P9-BZZ-592

Processes and Materials of Manufacture

THIRD EDITION

Roy A. Lindberg

Emeritus Professor of Production Engineering
University of Wisconsin

ALLYN AND BACON, INC.

Boston   London   Sydney   Toronto

This book is part of the
ALLYN AND BACON SERIES IN ENGINEERING
Consulting Editor: Frank Kreith
University of Colorado

Copyright © 1983, 1977, 1964 by Allyn and Bacon, Inc.,
7 Wells Avenue, Newton, Massachusetts 02159. All rights
reserved. No part of the material protected by this copyright
notice may be reproduced or utilized in any form or by any
means, electronic or mechanical, including photocopying,
recording, or by any information storage and retrieval
system, without written permission from the copyright
owner.

Library of Congress Cataloging in Publication Data

Lindberg, Roy A.
    Processes and materials of manufacture.

    (Allyn and Bacon series in engineering)
    Includes index.
    1. Manufacturing processes.   2. Materials.
I. Title.   II. Series.
TS183.L56   1983   671   82-22601
ISBN 0-205-07888-5

Printed in the United States of America

10 9 8 7 6 5 4 3   88 87 86 85 84 83

Contents

iii

Chapter Three

METAL-CUTTING THEORY AND PRACTICE 129

Chapter Four

TURNING MACHINES   165

Chapter Five

HOLE MAKING AND SIZING OPERATIONS   193

Chapter Six

STRAIGHT AND CONTOUR CUTTING   219

Chapter Nineteen

GROUP TECHNOLOGY   695

Chapter Twenty

**NONTRADITIONAL MATERIAL REMOVAL
PROCESSES   723**

Preface

The manufacturing or production engineer has the primary objective of producing quality parts at an economic price following a prescribed schedule. To accomplish this objective requires a broad background in many areas of engineering, but particularly in materials and manufacturing processes. This text is written to provide the manufacturing engineer, whether a student or practicing professional, with certain desirable fundamentals.

Practical relationships are stressed. For example: Why are some carbon steels more machinable than others? When, if ever, is it feasible to have a four-minute tool life? Where can composite materials take over for steel and other materials?

From a fundamental viewpoint, both machines and processes are stressed, as well as the more theoretical aspects of metal cutting and forming.

In recent years great strides have been made in automated manufacture and therefore more emphasis has been placed on automatic control systems, numerical control (NC), robots, computer-aided design (CAD), and computer-aided manufacture (CAM). Group technology, a relatively new manufacturing concept, is also discussed.

Case studies (new in this edition) have been designed to help the student become acquainted with practical problems faced in industry and their solutions. The case studies can be used to tie various aspects of manufacturing and materials together and can serve as a basis for class discussions.

It is a pleasure to acknowledge the help of my colleagues and their constructive criticisms; in particular, Professor Norman Braton for his review of the chapters on welding, Professor Marvin DeVries for his comments and help on numerical controls, and Professor Jay Samuel for the chapter on materials.

CHAPTER ONE

Environment, Energy, Safety, and Challenges of Manufacturing

In the peaceful era between Worlds War I and II, visionaries extolled the coming "Age of Technocracy," when the work week would be three days long and all the dull, routine, monotonous tasks would be performed by robot-type machines. Production would be high enough to allow sufficient income for the average individual to truly enjoy leisure living.

Obviously this utopian dream that was to be a gift of science and technology has not materialized. On the contrary, rather than being hailed as the great benefactor of mankind, technology is being singled out by some critics for many of the present ills of society, particularly those concerned with our environment.

Since its inception, technology has been largely self-directed and profit-motivated. It has enjoyed an era of laissez-faire. Now an alert citizenry is demanding many changes.

CHANGES AND SAFEGUARDS

Industry, the Corps of Engineers, pipeline builders, utilities, and others must now answer to an informed citizenry before making changes that have an environmental bearing. Technological and scientific programs that affect a great segment of society are now aired and debated. In retrospect, it is hard to believe that a huge program such as "National Goals in Space" was launched with little public debate. On the other hand, an informed constituency helped Congress debate and defeat the issue of a supersonic jet transport plane.

It is reasonable to expect that the procedures for safeguarding our environment have now been set up and that they will, in the aggregate, operate to the overall good of the country.

Industry and Environment

The predominant ecological problems that relate directly to manufacture are water and air pollution.

Water pollution

Industrial water pollution is due largely to chemical wastes discharged into sewers or into adjoining lakes, rivers, and streams. Of major concern in the industrial sector have been plating, pickling (chemical cleaning of oxide scale), and coolant wastes.

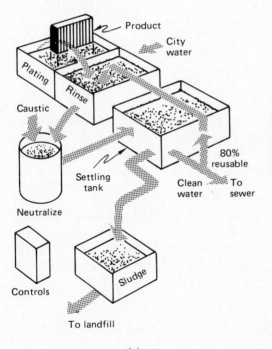

(a)

(b)

Figure 1.1. Liquid wastes from plating operations are treated as shown in the diagram before discharging into the sewer system (a). Dry sludge used for landfill (b). (*Courtesy Chemfix Division of Environmental Sciences, Inc.*)

Plating and Pickling Wastes.   Typical plating operations utilize salt solutions of cadmium, nickel, copper, tin, indium, and chromium. Operations allied with plating that pose similar waste disposal problems include chemical cleaning, chemical etching (chemical stock removal), and chemical stripping.

Various solutions have been worked out for the chemical treatment of effluent from plating and cleaning processes, as shown schematically in Fig. 1-1(a).

Some appreciation for the magnitude of the waste disposal problem can be gained by citing two examples. The Kodak plant at Rochester, New York, processes about 28 million gallons of waste water per day and more than 1500 tons of solid waste per week. Ford Motor Company at Lorraine, Ohio, processes about 55,000 gallons of slurry per day. Now, by the use of additives and settling tanks or large lagoons, the processed slurry becomes a gelatin in about three days. This claylike solid, Fig. 1-1(b), can then be used for landfill.

Coolant Wastes.   Considerable heat is generated in normal metal cutting and forming operations. Cutting fluids, often referred to as coolants, are used to carry off this heat and prolong the life of the tool, as shown in Fig. 1-2. Water provides an excellent cooling quality but does not provide lubrication for the cutting tool or prevent rust. Water-soluble oils are the most frequently used coolants and *straight oils* or undiluted oils are used for some applications.

In recent years, coolants have been developed that are entirely biodegradable and thus more easily disposed of. However, in some metalworking industries, various oils and greases contaminate the coolant so that it is not easily disposed

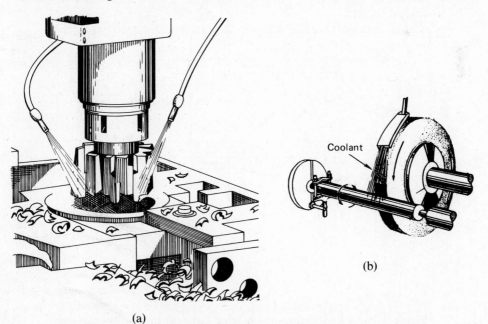

Coolant

(b)

(a)

Figure 1.2.   Coolants and cutting oils are used primarily to dissipate the heat in metal cutting (a), and grinding (b).

of. The process of incineration is used in this case. It operates on the principle that at a given temperature, in the presence of enough oxygen, chemical substances are converted to CO_2 and water. Hydrocarbons can be disposed of in this way. The incinerator itself is often fired from solid wastes collected from the plant. A schematic presentation of the incinerator principle is shown in Fig. 1-3.

Air pollution

Although a great deal of attention is directed toward air pollution, industry contributes only about 17% of it. The major source is transportation, which contributes 60%. Power plants contribute 14%, heating 6%, and refuse disposal 3%. The Clean Air Act of 1963, with further amendments in 1967 and 1970, has established emission standards of performance for hazardous pollutants from both moving and stationary sources.

The most reliable information on air-pollutant concentrations is obtained by continuous sampling with a wet chemical. This method can be automated using a continuous-flow technique. Reagents are added in their proper sequence, a reaction follows and is detected, and results are read out in a prescribed form.

In metal cutting, smoke is sometimes generated when straight oils are used. This occurs when metal is removed at relatively high volume rates. Even though the ventilation system is such that the concentration of smoke particles in the

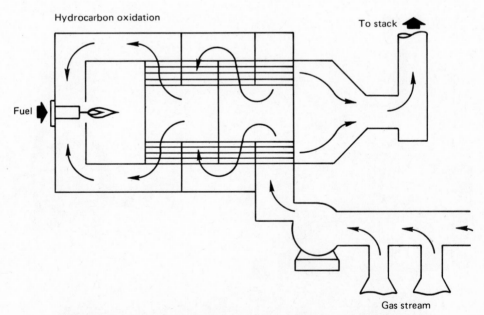

Figure 1.3. A gas incinerator is used to raise the temperature of the hydrocarbons to 1205–1500°F (649–816°C) in the presence of oxygen. The hydrocarbons are oxidized to CO_2 and water.

working environment is not excessive or does not cause discomfort, the exhaust of this smoke to the atmosphere is not permitted. The smoke is usually prevented from entering the atmosphere by *electrostatic precipitators*, or *scrubbers* (Fig. 1-4), or by some other means of removal.

An electrostatic precipitator is a means of putting an electric charge on the particles. The charge difference between the particle and the collecting plate causes attraction and collection.

Where the air becomes laden with large particles, gravity or simulated gravity provides a good means of separation. A gravity settling chamber, as shown in Fig. 1-5(a), is merely a "wide spot in the line" where the velocity of the smoke or gas is reduced to allow the particles to settle out. A cyclone separator subjects the particles to forces several times that of gravity as shown in part (b) of Fig. 1-5.

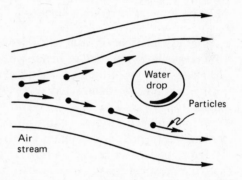

Figure 1.4. The air stream may be subjected to a scrubber for cleaning. Particles in the air are impinged against the water droplets. The gas velocity or opposing water velocity must be high enough to prevent particles from following the slipstream around the water droplets.

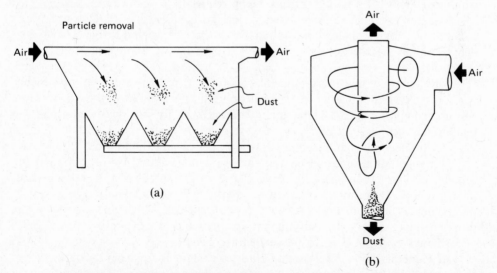

Figure 1.5. Where air is contaminated by large, heavy particles, a gravity-type settling chamber (a) will do a good job of separating them out. A cyclone separator (b) forces the particles to external walls where they are collected in a dust chamber.

Industry and Energy

Until recent years the supply of energy was not seen as a crucial item in manufacturing. A survey made by the Battelle Columbus Laboratories found that industry consumes about 40% of all the energy used in the United States, and that 22% of this consumption is returned to the atmosphere. A large portion of the energy used in industry is used in producing the basic materials of steel and aluminum. Next in order of energy consumption are the "hot-working" processes such as casting, forging and other hot-forming processes, welding, and heat treatment. Paint drying also uses considerable energy. Since energy, in whatever form, adds to the overall cost of the product, it has not been wasted. However, only recently has attention been seriously directed to effective energy conservation. The results of some of these studies will be presented to indicate what is being done in some manufacturing areas.

Industrial Heat Treatment.   Heat treatment, used extensively in industry to change the properties of metals, ranks high on the energy consumption list. Furnaces have been found to be extremely inefficient. Typically, less than 15% of the heat energy (natural gas or electric power) ends up in the form of heated parts.

Considerable strides have been made in bringing an aura of sophistication to heat treatment, for example, the benefits of the microprocessor-based control systems, of maximized efficiency furnace systems, of carefully blended alternative atmospheres, and of selective hardening options such as induction, electron beam, and laser processing are being realized.

Systems are available where much of the heat used in the furnace is recovered, as, for example, hot, flue gas from the carburizing furnace is used to heat a tempering furnace—thereby saving 90% of the energy needed for the tempering unit.

Although new and more efficient equipment will constantly be introduced, perhaps even greater energy savings will go to the alert and innovative processor. As an example, one casting company found that by raising the normalizing temperature for carbon and low-alloy steel from 1650°F to 1700°F (900°C to 925°C), the needed time at temperature dropped from 2 hr to $\frac{1}{2}$ hr.

Other potential areas of energy savings can be found by reexamining the product design. Often parts are made where various processes have been specified but are no longer needed or new and better ways are available. For example, one company developed an automatic process for case-hardening castings using nitrogen, methanol, and small amounts of methane to replace generated gases at a savings of from 15 to 25% over natural gas.

At General Motors Research Center, a system for plasma carburizing has been developed that reduces the use of natural gas by 99% and the total energy costs by 50%.

Selective-area heat treatment can now be accomplished by the use of the electron beam (EB). In most heat-treatment operations, where hardening is required in a specific area, the heating cannot be done fast enough. The result

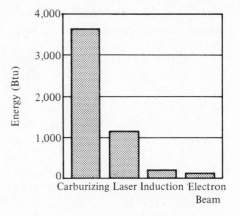

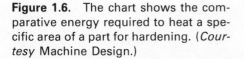

Figure 1.6. The chart shows the comparative energy required to heat a specific area of a part for hardening. (*Courtesy* Machine Design.)

is that an area larger than necessary is heated and a liquid quench is required. With the EB process, so much heat can be put into the surface of a part so fast that the rest of the material, being cold, acts as a heat sink, which is sufficient to quench the heated area, thus making it hard. Figure 1-6 is a chart showing comparative energy requirements for hardening by carburizing, laser, induction, and electron beam.

Paint Finishing. New kinds of paints and processes offer significant reductions in curing times and eliminate or reduce afterburning. Powder coatings, for example, contain no solvents, require no afterburning, and can be cured rapidly in infrared ovens that require 42% less energy than standard hot-air systems. The powdered paint particles are electrostatically charged and attracted to the base part, so that only a small amount leaving the gun misses the target. Even this overspray is recoverable when one color is used at a station. There is also a trend to go from solvent-type to water-based paints. This reduces energy requirements significantly.

In-plant painting may even be eliminated on some items by using precoated sheet and coil stock. Energy is reduced since the coating process is done much more efficiently at the mill. A more complete discussion of this and related topics is given in Chapter 22.

Welding. The energy consumed during welding is a function of welding time and volume of deposited filler metal. Joints should be designed to permit the fastest possible welding. Automated welding systems require at least 50% less energy than manual welding. Manual welders deposit metal only 15 to 30% of the machine's running time; during the remaining time, the equipment draws wasted no-load power. Automated wire processes reduce no-load time to 40% or less.

Material Selection. Material selected for a part contributes to the total energy consumed in two ways. First, energy is required to create the raw material. For some materials, such as aluminum, this is substantial. The material also dictates how much energy is consumed in manufacturing the product. Some metals are easily machined or formed, while others require several stages to obtain the final size.

Plastics not only require energy to create the raw material, but their product is also directly derived from petroleum and natural gas. Plastics are selected over other material options because of their unique properties. The cost in terms of "energy in a pound of plastic" is not the only consideration for an energy-conscious engineer. Other factors such as lighter weight or ease of fabrication contribute to the overall energy consumption picture. The long-range outlook suggests that new sources for polymer materials should be explored, such as cellulose, which occurs naturally in plants and trees.

Aluminum requires a great deal of energy to be produced. According to Alcoa, 108,000 Btu are required to produce one pound of primary or "virgin" aluminum. However, recycled aluminum can be produced for 13.8% of the thermal energy required for the primary metal.

A pound of steel, including a typical quantity of remelted scrap, can be produced for 16,500 Btu. Thus aluminum requires 6.54 times more energy by weight, or 1.57 times more energy by volume, to produce. But, as with other materials, comparisons on an energy-per-pound basis are misleading. Many steels, for example, require heat treatments to bring them up to maximum strength or toughness. Other steels may be difficult to work. In such cases, optional materials may compare more favorably with steels on an energy-per-product basis.

Design and Manufacturing Engineering Responsibility.   Many designs over-specify, either by habit or for "good measure." Overweight parts waste material and are more difficult to produce. Energy is not only wasted in the manufacture of the part, but also in the operational life of the product.

The manufacturing engineer must be alert to the use of optimum cutting speeds, feeds, and depth of cut. As will be shown in this text, the use of the proper cutting-tool materials can result in substantial savings in the time required for machining. By working together, design and manufacturing engineers can plan the products to consume as little energy as possible. This will usually result in good practice and in overall savings to the manufacturer as well as in con-servation of energy. Figure 1-7 is a diagram showing how overdesign adds up to energy waste.

Industry and Safety

In December 1970 the U.S. government stepped forcefully into the arena of insuring safety and healthful conditions for workers throughout industry by pass-ing the Occupational Safety and Health Act (OSHA). Since the signing of the act, the Secretary of Labor (under whose jurisdiction the act resides) has in-corporated standards that had previously been adopted by the National Fire Protection Association (NFPA) and the American National Standards Institute (ANSI). Under the provisions of this act, all employers must provide their

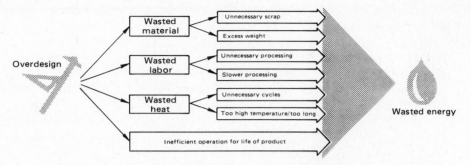

Figure 1.7. Overdesign results in wasted energy in many ways.

employees with safe working conditions and protect them from their own carelessness.

The act allows states to assume full responsibility for administering and enforcing their occupational safety and health laws. However, state standards must be equal to or better than the federal laws and must be accepted by the federal government. The act may be divided into three parts: investigation, enforcement, and penalization.

Investigation. Investigation by OSHA may take place as an unannounced visit or it may come at the request of one of the employees. Any employee who believes a violation of health or safety standards exists that may cause him serious physical harm can request an investigation by a federal representative of OSHA.

The federal representative can take one of two courses of action: (1) he may proceed with the investigation accompanied by an authorized representative of the employee or, (2) he may feel the complaint does not merit investigation and will give a written explanation of his reasons to the employee. If an investigation is made, it may involve private audiences with the person or persons involved. The private examination of witnesses is provided to maintain the employee's anonymity from his employer, thus guaranteeing his protection under the act.

Enforcement. The Secretary of Labor or any one of his representatives may issue a citation if the results of his investigation show noncompliance with the standards established under OSHA or if there are any proven "recognized hazards," even though there are as yet no standards covering them.

Citations will contain a written explanation of the violation or violations and set a realistic deadline for their elimination. The employer will have 15 days to respond to the Secretary of Labor's office; failure to do so eliminates any future review by the court.

Should the employer be unable to correct the violation by the date set because of extenuating circumstances, an extension may be granted. This is called a *variance*.

Penalization.   In order to make the law effective, a system of penalties was set up as follows:

1. A serious violation, fine of up to $1000.00.
2. Failure to correct violations, fine of up to $1000.00 per day until corrected.
3. Willful or repeated violations, fine of up to $10,000.00 for each violation.
4. Willful violation resulting in death, fine up to $10,000.00 and/or imprisonment up to 6 months.
5. Failure to display requirements, fine up to $1000.00.
6. Falsification in any application or report, fine up to $10,000.00 and/or up to 6 months imprisonment.

In addition to the stated penalties, the U.S. District Court has the right to shut down any operation or place of business found to be under imminent danger until that danger has been eliminated.

Additional Legal Powers.   Under the act several clerical duties must be performed:

1. All work-related deaths, as well as injuries and illnesses requiring medical care, must be reported.
2. The employer must monitor employee exposure to health hazards.
3. The employer must provide for physical examinations at regular intervals if environmental conditions warrant it.
4. The employer must disclose to all employees all hazards encountered in his work and provide educational programs and safety equipment to help guard against them.

OSHA and Industrial Noise

Tight standards have been set by OSHA in regard to industrial noise. Noise, which has been defined as unwanted excessive sound, is considered a major industrial health problem.

Noise-induced hearing losses are often insidious, usually occurring in the higher sound frequencies without discomfort or warning. Since these are not within the speech frequencies, the loss goes unnoticed. Although some people claim they become accustomed to noise, continued exposure to high levels spreads losses to speech frequencies, and the ability to understand and communicate is impaired. There is a growing concern that irritability, nervousness, headaches, stomach upsets or ulcers, high blood pressure, fatigue, poor morale, and other effects can be traced to noise pollution.

Fundamentals of Sound.   Sound is made up of variations in pressure that radiate from a vibrating object. The variations alternately compress and rarify the atmosphere and are referred to as sound waves. The frequency of the sound waves is commonly measured in hertz (Hz), or cycles per second. The lower the frequency, the longer the wavelength. Sound can generally be detected by the human ear when frequencies are in the range of 20 to 20,000 Hz, but the ear is less responsive to very low or very high frequency sounds than to sounds of medium frequency.

Decibels.   Sound is generally measured in decibels (db). The decibel is a dimensionless unit used in comparing the magnitude of sound pressures or powers. Shown in Table 1-1 are various sound pressure levels with the corresponding decibels. The sound of 80 db is shown as the sound pressure of only 0.00003 psi, or that of a busy office. The threshold of pain is generally considered to be near 140 db.

Decibels are logarithmic ratios and cannot be added or subtracted directly. To predict the noise from several different sources, the decibels can be converted to either sound pressures or sound powers, added or subtracted, and then converted back to decibels. Graphs are available to simplify this procedure.

Noise Measurement.   To control noise, it is often necessary to measure both sound and vibration near the source. Sound is picked up by microphones and

Table 1-1.   Sound pressure levels in decibels and psi. (*Courtesy Sperry Vickers.*)

Sound Pressure Level in Decibels (db)	Sound Pressure in Pounds per Square Inch (psi)	Common sounds (depending on distance)
160	3×10^{-1}	Medium jet engine
140	3×10^{-2}	Large propeller aircraft Air raid siren Riveting and chipping
120	3×10^{-3}	Discotheque
100	3×10^{-4}	Punch press (impact) Canning plant Heavy city traffic; subway Average machine shop
80	3×10^{-3}	Busy office
60	3×10^{-5}	Normal speech Private office
40	3×10^{-7}	Quiet residential neighborhood
20	3×10^{-8}	Whisper
0	3×10^{-9}	Threshold of hearing

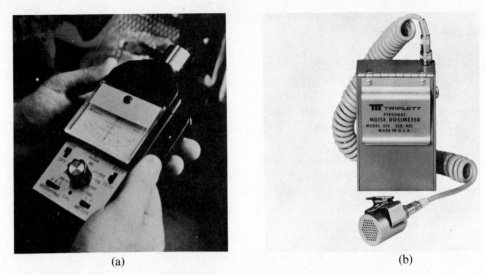

(a)   (b)

Figure 1.8. A sound-level meter being used for noise analysis at a punch press (a). An audio dosimeter that can be worn in the shirt pocket with a microphone clipped to the shirt collar (b). A memory cell is used to store the data. (*Courtesy Triplett Corp.*)

vibration by accelerometers. These meters are generally provided with slow or fast response. The slow response facilitates reading the meter by reducing the fluctuating speed of the indicating needle. Figure 1-8(a) shows a sound-level meter being used for noise analysis at a punch press. When fixed-position noise measurement is impractical, lightweight docimeters (noise exposure monitors) may be worn by the individual (b). The docimeters accumulate noise exposure data for the workday.

Legal Requirements.   OSHA standards, which specify protection against the effects of noise exposure, are shown in Table 1-2. When employees are subjected to sound exceeding the limits shown, effort must be made to reduce the sound levels or to provide personal protective equipment.

Controlling Industrial Noise.   Engineers, in compliance with the OSHA Act, have been seeking ways of controlling industrial noise. The control of noise is based on a knowledge of vibration isolation and damping. One method used to change the frequency of vibration is that of changing the mass or stiffness. Dampers consisting of a spring mass system, which can be tuned by altering the mass or the spring constant, are being used to improve the dynamic stiffness of machine tools.

Many materials are used for vibration damping, including sand, lead, asphalt, and impregnated felts. Materials that provide intermolecular friction to convert vibrational energy into heat are good for damping. Examples of this type of damping material are leaded vinyl sheets, sheet lead, polyurethane sandwiches,

Table 1-2. Permissible noise exposure.

Duration Per Day (hours)	Sound Level, db, Slow Exposure
8	90
6	92
4	95
3	97
2	100
1½	102
1	105
½	110
¼ or less	115

and thermoplastic materials sandwiched between steel sheets. A major plastics damping material is polyvinyl acetate that can be sprayed or troweled on surfaces.

Sound-absorbing materials reduce only a portion of the noise reaching them. Part of the energy is converted to heat, some is reflected, and some filters through. In addition to their use as an acoustic treatment for walls and ceilings, sound-absorbing materials are used for lining ducts and for enclosure panels.

Since the enactment of OSHA only a few areas of safety violation have gained significantly in litigation. Most of the contested cases have involved machine guarding, particularly as related to presses, shears, saws, and grinders.

The provision of proper guards by an employer is not enough. He must instruct employees in their use and make certain they are used. In many cases in which the employer was cited for lack of machine guarding, the proper guards were in the workplace, either on the floor or elsewhere.

Progressive welding shops have made considerable progress in cleaning atmospheres around the workmen by using exhaust fans and other devices for general ventilation, plus smoke eliminators and fume exhausters on welding guns. At present industry is uncertain about the actual effects of welding fumes and particulates on health. However, OSHA plans to institute stiffer rules, feeling they are necessary to insure the welder's well-being. Industry, meanwhile, looking at the cost, wants to be sure that the additional money will buy real, not imagined, benefits.

The American Welding Society sponsored a study of welding atmospheres at Batelle-Columbus Laboratories that led to a 170-page book, *The Welding Environment,* published in 1973. This book describes the nature of fumes, effects of toxic gases and particulates, studies of gases emitted by different welding processes, ventilation considerations, and techniques used for sampling and analyzing gases. Other health and environment books were also developed and are listed in the bibliography at the end of the chapter.

Arc-welding radiation hazards are not a priority item in welding safety and health since they have been handled well for years by industry. Improved understanding of radiation may lead to better protective clothing and to improvements in shop surroundings to reduce the amount of radiation.

Control Standards and Responsibility

There is little question that many changes will continue to be made in manufacturing technology that will help eliminate some of the pressing pollution, energy, and noise problems we now face. However, to expect industry to do it all is unrealistic. Everyone must be responsive, from the farmer who, in the interest of better crops, uses herbicides and fertilizer, to the consumer who strews cans and other debris along the countryside. As Pogo stated, "We have met the enemy and they is us."

Our environmental problems will require the cooperation and expertise of industry, government, and many individuals. Much of the work can be done by technically capable standards organizations, mentioned previously, with the Environmental Protection Agency (EPA) providing leadership in the perspective of costs for each increment of improvement. It must be pointed out that some of these agencies, such as EPA and OSHA, are still in their infancy, with major impacts yet to be worked out.

For further study of this topic, see the references at the end of the chapter.

THE CHALLENGE OF MANUFACTURING

The need for qualified manufacturing engineers has never been greater. To meet today's challenges, it is necessary to gain a broad background in both material sciences, design, and manufacturing processes. Considerable emphasis is also being placed on computer sciences and automatic controls.

The manufacturing engineer must be able to visualize the whole problem of production in the correct perspective: from the basic component materials and tooling through the step-by-step operational planning, final assembly, inspection, and shipment.

The manufacturing engineer is responsible for the expenditure of large sums of money for tooling and machinery to keep production on a competitive basis. He not only should be familiar with the theoretical aspects of fabrication, but also must be able to work out practical "bread and butter" problems of the production floor.

A few of the areas in which a manufacturing engineer can be expected to work are in planning process layouts, plant layouts, equipment specification, material specification, tool design, methods development, work standards, value analysis, and cost control.

As the manufacturing engineer gains experience, and if he has a high degree of creativity and innovative ability, he can expect to be involved in the generation of manufacturing systems concepts. He will also be involved in the development of novel and specialized equipment, research into the phenomena of fabricating technologies, and manufacturing feasibility studies of proposed new products.

Not all of the innovations that increase efficiency of manufacture are in the area of machine tools, inspection, or automated assembly per se. Increasing emphasis will be in the challenge of the individual worker. Men on production lines can become teams, collectively responsible for the end product rather than for just so many repetitious operations. Under this concept, it is the team's responsibility to help determine the distribution of work, the occupation of machines, and the speed and quality of work. It is assumed that the personnel will be given broader training so that they might have the flexibility and all-around skill to be able to step in where help is needed.

Modern manufacturing planning also includes the use of the computer to handle the vast amount of data needed to utilize raw information, such as sales forecasts and customer orders, in determining what is to be purchased, made, stored, etc. (Shown schematically in Fig. 1-9.)

The computer, including the minicomputer and programmable controllers (as discussed later in the text), will have increased impact in the areas of product, tool, and process design for optimum production. Computer-aided design (CAD) and computer-aided manufacture (CAM) have the potential of being the largest single development in manufacturing since the Industrial Revolution. Whether the full potential will be realized is dependent on its acceptance by industry and by colleges and universities where future manufacturing engineers are trained. This is not to imply that considerable learning does not take place outside of the formal classroom. On the contrary, because of today's flood of new scientific and technical knowledge, and the astonishing pace of technological change, the engineer risks obsolescence unless he makes some plans for self-study. However, even short courses and colloquiums may not be sufficient. For many areas of rapid technological change, a work cycle of four or five years followed by a year on campus may be a normal practice.

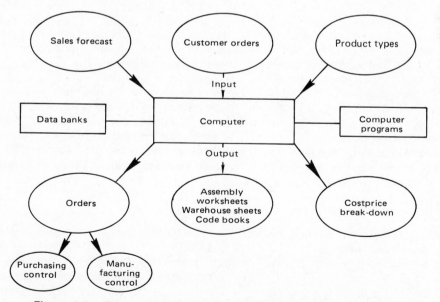

Figure 1.9.   The computer, with its data banks and programs, is an essential part of manufacturing planning. (*Courtesy* Engineering.)

Problems

1-1. Is industry the principal cause of air and water pollution? Explain.

1-2. (a) Does industry use most of the energy that is generated in the United States?
(b) Where is most of the energy used in industry?
(c) What are some ways that industry can conserve energy?

1-3. Does the OSHA Act supercede state safety laws? Explain.

1-4. Who is authorized to request an investigation by an OSHA inspector?

1-5. Does an OSHA inspector come out for every employee request? Explain.

1-6. Is there any protection from being fired for an employee if he turns in a request for inspection? Explain.

1-7. What happens if the employer refuses to recognize anything is unsafe in his plant?

1-8. Why may an individual not be aware of hearing losses?

1-9. According to OSHA standards, what is the daily limit of noise a person should be exposed to in the case of punch-press impact noise, or discotheque noise?

1-10. What is meant by a decibel?

1-11. What makes some materials good for vibration damping?

1-12. What may be a way of building employee morale in a production line situation?

1-13. Would you expect a supervisor whose department shows good production, low scrap, low turnover, and low grievance record to also have a low accident record? Explain.

1-14. In what areas of manufacture do most of the OSHA violations occur?

1-15. What is the employer's responsibility beyond insuring that a machine is properly equipped with guards?

1-16. Where can additional information be found about welding fumes?

Case Study

Company ABC had a record of 860,000 man-hours of work and 51 accidents, which caused a total of 209 lost workdays. Of the 250 people employed in the plant, 175 work in production. Operations include material and parts handling, flame cutting, positioning, manipulating, welding, assembling, and machining of heavy steel plate.

Unsafe material-handling practices accounted for 80% of the lost-time accidents.

The manufacturing engineer stated that back and foot injuries occurred most often, followed by lacerations and eye injuries.

Suggest a plan that would have as its goal a zero lost-time safety record. Include ideas on how the whole plant becomes involved on a day-to-day basis beyond putting up safety posters and signs stating number of hours without a lost-time accident. What outside agencies might be called on to help?

Bibliography

American Welding Society. *Arc Welding and Cutting Noise.* Miami, 1979.
———. *Effects of Welding on Health.* Miami, 1979.
———. *Fumes and Gases in the Welding Environment.* Miami, 1979.
———. *The Welding Environment,* Miami, 1973.
Betz, G. M. "How OSHA Welding Standards Are Interpreted." *Welding Design and Fabrication* 52 (January 1979): 119–122.
———. "OSHA's Challenge to Shop Man-

agement." *Modern Machine Shop* (June 1979).

Editor. "Trends in Heat Processing Technology." *Metal Progress* 77 (January 1970): 26–27.

Hankel, K. M. "Controlling In-Plant Noise."

Automation 21 (April 1974): 86–90.

OSHA. *The Who, What, Where, When, Why and How of the Occupational Safety and Health Act of 1970.* Washington, D.C.: U.S. Department of Labor, 1970.

CHAPTER TWO

Materials, Structures, Properties, and Fabricating Characteristics

Man discovered the art of smelting metal toward the end of the Stone Age. The first crude bronzes were probably the result of accidental mixtures of copper and tin ores. The Bronze Age was ushered in about 2500 B.C., when the art of extracting relatively pure tin had advanced to a point where intentional additions to copper were possible. Brass was not introduced until about 500 B.C., when copper and zinc ores were smelted. The Bible makes reference to Tubal Cain, a descendant of Cain, as an "instructor of every artificer in brass and iron" (Genesis 4:22).

Gold was used for jewelry and utensils as early as 3500 B.C. Silver was used about 2400 B.C. A wrought iron sickle blade was found beneath the base of a sphinx in Karnak near Thebes, Egypt, and a blade probably 5000 years old was found in one of the pyramids.

The origin of forced draft in the making of wrought iron is unknown, but prior to about 1500 B.C. the Egyptians had developed a bellows out of goat skin with a bamboo nozzle and an air inlet valve. Forced draft was the first major development in wrought iron manufacture. It was not until the end of the thirteenth century that water power was used to drive a bellows. Water power was also used at this time to drive a forging hammer.

Guilds sprang up in the early 1400s to perpetuate the skills and art of working metals. The iron and steel workers considered themselves very aristocratic and apprenticeships were reserved for legitimate sons of honorable parents. Children of night and gate watchmen, barbers, musicians, millers, tanners, weavers, shepherds and tax collectors were not eligible.

CRYSTAL STRUCTURE

The solidification of metal from the molten state requires that a sufficient number of atoms shall exist in the proper arrangement to form a crystal that will grow. The number of atoms that will form a crystalline nucleus depends on the decrease of energy when the liquid is replaced by a solid, and the increase of energy resulting from the formation of an interface between liquid and solid phases.

After nucleation, the structure grows until crystals, or grains as they are often called in metallurgy, are formed. The grains grow in a dendritic or "pine tree" columnar structure (see Fig. 2-1). This dendritic freezing mode is the pattern of ferrous alloys.

Space Lattices

Each grain consists of millions of tiny unit cells made up of atoms that are arranged in a definite geometric pattern of imaginary points. The atomic arrangements found in most metals are: (a) body-centered cubic (BCC), (b) face-centered cubic (FCC), and (c) hexagonal close-packed (HCP) (see Fig. 2-2).

Upon cooling, the space lattice structures expand in all directions of the axes of the lattice until development is stopped by contact with the container or by contact with adjacent growing grains. Where the crystals meet or intersect, it is impossible for another space lattice to develop. Thus, at the grain boundaries the atomic distances are not correct and the atoms are not in a stable position but are in a higher state of energy. The lower density of the atoms at the boundaries provides a place for impure atoms to reside. The size of the grain depends upon the temperature at which the metal is poured, the cooling rate,

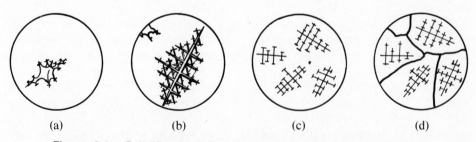

(a)   (b)   (c)   (d)

Figure 2.1. Solidification of pure metal. Nucleation and the first dendritic structure is formed at (a). The dendritic structures join together at (b) and (c). Development is stopped by interference with adjacent structures and the container or the grain boundaries.

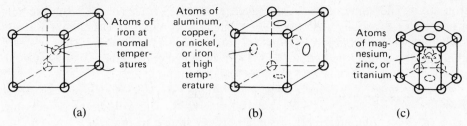

Figure 2.2.   The atomic structure of some metals.

and the nature of the metal. If the metal cools slowly, large grains will result; if the metal cools fast, smaller grains are obtained.

Crystal-Lattice Imperfections

Even if metal solidified very slowly, a large number of atoms would be laid down in the crystal structure each second. It is not surprising therefore that imperfections in the structure should occur. The most important of these are *vacancies, interstitials,* and *dislocations*.

Vacancies are, as the term implies, vacant sites in the crystal atomic structure [Fig. 2-3(a)]. Interstitials are extra atoms that fit into the interstices between the normal atom structure [Fig. 2-3(b)]. Both vacancies and interstitials produce local distortion and interrupt the regularity of the space lattice structure.

Vacancies help explain the phenomenon of *solid-state diffusion*. This is why one metal, such as copper, will diffuse through a second metal, such as nickel, when placed in intimate contact and heated to a high temperature. It is assumed that vacancies move through the lattice producing random shifts of atoms from one lattice to another.

An example of an interstitial solid solution is the introduction of carbon atoms into an iron atomic lattice. The carbon atoms are much smaller than the

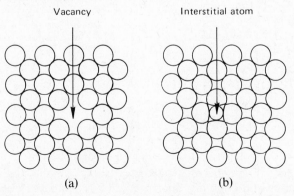

Figure 2.3.   Two types of crystal-lattice imperfections: (a) vacancy, (b) interstitial.

iron atoms and some of them can squeeze into the structure. At room temperature, iron has a body-centered cubic (BCC) lattice and can contain only a few carbon atoms in "solution." However, at elevated temperatures, the structure changes to face-centered cubic (FCC) and there is a considerable increase in the number of carbon atoms that form the solid solution.

Dislocations, a third defect, may be thought of as imperfections caused by the mismatching of atomic planes within a crystal. These line-type defects are referred to as *edge* and *screw* dislocations, as shown in Fig. 2-4.

The Crystal Structure Under Stress

One of the mysteries surrounding the enormous discrepancy between the theoretical and observed shear strength of metals was clarified with the aid of the electron microscope. Theoretical calculations show that a pure crystal of magnesium, for example, should have a yield strength of approximately 1,000,000 psi, whereas it may actually yield at 1000 psi. The advent of the electron microscope made it possible to see that it was not necessary for the whole plane of atoms to move at once, but that the movement could start at one place and only a few atoms shift progressively, as in the screw-type dislocation.

Elastic and Plastic Deformation.   When a crystal lattice is subjected to a stress below its elastic limit, the crystal structure will temporarily yield a small amount but will recover when the load is released. If a sufficiently large load is applied, plastic deformation takes place. This means the atomic structure has to slip. Slip takes place along certain crystal planes called *slip planes*. When slip planes are readily available, the metal is considered very formable. With few slip planes and directions of slip available, the metal is thought of as being

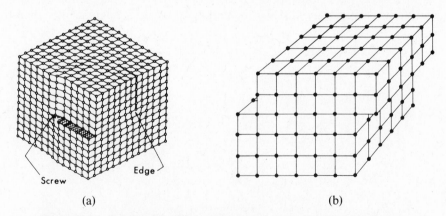

| (a) | (b) |

Figure 2.4.   Two types of line defects that occur in the crystal structure: screw and edge (a). Shown at (b) is an enlarged view of the screw dislocation that has passed through the crystal, causing unit slip.

difficult to form. FCC metals, for example, have 12 systems of slip, BCC have 48, and HCP have only 3.

Twinning takes place in some metals when they are subjected to external stresses, as shown in Fig. 2-5. In the case of slip, the offset is a multiple of the interatomic spacing; but in twinning, the offset produced by sliding of one plane against its neighbor is a fraction-of-unit slip. This causes a difference in the orientation between the twinned and untwinned regions in the crystal.

Deformation by twinning is most common in BCC and HCP metals and its effect on others is to move parts of grains to a more favorable position for slip to occur. Thus, all metals deform by slip and others by twinning and some by both twinning and slip. Twinning always takes place in pairs of lattices, hence the name.

Strain Hardening.   As external forces are applied to the metal, slip occurs in the atomic structure. It begins at the points of imperfection or dislocation and twinning. As the force continues and the crystals deform, more dislocations are formed along the slip planes. Strong interactions arise as more and more dislocations are forced to intersect on the various slip planes. The net result is increasing resistance to further deformation, or strain hardening. In this con-

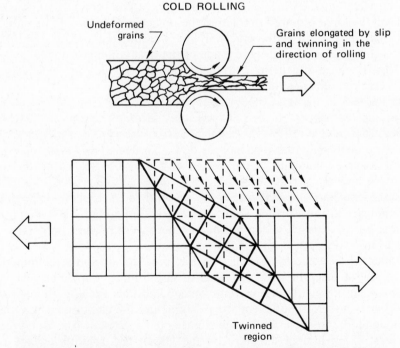

Figure 2.5. Plastic deformation results in a twinning action. All the atoms in the twinned region move a given amount and change orientation as shown.

dition, the metals are harder, stronger, less ductile, and have reduced electrical resistance.

Fracture.   Every solid material, from the most frail eggshell china to the toughest steel, has its breaking point. It resists stress up to a point, perhaps yielding somewhat and stretching, but then suddenly it fractures. Just what happens at the instant of fracture has been difficult to investigate due to the speed and finality of the reaction. For the engineer, fracture was simply a calamity to be avoided by careful design. However, the fracture of entire ship hulls during World War II brought the subject under concerted investigation.

Brittle fracture is characterized by the small amount of work absorbed and by a crystalline appearance of the surfaces of the fracture.

Even though most metallic and some nonmetallic crystals have no cracks initially, they do have dislocation defects which make them vulnerable to crack formation. If slip takes place freely, the result is just a change in shape of the crystal. However, if sliding is blocked by a hard particle inside the crystal or at the boundary between crystals or grains, a high concentration of stress collects. The atomic bonds in this situation are under great stress. They stretch beyond their limit and rupture, forming a tiny crack. Once a crack is started, it takes comparatively little energy to carry it through to a complete fracture. The energy is of course dependent on the specific case. As an example, consider a plate of steel 6 in. wide and 0.25 in. thick. Assume it has a 2-in. crack running into one side. Then the force required to fracture the remaining 4 in. would be only about 400 lb. Without the aid of the crack, it would require a force of about 450,000 lb to pull the plate apart if it were made of the best commercial steel available. It is a large leverage effect (Fig. 2-6) that makes it possible for a relatively small force to crack an entire ship in two.

Plastic flow or slip can be both detrimental and beneficial to the strength of a material. If slip is blocked, it may lead to crack formation. If it is not blocked, it may act to relieve the stress concentrations at the tip of a crack. As temperature decreases, the tendency to flow also decreases. With plastic flow inhibited, stresses build up at the vertex of the crack, stretching the atomic bonds to the breaking point. Some materials, such as ordinary steel, are normally resistant to cracking, but become highly susceptible to it at low temperatures. Referring again to the wartime ships: those used in southerly waters posed no problems, but those used in the northern waters were prone to cracking under shocks from heavy seas in the winter season.

Sudden and seemingly inexplicable fractures are often the culmination of a long series of discrete steps. A sheet of glass, for example, may support a load for a long time without apparent damage and then break without warning. The load does not exert sufficient stress to fracture the glass outright but exerts only enough force to open tiny invisible cracks in its surface. Atmospheric water vapor lays a thin film of moisture on the glass. The smooth surface resists penetration, but as the water reacts with silicon–oxygen bonds within the crack, it breaks those bonds and forms reaction products that have a larger volume

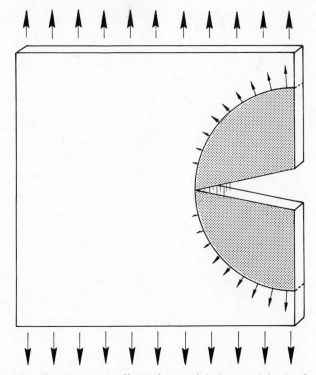

Figure 2.6. The leverage effect of a crack helps explain the fracture mechanism. Forces applied to the top and bottom of the steel plate are transmitted to the vertex of the crack through schematically depicted lever arms. As the crack grows the lever arms have a multiplying effective force at the crack vertex.

than the glass. The reaction products act as a wedge, spreading the crack faces apart. This stretches the bonds at the vertex of the crack, encouraging the next reaction. Thus the process is both self-sustaining and self-accelerating. The crack which at one time was advancing only a few atomic diameters at a time now surges forward at a speed approaching that of sound, and complete fracture is the result. The same so-called static fatigue fractures take place in other materials.

Ductile fractures evidence a substantial plastic deformation prior to failure. The fractured surfaces are characterized by a cup and cone appearance (Fig. 2-7). Microcracks form in the necked-down region that is progressively strain-hardened. The cracks join together and fracture occurs. The size of the cup depends on the relative shear and cleavage strength values. High-strength metals produce a smaller cup, a topic discussed in connection with metal forming in Chapter 4.

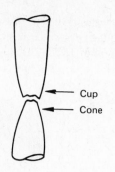

Figure 2.7. Ductile fracture. The final shear slip action produces a cup and cone fracture.

PROPERTIES OF METALS

Metal properties are often spoken of as separate entities; however, they are closely interwoven. For example, if a specified hardness is desired, it will be accompanied by a corresponding strength.

Before the specific properties of metals can be defined, it would be advantageous to be familiar with three basic terms: stress, strain, and elasticity.

Stress.   Stress may be defined as the load per unit area of $S = P/A_o$. Thus a rod whose original area is 0.5 sq in., subjected to a load of 2000 lb, would have the following stress:

$$S = \frac{2000 \text{ lb}}{0.5 \text{ sq in.}} = 4000 \text{ psi}$$

Stresses may be either tensile (tending to pull apart), compressive (tending to make shorter), or shear (tending to divide the material in layers). Bending stresses and torsional stresses (twisting) are combinations of the three main stresses.

Strain.   Strain is the percent change in the unit length during elongation or contraction of the specimen and is expressed as a measure of deformation under load. As an example, a weight of 2000 lb is suspended by a 0.25-in. diameter wire. The wire changes in length from 25 in. to 25.05 in.

$$n = \frac{(l_f - l_o \times 100)}{l_o} = \frac{(25.05 - 25) \times 100}{25} = 0.2\% \text{ strain}$$

where: n = strain
l_f = final length
l_o = original length

Elasticity.   Most engineering materials are used under what is known as elastic conditions, which means that when a material is deformed, the deformation is not permanent. Elasticity was expressed as a theory by Robert Hooke and therefore bears his name. Hooke's Law states that the degree to which a body bends or stretches out of shape (strain) is in direct proportion to the force acting on it (stress). This law is applicable within a given range of stresses, this range being termed the *elastic limit*. Loads beyond the elastic limit cause permanent deformation.

MECHANICAL PROPERTIES OF MATERIALS AND DESIGN

The basic modes of loading a component are tension, compression, bending, torsion, and shear. Mixed modes may also occur in service. Each of these five modes of loading is the key to determining certain mechanical properties of a component during its design stage.

Tension

Several basic properties of a material can be determined from a tensile test. A standard-type specimen (Fig. 2-8) is shown before and after a tensile test. The specimen is gripped at both ends and stretched at a slow, controlled rate of extension until rupture. Usually, frequent or continuous measurements of load and extension are made, tensile stress and strain are calculated, and a *stress–strain diagram* is constructed. From the diagram (Fig. 2-9) may be calculated the modulus of elasticity, proportional limit, tensile strength, and yield strength.

Modulus of Elasticity (*E*).   The modulus of elasticity is a measure of stiffness of the material and is represented by the area under the stress–strain curve up

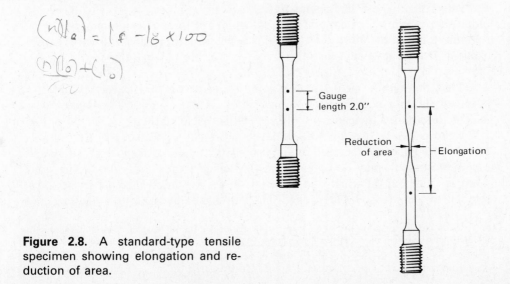

Figure 2.8. A standard-type tensile specimen showing elongation and reduction of area.

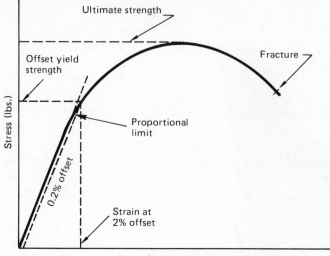

Figure 2.9. A simple stress–strain plot.

to the proportional limit. Steels have an average modulus of elasticity of 30 × 10⁶.

$$E = S/n$$

Proportional Limit. The proportional limit is the highest point on the stress–strain curve, where the strain is proportional to the stress.

Tensile Strength. Tensile strength is defined as the maximum load in tension that a material can withstand prior to fracture. This is the value most often used for listing the strength of a material and is given in pounds per square inch. In older metric terms, it is expressed as kilograms of force per millimeter squared (kgf/mm^2); but in the newer SI* units it is expressed as megapascals (MPa). (See Table 21-3 for conversion factors.)

Yield Strength. Yield strength is the stress at which a material exhibits a specified limiting permanent deformation. A practical approximation of it is usually determined by using an "offset." Offset yield strength, as shown in Fig. 2-9, corresponds to the intersection of the stress curve to a straight line parallel to the curved line at a specified offset, usually 0.2%. In the absence of a diagram, the yield strength may be obtained at a specified strain under load. The *yield point* is that point at which the material shows a marked increase in strain in relation to the stress. Yield strength may be calculated from the equation:

* SI is the abbreviation for "Le Systeme International d'unites," which is the internationally adopted metric system of weights and measures.

$$S_y = \frac{L}{A_o}$$

where: L = load
          A_o = original area

True Stress and Strain.   The relationship of properties just described is based on the original cross-sectional area of the specimen and is referred to as *nominal*. If each time the load was recorded it represented the instantaneous cross section of the specimen or the instantaneous length, it would be termed *true* stress and *true* strain.

$$\sigma = \frac{L}{A_i} \qquad \varepsilon = \int_{l_o}^{l_f} \frac{dl}{l} = \ln \frac{l_f}{l_o}$$

where: σ = true stress
          L = load
          A_i = instantaneous area
          ε = true strain
          l_f = final length; l_o = original length
          ln = natural log

The true-stress–true-strain curve exhibits no maximum; stress continues to rise until fracture (Fig. 2-10).

Percent Elongation.   The percent elongation refers to the increase in length over the original gage length.

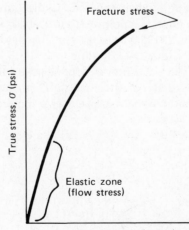

Figure 2.10.   True stress and true strain plot.

$$\% \text{ Elongation} = \frac{l_f - l_o}{l_o} \times 100$$

Percent Reduction of Area. The percent reduction in area is the original cross-sectional area compared to the final cross-sectional area taken at fracture.

$$\% \text{ RA} = \frac{A_o - A_f}{A_o}$$

Ductility. Ductility may be defined as the ability of a material to withstand plastic deformation without rupture. It is expressed in terms of percent elongation and reduction of area.

Ductility is also thought of in terms of bendability and drawability. The forming of cups as discussed in Chapter 8 or the drawing of wire to a smaller size is also a measure of ductility. Brittleness is the opposite of ductility. Usually, if two materials have approximately the same strength, the one that has the higher ductility is the more desirable.

Compression

The compressive strength of a material is a measure of the extent it deforms under a compressive load prior to rupture. The total strain (%) of a specimen immediately before rupture is indicated by direct measurement on a universal testing machine or by the corresponding stress–strain diagram (Fig. 2-11).

Buckling of a compression specimen may take place if the length-to-diameter ratio is greater than 2.5, as shown in Fig. 2.12.

Bending

Bending is characterized by the outside fibers being placed in tension and the inner fibers in compression. The stress goes to zero at the neutral axis, as shown in Fig. 2-13. The stress on the outer fibers depends on the section geometry, bend radius, and loading.

The deflection is dependent on the loading, section geometry, and modulus of elasticity of the material. To decrease the deflection at a given load, the section may be increased or a material with a higher modulus of elasticity may be selected.

The bending stress (σ) in a cantilevered shaft, as shown in Fig. 2-13(b), may be found by:

$$\sigma = \frac{My}{I}$$

where: M = moment (in.-lb)
  y = half the shaft diameter or the distance from the neutral axis to the outer edge (in.)
  I = area moment of inertia (in.^{4})

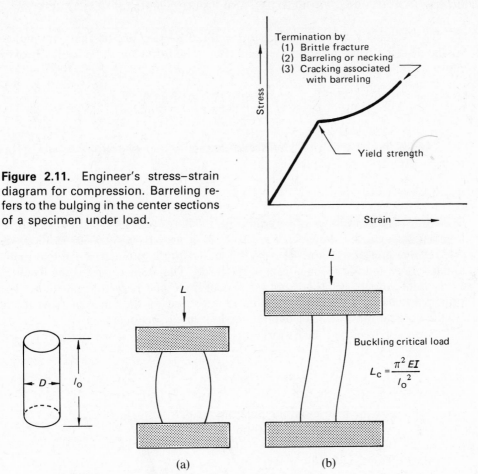

Figure 2.11. Engineer's stress–strain diagram for compression. Barreling refers to the bulging in the center sections of a specimen under load.

Termination by
(1) Brittle fracture
(2) Barreling or necking
(3) Cracking associated with barreling

Yield strength

Stress

Strain

Buckling critical load

$$L_c = \frac{\pi^2 EI}{l_o^2}$$

(a)   (b)

Figure 2.12. Conditions for barreling (a) when $l_o = 2d_o$ and buckling (b) when $l_o = 2.5d_o$.

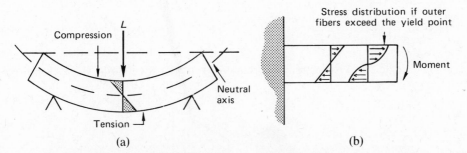

Compression

L

Neutral axis

Tension

(a)

Stress distribution if outer fibers exceed the yield point

Moment

(b)

Figure 2.13. Structural members subjected to transverse loads are classed according to the method of support. A simple beam supported at each end is shown at (a) and a cantilever beam is shown at (b). The moment at a section is resisted by the internal couple of compressive and tensile stresses.

31

The bending moment is equal to the force exerted on the shaft, multiplied by the distance from the supported end (PL). The formula can be further broken down as follows:

$$\sigma = \frac{My}{I} = \frac{(PL)d/2}{\pi d^4/64} = \frac{(PL)(32)}{\pi d^3}$$

One of the principal problems encountered in bending is *springback*. This topic is discussed in Chapter 4.

Torsion

Torsion is the application of torque to a member to cause it to twist about its longitudinal axis, as shown in Fig. 2-14. It is usually referred to in terms of torsional moment or torque (T), which is basically the product of the externally applied force and the moment arm of the force. The moment arm is the distance of the centerline of rotation from the line of force and perpendicular to it. The principal deflection caused by torsion is measured by the angle of twist (θ) or by the vertical movement of one corner in framed sections.

For a solid circular shaft, torque is related to stress at the outer fibers as:

$$T = .196\tau d^3$$

where: T = torque (lb in.)
τ = shear stress at the outer fibers (psi)
d = diameter of the shaft (in.)

For a hollow shaft:

$$T = .196\tau \times \frac{(d_o^4 - d_i^4)}{d_o}$$

where: d_o = outside diameter (in.)
d_i = inside diameter (in.)

The angle of twist (θ) may be related to torque for a solid shaft as follows:

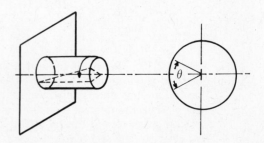

Figure 2.14.   A shaft in torsion. The left end is stationary. As the right end is rotated, internal shear stresses, known as torsional stress, are set up in the shaft.

$$\theta = 584\, \frac{Tl}{Gd^4}$$

where: l = length of shaft (in).

   G = modulus of elasticity in shear (Note: for steel—12 × 10⁶).

For hollow circular shafts:

$$\theta = 584\, \frac{Tl}{G(d_o^4 = d_i^4)}$$

Example: What is the minimum diameter of a steel shaft that can be used if it is 3 ft long and subjected to a torque of 50,000 lb in.? The maximum shearing stress should not exceed 8000 psi and the angle of twist will not be greater than 2°.

Solution:

$$T = .196\tau d^3$$
$$50,000 = .196 \times 8000 \times d^3$$
$$d = 3.17 \text{ in.}$$

and

$$\theta = 584 \left(\frac{Tl}{Gd^4} \right)$$

$$l = 584 \left(\frac{50,000 \times 12 \times 3}{12,000,000 \times d^4} \right)$$

$$d = 3.17 \text{ in.}$$

Therefore, to satisfy both conditions given in the problem, the shaft should be at least 3.17 in. in diameter.

Shear Strength

The ultimate shear strength of a material is the maximum load it can withstand without rupture when subjected to a shearing action. A method of determining the shear strength is by using a small sheet or disk of the material of known thickness which is clamped over a die. A corresponding punch is brought down on the material. A gradually increasing load is applied until the material is completely punched through. The shear stress (τ) may be calculated as follows:

$$\tau = \frac{P}{\pi dt}$$

where: P = the punch load in pounds
   d = punch diameter in inches
   t = the specimen thickness in inches

The shear strength of mild steels ranges from about 60 to 80% of the tensile strength.

Testing

Standardized tests have been made to determine the properties of materials. Discussed here are those used for determining hardness, tensile strength, impact strength, and fatigue.

In selecting a material to withstand wear or erosion, the properties most often considered are hardness and toughness. Hardness is the property of a material that enables it to resist penetration and scratching.

Hardness testing can be done by several standard methods such as Brinell, Rockwell, Vickers, and scleroscope.

Brinell. Brinell tests are based on the area of indentation a steel or carbide ball 10mm in diameter makes in the surface for a given load (Fig. 2-15). Loads used are 3000, 1500, or 500 kg. The load is applied for 15 sec on ferrous metals and at least 30 sec on nonferrous metals. When the load is released, the diameter of the spherical impression is measured with the aid of a Brinell microscope. From the diameter, a Brinell hardness number is obtained by consulting standard tables such as that made by the Society of Automotive Engineers (SAE), Table 2-1.

Brinell hardness tests are especially good for materials that are coarse-

Table 2-1. Hardness conversion table for hardenable carbon and alloy steels. Reprinted with permission, Copyright © 1966, Society of Automotive Engineers.

Brinell, 10 mm Carbide Ball, 3000 kg Load		Diamond Pyramid Hardness Number	Rockwell		Shore	Tensile Strength, 1000 psi
Indentation Diameter, mm	Hardness Number		C Scale 150 kg Brale	B Scale 100 kg 1/16-in. Ball		
—	—	940	68	—	97	—
—	767	880	66.5	—	93	—
2.25	745	840	65.5	—	91	—
2.30	712	—	—	—	—	—
2.35	682	737	61.5	—	84	—
2.40	653	697	60	—	81	—
2.45	627	667	58.5	—	79	—
2.50	601	640	57.5	—	77	—
2.55	578	615	56	—	75	—
2.60	555	591	54.5	—	73	298

Table 2-1 (continued).

Brinell, 10 mm Carbide Ball, 3000 kg Load		Diamond Pyramid Hardness Number	Rockwell		Shore	Tensile, Strength, 1000 psi
Indentation Diameter, mm	Hardness Number		C Scale 150 kg Brale	B Scale 100 kg 1/16-in. Ball		
2.65	534	569	53.5	—	71	288
2.70	514	547	52	—	70	274
2.75	495	528	51	—	68	264
2.80	477	508	49.5	—	66	252
2.85	461	491	48.5	—	65	242
2.90	444	472	47	—	63	230
2.95	429	455	45.5	—	61	219
3.00	415	440	44.5	—	59	212
3.05	401	425	43	—	58	202
3.10	388	410	42	—	56	193
3.15	375	396	40.5	—	54	184
3.20	363	383	39	—	52	177
3.25	352	372	38	(110)*	51	171
3.30	341	360	36.5	(109)	50	164
3.35	331	350	35.5	(108.5)	48	159
3.40	321	339	34.5	(108)	47	154
3.45	311	328	33	(107.5)	46	149
3.50	302	319	32	(107)	45	146
3.55	293	309	31	(106)	43	141
3.60	285	301	30	(105.5)	—	138
3.65	277	292	29	(104.5)	41	134
3.70	269	284	27.5	(104)	40	130
3.75	262	276	26.5	(103)	39	127
3.80	255	269	25.5	(102)	38	123
3.85	248	261	24	(101)	37	120
3.90	241	253	23	100	36	116
3.95	235	247	21.5	99	35	114
4.00	229	241	20.5	98	34	111
4.05	223	234	(19)	97.5	—	—
4.10	217	228	(17.5)	96.5	33	105
4.20	207	218	(15)	94.5	32	100
4.30	197	207	(12.5)	93	30	95
4.40	187	196	(10)	90.5	—	90
4.50	179	188	(8)	89	27	87
4.60	170	178	(5)	87	26	83
4.70	163	171	(3)	85	25	79
4.80	156	163	(1)	83	—	76
5.00	143	150	—	78.5	22	71
5.20	131	137	—	74	—	65
5.40	121	127	—	70	19	60
5.60	111	117	—	65.5	15	56

* Values in parentheses are beyond normal range and are for information only.

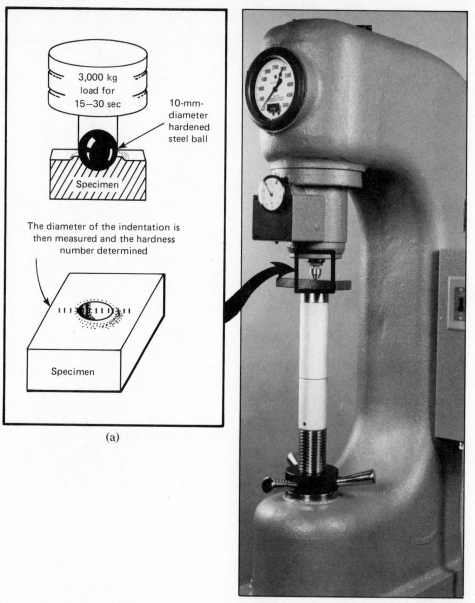

Figure 2.15. Principles of Brinell hardness testing (a). A Brinell hardness-testing machine (b). (*Courtesy Acco, Wilson Instrument Division.*)

grained or nonuniform in structure, since the relatively large indenter gives a better average reading over a greater area.

Rockwell.   The Rockwell hardness tester (Fig. 2-16) uses two types of penetrators, steel balls and a diamond cone or Brale. The ball indenter is normally $\frac{1}{16}$ in. diameter, but larger diameters such as $\frac{1}{8}$, $\frac{1}{4}$, or $\frac{1}{2}$ in. may be used for soft materials. The Brale is used for hard materials. The principle of this tester is based on measuring the difference in depth of penetration between a minor and a major load. The minor load is 10 kg and the major load varies with the material being tested. If the material is known to be relatively soft, the ball penetrator is used with a 100-kg load in what is known as the "B scale," (R_b). If the material is relatively hard, the diamond Brale is used with a 150-kg load on what is known as the "C scale," (R_c). Other scales are available and are useful for checking extremely hard or soft surfaces. Very thin sections such as razor blades or parts that have just a thin, hard outer surface may be checked with a Rockwell Superficial hardness tester. The loads range from 15 to 45 kg on what is known as the "T scale."

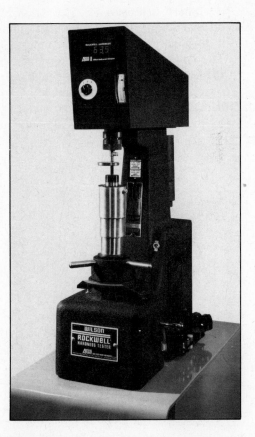

Figure 2.16.   Rockwell hardness tester with Brale penetration used in testing hard materials. (*Courtesy Acco, Wilson Instrument Division.*)

Microhardness Tests. Microhardness tests usually refer to tests made with loads ranging from 1 to 1000 grams. The indenter is either a 136° diamond pyramid or a Knoop diamond indenter (Fig. 2-17). The Knoop indenter is a diamond, ground to a pyramidal form that makes an indentation having an approximate ratio between the long and short diagonals of 7:1.

Prior to the advent of the microhardness tester, it had been assumed that the 136° diamond indenter produced a hardness number that was independent of the indenting load. In general terms, this can be accepted for loads of 1 kg and up. However, microhardness tests (performed with loads of 500 g or lighter with the Knoop indenter, and 100 g or lighter with the diamond pyramid indenter) are a function of the test load.

The Knoop hardness number is the applied load divided by the unrecovered projected area of the indentation. Both hard and brittle materials may be tested with the Knoop indenter.

The diamond pyramid hardness number (DPH) is also the applied load divided by the surface area of the indentation. The 136° diamond pyramid is often referred to as a "Vickers" test.

Shore Scleroscope. The scleroscope presents a fast, portable means of checking hardness. The hardness number is based on the height of rebound of a diamond-tipped metallic hammer. The hammer falls free from a given height. The amount of rebound is observed on a scale in the background. The harder the material, the higher the rebound, and vice versa. Thin materials may be checked if a sufficient number of layers are packed together to prevent the hammer from penetrating the metal to the extent that the rebound is influenced by the steel anvil. This is known as *anvil effect*.

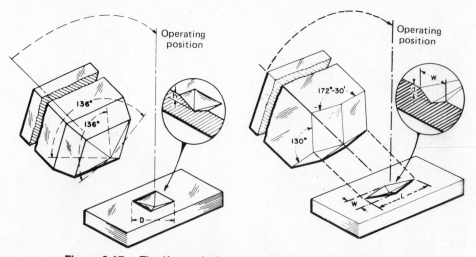

Figure 2.17. The Knoop indenter with the long impression is shown at the right. At the left is the diamond-pyramid or Vickers indenter. Hardness in both systems are in units of kg/mm². (*Courtesy Acco, Wilson Instrument Division.*)

Tensile Testing.   Tensile tests, when properly conducted, come closer to evaluating fundamental properties of a material for use in design than any other tests. This implies the use of standardized specimens with regard to shape and size, and a standardized testing procedure. Both tension and compression tests are made on a universal testing machine such as that shown in Fig. 2-18.

Although standard shapes have been designated for tensile or compression tests, this does not preclude making tests on full-size manufactured parts. In many cases, tests of this type are essential and are the best type to determine the properties of tubing, selected wires, reinforcement bars, fibers, fabrics, brick, tile, metal castings, etc.

Specimens are usually round; however, flat sheet stock is also used. The central portion is usually of smaller cross section than the ends in order to cause failure where it is not affected by the gripping device. The round specimen is often made 0.505 in. in diameter in order to have an even 0.200-sq in. cross section.

Impact Tests.   Many machines or parts of them must be designed to absorb impact loading. The aim of the designer is to provide for the absorption of as much of the energy as possible through *elastic* action and then to provide for some type of damping to dissipate it.

Standardized tests do not determine the energy-absorption characteristics

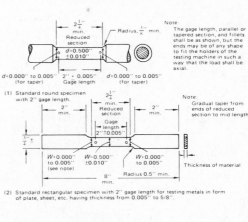

(a)   (b)

Figure 2.18.   A universal testing machine being used to pull a tensile specimen (a). Standard ASTM round and flat specimens for ductile materials are shown at (b). The machine may also be used for compression tests. (*Courtesy Acco, Wilson Instrument Division.*)

of a whole mechanism, but they can provide information on a given material by subjecting it to impact loading. Two tests developed to provide a measure of rupture strength of the material or *toughness* are the Izod and Charpy tests as shown in Fig. 2-19.

The tests are conducted on a machined specimen that is usually notched and struck a single blow. The energy absorbed in breaking the specimen is measured and converted into engineering design calculations. If the material distributes the stress uniformly, the impact value in terms of foot-pounds of energy absorbed will be high. Three tests are normally performed and the average of the three is used as the impact value.

Impact tests on ferrous alloys are usually carried out over a range of temperatures (−197 to 100°C) so that the transition temperature range can be determined. Above the transition temperature the specimen fractures ductilly while below this temperature the specimen fractures in a brittle manner. The transition temperature is especially important in the case of structural steels since there is a marked change in behavior over a relatively narrow temperature range.

Fatigue. Fatigue is a phenomenon that begins with minute cracks in a material that develop into fractures under repeated or fluctuating stresses. The maximum stress value is less than the tensile strength of the material. Normally the fatigue failures start at the surface, since the stresses are higher at that point.

The most common type of fatigue test is the rotating beam test. In this test, the specimen is rotated while being subjected to a bending moment. The purpose

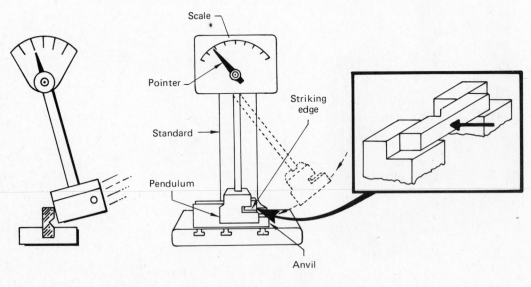

(a)   (b)

Figure 2.19. The Izod (a) and Charpy (b) impact-toughness tests.

of the rotation is to cause an alternate shift of the uniform bending stress from tension to compression every 180° of rotation (Fig. 2-20).

Great care must be taken in making the specimen, especially in machining the gage surface, since fatigue failure is very sensitive to surface influences.

To determine the endurance limit of a metal, it is necessary to prepare a number of similar specimens that are representative of the material. The first specimen is tested at a relatively high stress so that failure will occur with a relatively small number of stress applications. Succeeding specimens are then tested, each at a lower stress. Specimens stressed below the endurance limit will not rupture. The test results are usually plotted on S–N diagrams as shown in Fig. 2-21. For all ferrous metals tested and for most nonferrous metals, the S–N diagrams become horizontal, as nearly as can be determined for values of

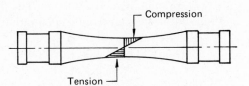

Figure 2.20. A rotating-beam specimen. A uniform bending moment is applied so that the upper fibers are always in compression and the lower fibers in tension. Thus in one cycle all the fibers have been subjected to both compression and tensile stresses. The stress ratio R, which is the algebraic ratio of minimum to maximum stress, is -1.

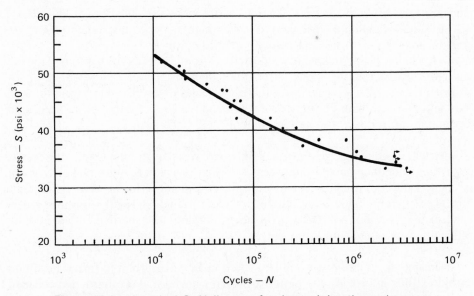

Figure 21.1. A typical S–N diagram for determining the endurance limit (fatigue) of metals under reverse flexural stress. The metal used for this curve was 7075–T6 aluminum.

N ranging from 1,000,000 to 50,000,000 cycles, indicating a well-defined endurance limit.

As discussed later in this chapter, alloying elements and heat treatment have considerable effect on metals. When the static strength of a metal is increased by these methods, so also is the fatigue life.

A general rule in design is to avoid, as much as possible, stress concentrations or abrupt changes in geometrical continuity such as holes, notches, threads, and rough machine marks.

THE ROLE OF ALLOYING ELEMENTS IN STEEL

General.   Every alloying element has a particular effect on a metal. Where a combination of two or more elements exist in the metal, the total result is usually an increase in each of the characteristics that is greater than the sum of their individual effects. Common alloying elements used in steel are carbon, nickel, chromium, molybdenum, vanadium, tungsten, manganese, copper, sulfur, boron, aluminum, and phosphorous. The effect of each of these elements on steel will be treated from a qualitative rather than a quantitative basis.

Carbon.   Technically, carbon should not be considered an alloying element of steel, since without it steel would not exist, but would remain iron. Varying amounts of carbon in the steel have a profound effect on its properties. Therefore, plain carbon steels are usually referred to by their carbon content as being low-, medium-, or high-carbon steels. These classifications are: 0.005 to 0.30% (low), 0.30 to 0.60% (medium), and 0.60 to 1.4% (high). Above this range of carbon content are the cast irons and below it are the wrought irons.

Among the properties influenced by the carbon content are hardness, tensile strength, yield strength, impact strength, reduction of area, and elongation, as shown in Fig. 2-22. It can be seen that tensile strength, yield strength, and hardness increase with carbon content. Impact strength, reduction of area, and elongation are reduced by carbon content.

Nickel.   Nickel increases toughness and resistance to impact. It lessens distortion in quenching and improves corrosion resistance. With nickel, a given strength value often can be obtained with considerably lower carbon content. Nickel is often included as an alloying element in high-strength structural steels used in the as-rolled condition or in heavy forgings that harden by air cooling rather than by quenching in oil.

Chromium.   Chromium, unlike nickel, joins with carbon to form chromium carbides, thus improving the depth to which a metal may be hardened (termed *hardenability*) and increasing its resistance to abrasion and wear. Chromium steels are stable at relatively high temperatures and are therefore used in steels where heat is a consideration. Chromium is also useful in resisting corrosion.

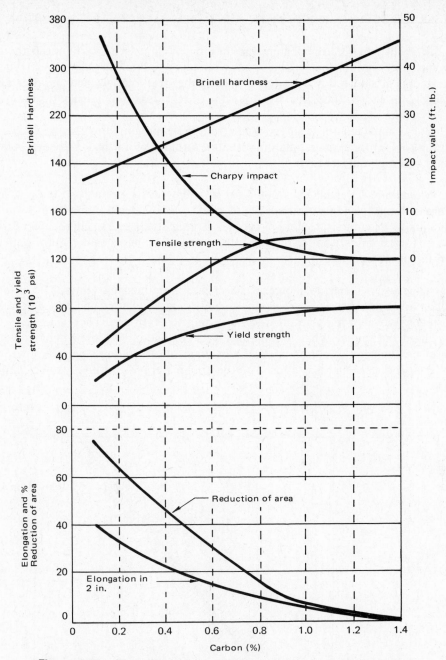

Figure 2.22. The effect of the carbon content on the mechanical properties of steel. The data represented are from 1-in. dia. rolled-steel bars.

Molybdenum.   Molybdenum, like chromium, joins with carbon and therefore promotes hardenability. It also has a tendency to hamper grain growth when the steel is at elevated temperatures, making the steel finer grained and tougher. Figure 2-23 shows the direct relationship of hardness and tensile strength of molybdenum steels.

Some familiar items that are made from molybdenum alloys are high-speed cutting tools, forged gears and crankshafts, turbine rotors, high-pressure boiler plate, and high-quality tubing.

Vanadium.   Vanadium is noted for providing a fine-grained structure over a broad range of temperatures. Parts made out of steels containing various amounts of vanadium are certain types of spring steel, gears, high-temperature steels, forged axles, shafts, turbine rotors, and other items that require impact and fatigue resistance.

Tungsten.   Tungsten increases hardness, promotes fine grain, and is excellent for resisting heat. Tungsten has a body-centered lattice and thus dissolves in steel at elevated temperatures, forming tungsten carbides. These carbides are very hard and stable. When used in higher percentages (18%) and combined with lesser percentages of chromium and vanadium, the resulting steel is known as a high-speed steel that is used as a cutting tool material for drills, reamers, lathe tools, etc.

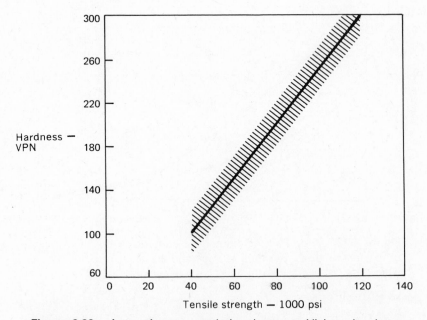

Figure 2.23.   Approximate correlation between Vickers hardness and tensile strength of molybdenum-based steels at room temperature. (*Courtesy Climax Molybdenum Company.*)

Manganese. Manganese is one of the basic alloying elements of steel. It is second only to carbon in frequency of use. Manganese contributes markedly to strength and hardness, but to a lesser extent than carbon. Actually the effectiveness of manganese is dependent on carbon. High manganese content, about 13%, causes the steel to strain or "work harden" easily, that is, it possesses the property of increasing hardness as the metal is worked. Thus the wear life of such items as power shovel teeth, crushing machinery, and railroad switch frogs is greatly increased.

Copper. Copper is added to steel in varying amounts, generally from 0.2 to 0.5%. It is used primarily to increase resistance to atmospheric corrosion, but it also acts as a strengthening agent.

Sulfur. Sulfur is considered an impurity that unites with steel to form iron sulphides, which contribute to cracking while being hot-worked. With manganese, this disadvantage is largely overcome due to the formation of manganese sulphide. Sulfur increases the free-cutting action of low-carbon steels because it breaks up the ferrite structure and helps the chips break free of the tool. Thus short chips are formed rather than long, hazardous types.

Boron. Boron is used to increase the hardenability of the steel. It is very effective when used with low-carbon alloy steels, but its effect is reduced as the amount of carbon increases.

Aluminum. Aluminum has an affinity for oxygen, therefore it is used in steel as an effective deoxidizer. It is also used to help form more and smaller grains.

Phosphorus. Phosphorus is also considered an impurity in steel. In large amounts, it increases the strength and hardness of steel but reduces ductility and impact resistance. Phosphorus in the low-carbon steels improves machinability.

HEAT TREATMENT

Heat treatment is a term used to denote a process of heating and cooling materials in order to obtain certain desired properties, often to obtain the best properties the material can offer. As an example, a shear blade must have its structure controlled to produce just the right combination of hardness, wear resistance, and toughness in order for it to successfully cut other metals. The reasons for heat treating ferrous metals may be listed briefly as:

1. To relieve internal stresses.
2. To refine grain size or to produce uniform grains throughout the metal structure.
3. To alter the microstructure.
4. To change the surface chemistry by adding or deleting elements.

Normalized Steel

The most common condition of steel is as a *normalized* structure. After the steel is produced in the furnace, it is poured into ingots for cooling and then rolled into sheets, plates, rods, etc. Thus the steel has proceeded from an elevated temperature to room temperature in a normal process. This process, along with other common heat treating processes such as annealing, hardening, tempering, stress relieving, and spheroidizing, can more easily be understood by following the iron–carbon equilibrium diagram shown in Fig. 2-24.

As steel cools from the elevated temperature, it passes through several stages, as shown on the diagram: from liquid, to liquid and austenite, and then to austenite. On the left, under the A$_3$ line, is austenite plus ferrite and on the right austenite plus cementite. Finally, below the A$_1$ line is ferrite plus pearlite, pearlite, and pearlite plus cementite. The formation of each of these structures will be discussed.

Austenite.   Austenite is shown in the area above the line GSE and is a solid solution of carbon in a face-centered cubic iron. The A$_3$ line represents the initial precipitation of ferrite from the austenite. The line SE indicates the primary deposition of cementite (Fe$_3$C) from austenite.

Ferrite.   Ferrite is an alpha iron with a body-centered cubic lattice that exists in the very narrow area at the extreme left of the diagram below 1674°F (911°C).

Cementite.   Cementite is a very hard, brittle compound of iron and carbon, Fe$_3$C, containing 6.67% carbon.

Pearlite.   Pearlite is a two-phase structure consisting of thin, alternate layers of iron carbide (cementite) and ferrite, as shown in Fig. 2-25 and in the enlarged structural views in the diagram in Fig. 2-24.

Normally, as the metal cools slowly from austenite, there is an automatic separation of ferrite and the ferrite–cementite mixture (pearlite), as shown in the diagram. (The white areas represent ferrite and the lined areas pearlite.) As the carbon content increases, it unites with greater amounts of ferrite, thus increasing the pearlite. At a point where all of the ferrite is in combination with carbon, the structure is entirely pearlite, as shown by the centered microstructural view. Theoretically, this is at 0.83% carbon but may range from 0.75 to 0.85% in plain carbon steels. It is represented on this diagram at 0.80% carbon. This combination of iron and carbon is known as a *eutectoid* steel. Eutectoid, taken from the Greek, means "most fusible." This combination occurs in binary alloys when a complete solid solubility does not exist. A eutectic transforms at a lower temperature than the melting point of either of the components. Steels with more than 0.80% carbon are called *hypereutectoid* and those below 0.80% carbon are

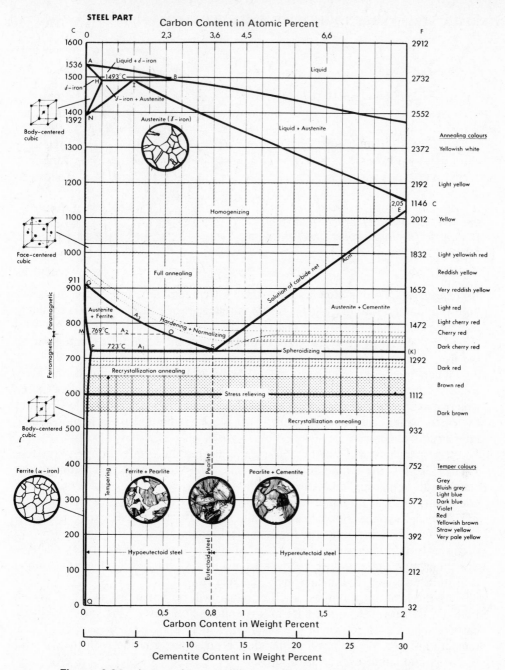

Figure 2.24. Iron-carbon equilibrium diagram. (*Courtesy Struers Scientific Instruments.*)

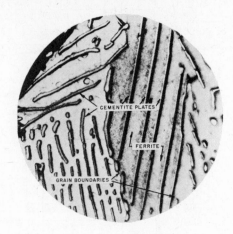

Figure 2.25.   Pearlite structure of steel.
(*Courtesy The International Nickel Co. Inc.*)

hypoeutectoid. During slow cooling, the Fe_3C precipitates out of the austenite until the temperature of 1333°F (723°C) is reached, at which time the remaining austenite changes to pearlite. The result is a pearlite structure surrounded by a network of cementite at the grain boundaries, as shown in the enlargement on the iron–carbon diagram.

When steel containing more than 2.11% carbon is slowly cooled, the alloy freezes out into iron and graphite. The fine, soft graphite forms flakes in the iron matrix, producing a gray cast iron. If cooling is more rapid, a brittle structure known as white cast iron will result. Cast irons will be discussed in more detail later in this chapter.

If a specimen of hypoeutectoid steel is heated uniformly, at approximately 1333°F (723°C), the temperature of the steel will stop rising, even though the heat is still being applied. After a short time, the temperature will continue to rise again. Ordinarily, metals expand as they are heated, but it is found that at 1333°F a slight contraction takes place and then, after the pause, expansion again takes place. This indicates the lower critical temperature of the steel, shown in the iron–carbon diagram as line A_1. Actually, an atomic change takes place at this point, and the structure changes from a body-centered arrangement (alpha iron) to a face-centered arrangement (gamma iron).

As the heating continues, another pause will be noted at line A_3, at which time the transformation of ferrite to austenite is complete. This varies with the carbon content, as shown in the diagram, and is known as the upper critical temperature. At this point all of the carbon goes into solution with the iron, so that it is now more evenly distributed.

Hardening Steel

Three requirements are necessary in order to successfully harden a piece of steel: first, the steel must contain enough carbon; second, it must be heated to the correct temperature; third, it must be cooled or quenched.

Carbon Content. In order to get extreme hardness in a piece of steel, it is necessary that the steel contain 80 points or more of carbon (one point of carbon = 0.01% carbon). Low-carbon steels (less than 25 points carbon) will not be materially affected by heat treatment. Medium-carbon steels (30 to 60 points carbon) may be toughened considerably by heat treatment, but they will not be hardened to a very great extent. High-carbon steels (60 to 150 points carbon) may be successfully hardened by simple heat-treating methods.

Heating. The steel must be heated above the upper critical (A_3) temperature. The actual hardening range is about 100 to 200°F above the upper critical temperature. Notice that this curve levels out beyond 0.80% carbon.

After the steel has been heated at the hardening range for a sufficient period of time to equalize the heat (about 30 minutes per inch of cross section), it is taken out of the furnace and quenched in a cooling medium, which may be water, brine, oil, molten salts, or lead baths. The rate at which the material cools will largely determine its hardness. In order to avoid undue stress caused by uneven cooling during the quenching action, the part should be vigorously agitated to prevent steam pockets from forming on the surface.

Given the proper heat treatment, the hardness will be determined largely by the carbon content. However, the depth of hardness, known as *hardenability*, will be determined largely by the alloying elements. The effect of alloying elements on the hardenability is shown in Fig. 2-26.

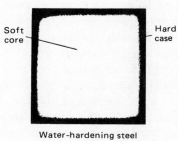

Soft core   Hard case

Water-hardening steel

(a)

Figure 2.26. The effect of alloys on the hardenability of steel. A plain high-carbon steel gets an outer layer of hardness when heat-treated (a). If manganese is added to plain carbon steel, it can be cooled at a slower rate, as in oil, and the hardness penetration will be deeper (b). As more alloys are added, the steel can be cooled more slowly, as in air, and the hardness will penetrate through the section (c).

Type analysis
Carbon, plus
Manganese 1.60%

(b)

Type analysis
Carbon, plus
Manganese 2.00%
Chromium 1.00%
Molybdenum 1.35%

(c)

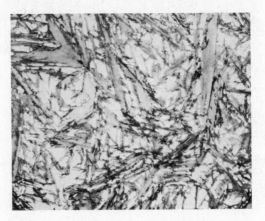

Figure 2.27. Martensite structure
$(500\times)$.

When austenite is immersed in a water quench, it does not have time to separate out and form pearlite. Some of the austenite transforms almost instantaneously to *martensite,* an interlaced, needlelike structure as shown in Fig. 2-27 that is hard and brittle. How this change takes place can be shown by means of another diagram, termed a time–temperature–transformation diagram or TTT diagram (Fig. 2-28). The diagram shows the various structures that occur as the metal is cooled and the times at which they occur. As an example, if a high-carbon steel is quenched in water such that it reaches a temperature of 550°F (288°C) in about 50 sec or less, a martensite transformation would start (M_s). The martensite transformation would be finished (M_f) at about 325°F (160°C). The martensite that results is a supersaturated solid solution of carbon trapped in a body-centered tetragonal structure. It involves no chemical change from the austenite stage. The unit cells of the martensite structure are highly distorted because of the entrapped carbon, which is the principal hardening mechanism. The body-centered cubic tetragonal structure is less densely packed than the FCC of austenite, hence there is about a 4% expansion causing still further localized stresses that distort the austenite matrix. Normally these internal stresses would have to be relieved as soon as possible to prevent cracking by a tempering process.

If the high-carbon steel had been cooled slightly slower, other structures may have formed in addition to martensite. This can be shown by a continuous-cooling-transformation diagram (C-T), Fig. 2-29. An end-quench hardenability bar is superimposed on the top of the diagram. The letters *A, B, C,* and *D* represent various distances from the quenched end and the corresponding cooling curves are shown on the transformation and C-T diagrams. Thus, the hardness of the bar at *A* will be the highest, as it has missed the "nose" of the transformation curve. At points *B, C, D,* and *E* the cooling will be progressively slower, so that it passes through the austenite–ferrite, austenite–bainite, and austenite–martensite transformation curves. The metal will consist of martensite, ferrite, and bainite, as shown by the microstructures below the diagram.

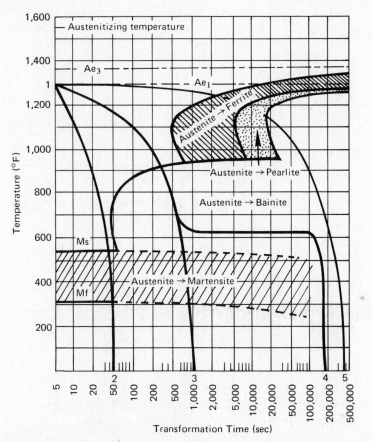

Figure 2.28. An isothermal-transformation diagram showing the transformation temperatures in relation to time as austenite is cooled. (*Courtesy United States Steel Corporation.*)

Bainite. Bainite is an intermediate-type structure, feathery in appearance, that forms between pearlite and martensite when the cooling rate is slowed to more than the critical rate. If the critical cooling rate is exceeded martensite will form.

Tempering

As mentioned previously, after quenching the steel is in a highly stressed, unstable condition. To avoid, or minimize, problems of cracking and distortion, the metal is reheated or tempered. There are several different ways this may be done: conventional reheating after quenching, martempering, and austempering (Fig. 2-30).

The conventional method consists of reheating the metal immediately after

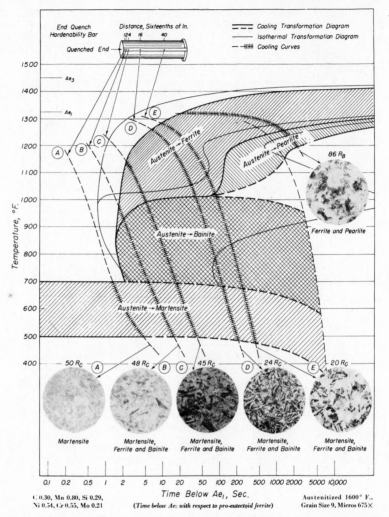

Figure 2.29. Jominy end-quench hardenability tests of a AISI 8630 steel correlated with continuous-cooling-transformation and iso-thermal-transformation diagrams. (*Courtesy United States Steel Corp.*)

quenching to a temperature less than critical. Since tempering softens the steel, the relationship of hardness to strength should be known. As an example, a steel that may have a hardness reading of R_c 62 (file hard) after quenching may be reheated to a temperature of 450°F (232°C) and then have a hardness reading of R_c 56. This would be hard enough to have high strength and good wear resistance, with the advantage of having a more stable structure.

In martempering, air-hardening steels may be quenched in a bath of molten

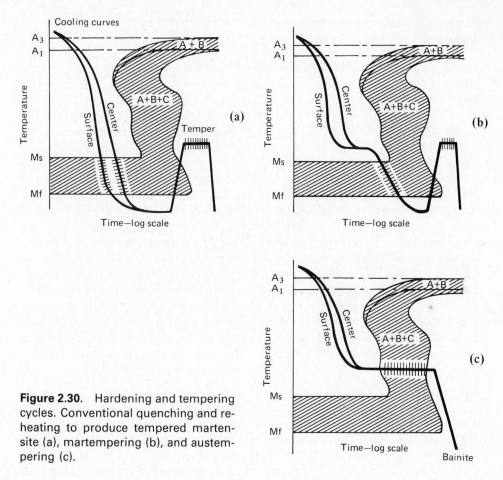

Figure 2.30. Hardening and tempering cycles. Conventional quenching and reheating to produce tempered martensite (a), martempering (b), and austempering (c).

salts at a temperature just above the M_s line. This allows the inside of the steel to arrive at the same temperature as the outside. The steel is then removed from the bath and allowed to cool at room temperature. For straight-carbon and some alloy steels a quench after the hold time is necessary. This process has the advantage of reducing the stress on the steel as it changes to martensite, since the temperature has already been equalized throughout. Tempering follows [see Fig. 2-30 (b)].

Austempering also pauses in the cooling curve to equalize the outside temperature of the part with that of the inside. The part then is kept at that temperature to produce a bainite structure. This eliminates the additional step of reheating the metal. The hardness can be the same as that of tempered martensite, depending on the transformation temperature. If a temperature of 450°F (232°C) is used as in the example of conventional tempering, the same relative hardness of R_c56 would be expected.

Annealing

Annealing in general refers to softening a metal by heating and cooling. It is generally of two types, *full annealing* and *stress-relief annealing*, as shown in the iron–carbon diagram, Fig. 2-24.

Full annealing consists of heating the steel about 100°F (38°C) above the hardening and normalizing range. It is held at that temperature for the desired length of time and then allowed to cool very slowly, usually in the furnace. This produces a soft, coarse, pearlitic structure that is stress-free.

In stress-relief annealing, the metal is heated to a temperature close to the lower critical temperature, followed by any desired rate of cooling. Stress-relief annealing is usually done to soften a metal that has become strain-hardened or work-hardened during a forming operation. As shown on the iron–carbon diagram, there is a recrystallization zone. At this temperature, the grains of a cold-worked steel will recrystallize and form fine grains. There are two zones shown for recrystallization. This does not mean that recrystallization occurs only at these temperature ranges, but rather that it can occur at different temperatures depending upon the amount of prior cold work. It takes energy to make the grains recrystallize, and if they have a considerable amount of energy already in the structure, it requires less to make them recrystallize. If there is no stress on the metal, the grains will not recrystallize until just below the lower critical temperature.

Spheroidizing

When steel is tempered at a temperature just below the lower critical or A_1 line, as shown in the iron–carbon diagram, the cementite will consist of small spheroids surrounded by ferrite (Fig. 2-31). Prolonged heating (16 to 72 hr) at this temperature may be required. The process may be speeded up, particularly on smaller items, by alternately heating them to temperatures slightly below and then slightly above the lower critical. The length of time and the number of cycles will be dependent on the original structure of the steel. A fine pearlite is preferred.

Spheroidizing results in greater ductility, which improves the forming qualities as well as machinability.

Surface Hardening of Steels

The heat-treating processes discussed were generally applicable to medium- and high-carbon steels. The depth of hardness was based on the alloy content and the quench rate. It is often desirable to have a hard surface accompanied by a softer, tougher core. This can be accomplished by flame hardening or induction hardening on steels of medium-carbon content or higher and cast irons of suitable composition. Carburizing and nitriding processes are used on low-carbon steels.

Flame Hardening.   Flame hardening consists of moving an oxyacetylene flame over a part, followed by a quenching spray. The rate at which the flame

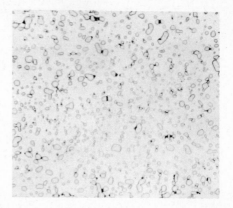

Figure 2.31. Spheroidized iron carbides in a matrix of annealed steel, at a magnification of 750×. (*Courtesy The International Nickel Co. Inc.*)

is moved over the part will determine the depth to which the material is being heated to a critical temperature or higher. Quenching can be built into the burner as shown in Fig. 2-32.

In flame hardening, there is no sharp line of demarcation between the hardened surface zone and the adjacent layer, so there is little likelihood that it will chip out or break during the service. The flame is kept at a reasonable distance from sharp corners to prevent overheating and drilled or tapped holes are normally filled with wet asbestos for edge protection. Stress relieving at about 400°F (204°C) is recommended for all flame-hardened articles except those made from air-hardened steel.

Some advantages of flame hardening are:

1. Large machined surfaces can be surface hardened economically.
2. Surfaces can be selectively hardened with minimum warping and freedom from quench cracking. Examples of selective hardening are gear teeth, machine ways, cam surfaces, and engine push rod ends.
3. Scaling of the surface is only superficial because of the relatively short heating cycle.
4. The equipment may be controlled electronically to provide precise control of the hardness.
5. The depth of case may vary to suit the part, from $\frac{1}{8}$ to $\frac{1}{2}$ in.

Disadvantages include:

1. To obtain optimum results, a technique would have to be developed for each design.
2. Overheating can cause cracks, or excessive distortion, especially where thin sections are involved.

Induction Hardening. Induction hardening is done by heating the metal with a high-frequency alternating magnetic field. Heat is quickly generated by high-frequency eddy currents and hysteresis currents on the surface layer. The primary current is carried by a water-cooled copper tube, the workpiece serves

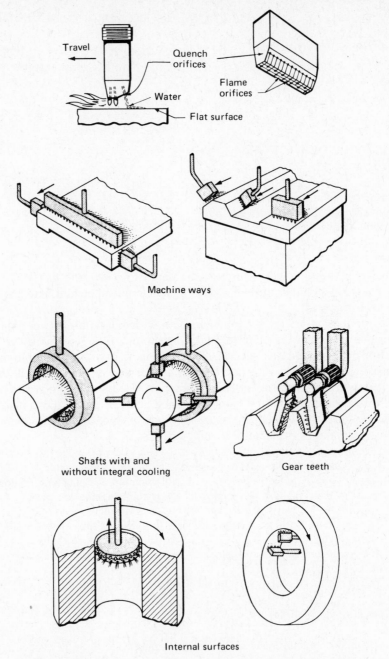

Travel

Quench
orifices

Water

Flat surface

Flame
orifices

Machine ways

Shafts with and
without integral cooling

Gear teeth

Internal surfaces

Figure 2.32. Flame hardening with various-shaped burners.

as the secondary circuit. The depth of penetration decreases as the frequency of the current increases, for example, the approximate minimum hardness depth for 3000 Hz is 0.060 in. and for 500,000 Hz is 0.020 in.

Induction hardening is fast, even on comparatively large surfaces. As an example, a large truck crankshaft can be brought to the proper temperature and spray-quenched in 5 sec. There is very little distortion due to the short cycle time.

Evaluation of Surface-Hardening Treatments.   If the prime consideration of surface hardening is wear resistance, then it is best to choose processes such as nitriding and cyaniding, where sufficient carbides are developed at the surface. If impact or torsion is involved, excess carbides will be detrimental, since cracks, chipping, and spalling will develop. As the case depth increases, the ability to withstand this type of loading also increases. Surface hardening procedures and applications are summarized in Table 2-2.

FABRICATING CHARACTERISTICS OF FERROUS METALS

Fabricating characteristics of a metal refer to the ability or ease with which it can be formed, machined, cast, or welded.

Before considering the fabricating characteristics of ferrous metals, it is important to know how they are classified, so that references can be made to the qualities of a specific metal.

The common ferrous alloys that will be discussed are plain carbon steels, low-alloy steels, tool steels, steel castings, cast iron (gray, malleable, and ductile), and stainless steel.

Classification of Alloy Steels

The first coded numbering system for alloy steels was made by the Society of Automotive Engineers (SAE). It consists of a four- or five-digit system: the first two digits refer to the type and percent of alloy and the last two (or three) refer to the percent or points carbon. The American Iron and Steel Institute (AISI) developed a code similar to that of SAE, with some prefixes and suffixes as shown in Table 2-3. Identification of steels in this code system is relatively simple; for example, AISI 4340 steel is of a nickel–chromium–molybdenum type with 0.40% carbon.

In addition, steels are often referred to as hot-rolled steels (HRS) or cold-rolled steels (CRS). Hot-rolled steels are easily identified because of an oxide scale that is formed during the heating process. Cold-rolled steels have the scale removed by acid baths or other methods and are then rolled to final size. The CRS are generally easier to machine, especially in the low-carbon range, and the dimensions are held within closer tolerances.

In addition to the standard code given, other organizations that have issued

Table 2-2. Summary of surface-hardening methods and applications.

Case Depth (in.)	Hardness, Rockwell C*	Remarks and Applications
Carburizing		
To 0.020	S 60+ C 18+	Light case depths used for high wear resistance and low loads. Typical applications are push-rod balls and sockets, shifter forks, small gears, and water-pump shafts.
0.020–0.040	S 60+ C 18+	Moderate case depths used for high wear resistance and moderate to heavy service loads. Applications include steering-arm bushings, valve rocker arms and shafts, gears, and brake-pedal shafts.
0.040–0.060	S 60+ C 18+	Heavy case depths used for high wear resistance to sliding, rolling, or abrasive action, and for high resistance to crushing or bending loads. Applications include ring gears, transmission and slide gears, piston pins, gear shafts, roller bearings, and kingpins.
0.060+	S 60+ C 18+	Extra heavy case for maximum wear and shock resistance. Typical applications are camshafts, armor plate, and cam surfaces.
Carbonitriding		
0.003–0.020	S 62–65 C 32–35	Produces a hard, wear-resistant, clean surface. Used on thin-wall tubing, rachet wrenches, bolts, screws, small gears, and pneumatic cylinders.
Nitriding		
To 0.030	...	Cycle of 50 hr produces 0.015-in. case, where initial 0.006 in. has Vickers hardness over 900. Cycle of 100 hr produces 0.030-in case where initial 0.011 in. has 900+ Vickers hardness. Applications include aircraft exhaust valves, instrument shafts, pump shafts, and steam valves.
Flame Hardening		
0.030–0.125	S 37–55 C 20+	Produces high surface hardness with unaffected core. Surface is relatively free from scaling and pitting. Used on sprocket and gear teeth, track rails, and lathe beds and centers.
Induction Hardening		
0.030–0.125	S 60+	Produces high surface hardness with ductile core. Parts have good fatigue resistance. Applications include camshafts, sprocket and gear teeth, rocker-arm shafts, mower and shear blades, lathe beds, and bearing surfaces of axle shafts and crankshafts.

* S = surface; C = core.

metal specifications are American Society for Testing Materials (ASTM), Aerospace Materials Specifications (AMS), American Society of Mechanical Engineers (ASME), Index of Federal Specifications and Standards, and Department of Defense. An excellent Information Guide on Metals is given in the May 1970 issue of *Materials Research and Standards*. The one most comprehensive publication is the *Annual Book of ASTM—Standards*. It contains over 1200 standards relating to metals and metal products published in 10 of the 33 parts of the book. These standards are widely used by engineers and, for purchasing specifications, by the major metal industries.

Fabricating Characteristics

The fabricating characteristics of each of the metals that follow will be discussed, where applicable, from the standpoint of machinability, formability, weldability, and castability.

Machinability

Machinability is an involved term with many ramifications. Simply stated, however, it refers to the ease with which the metal may be sheared in such operations as turning, drilling, reaming, threading, sawing, etc.

Ease of metal removal implies, among other things, that the forces acting against the cutting tool will be relatively low, that the chips will be easily broken up, that a good finish will result, and that the tool will last a reasonable period of time before it has to be replaced or resharpened. Another way of expressing this is to give each material a *machinability rating*. This has been done for most ferrous metals, using AISI B1112 as the basis of 100% machinability. Thus another metal may be said to have a machinability rating of 60%, as in the case of one type of stainless steel. Factors that increase or decrease machinability of a metal are shown in Table 2-4.

Formability

The ability of a metal to be formed is based on the ductility of the metal, which, in turn, is based on its crystal structure. Metals that have a face-centered cubic crystal have the greatest opportunity for slip—four distinct nonparallel planes and three directions of slip in each plane.

Other factors that govern, to a large extent, the flowability or ductility of the material are grain size, hot and cold working, alloying elements, and softening heat treatments such as annealing and normalizing.

Grain Size. If all metals consisted of a single grain or crystal, the tendency to slip in any pure metal would depend solely upon the number of slip planes in the crystal and upon the directions of ready slip in each plane. However, every metal contains many separate grains, and the planes of slip or direction of slip rarely coincide with each other. The tendency to slip in any single grain, therefore, is obstructed, to a certain extent, by the resistance of opposing slip planes in adjacent grains. As will be seen later, small grain sizes are recommended

Table 2-3. AISI (American Iron and Steel Institute) standard steels.

AISI Series	Nominal Composition or Range
Carbon Steels	
10XX Series	Non-resulphurized carbon steels with 44 compositions ranging from 1008 to 1095. Manganese ranges from 0.30 to 1.65%; if specified, silicon is 0.10 max. to 0.30 max., each depending on grade. Phosphorus is 0.040 max., sulphur is 0.050 max.
11XX Series	Resulphurized carbon steels with 15 standard compositions. Sulphur may range up to 0.33%, depending on grade.
B11XX Series	Acid Bessemer resulphurized carbon steels with 3 compositions. Phosphorus generally is higher than 11XX series.
12XX Series	Rephosphorized and resulphurized carbon steels with 5 standard compositions. Phosphorus may range up to 0.12% and sulphur up to 0.35%, depending on grade.
Alloy Steels	
13XX	Manganese, 1.75%. Four compositions from 1330 to 1345.
40XX	Molybdenum, 0.20 or 0.25%. Seven compositions from 4012 to 4047.
41XX	Chromium, to 0.95%, molybdenum to 0.30%. Nine compositions from 4118 to 4161.
43XX	Nickel, 1.83%, chromium to 0.80%, molybdenum, 0.25%. Three from 4320 to E 4340.
44XX	Molybdenum, 0.53%. One composition 4419.
46XX	Nickel to 1.83%, molybdenum to 0.25%. Four compositions from 4615 to 4626.
47XX	Nickel, 1.05%, chromium, 0.45%, molybdenum to 0.35%. Two compositions, 4718 and 4720.
48XX	Nickel, 3.50%, molybdenum, 0.25%. Three compositions from 4815 to 4820.
50XX	Chromium, 0.40%. One composition, 5015.
51XX	Chromium to 1.00%. Ten compositions from 5120 to 5160.
5XXXX	Carbon, 1.04%, chromium to 1.45%. Two compositions, 51100 and 52100.
61XX	Chromium to 0.95%, vanadium to 0.15% min. Two compositions, 6118 and 6150.
86XX	Nickel, 0.55%, chromium, 0.50%, molybdenum, 0.20%. Twelve compositions from 8615 to 8655.
87XX	Nickel, 0.55%, chromium, 0.50%, molybdenum, 0.25%. Two compositions, 8720 and 8740.
88XX	Nickel, 0.55%, chromium, 0.50%, molybdenum, 0.35%. One composition 8822.
92XX	Silicon, 2.00%. Two compositions, 9255 and 9260.
50BXX	Chromium to 0.50%, also containing boron. Four compositions from 50B44 to 50B60.
51BXX	Chromium to 0.80%, also containing boron. One composition, 51B60.
81BXX	Nickel, 0.30%, chromium, 0.45%, molybdenum, 0.12%, also containing boron. One composition, 81B45.

Table 2-3 (continued).

AISI Series	Nominal Composition or Range
94BXX	Nickel, 0.45%, chromium, 0.40%, molybdenum, 0.12%, also containing boron. Two compositions, 94B17 and 94B30.

Additional notes: When a carbon or alloy steel also contains the letter "L" in the code, it contains from 0.15 to 0.35% lead as a free-machining additive, i.e., 12L14 or 41L40. The prefix "E" before an alloy steel, such as E4340, indicates the steel is made only by electric furnace. The suffix "H" indicates an alloy steel made to more restrictive chemical composition than that of standard steels and produced to a measured and known hardenability requirement (i.e.) 8630 H or 94B30H. XXs indicate nominal carbon content within range.

Table 2-4. Factors affecting machinability of metals.

	Factors That Increase Machinability	Factors That Decrease Machinability
Structure	Uniform microstructure Small, undistorted grains Spheroidal structure in high-carbon steels	Nonuniformity Presence of abrasive inclusion Large, distorted grains
	Lamellar structure in low- and medium-carbon steels	Spheroidal low- and medium-carbon steels Lamellar high-carbon steels
Treatment	Hot-working of alloys that are hard, such as medium- and high-carbon steels	Hot-working of low-carbon steels
	Cold-working of low-carbon steels Annealing, normalizing, tempering	Cold-working of higher-carbon steels Quenching
Composition	Small amounts of lead, manganese, sulphur, phosphorus Absence of abrasive inclusions such as Al_2O_2	Carbon content below 0.30% or above 0.60% High-alloy content in steels

for shallow drawing of copper, and relatively large grains for heavy drawing on the thicker gages.

Hot and Cold Working. The tremendous pressures encountered in hot working tend to reduce the size of the crystals, either by preventing the growth of the crystals at elevated temperatures or by breaking up the existing crystals. Generally, the grains are distorted in the process. The amount of distortion will be a determining factor in the ductility of the metal.

Cold working also results in varying degrees of distorted crystals. Generally, cold-worked crystals are more distorted than are the hot-worked, and therefore cold-worked metals are usually less ductile than the hot-worked.

Alloying Elements.   Most alloying elements in a pure metal reduce its ductility. Whether the alloying element has replaced atoms of pure metal or has found room for itself in the spaces between the atoms of pure metal, the effect is to reduce the number of slip planes in which ready slippage can occur. For example, steel, which is an alloy of carbon and iron, is less ductile than iron. As steel solidifies, it can hold only an extremely small amount of carbon in solution. The excess carbon that is forced out of the individual cells of the atomic structure combines immediately with some iron to form iron carbides. Not only are the slip planes somewhat distorted by the presence of the alloy, but, as in the case of steel, the iron carbides offer increased resistance to slip. Therefore, the ductility of the steel decreases as the amount of carbon increases.

PLAIN CARBON STEELS

Plain carbon steels are steels whose principal alloy is carbon with no other special alloying elements.

Plain carbon steels are allowed to solidify under varying conditions which alter their characteristics. Under this classification the terms *killed, semikilled,* and *rimmed* are often used.

Killed.   Killed steels are those that have been chemically deoxidized so that they lie quietly in the mold as they cool. They are characterized by relatively uniform chemical composition and properties. Strips and sheets made of killed steel have excellent forming and drawing qualities.

Semikilled.   These steels are partly deoxidized before pouring into the mold. Thus, some gas evolution takes place during solidification. These steels are satisfactory for all but the most severe drawing operations.

Rimmed.   Rimmed steels get their name from the fact that the outer rim of the ingot has considerable gas evolution during the initial solidification period from the liquid state. During solidification, a layer of iron about three inches thick forms around the outer edge of the ingot. The liquid in the center retains the carbon, phosphorus, and sulfur. This rimmed steel is often known as drawing quality (DQ) steel. However, where draws or forming are more severe, killed steel may be required.

Temper.   When cold-rolled sheets are specified to a hardness range, they are designated by tempers as follows: quarter-hard (R_b 60 to 70 range), half-hard (R_b 70 to 80 range), and full-hard (R_b 84).

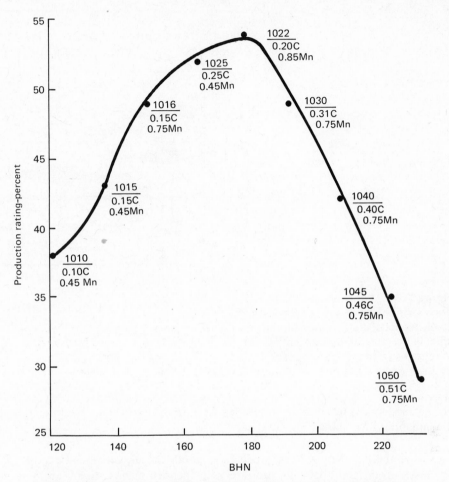

Figure 2.33. Small amounts of carbon improve the machinability of plain carbon steels up to a point. Beyond this point, carbon additives increase hardness but are detrimental to machinability. (*Courtesy* American Machinist.)

Machinability.   The machinability of plain carbon steels may be approximated by the curve shown in Fig. 2-33. Very low carbon steels do not machine well because the structure is mostly ferrite, which is too soft to produce a good shearing action. These steels must have manganese sulphides to help break up the ferrite structure, as shown in Fig. 2-34. Lead will also help break up the ferrite structure.

A number of grades of lead-bearing steels are available and are classified as free-machining steels. A small percentage of lead, 0.15 to 0.35%, is added to the steel just shortly before it solidifies. It is immediately dispersed throughout the metal. Leaded steels can be used at 50% higher cutting speeds than corresponding plain carbon steels. However they are more expensive and should

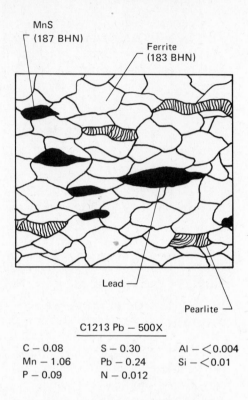

Figure 2.34. Lead and sulphur additions form inclusions in the low-carbon steel microstructure that promote machinability. The sulphur combines with manganese and the lead appears either by itself or as a tail in MnS inclusions.

C1213 Pb — 500X

C — 0.08	S — 0.30	Al — <0.004
Mn — 1.06	Pb — 0.24	Si — <0.01
P — 0.09	N — 0.012	

not be used unless production conditions warrant the extra cost. Factors affecting machinability are shown in Table 2-4.

Formability.   Low-carbon steels have good forming qualities because there are less carbon and alloys to interfere with the slip planes. Steels in this class can be given a class-2 bend, which is a 90° bend with a radius of *t,* or the thickness of the metal. Classes of bends have been defined by the American Society for the Testing of Metals (ASTM).

Medium-carbon cold-rolled steels are too low in ductility for any practical degree of cold forming. Hot rolled, medium-carbon steels are somewhat more ductile and can be bent to a 1-*t* radius up to 0.09 in. thick.

Weldability.   Plain carbon steel is the most weldable of all metals. It is only as the carbon percentages increase that there is a tendency for the metal to harden and crack. Fortunately, 90% of the welding is done on low-carbon (0.15%) steels. This amount of carbon presents no particular difficulties in welding. Near the upper end of the low-carbon range (0.24 to 0.30%), there may be some formation of martensite when extremely rapid cooling is used.

Medium- and high-carbon steels will harden when welded if allowed to cool at speeds in excess of the critical cooling rates. Preheating to 500 or 600°F (260–316°C) and postheating between 1000 and 1200°F (538–649°C) will remove any of the brittle microstructures.

The extra-high-carbon steels, or tool steels, having a carbon range of 1.00 to 1.70%, are not recommended for high-temperature welding applications. These metals are usually joined by brazing with a low-temperature silver alloy. Because of the lower temperature of this process, it is possible to repair or fabricate tool-steel parts without affecting their heat-treated condition.

ALLOY STEELS

Classification

Alloy steels may be defined as steels with alloying elements that exceed one or more of the following limits: 1.65 Mn; 0.60 Si; 0.60 Cu or aluminum, boron, chromium up to 3.99%; and containing one or more of the following elements: cobalt, columbium, molybdenum, nickel, titanium, tungsten, vanadium, and zirconium.

The nominal chemical compositions of the standard AISI alloy steels were given in Table 2-3. The carbon content was not given; however, the highest nominal carbon content is 0.95%.

In addition to the SAE–AISI code classification discussed, alloy steels may be classified as through-hardenable and surface-hardenable. Each type includes a broad family of steels whose chemical, physical, and mechanical properties make them suitable for specific product applications.

Through-hardening types are used when maximum hardness and strength must extend deep into the part.

Surface-hardening grades are used where a tough core and relatively shallow hardness are needed. These steels, after nitriding or carburizing, are used in truck transmission gears and steering worm gears—parts that must withstand wear as well as high stresses. Shown in Table 2-5 is a classification of the various types and uses of through-hardenable and surface-hardenable steels.

Fabricating Characteristics

Machinability. One reason for the widespread use of the alloy steels in the lower carbon range, such as 4024 and 4028, is that good machinability is obtained. The machinability drops off as the carbon content increases. For example, the machinability rating of 4023 is 70% (based on B1112 as 100%), 4032 is 65%, and 4047 is 55%. This holds true for the other alloy steels also.

Formability. Alloy steels are not usually used where forming operations other than forging are required. Here they are employed extensively for gears, bearings, crankshafts, connecting rods, axle shafts, and any other uses where good strength and toughness are necessary. These steels are often induction-hardened to provide a surface that can take high compressive loads and still have a core with great toughness.

Weldability. Weldability for alloy steels is generally good as long as the carbon content is in the low range. As the carbon content increases, preheating

Table 2-5.   Classification and use of plain-carbon, through, and surface-hardening steels.

Type	Carbon (%)	Hardness Range	General Characteristics and Applications
Through Hardenable			
Medium-carbon	0.3 to 0.5	250 to 400 BHN	Strength and toughness; for shafts, miscellaneous forgings, some gears (frequently cyanided), bolts, nuts, flanges, bearing cages.
High-carbon	0.5 to 0.7	375 to 500 BHN	Strength and moderate wear resistance; for springs and collets.
Bearing	1.0	60 to 64 R_c	High strength and high resistance to wear and scuffing; for bearings, balls, rollers, spacers, pins, and bushings.
Surface Hardenable			
Carburized	0.15 to 0.3	Case, over 60 R_c Core, to 45 R_c (case, 0.15 to 0.125 in. thick)	Wear-resistant case, high endurance strength; used for gears, shafts, bearings, and inspection fixtures.
Flame- or induction-hardened	Over 0.4	Case, over 55 R_c Core, to 45 R_c (case over 0.050 in. thick)	Wear-resistant case; used for shafts, pins, some gears, cages, and bushings.
Nitrided	0.3 to 0.5 and Nitralloys	Case, 500 to 1000 DPH Core, over 30 R_c (case, to 0.030 in. thick)	Wear-resistant case, high endurance strength, some resistance to corrosion and elevated temperatures; used for shafts, gears, couplings, and bushings.

and postheating are often used to reduce the stress. Heat treatment after welding will help produce a uniform structure in the weld and parent metal. For most of these steels, the best results are obtained by arc welding using a low-hydrogen electrode. Reducing the hydrogen content of the weld helps to eliminate brittleness.

Castability.   An important characteristic of alloy steels is their ability to air-harden. Thus, complicated castings can be hardened when using these alloys, and tensile strengths from 70,000 to 100,000 psi can be obtained without quenching. Nickel and molybdenum with manganese increase the capacity to air-harden.

Combinations of chromium, nickel, manganese, and vanadium are used to produce wear resistance and high strength.

Table 2-6. Representative tradenames of HSLA bar steels. (*Courtesy* Machine Design.)

Producer	ASTM Specification (Minimum Yield Strength)			
	A242 (To 50 ksi)	A441 (To 50 ksi)	A572 (42 to 70 ksi)	A588 (to 50 ksi)
Armco Steel Corp.	High-Strength A	High-Strength B	High-Strength C	High-Strength A-588
Bethlehem Steel Corp.	Mayari R	Manganese Vanadium	V Steels	Mayari R 50
Inland Steel Co.	Cor-Ten*	Tri-Steel	INX Steels	Cor-Ten*
Jones & Laughlin Steel Corp.	J&L Cor-Ten A*	Jalten #1	JLX Steels	J&L Cor-Ten B*
Republic Steel Corp.	Republic 50 & 60	Republic A-441	X-W Steels	Republic 50 & 60
U.S. Steel Corp.	Cor-Ten A*	Tri-Ten	Ex-Ten	Cor-Ten B*
Wisconsin Steel Co.	—	—	IHX Steels	—

* Cor-Ten is a registered tradename of U.S. Steel Corp., and other steel companies are licensed to produce the product.

High-strength low-alloy steels

HSLA steels have become increasingly important in recent years because of the weight consciousness of automobile manufacturers and of others who do not build stationary structures. These steels have been used where fatigue performance is critical; however, it is anticipated that they will be used extensively in parts, such as appliance frame members, in thicknesses of about 0.100 in. (2.54 mm), which will provide a significant weight savings.

To be competitive with mild steel in the automotive industry, the use of hot-rolled 50 ksi (345 MP$_a$) S_y HSLA steel must result in 10–20% gage reduction. An 80 ksi (550 MP$_a$) S_y steel must result in a 30% gage reduction. Table 2-6 indicates representative HSLA steels, listing their trade names and yield strengths.

Basically HSLA steels are low-carbon steels (maximum carbon content is 0.28%) containing small amounts of vanadium, columbium, copper, and other alloying elements. The steels gain their strength through controlled cooling and inclusion shape control. Controlled cooling is done by the steel mills as the hot sheets or bars are laying on the runout table. The inclusion shape control changes the inclusions in hot-rolled steels from stringers to round globules through the addition of zirconium or rare earths to the steels during deoxidation.

Forming HSLA Steels. HSLA steels can be fabricated in much the same way as structural carbon steels; however, greater forces are required to produce a permanent set. Since thinner sections are used, the forming pressures are only slightly higher. Liberal bend radii and slightly increased clearances are usually required. Provision must be made for greater springback, especially in the 80 ksi (550 MP$_a$) S_y materials.

Improved deep-drawing characteristics have resulted from adding alloys that remove interstitial elements from the solid solution. This results in no-yield point elongation, making it nonfluting and nonaging. Ordinarily a cold-rolled steel sheet in the "dead soft" annealed condition is not suitable for some forming operations since "fluting" or "stretcher strains" (localized yielding) occur. This condition is corrected by light cold-rolling or "temper rolling" but only for a time because of the aging phenomenon in which steels decrease in ductility and increase in hardness. See the section on "Formability" in Chapter 8.

Welding HSLA Steels. HSLA steels can be welded by any of the processes used for carbon steels. The welding can usually be done without preheat or postheat because the carbon content is low. Weldability decreases with carbon content. With constructional HSLA plate, the use of low-hydrogen electrodes (see Chapter 12) and other precautions to minimize hydrogen pickup are advised. Preheating is generally required for all thicknesses over 1 in. (25 mm) and for highly restrained joints.

Ultrahigh-strength steels

The term "ultrahigh-strength" is an arbitrary term since no universally accepted strength level has been set. However, it generally refers to the class of steels that has minimum yield strengths of 180 ksi (1240 MP$_a$)

Ultrahigh-strength steels can be divided into several groups as follows: medium-carbon low-alloy steels, chromium–molybdenum–vanadium steels, stainless steels, and high-alloy steels.

The medium-carbon low-alloy family includes AISI–SAE 4130 and 4340. In one modification (300M) the silicon content is increased to prevent embrittlement when the steel is tempered at the low temperature required for very high strength.

The forms in which the ultrahigh-strength steels are used are sheet, strip, bar stock, and forgings and, in some cases, castings. Table 2-7 lists the common ultrahigh-strength steels along with tensile strengths, yield strengths, and applications.

Strength Formation. Most ultrastrong steels derive the major part of their strength from the formation of martensite. In low-alloy steels, martensite is formed by oil or water quenching from an austenitizing temperature. Some highly alloyed steels form martensite even when slowly cooled from high temperatures.

The strength of martensite depends on the content of the interstitial elements—carbon and nitrogen. "Iron" martensites seem to be unique in that their "fine structure" limits the length of dislocations to the point that optimum interaction can be achieved between them and the intense stress fields surrounding each interstitial solute atom. The result is maximum resistance to dislocation movement and hence extraordinary strength.

Table 2-7. Approximate upper limit of strength at which selected steels are currently used. (*Courtesy American Society of Metals.*)

Type of Steel	Yield Strength, Psi	Tensile Strength, Psi	Applications
Low-alloy steel	250,000	300,000	Landing gear components
5Cr–Mo–V	240,000	290,000	Airframe structures
Stainless steels			
Martensitic	160,000	180,000	Bolts and rivets
Age-hardenable martensitic	170,000	190,000	Clamps and brackets
Extra-low carbon, age-hardenable martensitic	245,000	257,000	
Cold rolled austenitic	180,000	200,000	Missile tankage for liquid propellant
Semi-austenitic	220,000	235,000	Honeycomb structures
High-alloy steels			
18% Ni maraging	295,000	300,000	Fasteners and shafts
9Ni-4Co	240,000	280,000	
Matrix	290,000	360,000	Axles and shafting

Fabrication. When ultrahigh-strength steels are heated to suitable temperatures, they can be readily hammer-forged, press-forged, rolled, or extruded. In general, the low-alloy and 5 Cr–Mo–V steels are machined in the annealed condition with high-speed steel or carbide tools.

The gas-tungsten arc process, as discussed in Chapter 12, is favored for welding the ultrahigh-strength, low-alloy hardenable and the 5 Cr–Mo–V steels and many of the other types. The steels are preferably welded in the annealed condition and then heat-treated to the desired strength. The composition of the filler wire is chosen on the basis that it will respond to heat treatment in approximately the same way as the base metal. To avoid martensite formation during welding, preheating and interpass heating above the M_s are required. Complex structures must be heat-treated immediately after welding (usually before the weldments cool below a predetermined temperature) to avoid cracking.

Advantages and Limitations. Ultrahigh-strength steels provide weight savings and reduce bulk. The steels are often selected so that a component will fit into an envelope of a fixed size.

In competition with other high-strength materials, steel will usually win out on the basis of cost, not only initial cost but fabrication costs as well.

Ultrahigh-strength steels have an advantage in reducing weight for "moving structures" where overcoming inertia is a factor as in rotating or reciprocating mechanisms such as pumps, engines, actuators, and valves.

The high hardness that accompanies strength promotes wear resistance. However, the regular strength levels have a tendency for a sensitivity to sharp stress raising notches or defects. The present nondestructive techniques have not proved adequate in locating defects that may cause notch sensitivity.

The ratio of endurance limits to tensile strength decreases markedly as tensile strength rises above 200 ksi (1379 MP$_a$).

TOOL STEELS

Tool steels are so named because they have the properties needed in making tools such as dies for cutting and forming metal, for jigs and fixtures, and for precision gages and molds. In general, they have a higher alloy content and thus have high wear resistance, stability at high temperatures, and toughness.

There are about 75 types of tool steels in the AISI list, which is divided into seven main types. Tool steels are identified by letter and number as shown in Table 2-8. The letters are used to indicate the type of quench required for hardening or the main characteristic of the steel: O, for oil-hardening; A, for air-hardening; and S, for shock-resisting, the exception being L, D, P, and the 400 series. L is in the oil-hardening group and is sometimes referred to as a low-alloy tool steel. D is shown along with A in the air-hardening group and is often called a die steel. P and the 400 series are for plastic and die-casting molds.

Water-Hardening Steels (W).   Water-hardening steels are the least costly of the tool steels and have the best machinability. They are also more easily and safely welded than the other tool steels. Problems of distortion in heat treatment and shallow hardening limit the use of W1. It is not used where uneven die sections occur, length-to-thickness ratios exceed 10 : 1, high localized pressures occur, or severe wear conditions exist.

Oil-Hardening Steels (O).   Oil-hardening tool steels are generally used where more depth of hardening and more safety from distortion or cracking is desired than is obtainable in the water-quench type. Oil-hardening steels can tolerate holes near edges and great changes in die sections. These steels are used for forming dies, dies where high pressures are encountered but not where resistance to galling (pickup of metal particles on the die) is required.

Shock-Resisting Steels (S).   Shock-resisting steels are not only capable of taking a great deal of shock but also have high fatigue resistance and good wearing qualities. The most common type is probably S1 with 0.60% carbon and tungsten, chromium, or vanadium. Where shock is severe but heating is intermittent or nominal, a 0.60% carbon with 2% or more tungsten, 1.2% chromium, and 0.20% vanadium is often used.

These steels are generally hardened to 58 to 60 R$_c$ by oil quenching. They give extended performance where higher-carbon types of die steels fail.

Table 2-8. Comparison of tool steel properties. (*Courtesy Machine Design.*)

		Wear resistance / Red hardness / Toughness
Water Hardening	W1	------------------------------- . . . .
	W2	------------------------------- . . . .
Oil Hardening	O1	------------------------------- . . . .
	O6	--------------------------- . . . .
	L6	--------------------------- . . . .
Shock Resisting	S1	--------------------------------------- . . . .
	S5	--------------------- . . . .
	S7	------------------------------------- . . . .
Air Hardening	A2	--- . . . .
	D2	--- . . . .
	D4	--- . . . .
Hot Work	H11	--------------------------------- . . . .
	H12	--------------------------------- . . . .
	H13	------------------------------- . . . .
	H21	--- . . . .
Plastic Die Casting	P20	----------------- . . . .
	414	----------------- . . . .
	420	------------------------------- . . . .

Wear resistance --------------------
Red hardness . . . .
Toughness . . . .

71

Where toughness is a major consideration, S1 is useful. S series steels are excellent for rivet sets, hammer blocks, and long, slender punches, but are not used to make dies for cutting heavy-gage material or for forming dies where severe wear or pickup can develop. Where warranted, the tools can be case hardened to increase resistance to wear and pickup.

Air-Hardening Steels (A).   A2 steels are used in high-production cutting and forming dies that may be troubled with moderate wear or pickup. Nitriding may be used to improve wear and pickup resistance. The A steels can be hardened in large intricate sections with very little distortion. They are not used where toughness is required.

Air-Hardening (D) Steels.   D2 steels are used for forming blanking, and trimming dies where galling or wear is a problem in high production. These steels may be gas or liquid nitrided to increase wear resistance. Where alignment is difficult to maintain, as in slender punches or cutting dies, D2 should not be used.

Hot-Work Steels (H).   Hot-work steels are able to resist heat and are used in hot-blanking dies, hot-extrusion dies, hot-heading dies, and forging and casting dies. These steels contain varying amounts of tungsten, chromium, and molybdenum that have high resistance to softening at temperatures of 1100 to 1200°F (593–649°C).

Plastic and Die-Casting Mold Steels (P and 400).   The P steels are low-carbon steels with molybdenum, chromium, and cobalt that permit hobbing and carburizing. Hobbing is a process of making a mold by pressing the male form into the die to form the cavity. The dies made by hobbing are often carburized for wear resistance.

High-Speed Steels (T and M series).   High-speed tool steels are used extensively for metal cutting such as lathe tools, milling cutters, drills, reamers, and taps. These steels have excellent heat and wear resistance. They may be used in metal cutting operations and die operations at temperatures up to 1100°F (593°C) without softening below R_c 60.

The M or molybdenum type is somewhat inferior in heat resistance to the T or tungsten type but not in wear resistance.

Powder metallurgy tool steels

The powdered metallurgy (PM) process is discussed in Chapter 11. Basically the process consists of using finely divided powders, pressing them into a mold, and after removing the compact from the mold sintering or firing them to obtain the desired hardness and strength. High-speed steel tools are now being made by this process. The advantages claimed are superior strength and toughness, finer grain structure, a better distribution of carbides, better size stability after heat treating, and less chance of distortion and cracking during heat treatment. The tools are said to have longer life and are more easily ground with less chance of burning.

The cost for the PM tools is about 5 to 10% more than the conventional high-speed steels. However, this factor is rapidly outweighed by the increased performance benefits.

A limitation is that the PM tool steels are less machinable because they have a higher annealed hardness and greater carbide content, although PM parts require only very little machining since they are molded to shape.

CAST STEEL

Steel, although often associated only with the wrought form, as in plates, rods, bars, and tubes, is also extensively used in the cast form. Just as with other metals, casting steel provides a way of obtaining the desired product easily. The physical properties of the same steels, whether wrought or cast, are comparable. One advantage a cast steel has over the wrought is that the properties are the same in any direction, whether longitudinal or transverse.

CAST IRON

In order to understand the fabricating characteristics of cast iron, it is necessary to become familiar with the characteristics of the metal and the various types and classifications that are available.

One of the distinguishing features of all cast irons is that they have a relatively high carbon content. Steels range up to about 2% carbon. Cast irons overlap with the steels somewhat and range from about 1.5 up to 5% carbon. It is principally the *form of the carbon,* which is governed by thermal conditions and alloying elements, that provides various structures that may be classified into the following main types:

Gray cast iron
White cast iron
Malleable iron
Ductile (nodular) iron
Compacted graphite iron

Gray Cast Iron. The terms *gray* and *white* cast iron refer to the appearance of the fractured area. The gray cast iron has a grayish appearance because of the large amount of flake graphite on the surface. Typical matrix structures are shown in Fig. 2-35. The dark sections show the graphite flakes. The pearlitic

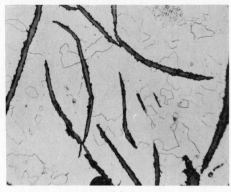

(a)

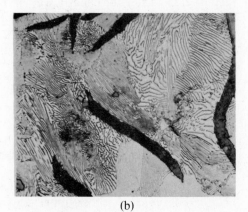

(b)

Figure 2.35. The microstructure of three variations of gray iron. The matrix surrounding the graphite flakes is ferritic (a), pearlitic (b), and bainitic or acicular (c). The corresponding Brinell hardness is also shown. Etched, at a magnification of 500×.

(c)

type may be made fine by faster cooling, or coarse by slow cooling. The size of the section will also determine the structure of the metal: the thinner the section the faster it cools. Thus a casting having large variations in section will also have large variations in hardness and strength unless special precautions are taken to ensure uniform cooling.

The basic composition of gray cast irons is often described in terms of carbon equivalent (CE). This factor gives the relationship of the percentage of carbon and silicon in the iron to its capacity to produce graphite. Thus

$$CE = C_t + \tfrac{1}{3}(\%Si + \%P)$$

where: C_t = total percentage carbon

Example: What is the CE of a gray iron containing 3.35% carbon and 0.65% silicon?

Solution:
$$CE = 3.35 + \tfrac{1}{3}(0.65)$$
$$= 3.57$$

The CE value may then be related to the tensile strength of the metal, as shown in Fig. 2-36. Irons with a carbon equivalent of over 4.3 are called *hypereutectic* and are particularly good for thermal-shock resistance, such as for ingot molds. The higher strength gray irons, with less than 4.3 CE, are termed *hypoeutectic*.

The compressive strength of gray iron is one of its outstanding features. In general, it ranges from 3 to 5 times the tensile strength. As an example, a class-20 gray iron which has a tensile strength of 20,000 psi has a compressive strength of 83,000 psi.

Gray-Iron Classification.   The most-used classification of gray irons is that of ASTM Specification A48-48. The last number indicates the year of original adoption as standard, or the year of last revision. This specification covers seven classes that are numbered in increments of 5, from 20 to 60. The class number

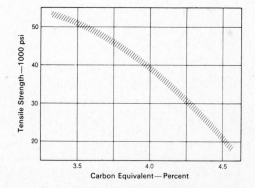

Figure 2.36.   A typical relation between the carbon equivalent and tensile strength of gray iron as cast in 1.2-in. dia. bars.

refers to its tensile strength. For example, a class-40 cast iron refers to one having a minimum tensile strength of 40,000 psi. The carbon content becomes less as the strength requirement goes up. Alloys in increasing quantities are used to achieve strengths higher than 45,000 psi. Although high-strength cast irons have been developed, they are not widely used because they are still quite brittle.

Gray iron is the most widely used of all cast metals. Typical applications are engine blocks, pipes and fittings, agricultural implements, bathtubs, household appliances, electric motor housings, machine tools, etc.

White Cast Iron.   White cast iron is produced by cooling the cast iron rapidly so that the carbon is in the combined form as cementite. It is often referred to as *chilled* cast iron. The structure is very hard and brittle and the fractured surface produces a shiny appearance. If the structure contains alloying elements such as nickel, molybdenum, or chromium, martensite will be formed. Irons of this type are extremely hard and wear-resistant, and are used on hammer mills, crusher jaws, crushing rollers, balls, wear plates, etc.

Malleable Iron.   Malleable iron is made from white cast iron, which has the carbon in the combined form. The white cast iron is subjected to a prolonged annealing process. It may take two days to bring the castings to a temperature of about 1600°F (870°C). The castings are held at this temperature for a period of 48 to 60 hr. They are then cooled at a rate of 8 to 10 degrees per hour until the temperature is about 1300°F. The temperature is held in the 1250 to 1300°F (677–704°C) range for a period of up to 24 hr. The furnace is then opened and allowed to cool to room temperature. During this time all the combined carbon has separated into graphite and ferrite. The graphite "flake" shown in Fig. 2-37 is often referred to as a rosette or merely temper carbon. It consists of an aglomeration of many very small graphite flakes.

A pearlitic malleable iron can be made by allowing the cooling rate to be fast enough to retain the desired amount and type of combined carbon in the matrix. In most grades, the matrix is later modified by further treatment which

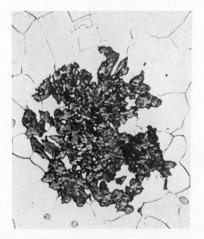

Figure 2.37. Malleable iron showing the graphite flakes in a ferrite matrix.

tempers the asquenched structure, raises its ductility, and improves its machinability.

As the name implies, malleable irons are softer than gray irons and are therefore used where greater toughness and shock resistance are required. Malleable irons are used extensively on farm implements, automobile parts, small tools, hardware, and pipe fittings.

Classification of Malleable Irons. About 90% of all the malleable iron made is classified as 32510, which means that it has a yield point of 32,500 psi and an elongation of 10%. Grade 35018 has somewhat lower carbon content, but it has somewhat higher strength and better ductility. Grades 32510 and 35018 are known as standard types of malleable iron. There is an increasing demand for a newer type termed *pearlitic malleable*. Instead of having a ferrite matrix, it is pearlitic, which is stronger and harder than ferrite. These irons are obtained by interrupting the second-stage annealing process, by introducing larger quantities of manganese which prevents graphitization of the pearlite, or by air cooling followed by tempering.

Pearlitic malleables have yield strengths up to 80,000 psi, but low ductility. They are used where higher strengths and wear resistance are needed, along with less resistance to shock. Class number examples for pearlitic malleables are 45010, for the lowest yield strength, and 80002 for the highest yield strength.

Ductile Iron. For generations, foundrymen and metallurgists have searched for some way to transform brittle cast iron into a tough, strong material. Finally, it was discovered that magnesium could greatly change the properties of cast iron. A small amount of magnesium, about one pound per ton of iron, causes the flake graphite to take on a spheroidal shape (Fig. 2-38). Graphite in a sphe-

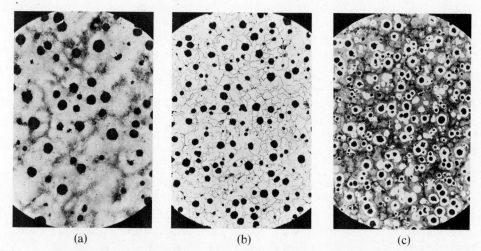

(a) (b) (c)

Figure 2.38. The matrix structure surrounding the spheroidal graphite of ductile iron may be pearlitic (a), ferritic (b), or a ferritic band surrounding the spherulite, termed a bulls-eye structure (c).

roidal shape presents the minimum surface for a given volume. Therefore there are less discontinuities in the surrounding metal, giving it far more strength and ductility. The processing advantages of steel, including high strength, toughness, ductility, and wear resistance, can be obtained in this readily cast material.

Ductile iron can be heat-treated in a manner similar to steel. Gray irons can too, but the risk of cracking is much greater.

There has been a rapid growth in the number of foundries licensed to produce this relatively new material, which goes by other names such as nodular iron and spheroidal iron. The combination of excellent casting qualities, excellent machinability, high strength, toughness, and wear resistance have led to wide acceptance of ductile irons for both small intricate castings, as used on a car door lock, to huge press rolls weighing 150 tons or more.

Classification of Ductile Irons.   The general classification numbers for ductile irons are similar to those for malleable cast iron, but a third figure is added which represents the minimum percent of elongation that can be expected. For example, an 80-60-03 has 80,000 psi minimum tensile strength, 60,000 psi minimum yield strength, with 3% minimum elongation. There are four regular types of ductile irons: 60–45–10, 80–60–03, 100–70–03, and 120–90–02. In addition, there are two special types: one is for heat resistance, the other (Ni-Resist*) has resistance to both corrosion and heat. The Ni renders the matrix austenitic so this alloy is somewhat similar in a sense to austenitic stainless steel.

Compacted Graphite Cast Irons.   A relatively recent addition to the field of cast irons is that of *compacted graphite*. In 1976 a commercial treatment alloy made the process control possible. The structure and properties are generally considered to be intermediate to those of gray and ductile cast irons.

Compacted gray cast iron is characterized by interconnected graphite that is intermediate to the flake graphite of gray iron and the nodular graphite of ductile iron. (This structure is often referred to as vermicular iron.) The physical properties of compacted graphite irons are to a large extent related to the interconnected graphite. While individual properties are generally intermediate to those of gray and ductile cast iron, some of the better properties of both gray and ductile are combined in compacted graphite irons.

Figure 2-39(A) shows a cooling curve with a series of microstructures observed from successively quenched specimens during solidification. The microstructures (a through f) correspond to the temperature shown on the cooling curve and show how compacted graphite develops from the nodular form. Figure 2-39(B) shows an enlarged view of both nodular and compacted graphite.

In tensile strength, compacted graphite iron is greater than that of alloyed high-strength gray cast iron, and tensile and yield strengths approach those of ductile cast irons—depending upon the base chemistries and section size. Ductility varies from 1 to 6% for the higher- and lower-strength graphite irons, respectively.

* Registered trademark of the International Nickel Company, Inc.

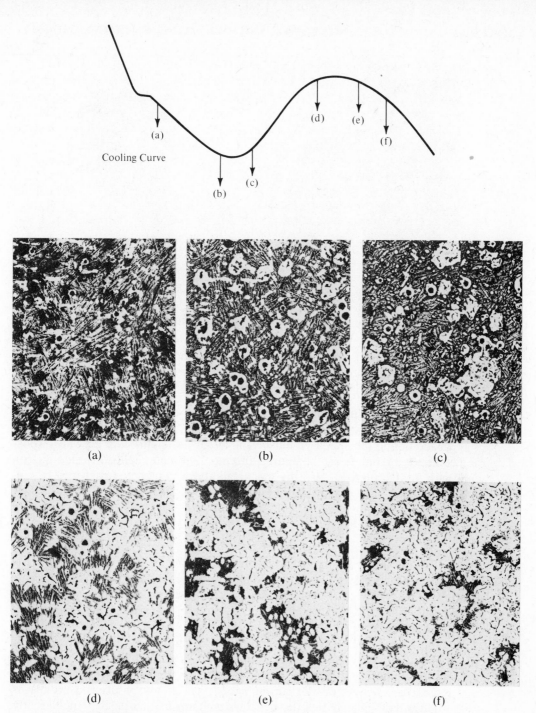

Cooling Curve

(a)

(b)

(c)

(d)

(e)

(f)

Figure 2.39(A). A series of microstructures observed from specimens quenched at various stages in the cooling curve. The progression shows the formation of vermicular graphite from the graphite nodule. Sodium Picarate etched, ×50.

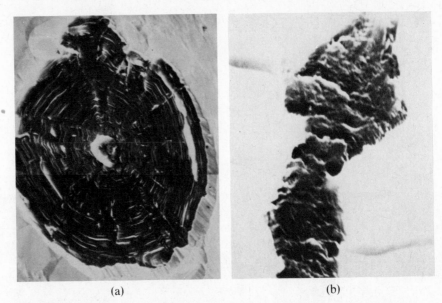

(a)   (b)

Figure 2.39(B). An enlarged view of the nodular and vermicular forms of graphite.

Impact and fatigue properties of compacted graphite irons are substantially better than for gray cast irons although lower than for ductile cast iron.

In addition to the increased tensile and fatigue strengths over gray cast irons, the compacted graphite can be poured into more intricate molds than ductile iron and with less shrinkage. Also the machinability of compacted cast iron is substantially better than that of ductile iron for comparable hardness.

Fabricating Characteristics

Gray Iron. The usual range of carbon in gray iron is between 2.50 and 3.75%. The graphite flakes cause discontinuities in the ferrite. This helps the chips to break up easily. The graphite also furnishes lubricating qualities to the cutting action of the tool. Although gray cast iron is considered quite machinable, it varies considerably because of the microstructure. The ratings may range from 50 to 125% of that of B1112.

One of the easiest ways to determine the machinability of a given piece of material is the Brinell test. The normal hardness range of 130 to 230 Brinell will present no machining difficulties. Beyond this range, however, even a few points will have a large effect on the cutting speed that should be used. Hardnesses above 230 Brinell indicate that the structure contains free carbides, which greatly reduce tool life.

Tungsten-carbide tools are usually used to machine cast irons because of the abrasive qualities of the material.

Machinability of gray cast irons can be improved through annealing. When this is done, its machinability exceeds that of most other ferrous materials.

Table 2-9. Machinability ratings for cast irons.

Type of Iron	Machinability Index
Malleable (standard)	110–120
Gray iron (flake graphite)	110
Ductile iron	90–110
Gray iron (pearlitic)	68

Malleable Iron. Malleable iron has a machinability rating of 120% and is considered one of the most readily machined ferrous metals. The reason for this good machinability is its uniform structure and the nodular form of the tempered carbon. The pearlitic malleables, which have a different annealing that leaves some of the carbon content in the form of combined carbides, have machinability ratings of 80 to 90%. (The term pearlitic is just a convenient term and does not mean that the microstructure is necessarily in pearlite form.)

Ductile Iron. The machinability of ductile iron is similar to that of gray cast iron for equivalent hardnesses and better than that of steel for equivalent strengths. Type 60–45–10 has both maximum machinability and toughness. Although the cutting action is good, the power factor is higher owing to the toughness of the material.

The machinability ratings for each of the main types of cast iron are given in Table 2-9.

Formability

Cast irons are usually not considered for forming operations; however, the ductility of malleable irons makes it possible to use many production shortcuts that add up to time and material savings. Some typical examples are:

1. Sizing and straightening the casting in a press.
2. Coining—striking the part either hot or cold in a press. The operation can often bring the dimensions within ±0.010-in. tolerance when done cold and within ±0.004 in. when done hot.
3. Staking (heading over) and crimping can be used to eliminate the need for fasteners.

Weldability

Gray Iron. Gray cast iron can be welded with the oxyacetylene torch or with the electric arc. However, because of its low ductility, special precautions should be taken to avoid cracking in the weld area. Preheating is done, in the case of oxyacetylene welding, to make expansion and contraction more uniform. The opposite effect is used in arc welding, where short welds are made in order to keep the heat at a minimum. Slow cooling is essential to permit the carbon to separate out in the form of graphite flakes. Failure to do this will result in mottled or chilled areas.

Generally, the welding of gray cast iron is limited to repair work rather than fabrication. However, braze welding with bronze or nickel copper is frequently used in fabrication.

Malleable Irons. Malleable irons are not considered weldable in the same sense that gray cast iron is weldable. The heat necessary to melt the edges of the break will completely destroy the malleable properties. Due to the long-term anneal required to produce these properties originally, a simple annealing process would not restore them. There are times, however, where stresses are low or are in compression only; in this case, successful arc welding can be accomplished. A commercially pure nickel rod or a 10%-aluminum bronze rod is used.

Since brazing can be done at a temperature of 1700°F (927°C) or less, it is a preferred method of repair. Silver brazing is also used, since this procedure can be accomplished with even less heat.

Ductile Irons. Ductile iron, being a high-carbon-content material, should be given special consideration in welding, just as is gray cast iron. It can be welded by a carbon-arc process and most other fusion processes, either to itself or to other metals such as carbon steel, stainless steel, and nickel. The most easily welded types are 60–45–10 and the high-alloy variety. High-quality welds are made with flux-cored electrodes having 60% nickel and 40% iron. Noncritical applications may be welded with ordinary steel electrodes.

This material can also be brazed with silver or copper brazing alloys. Crack-free overlays can be made with commercial hard-surfacing rods. This will provide special abrasion and corrosion resistance.

Castability

Gray Iron. From the iron–carbon diagram (Fig. 2-24) it can be seen that carbon markedly lowers the melting point of iron. Many benefits accrue from this fact. The pouring temperature, for example, can be several hundred degrees higher than the melting point, making for excellent fluidity. This makes it possible to produce thin-section castings over a large area. Shrinkage is considerably less than that of steel because the rejection of graphite causes expansion during solidification. The amount of solidification shrinkage differs with the various casting alloys. Class-20 and class-25 irons contain sufficient graphite so that there is virtually no shrinkage at all. The higher classes have one-half to two-thirds the solidification shrinkage of any ferrous alloys.

Malleable Irons. The molten white irons from which malleable irons are made have high fluidity, allowing complicated shapes to be cast. Good surface finish and close dimensional tolerances are possible. Since malleable irons require a prolonged heat treatment, internal stresses, which may have occurred as the casting solidified and cooled, are removed.

Ductile Irons. Ductile iron is similar to gray cast iron when it comes to casting qualities. It has a low melting point and good fluidity in the molten state. Therefore, it can be poured in intricate shapes and thin sectional parts. The metal will flow several inches, in sections as small as $\frac{1}{16}$ in., in greensand molds at normal foundry pouring temperatures. Many parts are now being cast from ductile iron that formerly could not be cast, because gray iron did not possess adequate properties and steel could not be cast in such intricate shapes.

STAINLESS STEELS

The principal difference between stainless steels and mild steels is, of course, corrosion resistance. The mechanical properties of stainless steels are also superior to those of ordinary steels. Ultimate strengths of the annealed 300-series stainless steels vary from 80,000 to 100,000 psi, whereas structural steel ASTM A36 has a yield point of 36 ksi (250 MPa). Figure 2-40 shows a representative stress–strain curve for half-hard stainless steel and A36 mild steel. Stainless steel does not show a definite yield point so it is calculated at 0.2% offset as shown. In a fully hardened condition, the strength of the stainless steel may exceed 300 ksi (2068 MPa).

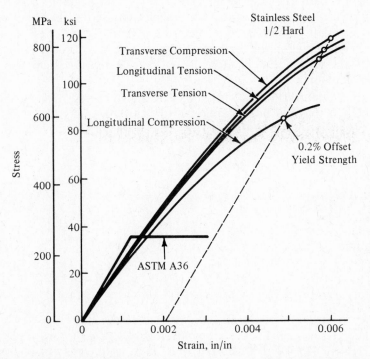

Figure 2.40.   Representative stress–strain curve for half-hard stainless steel and mild steel (2).

What is it that makes stainless steel stainless? Current theory holds that a thin, transparent, and very adherent film forms on the surface. It is inert or passive and does not react with many corrosive materials that attack mild steel and other metals.

Unique to any other alloy system, the stainless steel family serves a temperature range from −454°F (−234°C) to above 1800°F (982°C). Within this range it exhibits strength, toughness, and corrosion resistance superior to most metals. Considerable processing and research is now being done at very low (cryogenic) temperatures. Stainless steel 18–8 (18% chromium–8% nickel) type 304 is a virtual "workhorse" for handling and storage of liquid helium, hydrogen, nitrogen, and oxygen that exist at cryogenic temperatures.

Classification

The three main types of stainless steels are *martensitic, ferritic,* and *austenitic* (Fig. 2-41). In addition there is a special class of stainless steels referred to as "precipitation-hardening." A brief description of each of these types of steels follows.

Class I: Martensitic Type (Hardenable).   The martensitic stainless steels are primarily straight chromium steels containing 11.5 to 18% chromium together with enough carbon to render them hardenable by heat treatment. As the carbon content goes up, so does the range of mechanical properties that can be obtained by heat treatment. However, the increased carbon content ties up additional amounts of chromium in the form of chromium carbides, which, in turn, lowers the corrosion resistance. Therefore, as carbon content goes up, so must the chromium content.

Tables 2-10 through 12-12 list the AISI and the New Unified Numbering System (UNS) that has been developed for all metals and alloys. Martensitic stainless steels are listed in Table 2-10.

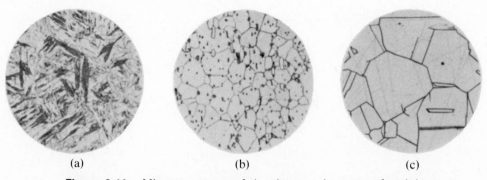

(a)   (b)   (c)

Figure 2.41.   Microstructures of the three main types of stainless steels, (a) martensitic, (b) ferritic, and (c) austenitic. (*Courtesy Republic Steel.*)

Class II: Ferritic Type (Nonhardenable). The ferritic stainless steels are straight chromium steels containing 12 to 27% chromium with carbon controlled to the lowest practical limit to minimize its harmful effect on corrosion resistance. The ferritic steels can be forged and hot-worked easier than the martensitic steels and after hot working may be air-cooled without danger or cracking. The resistance to corrosion and scaling at elevated temperatures is better than that of Class I.

The ferritic stainless series cannot be hardened by heat treatment, and only moderately by cold working. They are magnetic and have good ductility and resistance to corrosion. Type 430 is the general purpose type of this group. The ferritic stainless steel numbers are shown in Table 2-11.

Class III: Austenitic Type (Nonhardenable). The austenitic stainless steels are chromium-nickel steels containing from 16 to 26% chromium and from 6 to 22% nickel. As in the ferritic types, the carbon is there as a residual element and is controlled to the lowest practical limit. These steels can be work-hardened to high-strength levels. As a group, the austenitic steels have considerably better corrosion resistance than martensitic or ferritic steels and are characterized by excellent strength and oxidation resistance at elevated temperatures. However, these steels are susceptible to intergranular corrosion after prolonged heating within the temperature range of 750–1650°F (399–899°C). Chromium precipitates at the grain boundaries in the form of chromium carbides. Thus the corrosion resistance at the grain boundaries is greatly reduced. This problem can be avoided

Table 2-10.   Martensitic stainless steels.

AISI TYPE	Equivalent UNS	AISI TYPE	Equivalent UNS
403	S40300	420F	S42020
410	S41000	422	S422000
414	S41400	431	S43100
416	S41600	440A	S44002
416Se	S41623	440B	S44003
420	S42000	440C	S44004

Table 2-11.   Ferritic stainless steels.

AISI TYPE	Equivalent UNS	AISI TYPE	Equivalent UNS
405	S40500	430FSe	S43023
409	S40900		S43400
429	S42900		S43600
430	S43000		S44200
430F	S43020		S44600

Table 2-12.   Austenitic stainless steels.

AISI TYPE	Equivalent UNS	AISI TYPE	Equivalent UNS
201	S20100	310	S31000
202	S20200	310S	S31008
205	S20500	314	S31400
301	S30100	316	S31600
302	S30200	316L	S31603
302B	S30215	316F	S31620
303	S30300	316N	S31651
303Se	S30323	317	S31700
304	S30400	317L	S31703
304L	S30403	321	S32100
	S30430	329	S32900
304N	S30451	330	NO8330
305	S30500	347	S34700
308	S30800	348	S34800
309	S30900	384	S38400
309S	S30908		

by using types 321 and 347 steels, which are stabilized with titanium and columbium, respectively. These elements act to combine with carbon in the form of titanium or columbium carbides rather than chromium carbides.

Table 2-12 lists austenitic stainless steels. Alloys containing chromium and nickel are identified as the 200 series. Alloys containing chromium, nickel, and manganese are identified as the 300 series type.

Precipitation-Hardening (PH) Stainless Steel.   Precipitation, as the name implies, is a strengthening process that takes place during aging. One type, called 17–7 PH, contains 17% chromium, 7% nickel, 1% aluminum, and a small amount of titanium.

Table 2-13 lists the unified numbering system for these steels. Precipitation-hardening steels are solution heat-treated and aged to obtain high strength.

Machinability

Stainless steel is generally selected for its performance in service rather than for its machining qualities; however free-machining types have been developed in each of the three common types, as shown in Table 2-14.

Sulfur and selenium improve machinability of stainless steels, forming soft, nonmetallic sulfides and selenides in the steel. These appear in the microstructure as inclusions or stringers. They act to lubricate the cutting tool and produce nonclogging chips.

The machinability of stainless steels has improved significantly in recent years, primarily through melting and refining practices that permit tighter analysis control. Figure 2-42 shows a comparison of four frequently used stainless steels and their improved machining quality between 1965 and 1975.

Table 2-13. Precipitation-hardening stainless steels.

AISI UNS	AISI UNS
S13800	S17400
S15500	S17700

Table 2-14. Recommended cutting speeds for free-machining stainless steels.

AISI Type	Cutting Speed (sfpm) for HSS Tool
Austenitic, 303, 303 Se*	85–120
Martensitic, 416, 416 Se	110–160
Ferritic, 430F, 430 Se	80–115

* The addition of a small amount of selenium to improve machinability.

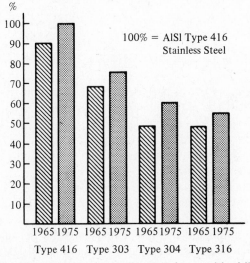

Figure 2.42. The relative improvement in machinability of four common stainless steels from 1965 to 1975.

Weldability

Stainless steels are not particularly difficult to weld if the basic properties of the metal are considered. This can be done, in part, by comparing some of the important physical properties of stainless steel with those of the most weldable of metals, mild steel.

The expansion and contraction of metals must be considered in all fabricating processes, but especially so in welding. The coefficients of expansion are given as a means of comparing one metal with another. You will note, by studying Appendix A, that stainless steel has a considerably higher rate of expansion than that of mild steel.

Also the thermal diffusivity of the metal (ability of the metal to absorb heat) must be considered. It is used in welding to estimate the amount of energy required to bring the metal to the fusion point. Metals with low thermal conductivity, such as stainless steel, tend to localize the stress. That is, the hot metal wants to expand but is hemmed in by the adjacent cold metal. When the heat is removed, the metal cools and contracts. However, the cooler adjacent metal does not yield and results in cracks at the edge of the weld. Methods of overcoming these problems are discussed in Chapter 15.

Austenitic chromium–nickel steels are readily welded by four commonly used processes: shielded metal-arc, gas metal-arc, gas tungsten-arc, and submerged arc. The molten weld metal is protected from atmospheric oxidation in all four processes by flux, slag, or inert gas. The complete protection from oxidation retains all the essential elements while excluding foreign elements that effect the strength and corrosion resistance, or weld quality. Due to the low thermal conductivity, the wire-melting rate (used in gas metal-arc welding), measured in in./min/amp, is approximately 30% higher with stainless steel than with carbon steel.

LIGHT METALS AND ALLOYS

The age of air and mass travel has focused the attention of engineers on light metals. About 17% of the aluminum produced in the United States is used in the manufacture of aircraft, automobiles, trucks, buses, ships, and trains.

Metals commonly classed as light metals are those having a density less than that of steel (7.8 gm/cm). Common metals in this classification are aluminum, magnesium, beryllium, and titanium.

Although aluminum is in abundant supply on the earth's crust, in all soils, clays, rocks, and minerals, it cannot be easily separated. It can only be profitably separated from bauxite. An electrolytic process is used to reduce the alumina. Commercially pure aluminum has about one-third the density of steel. It possesses excellent corrosion resistance and has good electrical, thermal, and reflectivity characteristics.

Classification of Aluminum Alloys

Aluminum is used extensively in both the wrought and cast forms. Casting alloys are designated by a four-digit number. The first digit identifies the alloy very much the same as that shown in Table 2-15 for wrought alloys. For cast alloys, number six is not used and number eight is tin. The second two digits, of the four-digit series, are assigned when the alloy is registered. A later modification of this alloy is indicated by a letter prefix before the numerical designation.

Table 2-15. Aluminum alloy specifications.

Type of Aluminum Alloy	Number Group
Aluminum—99.00% purity or greater	1XXX
Copper	2XXX
Manganese	3XXX
Silicon	4XXX
Magnesium	5XXX
Magnesium and silicon	6XXX
Zinc	7XXX
Other element	8XXX
Unused series	9XXX

Castings are designated by a suffix (0) which is separated from the first three digits by a decimal point. If it is an ingot of the same chemical composition, the number 1 will follow the decimal point. An example of an aluminum casting number is A514.0.

   A—Indicates it is the first of what may be a series of the same type of aluminum.
   5—Magnesium alloy.
   14—Aluminum purity.
   .0—Product form; .0 = casting; .1 = ingot.

For a more complete discussion see the book, *Aluminum Standards and Data,* listed in the bibliography at the end of this chapter.

Wrought aluminum alloys are also identified by a four-digit number as shown in Table 2-15. The first digit designates the alloy type, the second digit the alloy modification, and the last two digits the aluminum purity. For alloys in use before this classification system was devised, the last two numbers are the same as the old numbers. For example, a widely used type of aluminum was 24S. In the new specification, it is 2024.

   2—Copper alloy.
   0—No modification.
   24—Old number.

Temper Designations.   A letter following the four-number identification, separated by a hyphen, indicates the basic temper of the metal or the treatment used in hardening, strengthening, or softening the alloy. Temper designations are shown in Table 2-16.

Annealing of a work- or strain-hardened metal is accomplished by heating to a temperature at which recrystallization takes place, followed by slow cooling. The temperatures to which it is heated will depend upon the amount of prior cold work and the alloy content.

Table 2-16. Aluminum-alloy temper designations.

-O	Annealed
-F	As fabricated
-H	Strain hardened

-H1	Strain hardened only
-H2	Strain hardened followed by partial annealing
-H3	Strain hardened and stabilized

-T	Solution heat treated

-T2	Annealed (cast shapes only)
-T3	Solution treated followed by cold work
-T4	Solution treated, natural aging at room temperature
-T5	Artificially aged after quenching from hot-working operations such as casting or extrusion
-T6	Solution treated, artificially aged
-T7	Solution treated, stabilized to control growth of distortion
-T8	Solution treated, cold worked, artificially aged
-T9	Solution treated, artificially aged, cold worked

Solution heat treatment (combined with aging) is a hardening and strengthening process that is accomplished by heating an aluminum alloy, such as 2024, to a high enough temperature to affect a solution of the hardening constituents and then quenching it rapidly. For many aluminum alloys, this is 900 to 1000°F (482–538°C). During solution heat treatment, the soluble alloying constituents are dispersed uniformly in the matrix. When subsequently quenched, there is not enough time for these constituents to precipitate out of solution, which results in a "supersaturated" solid solution. That is, the matrix has more constituents than it can normally carry in solution. This is an unstable condition and certain constituents begin to precipitate from the aluminum matrix.

Age hardening or precipitation may take place at room temperature and is known as "natural aging." Certain other alloys must be heated slightly to bring the precipitation to completion. This is referred to as "artificial aging." In either case, precipitation is aimed at providing the correct size, character, and distribution of the precipitate. The precipitate should be of a "gritty" type to provide the maximum resistance to slippage along the grain boundaries, and, more importantly, along the slip planes of each crystal. Figure 2-43 shows the comparative tensile and yield strengths of common aluminum alloys.

Fabricating Characteristics

Machinability.   Pure grades of aluminum are too soft for good machinability. Work or strain hardening helps overcome some of the gumminess. The most easily machined alloy is 2011–T3. This metal is often referred to as the free-cutting aluminum alloy. It is used extensively for a broad range of screw-machine

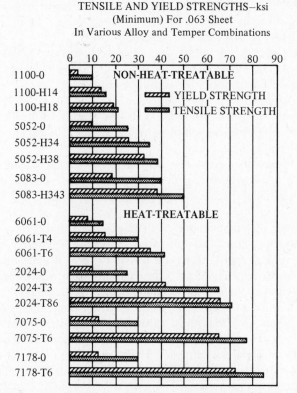

TENSILE AND YIELD STRENGTHS–ksi
(Minimum) For .063 Sheet
In Various Alloy and Temper Combinations

Figure 2.43. Mechanical properties of representative aluminum alloys.

products. Small amounts of lead and bismuth help produce a better finish and easy-to-handle chips.

The softer alloys sometimes build up on the cutting edge of the tool. However, the proper use of cutting oils and a highly polished tool surface will tend to minimize this condition.

Two important physical properties that should be kept in mind when machining aluminum are modulus of elasticity and thermal expansion, as shown in Appendix A.

Due to the much lower modulus of elasticity (about one-third that of steel), aluminum shows much greater deflection under load. Care should be taken, in making heavy cuts and in clamping the work, to avoid distortion.

When dimensional accuracy is necessary, thermal expansion of the metal is an important consideration. Since the thermal expansion of aluminum is almost twice that of steel, care should be exercised in keeping heat down. Overheating can be reduced to a minimum by using sharp, well-designed tools, coolants, and feeds that are not excessive. It may be necessary, on larger parts, to make a thermal allowance in measuring.

Tool angles vary widely for turning aluminum, depending on the type of material and the finish required. In general, large rake angles are employed for finishing cuts and for the softer alloys. Low rake angles are generally used for heavy cuts and for the harder, free-machining alloys.

Surface foot speeds for aluminum range anywhere from 400 to 20,000 fpm (feet per minute). The alloys 2011–T3, 2017–T4, and 2024–T4 permit the highest surface foot speeds, long tool life, and clean cutting.

Formability.   Most aluminum alloys can be readily formed cold. However, the wide range of alloys and tempers makes for considerable variation in the amount of forming possible.

Generally, depending upon the alloy and temper used, aluminum requires less energy and horsepower than does the forming of heavier metals. Consequently, machine life is increased, power costs are lower, and maintenance is minimized.

As the degree of temper increases, the plastic and elastic ranges decrease, which makes forming more difficult to accomplish and control. Material must be stressed beyond its yield strength in order to form it permanently, or it will tend to return to its original shape. Thus, material that is insufficiently stressed will spring back in proportion to the amount of stress applied. Of course, very little springback will be encountered in the soft (O—annealed) plate. The exact amount of springback is difficult to compute. Therefore, tools should be provided with adjustments for overbending, unless preliminary experiments are undertaken to determine the material's springback characteristic.

Each temper (degree of hardness) of an aluminum sheet alloy establishes a specific workability characteristic that must be considered in the planning of operations for bending it. Bending characteristics also vary with the thickness of the sheet.

The non-heat-treatable alloys—1100, 3003, 5005, 5050, and 5154—work-harden but have good forming qualities in the softer tempers. In the annealed state they can be bent to less than $1t$ radius up to 0.25 in. thick (t refers to the material thickness).

The heat-treatable alloys—2014, 2024, 6061, and 7075—can be formed readily in the annealed temper and are often worked in this state. Work can be done after solution heat treatment and quenching, but it must be done quickly before natural aging or precipitation starts. Sometimes the solution treatment is performed and the metal is placed in cold storage to delay the aging process until the parts can be formed.

Weldability

Physical Properties.   The physical properties that must be considered in welding aluminum are melting point, thermal conductivity, and the coefficient of expansion, as shown in Appendix A.

By comparing some of the physical properties of aluminum and steel, it can be seen that the melting point of aluminum is much lower and the thermal conductivity is much higher. Therefore, more heat will be required to weld a corresponding thickness of aluminum.

The thermal expansion of aluminum is almost twice that of steel. This, coupled with the fact that aluminum weld metal shrinks about 6% in volume upon solidification, may put the weld in tension. Some restraint is often necessary to prevent distortion during the welding operation. However, excessive restraint on the component sections during cooling of the weld may result in weld cracking.

The speed at which the weld is made is one of the determining factors in preventing distortion. A slow rate causes greater area heating with more expansion and subsequent contraction. For this reason, arc- and resistance-welding methods that make use of highly concentrated heat are the most used.

Inert-gas–shielded electrical-arc welding or gas-metal-arc welding (GMAW), gas-tungsten-arc welding (GTAW), and resistance welding (spot and seam) are the most widely used commercial methods of welding aluminum (Fig. 2-44).

Commercially pure aluminum and low-alloy aluminums such as 1100 and 3003 are easily formed and welded. These metals have little or no tendency to crack after welding, since they are quite ductile.

The heat-treatable aluminum alloys of the 6000 series (magnesium and silicon), the 2000 series (copper), and the 7000 series (zinc with magnesium and/or copper) gain their strength by solution heat treatment, quenching, and age hardening (precipitation). These metals are readily welded but the filler rod must be such that it will not produce hot-short crack sensitivity. Hot-short is a term used to describe metals that are very weak when they are near the molten state. Thus contraction cracks develop in the hot metal as it starts to solidify. Filler

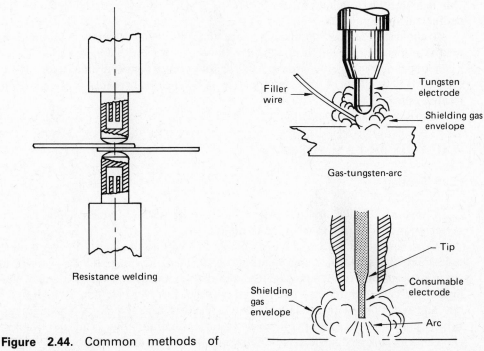

Figure 2.44. Common methods of welding aluminum.

Resistance welding

Gas-tungsten-arc

Gas-metal-arc (GMA)

metals can be chosen that will overcome these difficulties. As in the case of welding the aluminum–magnesium metals, a filler rod containing 5% magnesium will prevent weld cracking.

Casting.   Aluminum is considered one of the more easily cast metals. It requires less energy to bring it to the molten state, about 1200°F (649°C) and is therefore easier on the refractory furnace lining. In addition to sand casting, permanent metal molds and die casting are used extensively.

The combination of low density and low melting point combine to practically eliminate most problems of sand washes (sand moving out of place) that occur when heavier metals are poured.

Special attention must be given to casting shrinkage, hot shortness, and gas absorption. Due to the large shrinkage, from 3.5 to 8.5% by volume, it is desirable to have uniform sections wherever possible. If section changes are necessary, they should be gradual. The minimum section for sand castings is about $\frac{3}{16}$ in.

Hot shortness, the low strength of the metal immediately after solidification, makes it necessary for the casting designer to minimize shrinkage stresses.

Molten aluminum readily absorbs oxygen and hydrogen. Since most of the pinholes in aluminum castings are caused by hydrogen, precautions must be taken to minimize it. This is done by having the metal no hotter than necessary for pouring, and by keeping excess moisture from the molding sand.

Alloying elements intentionally added to facilitate aluminum casting and heat treatment are silicon, magnesium, and copper. Silicon increases the fluidity of the molten aluminum, allowing it to flow farther in thin walls of the mold cavity and produce finer detail. It also reduces internal shrinkage and reduces the coefficient of expansion.

Aluminum castings containing more than 8% magnesium will respond to heat treatment; however, it also makes the metal more difficult to cast.

Copper is one of the principal hardening constituents in aluminum. It increases the strength of the aluminum in both the heat-treated and non-heat-treated conditions.

COPPER AND COPPER-BASE ALLOYS

Classification

Coppers.   Unlike other common metals, the various types of copper are better known by name than by code number.

Commercially pure copper is available in several grades, all of which have essentially the same mechanical properties. *Electrolytic tough-pitch copper* is susceptible to embrittlement when heated in a reducing atmosphere, but it has high electrical conductivity. *Deoxidized copper* has lower electrical conductivity but improved cold-working characteristics, and it is not subject to embrittlement. It has better welding and brazing characteristics than do other grades of copper. *Oxygen-free copper* has the same electrical conductivity as the tough-pitch copper and is not prone to embrittlement when heated in a reducing atmosphere.

Modified coppers include *tellurium copper,* which contains 0.5% tellurium for free-cutting characteristic (selenium and lead are also used for this purpose), and *tellurium-nickel copper,* an age-hardenable alloy that provides high strength, high conductivity, and excellent machinability.

Brasses.   Brasses are principally alloys of copper and zinc. They are often referred to by the percentage of each; for example, 70-30 means 70% copper and 30% zinc. Small amounts of other elements such as lead, tin, or aluminum are added to obtain the desired color, strength, ductility, machinability, corrosion resistance, or a combination of these properties. The main classifications of brasses are given in Table 2-17.

Bronze.   Bronzes are usually thought of as copper–tin alloys. However, there are some types of bronze that contain little or no tin. The main classifications of bronzes are given in Table 2-18.

Fabricating Characteristics

Machinability.   Because of the wide variety of copper-base alloys, some general information will be given first, followed by machinability ratings according to groups.

General Speeds and Feeds.   For most copper alloys, it is generally good practice to use the highest practical cutting speed with a relatively light feed and moderate depth of cut. An exception is the machining of sand castings with high-speed tools, in which case low speeds and relatively coarse feeds are used to increase tool life. After the scale has been removed, the higher speeds and lighter feeds can be resumed. Tool life can be considerably increased if the castings

Table 2-17.   The main classifications of brasses.

Alloy	Meaning
Alpha brasses	Contain less than 36% zinc and are single-phase alloys.
Alpha-beta brasses	Contain more than 36% zinc and have a two-phase structure.
Leaded brasses	Contain up to 88.5% copper and up to 3.25% lead.
Tin brasses	Known as admiralty and naval brass when the percentage of tin is low, 0.75 to 1.0%.
Nickel silvers	Brasses containing high percentages of nickel. (The designation 65-18 indicates approximately 65% copper, 18% nickel, and the remainder zinc. Nickel is added primarily for its influence on color. When the percentage is high, a silvery white color is obtained. Nickel also improves mechanical and physical properties.)

Table 2-18.   The main classifications of bronzes.

Alloy	Meaning
Phosphor bronze	Most copper–tin alloys are deoxidized with phosphorus; the small amount left in the metal may range from a trace to over 0.35% phosphorus, increasing strength, hardness, toughness, and corrosion resistance.
Aluminum bronze	Contains up to 13.5% aluminum and small amount of manganese and nickel; has good antifrictional properties.
Silicon bronze	Contains up to 4% silicon; has strength similar to mild steel, and excellent corrosion resistance to brine and other nonoxidizing inorganic acids.
Manganese bronze	Contains up to 3.5% manganese, which imparts high strength.
Beryllium bronze	Contains from 2 to 2.75% beryllium, which makes it respond to precipitation hardening.

are sandblasted, pickled in acid, or tumbled to remove the extremely hard, abrasive surface scale.

Machinability Ratings.   Steel and steel alloys have machinability ratings based on B1112 as 100%. Copper and copper-base alloys use either free-cutting brass—consisting of 61.50% copper, 35.25% zinc, and 3.25% lead—as the base material, or B1112. The base must be designated.

To classify the relative machinability of copper-base alloys, they can be placed in three groups, based on the amount of tool life and amount of power required. The line of demarcation from one group to the other is not sharp but will act as a guide.

Group I.   The materials in this group are the alpha-beta or two-phase brasses. The beta phase is harder and more brittle, which makes the alloy less ductile, thus permitting heavier feed rates in turning. Lead is added to both single- and double-phase brasses. It is effective in reducing the shear strength of the metal and assists in chip breakage. Alloys and ratings in this group are:

> Free-cutting brass—100% (base)
> Selenium copper—80%
> Leaded naval brass—80%
> Leaded commercial bronze—90%

Group II.   The materials in this group are considered readily machinable, with a rating from 30 to 70%. The principal alloys in this group are:

> Yellow brass—40%
> Manganese bronze—30%

Admiralty brass—50%
Leaded phosphor bronze—50%
Leaded nickel silver—50%
Tin bronze—40%

Group III.   These materials have a machinability rating up to 30%. This group contains the unleaded coppers, nickel silvers, some phosphor bronzes, and beryllium bronze. These alloys usually produce tough stringy chips.

As with aluminum, the main angles to consider for cutting tools are the ones on top of the tool, or the rake angles. Moderate rake is suggested for group I to reduce the tendency of the tool to grab or "hog-in." More pronounced rakes are used in groups II and III to provide free chip flow. The side rake angles in group III exceed those for steel.

Best results are obtained in drilling group I and II materials with straight-fluted drills [Fig. 2-45(a)]. These drills have a natural zero-degree rake angle and are particularly good for automatic machine work. Regular drills may be modified to reduce the cutting-lip angle and the tendency to grab, as shown at (b) of Fig. 2-45.

Formability.   The forming qualities of copper-base alloys are dependent on three main factors: alloy composition, grain size, and temper.

Alloy Composition.   The most-used alloys for forming are the 70–30 cartridge brasses. However, all brasses from 95–5 to 63–37 are used for press-working operations. Trouble is encountered when less than 63% copper is used, because the beta phase is brittle and may produce fractures or waviness—surface defects.

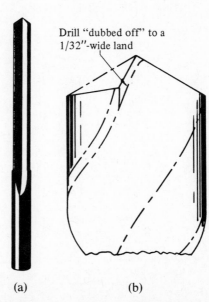

Drill "dubbed off" to a
1/32"-wide land

Figure 2.45.   A straight-flute drill used to drill soft metals (a). A standard drill modified by grinding a $\frac{1}{32}$ in. wide flat on the cutting lip (b) to keep the drill from grabbing or "hogging-in" on soft metals.

(a)   (b)

Pure coppers such as the electrolytic tough-pitch, oxygen-free, and deoxidized coppers are very ductile, and they work-harden less rapidly than do bronzes or brasses. The deoxidized and oxygen-free coppers can withstand more severe bending and forming than the tough-pitch coppers.

Grain Size.   The grain size of copper alloys is expressed as the average grain diameter in millimeters. Grain structure is related to tensile strength, elongation, hardness, and other properties.

Fine grain or ultrafine grain is material below 0.010-mm grain size. Such material is very smooth after bending, forming, or drawing and can economically be buffed to a high luster. Other grain-size recommendations for cold-working annealed copper alloys are given in Table 2-19.

In general, ductility increases with increasing grain size. For very severe draws, particularly of heavy-gauge strips, grain sizes as large as 0.120 mm are not uncommon.

Temper.   Copper and copper-base alloys may be obtained in the tempers shown in Table 2-20.

The most common tempers for rods are half-hard and hard.

The minimum bend radius varies with alloy and tempers. Annealed copper-base alloys can be bent to a radius equal to the metal thickness, and some of them to one-half the metal thickness.

Weldability.   All four of the commonly employed methods—gas, metal arc, carbon arc, and gas-shielded arc—are in general use for welding copper. Copper is also joined by brazing and by newer methods discussed later, such as laser

Table 2-19.   Grain size for cold-working annealed copper alloys. (*Courtesy* American Machinist.)

Nominal Grain Size (mm)	Typical Uses
0.015	Slight forming operations
0.025	Shallow drawing
0.035	For best average surface combined with drawing
0.050	Deep-drawing operations
0.070	Heavy drawing on thick gauges

Table 2-20.   Temper designations of copper-base alloys.

Temper Designation	Meaning
$\frac{1}{4}$ hard	Extra hard
$\frac{1}{2}$ hard	Spring
$\frac{3}{4}$ hard	Extra spring
Hard	

and ultrasonic welding. The gas method, because of the variety of atmospheres obtainable, finds wide use. Its disadvantages are cost and, in some cases, warpage and attendant high-conduction losses into the metal adjacent to the weld.

The most suitable of the coppers to use for welding is the deoxidized type. The fact that the coppers do not contain oxygen, plus the fact that there is some residual deoxidizer in the metal, reduces embrittlement to a minimum.

Resistance welds can be made on all brasses containing 15 to 20% zinc or more. The thermal conductivity is sufficiently low in these metals to make spot or seam welding of the lighter gauges practical.

Both soldering and low-temperature brazing are widely employed in joining coppers and brasses.

Castability.   Copper castings are made almost entirely in sand molds. The technique of casting high-conductivity copper is especially difficult, since considerable skill is required in handling the reducing agents. The molten metal does not flow well, making it difficult to fill molds that are intricate. Copper has the property of being hot-short, and will break as it solidifies if the design does not permit shrinkage.

The copper-base alloys are quite easily cast and handled in the foundry.

ORGANIC MATERIALS

Introduction

About 100 years ago there was no such thing as a commercial plastic material. Chemists and scientists had not yet realized that by the combination of such basic organic materials as oxygen, hydrogen, nitrogen, chlorine, and sulphur new materials could be created.

In 1868 a shortage of ivory for billiard balls developed. A determined and inventive young printer named John Wesley Hyatt mixed pyroxylin, made from cotton and nitric acid, with solid camphor. The resulting material was called Celluloid. The new material was soon adapted to many uses. Colored pink it was used by dentists as a replacement for hard-rubber denture plates. Other popular uses were for toothbrush handles, combs, and motion picture film.

About 40 years passed before another major step was made in the plastics industry. In 1909 Dr. Les Henrik Baekeland introduced phenolformaldehyde resins. Although others had experimented with this combination, Dr. Baekeland was the first to achieve a controllable reaction and to develop a commercial product. This first phenolic material was given the tradename Bakelite.

Cellulose acetate was the next large-volume plastic to be developed commercially. In 1927 it was available only in rod, sheet, or tube form and in 1929 it became the first plastic material to be injection molded. Other now well-known materials began developing at an ever-increasing rate, such as the vinyl resins, polystyrenes, and polyethylenes.

Organic Structures

Early investigators began to deduce the existence of large organic molecules from the insolubility of certain substances and their resistance to melting. Chemists named them by attaching "poly" to the basic monomer unit. A high polymer was one with a very large aggregation of units, i.e., compounds of high molecular weight. Molecular weight is used to indicate the size of a polymer. It represents the cumulative weight of all elements forming the giant molecule and may range between 1000 and 1,000,000.

Another parameter that is used to designate the size of a molecule is the *degree of polymerization* (DP). It refers to the number of repeat units in the chain. In commercial plastics the DP normally falls in the range of 75 to 750 mers per molecule.

Even though the polymer molecule appears to be very large by weight, it is still below the resolving power of an optical microscope and only under certain circumstances can it be resolved by an electron microscope. Consequently, molecular weight determinations are usually made indirectly by such physical means as viscosity, osmotic pressure, or light scattering, all of which are affected by the number, size, or shape of molecules in a suspension or in a solution.

Polymerization Mechanisms

The length that a polymer chain attains in a particular reaction is determined by such random events as which materials are available at the end of the growing chain during a *condensation reaction,* and how soon one free radical meets another during an *addition reaction.*

Condensation Reaction.   Where two or more reactive compounds are involved, there is a repetitive elimination of smaller molecules to form a by-product. Let us suppose, for example, that there are two monomers: one has a basic group, the amino group (NH_2), at each end; the other has an acid group, a carboxyl group (COOH), at each end. The two monomers will hook up end to end, the basic end of one linking up with the acid end of the other, splitting off a molecule of water in the chemical reaction. The condensation reaction takes place during the molding process. Unless the volatile by-products are completely driven off, dimensional instability and low strength will result.

The condensation method of building polymers is rather slow and tends to come to a complete stop before the molecules have attained really giant size. As the chains grow, they become less mobile and less numerous and therefore less likely to encounter free building blocks in solution. Fortunately, such products as nylon have acquired their valuable properties by the time they reach a molecular weight of 10,000 to 20,000.

Addition Polymerization.   The addition method, in contrast with the condensation method, can build molecules of almost unlimited size. The first stage is to activate a monomer by heat and pressure as shown in Fig. 2-46. An ethylene monomer, $CH_2 = CH_2$, is made by removing two hydrogens from ethane, $CH_3 —$

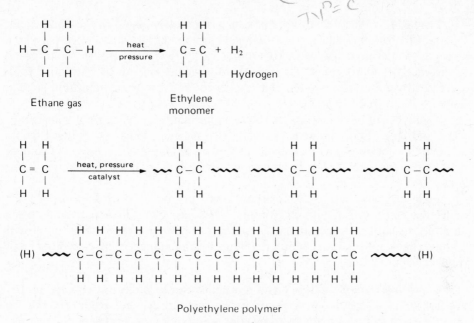

Polyethylene polymer

Figure 2.46. An example of addition polymerization.

CH_3. A redistribution of electrons occurs, and a stable double bond is formed. By counting the double bonds as two, each carbon still has four energy bonds satisfied. Every atom within a molecule must have all its energy bonds completely satisfied if the compound is to be stable.

Starting with billions of molecules of a monomer in a reactor, under heat, pressure, and catalysts, one of the monomer double bonds will rearrange into two "half bonds," one at each end. These half bonds quickly combine with half bonds of other rearranged monomer molecules, forming stable "whole bonds" between them. As each monomer joins with others, the chain length grows until it meets a stray hydrogen, which combines with the reactive end, stopping the chain growth at that point.

During the polymerization reaction, millions of separate polymer chains grow in length simultaneously, until all of the ethylene monomer is exhausted. By adding predetermined amounts of hydrogen (or other chain stoppers), the polymer chemists can manufacture a relatively consistent average chain length. Chain length (molecular weight) is important because it determines many of the properties of a plastic, and it also affects the ease of processing.

Factors Affecting Polymer Properties

Factors that largely determine the properties of a polymer are branching, co-polymerization (random block and graft), crystal orientation and size, degree of crystallinity, molecular weight, and molecular weight distribution.

Branching.   The density of a polymer is directly related to the number and length of the side branches. The greater the branching, the less the density. The strength of intermolecular attractive forces varies inversely with the sixth power of distance between chains. Thus, as the distance is halved, the attractive force increases by a factor of 64. For this reason chain shape is as important as chain length.

Two types of branches are formed from the main stem during polymerization: short and long. Short branches (usually two carbon atoms) are formed by the action of the catalyst (which controls the rate of initiation). Increasing the catalyst and temperature yields shorter chains and lower molecular average weights (Fig. 2-47).

Long branches, up to 2000 carbon atoms per branch, are formed by inter-molecular *chain transfer*. In chain transfer, the growing polymer radical attracts an atom, usually hydrogen, from an organic compound, terminating the growth of one chain and starting another. Thus, the monomer itself can be a chain transfer agent.

Copolymerization.   Large chain molecules can be built from monomers that are not alike. The monomers need not alternate on a regular pattern, but frequently do. A polymer made from two different monomers is called a copolymer; one made from three different monomers is called a terpolymer.

Final properties of a copolymer depend on the percentage of monomer A to monomer B and how they are arranged along the chain.

Block Copolymers.   Block copolymers have long chains of monomer A alternating with long chains of monomer B.

Graft Copolymers.   A graft copolymer consists of a main chain of one polymer with side groups of monomers. A copolymer with a flexible polymer for the main chain but with rigid side chains is very stiff, yet has excellent resistance to impact—a combination of properties not often found in the same plastic. Figure 2-48 shows a schematic representation of various copolymer arrangements.

Crystal Orientation and Size.   It has long been believed that single crystals could not be produced from polymer solutions because of the molecular entanglements of the chains. However, since 1953 single crystals have been reported for so many polymers that the phenomenon appears quite general. Electron-diffraction patterns show platelets about 100 Å thick in which perfect order exists. Dislocations also exist that are similar to those found in metals and low-molecular crystals.

The crystallites, or spherulites as they are called, can be oriented in two different ways to produce increased strength: one-dimensional and biaxial.

An example of the one-dimensional arrangement is polypropylene monofil-aments, or threadlike fibers. An example of biaxial arrangement is polypropylene film.

The size of the spherulites can be controlled by (a) rapid cooling from the melt condition, or (b) the addition of a nucleating agent. If the spherulites become too large, the polymers may become brittle. Nucleated resins normally have increased toughness.

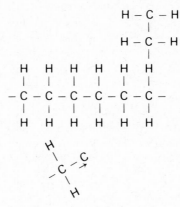

Figure 2.47. An example of how short branching takes place. The growing polymer has an unsatisfied valance that attracts the monomer.

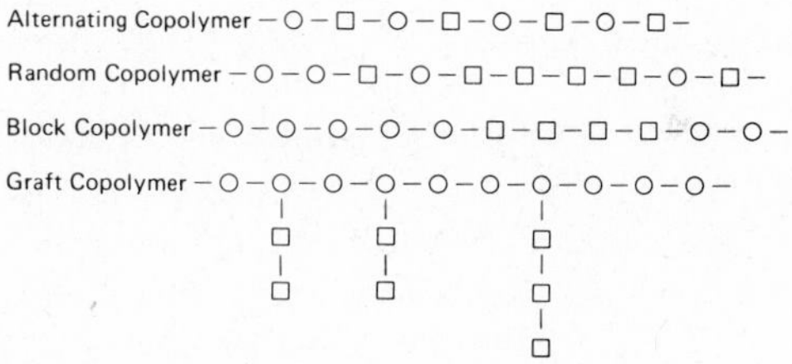

Alternating Copolymer — O — □ — O — □ — O — □ — O — □ —

Random Copolymer — O — O — □ — O — □ — □ — □ — □ — O — □ —

Block Copolymer — O — O — O — O — O — □ — □ — □ — □ — O — O —

Graft Copolymer — O — O — O — O — O — O — O — O — O — O —
                          | | |
                          □   □   □
                          | | |
                          □   □   □
                                                  |
                                                  □

Figure 2.48. A schematic of the basic arrangement of copolymers based on two monomers. Each copolymer has different properties and processing characteristics.

Degree of Crystallinity. Polymer structures range from partly crystalline to completely amorphous. The degree of crystallinity is largely determined by the geometry of the polymer chains. It must be remembered that the side groups attached to the main backbone of the molecular chain give it a bulkiness in three dimensions. The closer the polymer chains can be packed, the more crystalline it will become. Conversely, the farther apart the chains are held due to irregularities, the more amorphous the structure will be. If a polymer molecule has a symmetrical shape that can pack closely, the intermolecular forces will be very large compared to those of a nonsymmetrical shape, as shown by the schematic sketch of the structures of high- and low-density polyethylene in Fig. 2-49.

The shape of the polymer chain can be controlled by various modes of polymerization, as shown in Fig. 2-50. These modes are three chain arrangements and their corresponding cross-sectional views. The *isotactic* mode has very high symmetry, which allows close packing. The *syndiotactic* mode is of intermediate symmetry. The *atactic* mode is of random configuration. Considering these configurations, it is not surprising that polyethylene, with a high degree of regularity (isotactic), is closely packed and has a melting point of 338°F (170°C), while polypropylene, with a random (atactic) arrangement, flows at a temperature of 100°F (212°C).

A crystallized structure is stiffer, has higher tensile strength, has increased creep resistance, and has heat resistance, but also has a greater tendency to stress-crack than the polymers that are more noncrystallized or amorphous.

Crystalline polymers are more difficult to process, have higher melt tem-

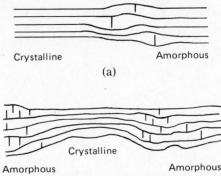

Figure 2.49. The chain packing of high-density polyethylene (a) and low-density polyethylene (b). In high-density polyethylene the chains are mostly linear, with one to three branches per thousand carbon atoms.

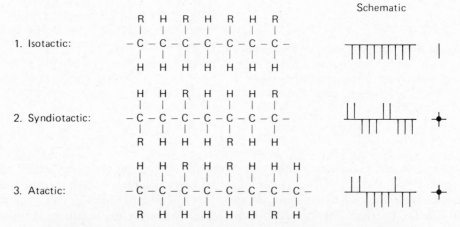

Figure 2.50. Typical polymer structures where R represents a side group. The term isotactic refers to a molecule of high symmetry, syndiotactic to intermediate symmetry, and atactic to a random configuration. The modes are shown as two-dimensional; however it should be realized that even as molecules exist in three dimensions, so do the side-group locations.

peratures and melt viscosities, and tend to shrink and weigh more than amorphous polymers.

The *glass transition* (T_g) temperature of a polymer is also related to crystallinity. Glass transition refers to the temperature at which a polymer changes from a melt to a solid. Amorphous polymers do not crystallize readily but change from a melt to a solid gradually over a range of about 80°F (45°C). The change in viscosity, when a crystalline structure goes from a melt to a solid, is more pronounced and is given as the T_g temperature. If side groups attached to the main chain have different transition temperatures, the effect is to produce rubbery or leathery stages before the solid state.

Table 2-21. Molecular weight of polymers is built up by adding CH_2 groups.

Chemical Formula	Common Name	Molecular Weight	Physical State
CH_4	Methane	16	Gas
C_2H_6	Ethane	30	Gas
C_3H_8	Propane	44	Gas
C_4H_{10}	Butane	58	Gas
C_5H_{12}	Pentane	72	Liquid
$C_{17}H_{36}$	Kerosene	240	Liquid
$C_{18}H_{38}$	Paraffin	254	Soft solid
$C_{50}H_{102}$	Hard waxes	702	Brittle solid
$C_{100}H_{202}$	LMW polyethylene	1402	Tough solid

Many engineering plastics such as nylon, polypropylene, and acetal share common characteristics of crystallinity and hence have similar properties.

Molecular Weight and Distribution.   As stated previously, polymers consist of a mixture of giant-size molecules. The size of the molecules is described in terms of *molecular weight* and the distribution of the sizes is referred to as *molecular-weight distribution* (MWD). Molecular weight of a given polymer refers to the average of all sizes of polymer chains produced during polymerization. In general, high-molecular-weight polymers have better physical properties than their low-molecular-weight counterparts. This is especially true in certain critical properties such as impact resistance (toughness), chemical resistance, and stress-crack resistance. However, because high-molecular-weight polymers also have a higher melt viscosity, they are more difficult to process.

Molecular-weight distribution has a somewhat smaller effect on the polymer properties and processing than molecular weight, but nevertheless it is significant. In general, polymers that have a narrow MWD have higher impact strength, better resistance to environmental stress cracking, and better low-temperature toughness. On the other hand, polymers with broad MWD have better processing properties, such as less resistance to flow (important in molding and extruding), and a better drawing rate (important in producing films). Therefore it is important in selecting a polymer compound to recognize both the endproduct qualities and the ease of processing. Table 2-21 is a classification of polyethylene by molecular weight.

Both high-density polyethylene (HDPE) and ultrahigh-molecular-weight polyethylene (UHMWPE) have interesting design properties. Recent developments in copolymerization techniques have eliminated the major problem of stress cracking that has plagued polyethylene since its discovery. Recent modifications also have produced polymers with narrow molecular weight distribution. Parts molded from these new resins are more warp-resistant than are their broad-distribution counterparts, and they permit injection molding of parts that are essentially flat. An application of narrow-molecular-weight-distribution HDPE is the injection molding of 5-gallon pails and drums that have excellent resistance to impact at low temperatures and are used to replace steel and other container materials. More recently, HDPE has been found to be the best material available for hip-joint replacements.

UHMWPE is the most impact-resistant thermoplastic known. It also has excellent resistance to abrasive wear. It will outwear stainless steel by a factor of 8 to 1 under certain conditions. Abrasion resistance apparently depends more on molecular weight *distribution* than on the molecular weight itself. The effect of molecular weight on the key properties of impact strength, tensile strength, and stiffness are shown in Fig. 2-51.

At this point there is no strict definition as to what constitutes a UHMWPE; however, in this discussion it has been assumed to have a molecular weight of 2 million or higher.

Thermoplastic and Thermosetting Structures

Figure 2-52(a) shows a sketch of an amorphous polymer with random-type attractions. The strength is limited because it depends on hit-or-miss "kissing"

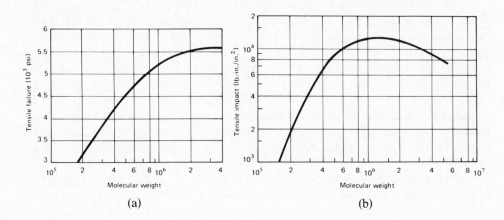

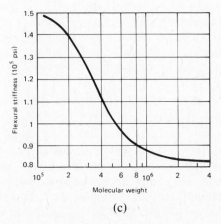

Figure 2.51. The effect of molecular weight on tensile failure (a), tensile impact (b), and flexural stiffness (c). (*Courtesy* Machine Design.)

among the polymer chains. These secondary bonds are easily broken with heat, so such polymers are called *thermoplastics*. Common examples of this type of plastic are polystyrene and acrylics. Thermoplastics are readily molded or extruded because they melt to a viscous liquid at a temperature slightly above that of boiling water.

When the molecular chains are connected by a network of primary bonds it is very much like a three-dimensional spider web [Fig. 2-52(b)]. Even though heat is applied, the chains will not separate completely from one another. These polymers are termed *thermosetting*. Common examples of this type of plastic are urea-formaldehyde and phenolformaldehyde or Bakelite. In the example of phenolformaldehyde, the reaction is stopped at a point where the chains are still mostly linear. Unreacted chain portions are capable of flowing under heat and pressure. The final stage of polymerization is completed at the time of molding. The additional heat and pressure liquifies the polymer, producing a crosslinking reaction between the molecular chains. Unlike the thermoplastic monomer, which has only two reactive ends for linear chain growth, a thermoset monomer must have three or more reactive ends, so the chains crosslink in three dimensions.

Once a thermoset plastic is molded, virtually all of its polymer molecules are interconnected with strong permanent bonds that are not heat reversible. If the thermoset plastic is heated too much or too long, chain breakage occurs, resulting in degradation.

High-Temperature Plastics

High-temperature plastics have in the past been much higher in cost and presented greater processing difficulties than other materials. Now, however, some of the newer resins are processable by injection molding and the cost has been considerably reduced.

Polysulfone plastics have been introduced in recent years that have excellent heat-resistance service temperatures at about 375°F (190°C). The parts are clear with a slight amber tint. In most cases, polysulfone products can be used continuously in air or steam at the rated temperatures and do not cloud, craze, or otherwise destroy the transparency of the polysulfones. A common example of

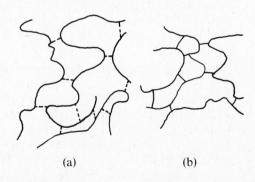

Figure 2.52. An amorphous, thermoplastic polymer (a) showing weak crosslinking. Shown at (b) is a thermoset polymer with strong crosslinks. In the latter, after crosslinking has been established (usually by heat) the polymer is said to be cured.

(a)   (b)

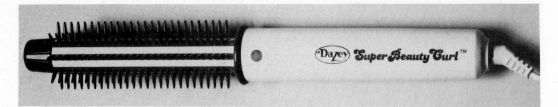

Figure 2.53. Polysulfone plastic parts are able to withstand prolonged services temperatures at about 375°F (190°C) without harmful effects. This comb contains eleven injection-molded polysulfone pieces. (*Courtesy Union Carbide.*)

the use of polysulfone where heat resistance is needed is the hair curler shown in Fig. 2-53.

Basic Structure. As stated at the end of the discussion of thermosetting plastics, too much heat causes degradation. In producing high-temperature plastics, the objective is to increase the number of strong, *inert* (resistant to oxidative and chemical attack) chemical bonds in the backbone so that, at intervals, atoms are joined together by two bonds instead of one. This structure is analogous to that produced by double-joining of links in a chain. To do this, "rings" are incorporated into the polymer backbone, referred to as "aromatic rings." In this way, six carbon atoms are joined together in a hexagon structure (Fig. 2-54). This provides (a) stronger bonds (aromatic bonds are stronger than normal C—C single bonds) and (b) more bonds in a given space. Thus, more energy is required to break down the structure. The ring-type structure is typified by a polyphenylene hexagon structure or by the polyimide five-numbered rings containing carbon, oxygen, and nitrogen regularly placed around the ring.

Fluorocarbons are examples of polymers whose bonds are particularly inert to chemical degradation. The polymer does not decompose chemically at high temperatures; but since it is a single-link polymer, it softens considerably at high temperatures.

A new and still experimental class of resins—called ladder polymers—are reported to have the greatest high-temperature strength yet achieved. In this structure, every atom is joined into a ladderlike structure. This type of structure is represented by the polybenzimidazoles. As might be expected, the ladder polymers are the most difficult of all to process.

Selecting Pertinent Polymer Properties

Unlike metals, the property selection of polymers can be quite misleading. Only those properties that are clearly defined as mechanical constants are useful in comparing one polymer with another. As an example, tensile strength comparisons can be misleading because plastics tend to creep and show stress relaxation (viscoelastic effect) even at room temperature. Two ABS (acrylonitrile butadiene styrene) plastics may have identical tensile strengths of 5500 psi, but under long-

Figure 2.54. Six carbon atoms are joined together in a hexagon structure. This increases strength because there are more bonds in a given space and increases heat resistance because more energy is required to break it down.

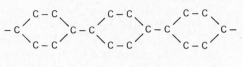

Polyphenylene

term loading these may change to 2150 psi for one and 1200 psi for the other. Thus, for parts that must support loads, the long-term strength must be known.

Performance properties are those that essentially characterize the effect that mechanical loading will have at the stress encountered in the application. These include measurements of rigidity, or modulus, in various loading configurations, toughness index, Poisson's ratio, evaluation of creep under long-term conditions, impact strength, fatigue limits, and others.

The evaluation of environmental effects is considerably more critical for plastics than for metals. Too often the effect of temperature on several properties is taken as the only criteria in making a material selection. Time and space prohibit the discussion of all environmental conditions but certain key elements should be examined.

1. *Temperature exposure.* The temperature the product will operate in, including extremes and number of cycles between those extremes, must be considered.

   A generalized comparison of the properties discussed in relation to temperature and modulus of elasticity of three different polystyrenes—crystalline, thermoset, and amorphous—are shown in Fig. 2.55. At room

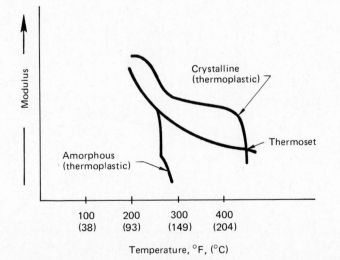

Figure 2.55. A comparison of crystalline, amorphous, and thermoset properties of three different polystyrenes in relation to temperature and modulus of elasticity.

temperature, all three have about the same stiffness. At 220°F (104°C) the modulus of the crystalline material drops only slightly because of the interchain alignment. Only at well above 400°F (204°C) does the modulus drop, a signal that the crystalline regions are breaking up. The other two materials soften at 220°F (104°C). This initial loss of properties takes place at the glass transition temperature.

2. *Chemical exposure.* Most polymers can be significantly damaged by many chemicals that act independently or in association with a mechanical load. Oxidation can be significant at elevated temperatures.

3. *Impact exposures.* In applications where the part is subject to frequent pounding, materials that have high impact strength or toughness must be selected. As an example, ABS plastics were chosen to replace fiber-reinforced plastic tote boxes for the post office because it would withstand the constant beating that boxes receive when traveling along conveyors.

Toughness varies with the rate at which the load is applied. Fig. 2-56 shows a schematic of a puncture test and a corresponding curve. The plastic is clamped between two plates, exposing a 1-in. diameter surface. The rounded-end rod is then driven into the plastic at speeds up to 10,000 in. per min. As shown at low speeds, rupture is relatively independent of speed and closely related to tensile strength. As speeds increase, the load required to puncture drops off sharply. In this case, the peak for an acetal resin was reached at about 500 in./min.

Comparison service tests show that a puncture resistance up to a speed of

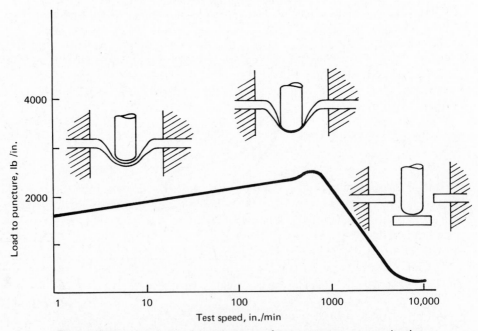

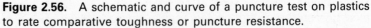

Figure 2.56.   A schematic and curve of a puncture test on plastics to rate comparative toughness or puncture resistance.

2000 in./min provides a sound basis for selecting plastics that must withstand considerable abuse.

An old cliché states that the best way to evaluate a material and design is to study the finished product. Although this is fundamentally true, few companies can afford the luxury of waiting. Therefore every step from conception to final processing must be studied carefully. A thorough knowledge of materials and how they are affected by processing is essential for both the designer and the manufacturing engineer. With this knowledge, effective evaluation can be made at each step so that modifications can be made either in the design or process, or both, as required to produce the best possible product.

To design with plastics it is necessary to know not only the various materials that are available, their characteristics, and their service life, but also the various molding processes that can be used. For example, a given part may be made by injection molding, blow molding, thermoforming, or rotational molding. These processes are discussed in Chapter 10.

ELASTOMERS

Thermosetting Elastomers

Elastomers are a special type of thermoset polymer that have long-chain molecular structures whose glass transition temperatures are well below room temperature, usually in the range of −20 to 80°F (−29 to 26.7°C). In their "green" state they are tacky, and cold-flow easily. To be useful they must be crosslinked. With rubbers this is termed vulcanization.

Elastomer molecules may be compared to a network of thin, soft coil springs. When this structure is stretched, the coils open until they are almost straight. The motion is limited by the number of junction points or crosslinks. When the stress is removed, the assembly returns to its original coiled position. The farther apart the crosslink points, the weaker the system and the greater the elongation.

Thus, an elastomer or rubber is basically a low-T_g thermoset with relatively wide spacing between crosslinks in the polymer chains. This structure enables the elastomers to readily absorb and dissipate energy.

Rubber Classification.   The standard that has been adopted for rubber classification is ASTM D2000 "A Classification System for Elastomeric Materials for Automotive Applications." While this standard was originally intended to cover automobile uses, it covers all potential rubber compositions. The following is a summary of some of the principal types.

R Class: R-class rubbers are those that have an unsaturated carbon chain and may be natural or synthetic.

Natural Rubber (NR).   Because of its relatively low cost and outstanding physical properties in both reinforced and unreinforced state, natural rubber is still the only material that satisfies many applications. It may be compounded to have good abrasion and high tear resistance. Special compounding is needed if good weathering is required. It is ideal for general purpose applications where "snap" or resilience is important.

Typical applications for natural rubber are pipe gaskets, respirators, windshield wiper blades, swimming goggles, drain boards, and pharmaceutical products.

Styrene Butadiene (SBR). SBR was the first elastomer to find wide acceptance as a substitute for natural rubber. It is a low-cost, high-volume rubber that is available in various ratios of the two monomers. If compounded with a high-styrene and low-butadiene content, it has high tensile strength, high modulus, and high hardness without the use of large amounts of filler materials. Due to its good abrasion resistance and leather-like properties, it is used in the manufacture of shoe soles and floor tile. High-butadiene, low-styrene copolymers offer good flexibility at low temperatures.

Typical applications for SBR rubbers include tires, pipe gaskets, automotive brake components, and other mechanical items.

Nitrile Butadiene (NBR). NBR is a copolymer of acrylonitrile and butadiene. It is made in various ratios of these monomers. Nitrile rubbers have good oil and gasoline resistance, tensile strength, elongation properties, heat resistance, and low compression set. Special compounding can produce good weatherability.

Typical applications are seals, O-rings, gaskets, diaphragms, pipe gaskets, tank linings, boots, and bellows.

Chloroprene (CR). Chloroprene, also known as neoprene, is made from chlorine and butadiene in several formulations. Chloroprene rubbers have good resistance to heat, oil, chemicals, ozone, weathering, flexing, compression, stain, and flame.

Typical applications are hoses used in handling fuels, bellows, gaskets, seals, tank linings, and O-rings.

M Class: M-class rubbers have a saturated chain of the polymethylene type.

Polyacrylic (ACM). ACM elastomers are based on ethyl or butyl acrylate or a combination of the two monomers, with other comonomers.

Polyacrylic rubbers are superior to nitrile rubbers in resistance to deterioration by high-aniline-point oils, extreme-pressure lubricants, and transmission fluids. They also have excellent resistance to ozone, sunlight, and weathering, and are suitable for use down to temperatures of $-40°F$ ($-40°C$) and intermittent use at temperatures up to 300°F (150°C).

Other M-class rubbers are chlorosulfonyl polyethylene (CSM), chloropolyethylene (CM), ethylene propylene (EPM), fluorocarbon (FKM), and epichlorohydrin (ECO). For further discussion of the various properties and uses, see the bibliography at the end of the chapter.

Q Class: The Q-class rubbers have a substitute group on the polymer chain and are indicated prior to the silicone (Q) designation.

Fluorosilicone (FVMQ). The outstanding characteristic of all types of silicone rubbers is the wide service-temperature range, -55 to 200°F (-48 to 93°C). In addition, they have excellent resistance to oils and fuels.

Silicones are relatively expensive rubbers which include several types. Typical applications are for seals, gaskets, tank linings, and diaphragms. Some versions require postcure for development of optimum properties.

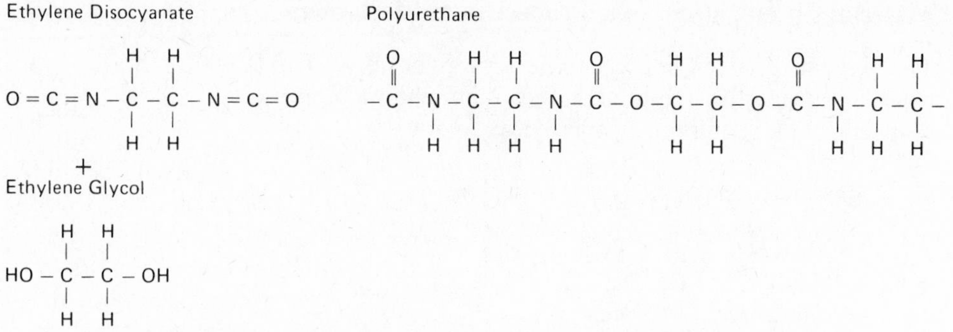

Ethylene Disocyanate   Polyurethane

```
        H   H                              O     H   H     O     H   H     O      H   H
        |   |                              ‖     |   |     ‖     |   |     ‖      |   |
O = C = N - C - C - N = C = O        - C - N - C - C - N - C - O - C - C - O - C - N - C - C -
        |   |                              |   |   |   |         |   |            |   |
        H   H                              H   H   H   H         H   H            H   H
       +
```

Ethylene Glycol

```
        H   H
        |   |
HO - C - C - OH
        |   |
        H   H
```

Figure 2.57. Polyurethane is made from two monomers, ethylene diisocyanate and ethylene glycol, by the condensation process. Two molecules combine, usually with the elimination of water (H–O–H), to form the repeating units of the copolymer chain.

U Class: U-class rubbers are those having carbon, oxygen, and nitrogen in the polymer chain, as shown in Fig. 2-57.

Polyurethane Rubbers. The two types of polyurethanes are polyester urethane (AU) and polyether urethane (EU). Both are available in viscous liquid and millable gum form. The gum forms are processed in conventional rubber equipment, and the viscous liquid form depends on the part being fabricated. Casting is frequently used for the liquid version. The properties generally exceed most other rubbers in higher hardness range, extremely high tensile strength, and abrasion resistance. In addition, these urethanes have high load-bearing characteristics, low compression set, good aging characteristics, and good oil and fuel resistance.

Typical applications include seals, bumpers, metal-forming dies (Fig. 8-19), valve seats, liners, rollers, wheels, and conveyor belts.

Thermoplastic Elastomers

Thermoplastic elastomers are similar to the thermoset type in service properties. However, they do offer greater ease and speed of processing. Molding or extruding can be done on standard plastic-processing equipment with considerably shorter cycle times than that required for compression molding of conventional rubbers.

The principal thermoplastic elastomers are polyester copolymers (*Hytrel*, Du Pont Co.), styrene–butadiene block copolymers (*Krayton*, Shell Chemical Co. and *Solprene*, Phillips Petroleum Co.), polyolefin (*TRR*, Uniroyal, Inc.), and polyurethane, from a number of producers.

Polyurethane. Polyurethane was the first major elastomer to be developed that could be processed as a thermoplastic. It does not have quite the heat resistance or the compression set resistance of the crosslinked types, but most other properties are similar.

Material Specifications and Properties of Elastomers

The system of specifying elastomers as given in ASTM D2000 or SAE 5200 is based on properties required, rather than material composition. An example of a specification requirement may be as follows:

> Applicable type of material: chloroprene
> Original properties:
> > Hardness   50 ± 5 (durometer)
> > Tensile   1500 psi min
> > Elongation   350% min
>
> Air oven, 70 hr at 100°C:
> > Hardness change   + 15 max
> > Tensile change   − 15 max
> > Elongation change   − 40% max
>
> ASTM #3 oil, 70 hr at 100°C:
> > Tensile change   − 70% max
> > Elongation change   − 55% max
> > Volume change   + 120% max
>
> Compression set, 22 hr at 100°C:
> > Set   80% max

THE RECLAMATION OF RUBBER AND PLASTIC

Rubbers.   To the ecologically and energy minded world the problem of rubber reclamation has become one of prime importance. The process is not new. Goodyear patented a process for grinding waste rubber into powder and mixing it with pure rubber in 1853. Some work has been done on "desulphurizing" rubber in an attempt to break down the crosslinked structure in order to be able to reprocess and revulcanize it.

One of the most used processes is the steam process, which does not necessitate prior grinding of the waste products. The rubber is cut and put into an autoclave with 40 to 70 psi (65 to 120 MPa) steam pressure or with superheated steam at 400 to 430°F (200–220°C) or with saturated steam at 670 to 700 psi (670 to 1175 MPa) after which it is dried and refined.

Somewhat the same means are employed in recovering synthetic rubbers. However, special plasticising oils or special catalysts are usually added. It is almost exclusively SBR and NBR that is recovered on an industrial scale.

Since more than 215 million old tires are discarded annually in the United States, considerable interest is being developed toward possible uses. At the present time only a small percentage (approximately 10%) is reclaimed. A process developed at the University of Wisconsin utilizes liquid nitrogen for freezing the rubber and then shattering it by impact. The result is small rubber granules that can be used in athletic tracks, industrial floor mats, and landfill.

Plastics. Thermoplastics can be ground up and remolded, but the thermosetting resins cannot. Since collection of used plastic containers poses an environmental problem, comprehensive studies have been carried out, and are still in process, to determine the extent of plastic pollution. Newer plastics are being developed, especially for the packaging industries, that are biodegradable or susceptible to biodeterioration. Evidence has now been introduced to show that polyvinyl chloride plasticized with dodecyl esters of aliphatic dibasic acids are much more biodegradable than those formulated previously. Modifications of polyolefin polymers by the introduction of carboxyl, amide, and other substituting groups has significantly increased their biological conversion.

New mutant strains of soil microorganisms are also being developed that show increased capacity for degrading the more resistant types of synthetic polymers. The use of specialized bacteria and fungi as inoculants in waste disposal areas will do much to help the problem of degrading waste polymers.

Attention has also been given to the incineration of plastic packaging materials. A new material, Barex 210FDA, is suitable for a wide variety of food, beverage, and commercial products that can be conveniently disposed of by incineration with no hazard to the environment. The new material is described as having good transparency, shatter resistance, and lightness.

COMPOSITE MATERIALS

The dramatic technological advances of the past few decades have, in part, been based on ever better material properties. Limitations on how far the material properties can be expanded may present equivalent limitations on further successes in aerospace, hydrospace, nuclear energy, defense, transportation, and other fields.

Composite materials that have the potential of integrating various metals, ceramics, polymers, and natural materials appear to be well on the way toward eliminating some of the obstacles that relate to strength and high-temperature performance.

Fibers, Filaments, and Whiskers

Fibers

Man has only to observe nature to learn that fibers, as in plants, increase the strength of a structure considerably. The word "fiber" is generally associated with threadlike materials like those found in wool, cotton, or hemp. The main characteristic of a fiber is that its length is considerably greater, at least 200 times greater, than its effective diameter. Two well known fibers used in today's composites are glass and high-strength wire, as shown in Fig. 2-58.

Filaments

Filaments are defined as "endless" or continuous fibers. Filaments are used in making glass, yarn, and cloth. Glass filaments are formed by jets of air that pull

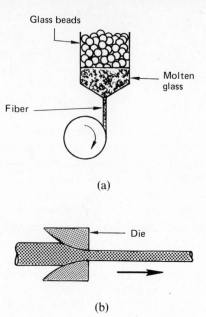

(a)

Figure 2.58. Glass fibers are made by continuous drawing of molten glass from an orifice and winding the fiber on a mandrel (a). Other fibers can be made in a similar manner. Fine wires can be made by drawing through a die (b). Wires used in composites include tungsten, molybdenum, steel, and superalloys. Diameters range from about 1 to 20 mm.

(b)

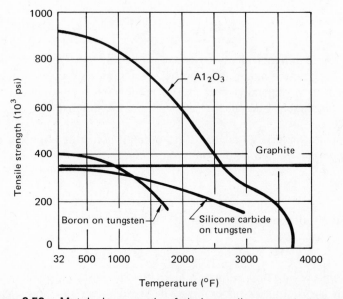

Figure 2.59. Metals lose much of their tensile strength with increasing temperature. Graphite offers obvious advantages with its flat strength curve, 350,000 psi to at least 4000°F (2250°C). Graphite unfortunately is incompatible with many metal matrix materials. Silicon carbide is relatively compatible with metals and its strength is over 200,000 psi at 2000°F (1093°C). Alumina (sapphire) has the highest strength above 2500°F (1371°C). (*Courtesy* Machine Design.)

glass strands from a spinneret onto a revolving drum. These filaments can be handled and processed on a mass-production basis.

Whiskers

Whiskers are single crystals that are grown with almost no defects. They are short, discontinuous fibers of polygonal cross sections that can be made from over 100 materials. Common examples are copper, iron, graphite, silicon carbide, aluminum oxide, silicon nitride, boron carbide, and beryllium oxide. The tensile strength of some of these fibers in relation to heat is shown in Fig. 2-59.

Generally, whiskers grow from vapor or metallic deposits of gases or liquids on a surface, as in the case of sapphire (Fig. 2-60). Oxide, carbide, and nitride whiskers are of primary interest in composite-materials technology. For a long time, whiskers were only a laboratory curiosity, but in 1952 Herring and Galt, of the Bell Telephone Laboratories, observed that the strength of metallic tin whiskers was 1,000,000 psi.

Multiphase Fibers.   Multiphase fibers consist of materials such as boron carbide and silicon carbide that are formed on the surface of a very fine wire substrate of tungsten (Fig. 2-61).

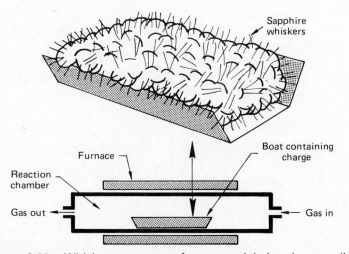

Figure 2.60. Whiskers are grown from materials in a boat or dish placed in a heated, controlled-atmosphere reaction chamber. For example, aluminum in a ceramic boat, in the presence of hydrogen, forms vapors of Al_2O or AlO. These vapors, in turn, react with oxygen (present in the form of water vapor or SiO vapor) that forms from the boat material. The result is solid, needlelike, single crystals of alumina that grow on the surfaces of the boat or its contents. Whiskers usually grow to about 0.5 in. (12.70 mm) long but sometimes may attain a length of about 2 in. (50.80 mm).

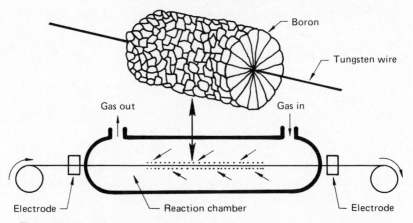

Figure 2.61. Multiphase fibers such as boron on a tungsten substrate are commonly produced by vapor deposition methods. Heated tungsten wire of about 0.0005 in. dia. is moved through a chamber in which boron is deposited on the wire from gases containing a boron compound. The final diameter of the multiphase fiber is about 0.004 in. (0.10 mm).

Fiber-Metal Composites

The primary purpose of making a composite is to incorporate the strength or the unusual properties of the fibers into usable shapes by bonding them into a continuous matrix, as shown in Fig. 2-62. The tensile strength and stiffness of a composite are directly related to the fiber content and to the strength of the fiber, assuming the fibers are aligned in the direction of loading.

Carbon-Fiber Composites

Carbon fibers of the type discussed here are often referred to as "graphite fibers" in the trade and other literature. Graphite, by its technical definition, implies a three-dimensional crystallographic order, which is not present in the commercially available fibers. Hence the term "carbon fiber" as used here is technically correct.

Carbon fibers are among the strongest and stiffest of the reinforcing materials being used today in resin-matrix systems for high-performance structures. Because their cost is still relatively high, various combinations, often including glass fibers, are used to achieve maximum cost effectiveness. Formerly used only for aerospace components and high-priced sports equipment, some molding compounds at the lower end of the performance range are now selling for well under $10.00/lb. The manufacture of graphite fibers is shown schematically in Fig. 2-63.

Design Considerations. Design considerations are entirely different from those generally used with homogenous, isotropic materials having well-defined elastic and plastic stress–strain behavior. For structural parts, all failure modes

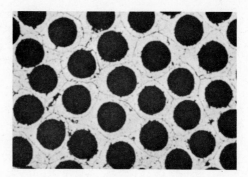

Figure 2.62. The cross section of coated fibers hot-pressed in an aluminum matrix.

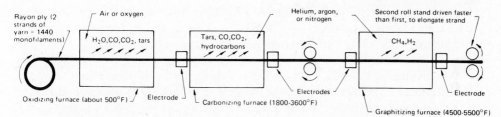

Rayon ply (2 strands of yarn = 1440 monofilaments)

Air or oxygen

H_2O, CO, CO_2, tars

Tars, CO, CO_2, hydrocarbons

Helium, argon, or nitrogen

Second roll stand driven faster than first, to elongate strand

CH_4, H_2

Oxidizing furnace (about 500°F)

Electrode

Carbonizing furnace (1800-3600°F)

Electrodes

Electrode

Graphitizing furnace (4500-5500°F)

Figure 2.63. Graphite fibers are made by a process that converts organic materials to carbon, then to graphite. Bundles of the rayon or polyacrylonitrile are heated to drive off most organic constituents, leaving a residue of carbon. The carbon strand is then heated to a higher temperature in controlled atmosphere chambers and a slight stress is applied to produce elongation of 40 to 60%. The strands elongate and the molecular structure is changed from carbon to graphite. The diameter of the graphite filaments is about 5 to 10 microns.

must be investigated and particular attentions must be given to interlaminar tensile and shear stresses because strengths in these modes depend principally on the matrix resin, not on the reinforcing fiber.

Components made from carbon–resin or carbon–glass–resin combinations, often called "advanced composites," are usually of laminated construction. They can be isotropic, quasi-isotropic, or very anisotropic, depending on the material form, layup configuration, and fabrication method. Most parts are designed to be anisotropic to utilize fully the highly directional properties of the fibers.

Fiber Properties. Carbon-fiber–resin matrix composites have very high strength-to-weight and stiffness-to-weight ratio. With proper selection and placement of fibers, composites can be stronger and stiffer than steel parts and weigh 40 to 70% less. Direct weight savings can in turn lead to additional indirect weight savings by requiring lighter support members.

Carbon-fiber composites show excellent fatigue and environmental resistance, being able to operate at temperatures over 300°F (149°C) for long periods of time.

An outstanding advantage of composites is that the stiffness can be varied significantly in different areas of a part by selecting the type and form of fiber, by judicious orientation, and by concentrations of fibers.

By incorporating carbon fibers into plastic resins, the compound becomes conductive, both electrically and thermally. Electrical conductivity is beneficial for static drain and RF suppression in electronic enclosures, automotive ignition system devices, and small appliance housings. Thermal conductivity provides for heat dissipation in such components as gears, bearings, brake pads, and other friction-related products and in cryogenic processing equipment.

The thermal expansion characteristics of composites are quite unique. Linear expansion coefficients range from slightly negative for 30-million-psi modulus fibers to approximately -0.5×10^{-6} in./in./°F for high modulus fibers. This property makes possible the design of structures with zero or very low linear and planar thermal expansion—a valuable characteristic for precision instrument components and where critical alignment applications are required.

Discontinuous Fibers.   Carbon fibers can be long and continuous, or fragmented; they can be directionally or randomly oriented and they can be dispersed in thermoplastic or thermosetting material. Each fiber form has its limitations. In general, short fibers cost the least, and fabrication costs are lowest; but properties of the resulting composite are also lower than those obtainable with longer fibers. The shortest fibers used for reinforcement are finely ground, or milled, ranging in length from 30 to 3000 microns, with an average of about 300 microns. Incorporating 30% (by weight) of these fibers in a nylon matrix substantially improves many properties. Tensile strength increases 43%, tensile modulus 217%, flexural modulus 263%, compressive strength 55%, and notched Izod impact strength 77%. Heat-deflection-temperature is higher by 338°F (170°C). There is a large drop in ductility and a lower unnotched Izod impact strength.

As the chopped carbon fibers increase in length, both strength and modulus increase significantly; however, the molding compound costs also increase. High flexural modulus can be obtained with ¼-in. (6.35 mm) chopped carbon fibers used to reinforce matrix material. Chopped reinforcing fibers ½ to 2-in. (12.70–50.80 mm) long are commonly added to thermosetting polyester resin (usually 25 to 50% resin) in manufacturing sheet molding compound (SMC), which is normally processed by compression molding (see Chapter 10).

Fibers over 2 in. (50.80 mm) long can be placed in a preferred direction. Even allowing for some disorientation due to flow mold closure, such composites offer greatly increased resistance to load deformation in a specified direction.

Continuous Fibers.   The ultimate in performance in weight reduction is achieved with the continuous fibers. These fibers are available in a variety of forms, including yarns or tows containing from 400 to 160,000 individual filaments; unidirectional tapes up to 48 in. wide (1888 mm); multiple tape layers with individual layers or plies, at selected fiber orientation; and fabrics of many weights and weaves (Fig.2-64).

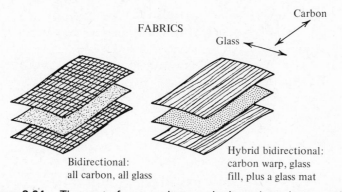

FABRICS

Carbon

Glass

Bidirectional:
all carbon, all glass

Hybrid bidirectional:
carbon warp, glass
fill, plus a glass mat

Figure 2.64. The cost of composites can be kept down by a number of combinations, such as the hybrid structure shown at the right where glass fill and glass mat are used. (*Courtesy* Machine Design.)

The individual carbon fibers are usually 10 microns or less in effective diameter. Tensile strengths can exceeed 400,000 psi (2758 MPa).

Properties of composite structures in the transverse direction depend largely on the matrix material but, in any case, are very low compared to the longitudinal properties. In designs where both longitudinal and transverse loadings are expected, the fibers must be oriented in specific directions to withstand these loads. Further information can be obtained from *Advanced Composite Design Guide* published by the Air Force Materials Laboratory, Wright-Patterson Air Force Base, Dayton, Ohio.

Hybrid Forms. The major drawbacks to carbon fibers are low-impact resistance and high cost. Both of these factors are improved by the use of hybrids (carbon fibers incorporated with lower-cost, higher-impact-strength reinforcements). Commercial fabrics are available containing a mix of fibers within the warp/or fill, as well as with carbon in one direction and the secondary fiber in the other (Fig. 2-65).

Joining Advanced Composites

The manufacture of parts using composite materials has been largely labor-intensive. At the present time, however, high-production methods are being developed. More uses will develop as more is learned about the special design requirements of advanced composites.

Adhesives or Bolts? Most structural composites are made from thermoset plastics and therefore cannot be welded as can thermoplastics and metals. This leaves a choice of joining with adhesives or mechanical fasteners. Both methods have advantages and disadvantages.

Mechanical fasteners have the advantage of not being adversely affected by humidity or thermal cycling. They can be disassembled without destroying the

TAPES

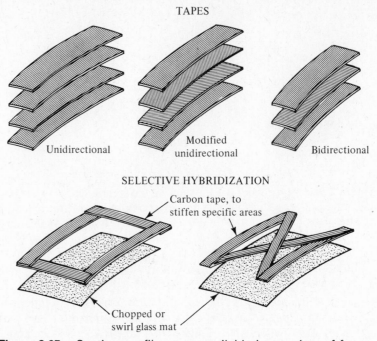

Unidirectional   Modified
                        unidirectional   Bidirectional

SELECTIVE HYBRIDIZATION

Carbon tape, to
stiffen specific areas

Chopped or
swirl glass mat

Figure 2.65. Continuous fibers are available in a variety of forms including unidirectional tapes with individual layers or plies, multiple layers at selected fiber orientation, and fabrics of many weights and weaves. (*Courtesy* Machine Design.)

substrate and little or no surface preparation is required. On the less favorable side, mechanical fasteners require machining the substrate to accommodate bolts or screws, thus weakening the structure and causing stress concentrations. Figure 2-66 shows a sketch of the design of a composite material joint.

Slight dimensional inaccuracies in bolt patterns seldom cause problems in materials that deform plastically such as aluminum or steel. The material around the hole can yield to distribute the load to adjacent bolts. This is not true of advanced composite materials since they have no yield point. If one bolt is tighter in its hole than others, the bearing stress at that hole remains higher. One way to avoid such problems is to soften the laminate in the bolt area through the use of staggered fabric plies. Another way is to coat the bolts with uncured resin prior to attachment. The resin fills the clearances between bolts and holes and, when cured, helps distribute the load evenly.

Adhesive-bonded joints have the advantage of light weight, distribution of the load over a larger area than mechanical joints, and elimination of the need for drilled holes that weaken the structural members. The disadvantages of the bonded joint is the difficulty of inspection for bond integrity, the possibilities of degradation in service from temperature and humidity, and the relative permanence that does not permit disassembly without destruction of the joint. In

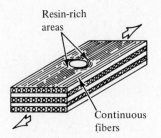

Avoid molded-in holes. In molding, continuous fiber is forced around the plug that forms the hole, leaving resin-rich areas, which are relatively weak.

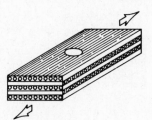

Drilled holes are better. The cross fibers, at 45° or 90° to the load, maintain transverse integrity, keeping cracks from forming.

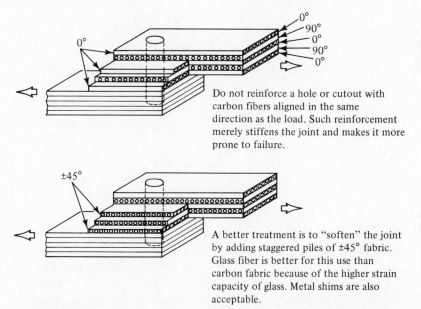

Do not reinforce a hole or cutout with carbon fibers aligned in the same direction as the load. Such reinforcement merely stiffens the joint and makes it more prone to failure.

A better treatment is to "soften" the joint by adding staggered piles of ±45° fabric. Glass fiber is better for this use than carbon fabric because of the higher strain capacity of glass. Metal shims are also acceptable.

Figure 2.66. Bonded joints in advanced-composite structures are preferred where possible. But when strength requirements exceed those available from adhesives alone, bolts are used, often in conjunction with an adhesive. Because of their low strain capacity and inhomogeneity, composites require special design treatment when they are to be joined by mechanical fasteners. (*Courtesy* Machine Design.)

addition, bonded joints generally require rigorous cleaning and preparation of the mating surfaces.

Multilayer composites should have the adjacent layers oriented so the fibers are in the direction of the load; otherwise there will be no fiber strength at the joint. Figure 2-67 shows a sketch of good and poor joint design.

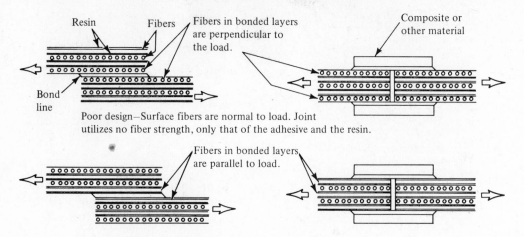

Resin Fibers Fibers in bonded layers are perpendicular to the load.

Composite or other material

Bond line

Poor design—Surface fibers are normal to load. Joint utilizes no fiber strength, only that of the adhesive and the resin.

Fibers in bonded layers are parallel to load.

Good design—All fibers at the bond lines are oriented in the same direction as the load.

Figure 2.67. Multilayer composites must not be stressed in a manner likely to cause delamination. Proper orientation of the surface fibers in a bonded joint—so the fibers are in the direction of the load—must be included in the design specifications. Otherwise, machining of bond-line areas can result in a fiber orientation that weakens the joint drastically. The sketches show this principle in use on components having 0° and 90° fiber orientation. (*Courtesy Machine Design.*)

Steel Laminate Composites

A newer type of laminate that holds considerable promise is sheet steel laminate. Two thin sheets of steel are heat-laminated to a solid plastic (polypropylene) inner core. Laminates with a total thickness ranging from 0.040 to 0.350 in. (1 to 9 mm) are being produced. Applications range from battery-powered golf cars to structural members for luggage.

The laminate can be formed like steel, is dent resistant, and has a very high stiffness ratio. It also resists perforation by corrosion and does not delaminate after extended periods of exposure to corrosive media. Strength is obtained via an I-beam effect.

The use of laminates can result in substantial weight savings. As an example, a simple oil pan cover made from sheet steel may weigh $25\frac{1}{2}$ oz. (725 g), while the laminate version weighs $15\frac{1}{2}$ oz. (440 g).

Problems

2-1. What are some of the major factors that design and manufacturing engineers would have to consider if management posed the problem of making an automobile hood out of aluminum that is now made out of steel? The present hood material is AISI 1010 steel, 0.035 in. thick.

2-2. Shown in Fig. P2.1 are three sketches of parts. Show how the design of each of these parts can be improved, assuming they will be subjected to stress loads or heat-treating stresses.

2-3. If a mild-steel butt weld in $\frac{1}{8}$ in. thick steel is normally made with $1t$ spacing between the plates, would the same rule apply to stainless steel? Why or why not?

2-4. A mild-steel cold-rolled rod 0.5 in. dia. is used to support a maximum tensile load of 20,000 lb. It was found after some time that a portion of the rod had "necked down" to a minimum of 0.400 in. dia. and the length had increased from 8 in. to 8.03 in. (a) What was the nominal ultimate stress for this rod? (b) What was the maximum natural strain?

2-5. If the load at yield in Problem 2-4 was 12,400 lb, what would the nominal yield stress be?

2-6. (a) What would the true stress be for the rod in Problem 2-4? (b)What would be the true strain? (Dia. at max. load = 0.45 in.)

2-7. Aluminum sheets are used in roofing. Three 12-ft long sheets are used end to end. What will the total difference in length be between the temperature range of $-20°F$ to $+95°F$?

2-8. A steel bar is to be subjected to a load of 12,000 lbs. What diameter should it be if the stress will not be in excess of 20,000 psi?

2-9. An aluminum bracket with an initial area of 0.5 sq in. is used to support a maximum tensile load of 10,000 lb. After several weeks in service it fractures. At the time of fracture it was supporting a load of 8000 lb. It was found that the area at fracture had reduced to 0.4 sq in. (a) What was the nominal stress at fracture? (b) What was the true stress at fracture?

2-10. The bracket, in Problem 2-9, was originally 2 in. long. It was found to be 2.38 in. long at the time of fracture. What were the nominal and true strains at fracture?

2-11. Ten-in. dia. blanks are to be made by shearing in a standard punch and die. The material is AISI 1015 plain carbon steel $\frac{1}{8}$ in. thick. What size press, tonnage wise, would be required for this operation? Assume a 100% safety factor. The shear strength of the material is 45 ksi.

2-12. A high-carbon steel rod of 0.5 in. diameter is hardened, quenched and tempered to R_c 54. How much difference would there be in

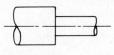

Figure P2.1

the maximum tensile strength of the bar if it were tempered to R_c 50?

2-13. (a) Why is it customary or advantageous to flame harden such parts as the bearing surfaces on top of a lathe bed or the guideways used for a large machine table to slide on? (b) What hardness, and to what depth, can be expected by this process?

2-14. The hardness of some razor blades needs to be checked. What types of hardness checking should be used and why?

2-15. What is the difference between nominal tensile strength and true tensile strength?

2-16. What are the advantages of using a standard-type specimen for tensile testing over a straight rod or strip?

2-17. Why is a high-carbon steel so much stronger after heating above the upper critical temperature and quenching in water?

2-18. (a) Compare the tensile and yield strengths of 1100-O and 1100-H18 aluminum. (b) Explain the reason for the change in tensile and yield strengths.

2-19. (a) What is the main difference between 1100 aluminum and 2024 aluminum as to content. (b) Explain the mechanism by which the high tensile strength is achieved in 2024-T6. (c) What aluminum has a tensile strength of over 80,000 psi?

2-20. What tool steels would be used for the following applications? (a) Bolt-header dies for cold heading. (b) Blanking dies. (c) Forging dies. (d) Milling cutters.

2-21. (a) What are the five types of cast irons discussed in this chapter? (b) What is the form of the carbon in each of the cast irons and what is the matrix form?

2-22. What would the approximate tensile strength be of a gray cast iron of the following chemical analysis: Si 1.14%, Mn 1.07%, P .18%, and total carbon 2.98%?

2-23. What would be the approximate tensile strength and the minimum compressive strength of ASTM class-40 gray cast iron?

2-24. Why is malleable cast iron generally considered nonweldable?

2-25. Which stainless steel would you recommend for each of the following? (a) Hardness and strength. (b) A very corrosive environment. (c) Good weldability. (d) Good machinability. (e) Good forming qualities. (f) Very high strength.

2-26. What modification is recommended for a standard drill if it is to be used in drilling copper or aluminum?

2-27. (a) Why does a metal with a large grain size form more easily than that with a small grain size? (b) What may be a disadvantage of a large grain size?

2-28. Why are composite materials able to achieve fantastic strengths, some even at elevated temperatures?

2-29. What fibers used in composites are being introduced into commercial applications?

2-30. What metal will compare favorably in tensile

strength to a graphite composite at room temperature?

2-31. What would be the approximate tensile strength of a 0.5-in. dia. rod of graphite composite at 3000°F (1643°C)?

2-32. (a) Why is martempering better than the conventional quench and temper process? (b) Why isn't martempering always used instead of the conventional quench and temper process?

2-33. What are the main differences between martempering and austempering?

2-34. What are two main advantages of using cast steels in the fabrication of equipment?

2-35. Identify each of the following materials: 1020, 5100, 1100–O, class 20, 4150, 2024–T3, 302, 80–60–03.

2-36. What is the difference in structure between a plastic and an elastomer?

2-37. What molecular weights constitute a low range, high range, and ultrahigh range for polyethylene?

2-38. What type of molecular structure makes it possible to achieve a very crystalline polymer structure?

2-39. What is the advantage of copolymerization?

2-40. What type of molecular structure is needed to resist high temperatures?

2-41. How does the evaluation of design properties of a plastic differ from that of a metal?

2-42. What are some ways that rubber can be reclaimed?

Case Study

A new oil refinery is to be built. Because refineries depend heavily on the choice of metals in vessels and piping, the equipment must resist heat, corrosion, pressure, shock, and fatigue, either separately or in combination, for a lifetime.

Refining equipment needs close inspection during shop and field construction.

Crude and vacuum towers are affected by sulfur compounds and chlorides. When

heated to fractionalizing temperature, the sulfur compounds in crude oil break down to form H_2S_1 chlorides from hydrochloric acid and, depending on the type of crude, concentrated naphthenic acid.

The vacuum towers operate at 500°F (260°C) and some of the piping that handles the heavy heating oil at 1300°F (704°C).

The elbows used in the piping system were fabricated by forging. Plugs are fitted into the pipes and elbows to facilitate inspection. The oil flows at high speed in some sections. Fine grain size steel is specified for heads and shell plates. State what material should be chosen for the vacuum towers, pip-ing, and fittings. What steps should be taken to insure that the material specified is the material received? State why the material specified is chosen and how it is inspected.

Bibliography

Allen, D. K. *Metallurgy Theory and Practice.* Chicago: American Technical Society, 1969.

AISI. *Design Guidelines for the Selection and Use of Stainless Steel.* Washington, D.C., 1977.

———. "Tool Steels Today." *Newsletter* (current issues).

American Welding Society. *Welding Aluminum,* 6th ed. Miami, n.d.

ASM. *Metals Handbook,* vol. 1. *Properties and Selection: Irons and Steels,* 9th ed. Metals Park, Ohio, 1978.

Bethlehem Steel Corporation. *An Analysis of Tool and Die Failures, Handbook 2828-A.* Bethlehem, Pa., n.d.

Dreger, D. R., ed. "Design Guidelines for Joining Advanced Composites." *Machine Design* 52 (May 8, 1980):89–94.

Dreger, Donald R., ed. "The Polysulfones, Heat-Resistant, Superstrong, and Ultra-stable." *Machine Design* 50 (January 12, 1978):114–120.

Dwyer, John J., ed. "Sheet Steel Today." *American Machinist Special Report 687* 120 (May 1976):65–88.

Eny-Ning Pan, and Carl Loper. "Research Into the Production and Formation of Compacted/Vermicular Graphite Cast Iron by Rare Earths." *Ph.D. Thesis,* University of Wisconsin Mining and Metallurgy Department, 1982.

Evans, R.; J. V. Dawson; and M. J. Lalich. "Compacted Graphite Cast Irons and Their Production by a Single Alloy Addition." *AFS Transactions* 84 (1976):215–220.

Hall, A. M. "If It Looks Impossible, Try High Strength Steel." *Steel* (August 22, 1966).

———. "The Status of Ultrahigh-Strength Steels Today." *Metal Progress* 108 (August 1975):37–46.

MacKenzie, B. A. "HSLA Bar Steel for Lighter, Stronger Parts." *Machine Design* 51 (August 23, 1979):93–95.

Palmer, F. R.; G. V. Luerssen; and J. S. Pendleton, Jr. *Tool Steel Simplified,* 4th ed. Publication sponsored by the Carpenter Technology Corporation. Radnor, Pa.: Chilton Co., 1978.

Tang, S. "Advanced Composites: Design Technology Transfer from Aerospace." *Mechanical Engineering* 99 (June 1977):36–39.

The Aluminum Association. *Aluminum Brazing Handbook,* 3rd ed. Washington, D.C., 1979.

———. *Aluminum Soldering Handbook,* 3rd ed. Washington, D.C., 1976.

———. *Aluminum Standards and Data,* 7th ed. Washington, D.C., 1980.

Van Vlack, L. H. *Elements of Material Science.* Reading, Mass.: Addison-Wesley, 1964.

Walton, C. F., ed. *The Properties of Gray Iron.* Cleveland: Gray and Ductile Iron Founders' Society Inc.

Wheeton, J. W. "Fiber Metal Matrix Composites." *Machine Design* 41 (February 20, 1969):141–156.

Wick, Charles, ed. "Better Tools from PM High-Speed Steels." *Manufacturing Engineering* 85 (September 1980):52–54.

CHAPTER THREE

Metal-Cutting Theory and Practice

The rapid growth of industrialization did not depend on scientific means as much as on empirical means. In metal cutting it was largely left up to the machinist how a workpiece was to be machined. He selected the tooling, determined the feeds and speeds and ground the tools to his own particular formula of satisfaction.

It was not until 1906 that Fredrick W. Taylor (after 25 years of patient research) delivered his famous thesis which contained the formula $T = CV^{-n}$. This was the real birth of modern industrial engineering. In this formula, Taylor displayed a clear mathematical relationship between tool life and the speed at which a tool was run. This chapter is devoted to the mechanics of efficient metal removal and the use of Taylor's principles.

MECHANICS OF METAL CUTTING

In any metal-cutting operation, the unit product of metal removed is called a "chip." The chip is usually thicker than the layer removed from the workpiece, thus the mechanism of plastic deformation is at work. Also the separation of the chip from the parent metal means shearing and/or cleavage is taking place.

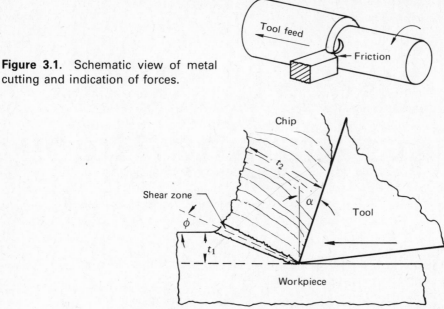

Figure 3.1. Schematic view of metal cutting and indication of forces.

Figure 3.2. Schematic of the chip-formation process.

Friction develops as the chip slides over the face of the tool which results in heat and wear (Fig. 3-1).

The cutting action can be more easily analyzed if the edge of the tool is set perpendicular to the relative motion of the material, as shown in Fig. 3-2. This two-dimensional type of cutting, in which the cutting edge is perpendicular to the cut, is known as *orthogonal*, as contrasted with three-dimensional *oblique* cutting as shown in Fig. 3-1, in which the cutting edge is inclined (at an oblique angle) to the cutting direction.

Orthogonal Cutting

In orthogonal cutting, a surface layer of constant thickness (t) is removed by the relative movement of the tool and workpiece. Some indication of the ease or difficulty of the machining process can be obtained by comparing the chip thickness (t_2) to the original depth of cut (t_1). The ratio of t_1 to t_2 is defined as the *chip-thickness ratio* (r). Usually, materials that do not shear well but "pile up" in front of the tool are considered less machinable than those materials that have only a small increase in chip thickness. Thus a measure of the ease of metal removal or *machinability* is expressed by the chip-thickness ratio. This ratio may also be expressed as the shear angle, ϕ

$$\tan \phi = \frac{r \cos \alpha}{1 - r \sin \alpha}$$

where: α = rake angle of the tool, as shown in Fig. 3-2.

The closer the shear angle approaches a 1 : 1 ratio, or 45°, the better the machinability is said to be.

Chip Formation

About 95% of the power expended for metal removal is used in the deformation taking place in the shear zone. This is the work required to form and remove the chip and the incidental plastic deformation of the surface layer of the finished workpiece. The remaining power consumed, about 5% of the total, is expended in stored elastic energy or *residual stresses* in the workpiece and friction.

The chips are formed largely by shearing action and compressive stresses on the metal in front of the tool. The compressive stresses are the greatest farthest from the cutting tool and are balanced by tensile stresses in the zone nearest the tool, hence the chip curls outwardly or away from the cut surface.

Chip Types

Chips, as formed in metal machining, have been classified into three basic types: segmented, continuous, and continuous with a builtup edge (Fig. 3-3).

Segmented Chip.   The segmented chip separates into short pieces, which may or may not adhere to each other. Severe distortion of the metal occurs adjacent to the tool face, resulting in a crack that runs ahead of the tool. Eventually the shear stress across the chip becomes equal to the shear strength of the material, resulting in fracture and separation. With this type of chip, there is little relative movement of the chip along the tool face.

Continuous Chip.   The continuous chip is characterized by a general flow of the separated metal along the tool face. There may be some cracking of the chip, but in this case they usually do not extend far enough to cause fracture. This chip is formed at the higher cutting speeds when machining ductile materials. There is little tendency for the material to adhere to the tool. The continuous chip usually shows a good cutting ratio, tends to produce the optimum surface finish, but may become an operating hazard.

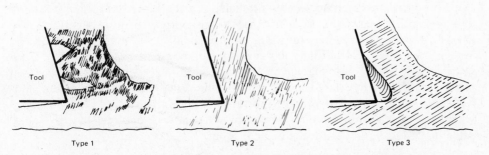

Type 1   Type 2   Type 3

Figure 3.3.   The three main types of chips: (1) segmented, (2) continuous, (3) continuous with a builtup edge.

Continuous with a Built-Up Edge.   This chip shows the existence of a localized, highly deformed zone of material attached or "welded" on the tool face. Actually, analysis of photomicrographs show that this builtup edge is held in place by the static friction force until it becomes so large that the external forces acting on it cause it to dislodge, with some of it remaining on the machined surface and the rest passing off on the back side of the chip.

The builtup edge (BUE) is closely related to cutting speed with its related heat and friction. It does not appear at very low speeds since there is neither sufficient heat or friction developed. At much higher speeds, the heat developed is sufficient to cause the material in the shear zone to be annealed, eliminating the strain hardening that takes place in the BUE. Studies show that the greatest BUE takes place in a temperature range of 750 to 800°F (399–427°C) for mild steel (intermediate cutting speeds).

The BUE often results in an inferior surface finish, since the effective tool geometry is altered.

SELECTION OF CUTTING-TOOL MATERIALS

Relatively few materials are able to withstand the heat, unit pressures, shock, and wear that are required of metal-cutting tools used in high production. The most common materials that meet these requirements in varying degrees are high-speed steels, cemented carbides, ceramics, cermets, diamonds, and borazon.

High-Speed Steels (HSS).   High-speed steels introduced at the turn of the century by Fredrick Taylor and Manuel White were a vast improvement over the plain high-carbon steels used previously. The first HSS used was referred to as 18–4–1 denoting the percentages of tungsten, chromium, and vanadium, respectively.

HSS are able to retain a high hardness up to the range of about 1000°F (528°C). The ability to hold a cutting edge at dull red heat is known as a *red hardness*. An examination of the microstructure of HSS reveals hard, brittle, refractory carbide particles held in a lower melting-point matrix. The most significant change in HSS in recent years has been brought about by the use of powder metallurgy (see Chapter 2). A comparison between structures of powder metallurgy tools and conventionally produced bar stock reveals a finer and more uniform carbide particle size. This process has permitted more highly alloyed HSS to be fabricated with a resultant improvement in wear resistance with no loss in toughness. These materials are often referred to as super HSS. Conventional HSS tool grades such as M2, M3, M4, T15, etc., are available in powder metallurgy stock. M and T indicate the principal alloying element, molybdenum or tungsten. Basically, the tungsten-based tools have high heat resistance while the molybdenum family shows greater shock resistance. Cobalt helps maintain the hot hardness of both groups.

Cemented Carbides.   Cemented carbides have the highest ratio of carbides to matrix of the cutting-tool materials—about 80%. Carbide tools are made by mixing pure tungsten powder under high heat with pure carbon (lampblack) in the ratio of 94% and 6%, respectively, by weight. The resulting compound, WC, is then mixed with cobalt until the mass is homogenous. The powder mixture is compacted at high pressure and sintered in a furnace at 2500°F (1371°C). After cooling, the carbide tools are ground and, in some cases, subjected to further finishing operations.

Although tungsten carbide tools were a boon to industry, it was found that in machining steels there was a tendency for the lower melting-point cobalt to wear first, exposing the carbide particles. This was followed by a seizure of both the high and low points of the cutting tool surface. To help prevent this condition, tantalum and titanium were added to the basic composition, hence the standard steel-cutting carbide tool contains WC–Co–TaC and TiC.

Cemented carbides cover a wide range of the tool material spectrum. The roughing grades approach high-speed steels in toughness while some finishing grades are nearly as hard as ceramics. Between these two extremes are literally hundreds of different compositions available in categories C–1 through C–8.

To simplify the selection task, categories C–1 through C–4 contain grades suitable for machining nonferrous materials along with cast irons and steels that yield short discontinuous chips. The C–5 through C–8 categories include grades having greater crater resistance (wearing a cavity on top of the tool) required for machining steels that produce longer chips (see Table 3-1).

Table 3-2 lists the ISO (International Standards Organization) classification of carbide cutting tools. The ISO group K roughly parallels the US designation C1–C4-cast iron and nonferrous metals; P parallels US designations C5–C8—steel applications; and M is an intermediate group. Within each group, numbers indicate the range of properties from maximum hardness (01) to maximum toughness (50), opposite to the direction of the US numbers which increase with increasing hardness.

Coated Carbides.   A newer type of carbide tool is one that has a thin 0.0002 to 0.0003 in. (0.005 to 0.008 mm) coating of titanium nitride (TiN), titanium carbide (TiC), or aluminum oxide (Al_2O_3), which is deposited uniformly over all surfaces by means of a vapor deposition process. Since these coatings have no binder they are substantially harder than the base carbide. Figure 3-4 shows a drawing of a carbide-coated insert and a photomicrograph of the coated tool. Test results show, in machining alloy steels, the TiN coated tool has a lower

Table 3-1. Carbide cutting-tool classification.

Cast Iron, Nonferrous, and Nonmetallic Materials	Steel and Steel Alloys
C-1 Roughing	C-5 Roughing
C-2 General purpose	C-6 General purpose
C-3 Finishing	C-7 Finishing
C-4 Precision finishing	C-8 Precision finishing

Table 3-2. ISO classification of carbides according to use.*

Group	ISO class	Material to be Machined	Uses, Working Conditions	Increasing Characteristic of Cut Carbide
P—Blue Ferrous metals with long chips	P 01	Steel; steel castings	Finish turning, boring; high sfm, thin chips; high accuracy, finish; no vibration	← Toughness → / ← Wear resist / ← Feed / Speed →
	P 10	Steel; steel castings	Turning, copying, threading, milling; high sfm; thin to medium chips	
	P 20	Steel; steel castings; malleable iron (long chip)	Turning, copying, milling; medium sfm and chip thickness; planing, thin chips	
	P 30	Steel; steel castings; malleable iron (long chip)	Turning, milling, planing; medium to low sfm, medium to thick chips; unfavorable conditions*	
	P 40	Steel; steel castings with inclusions, voids	Turning, planing, slotting; low sfm, thick chips. Large cutting angles; unfavorable conditions*; automatic machines	
	P 50	Steel; medium- to low-tensile steel castings with inclusions, voids	Operations requiring high toughness; turning, planing, slotting; low sfm, thick chips. Large cutting angles; unfavorable conditions†; automatic machines	
M—Yellow Ferrous metals with long or short chips; and nonferrous metals	M 10	Steel; steel castings; Mn steel; gray, alloy irons	Turning; medium to high sfm; thin to medium chips	← Toughness → / ← Wear resist / ← Feed / Speed →
	M 20	Steel; steel castings; austenitic, Mn steels; gray iron	Turning, milling; medium sfm and chip thickness	
	M 30	Steel; steel castings; austenitic steel; gray iron; high-temperature alloys	Turning, milling, planing; medium sfm, medium or thick chips	
	M 40	Mild, free-cutting, low-tensile steels; nonferrous; light alloys	Turning, cut-off, particularly on automatic machines	

	Toughness →	← Wear resist
	Feed →	← Speed

Grade	Material	Operations
K—Red Ferrous metals with short chips; nonferrous metals; and nonmetals	Very hard gray iron; chilled castings over 85 Shore; high silicon aluminum; hardened steel; highly abrasive plastics; hard cardboard; ceramics	Turning, finish turning, boring, milling, scraping
K 10	Gray iron over 220 Bhn; malleable iron (short chip); hardened steel; silicon aluminum; copper alloys; plastics; hard rubber; hard cardboard; porcelain; stone	Turning, milling, drilling, boring, broaching, scraping
K 20	Gray iron up to 220 Bhn; nonferrous; Cu, brass, Al	Turning, milling, planing, boring, broaching; demanding very tough carbide
K 30	Low hardness gray iron; low-tensile steel; compressed wood	Turning, milling, planing, slotting; unfavorable conditions†; large cutting angles
K 40	Soft or hard wood; nonferrous	Turning, milling, planing, slotting; unfavorable conditions†; large cutting angles

* Adapted from ISO recommendation 513.
† Raw stock or components in shapes awkward to machine; casting or forging skins; variable hardness or depth of cut; interrupted cut; vibration.

135

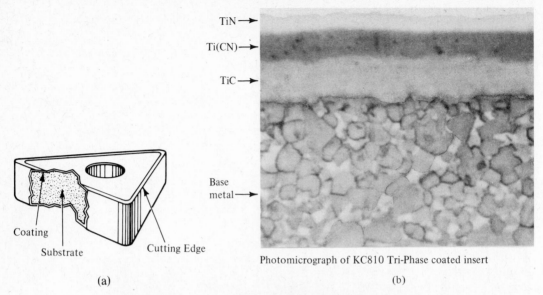

TiN →

Ti(CN) →

TiC →

Base metal →

Coating

Substrate

Cutting Edge

Photomicrograph of KC810 Tri-Phase coated insert

(a)

(b)

Figure 3.4. The coated carbide insert (a); the coating and tool microstructure (b). (*Courtesy Kennametal Inc.*)

coefficient of friction, hence the shear angle of the chip is greater, resulting in less force required and better surface finish. In addition, coated carbides can be used at higher cutting speeds and still maintain the same tool life.

There are some limitations for coated carbides. Since the coating is thin, it does not have the long-term abrasion resistance and should not be used where heavy-scale or other abrasive conditions predominate. Also the coating tends to reduce the transverse rupture strength and therefore should not be used for heavy, interrupted cuts. These are generalizations and since there are many grades of coated carbides it is best to consult with the supplier as to specific applications. Shown in Appendix C is a table of carbide tools and recommended speeds for given materials.

Ceramics. The term "ceramic tooling" is applied to tools made of aluminum oxide. The strongest tools are made by hot pressing a finely divided aluminum oxide powder in a graphite mold to obtain a solid piece porosity free within a fraction of 1%.

Of all tool materials, aluminum oxide best resists high temperatures. It is also hard, being surpassed only by TiC and diamonds. Its high hardness and high temperature resistance combine to provide superior wear resistance. This means excellent tool life on abrasive materials such as cast iron. Tool life is often three times that of other materials, even at three times the cutting speed. The high heat resistance also means better thermal stability with more accuracy of the machined surface.

The combination of high hardness and low toughness limits the usefulness of Al_2O_3 tools mainly to high-speed turning of cast irons and high-strength steels. Ceramic tools are at the point of awaiting better machine tools with greater rigidity and higher horsepower. Shown in Appendix D is a table of recommended cutting speeds for various materials using ceramic tools.

Cermets.   A more recent development in ceramic tooling has been the "straight" oxides. The new material consists of aluminum oxide with 30% titanium carbide as a fine dispersion, often referred to as *cermets*. Although these oxide–carbides are only slightly harder than oxides, they are stronger and more resistant to thermal shock. This unique combination of properties permits them to be used for milling steels as well as cast irons. They are frequently used at speeds of 2000 sfm in face milling gray irons, and are being increasingly used for rough turning, boring, grooving, facing, and other jobs when conventional oxides fail by thermal shock, chipping, or fracture.

Diamonds.   Single crystal diamonds have been used for many years in machining nonferrous materials, plastics, graphite, and ceramics. Their extreme hardness, coupled with low shock resistance, has limited their usage to the softer but very abrasive materials mentioned.

In the past it has been necessary to orient the diamond crystal to obtain the best hardness and wear-resistant surface. Recent developments have eliminated this process. Diamond powders can now be fabricated into solid tool shapes by using high temperatures and pressures. Some polycrystalline diamond tools are made as a layer over a carbide substrate.

One of the newest cutting tool materials is the cubic form of boron nitride, usually known by its General Electric tradename of Borazon. The first significant application of Borazon was in grinding wheels used to grind the hard, tough cutting-tool materials such as the tungsten-type tool steel T–15. Now it is used as a polycrystalline coating over carbide inserts and is used to machine nickel-rich alloys such as Inconel 718 and René 95. These tools are also used to turn hardened tool and die steels at 250–400 fpm (76–122 Mpm) at a depth of cut of up to 0.250 in. (6.3 mm), which represents a typical condition for cutting soft steel with conventional tools. The inserts, shown in Fig. 3-5, perform well when used to machine chilled cast iron and steel rolls with hardnesses up to 67 Rockwell C. Conversely these inserts do not do well on cutting materials that are less than Rockwell C 45.

CUTTING-TOOL COMPARISON

Figure 3-6 compares the various tool materials discussed and recommended surface foot speeds for each material. Surface foot speed means the number of feet of material that pass the cutting tool per minute. You will note there is considerable overlap between the recommended sfm of each of the tools. The overlap allows for other conditions that must be taken into account by the operator such as surface scale and interrupted or intricate cuts.

Figure 3.5. Alloy steels, tool steels, chilled cast iron, nickel and cobalt-base alloys can be machined at high speeds with CBN* compact tools. The layer of polycrystalline, Borazon (cubic boron nitride), and a tungsten carbide substrate are produced as an integral blank by General Electric's high pressure/high temperature process. (*Courtesy General Electric Specialty Materials Department, Worthington, Ohio.*)

*Trademark of General Electric Company USA.

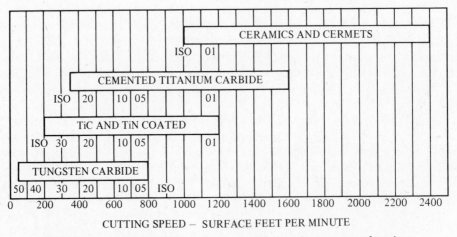

CUTTING SPEED – SURFACE FEET PER MINUTE

Figure 3.6. Approximate speed range for optimum use of various cutting tool materials.

TOOL GEOMETRY

The principle of tool geometry is to provide a sharp cutting edge that is strongly supported. The development of basic tool geometry is shown in Fig. 3-7. The purpose of each of the angles is discussed briefly.

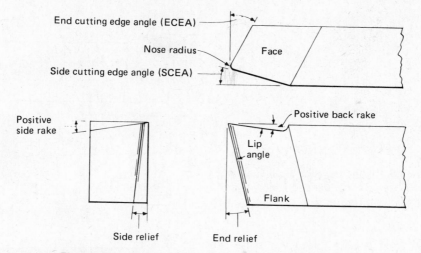

Figure 3.7. The development of basic tool geometry.

Relief Angles. As the name implies, relief angles, both side and end, provide relief for the cutting edge to keep the tool from rubbing on the cut surface. Relief angles are usually kept to a minimum to provide good support for the cutting edge. They may be increased to produce a cleaner cut on soft materials.

Side Cutting-Edge Angle. The side cutting-edge angle serves two purposes: it protects the point from taking the initial shock of the cut and it serves to thin out the chip by distributing the cut over a greater surface.

End Cutting-Edge Angle. The end cutting-edge angle reduces the tool contact area in respect to the metal being cut.

Rake Angles. Rake angles are ground on the top or face of the tool and may be either positive, neutral, or negative. Positive rake angles reduce the cutting force and direct the chip flow away from the material. Negative rake angles increase the cutting force required but provide greater strength at the cutting edge. Negative angles also make it possible to use six cutting edges of a triangular insert or eight cutting edges of a square insert, as shown in Fig. 3-8.

Newer tool geometry can provide positive rake cutting action even though the insert may be set at a negative angle (Fig. 3-9). A standard 5-degree negative rake may restrict the chip and exert excessive loads upon the tool face causing severe crater wear. The effect of negative rake, negative-rake grooved, negative-rake wide grooved, and a negative land are all shown in Fig. 3-10.

Nose Radius. The nose radius is the rounded end that blends the side cutting-edge angle with the end cutting-edge angle. It performs the same function as the side cutting-edge angle in that it decreases the thickness of the chip as it approaches the point of the tool (Fig. 3-11). The nose radius is also important in controlling surface finish. For a given depth of cut and feed, the tool with

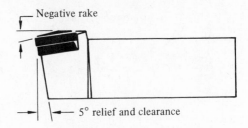

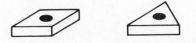

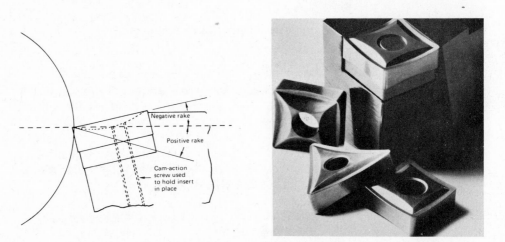

Figure 3.8. Negative rake angles allow a tool insert to be used on all six or eight of its cutting edges.

Figure 3.9. Negative-rake tool holders used with effective positive-rake inserts. Thus all cutting edges can be used with the added advantage of lower cutting forces provided by the positive rake at the cutting edge. (*Courtesy Kennametal Inc.*)

a large nose radius will produce a smoother finish than a tool with little or no nose radius. Too large a nose radius will tend to produce chatter due to excessive contact area. Also, pressures on the cutting tool and material being machined are proportionate to the contact area. As an example, small-diameter parts require a small nose radius with light cuts to minimize tool pressures and obtain better accuracy.

Chip Control

Long, stringy chips produced by some workpiece materials can be a hazard to the operator and troublesome on automated machines. It was fairly early in the history of inserts that mechanical chip breakers were introduced. A wedge-

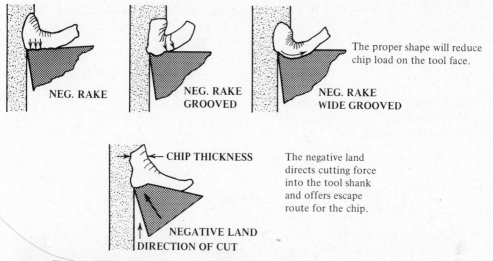

Figure 3.10. The effect of rake angle, grooves, and lands on chip load. (*Courtesy Adamas Carbide Corporation.*)

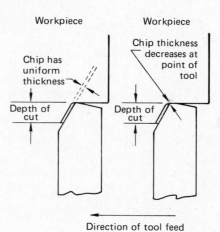

Figure 3.11. The nose radius acts to thin out the chip at the point of the tool. (*Courtesy Adamas Carbide Corporation.*)

shaped piece was simply clamped on top of the insert and set back a short distance from the cutting edge [see Fig. 3-12(a)]. When the chip moved against this obstruction it was forced to curl tightly and break.

The next chip breaker was the molded-groove type of chip breaker insert with a lockpin hole for securing the insert in the toolholder pocket [see Fig. 3-12(b)]. This led to the positive–negative insert already described.

Variations in both depth of cut and feed affect the performance of molded chip breakers. As a result new designs have been conceived in attempts to both broaden the application range of inserts and to produce inserts tailored for greater efficiency in specific applications such as one insert for roughing and another for finishing cuts. Figure 3-13 shows a graph and tool insert with the effective chip-breaking range.

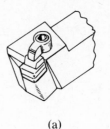

Figure 3.12. The clamped chip breaker (a). The lockpin insert with a molded groove chip breaker (b).

(a)   (b)

TNMM
5° negative
honed with
chip control

SAE 1045 STEEL

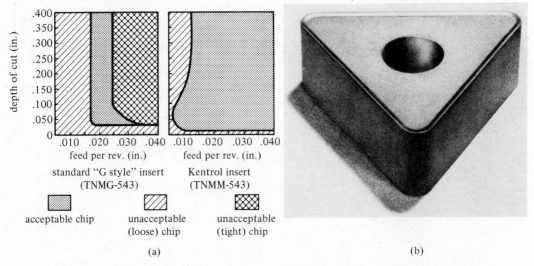

standard "G style" insert
(TNMG-543)

Kentrol insert
(TNMM-543)

acceptable chip   unacceptable (loose) chip   unacceptable (tight) chip

(a)   (b)

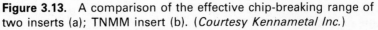

Figure 3.13. A comparison of the effective chip-breaking range of two inserts (a); TNMM insert (b). (*Courtesy Kennametal Inc.*)

TOOL FAILURES

Carbide cutting edges lose their usefulness through wear, breakage, chipping, or deformation. Wear, which accounts for most of the tool failures, is very complex and involves chemical, physical, and mechanical processes, often in combinations.

Wear

The study of wear between two metallic surfaces is complicated because it involves several unique mechanisms. Four of these mechanisms have been categorized as (1) abrasion, (2) adhesion, (3) diffusion, and (4) oxidation. Each of these wear mechanisms will be described briefly.

Abrasive Wear.   Abrasion takes place when hard constituents in the work-piece microstructure or underside of the chip "plow" into the tool face or flank as they flow over it. Highly strain-hardened fragments of a built-up edge or inclusions contribute to this type of wear. Abrasive wear is generally considered to be a mechanical wear mechanism that is predominant at low cutting speeds and at relatively low temperatures. It is generally more pronounced on the flank surface of the tool (Fig. 3-14).

Adhesive Wear.   The chip, as it slides over the face of the tool, is not smooth on a microscopic scale. It consists of hills and valleys, or asperities. Likewise, the cutting tool may have asperities. As these asperities slide past each other under heat and pressure, they tend to adhere. If the junction formed is weaker than both parts of the metal, a fracture occurs with little transfer of metal. However, a certain amount of plastic deformation takes place in the bonding asperities. In repeated action, work hardening takes place. After a time of deformation contact, the asperities become brittle and break off, taking with them minute amounts of tool material. Eventually a crater begins to form (Fig. 3-15). This action takes place because of the extremely high unit pressures. If, however, the temperature is near or beyond the recrystallization temperature, the bond between the chip and tool asperities is no weaker than the material

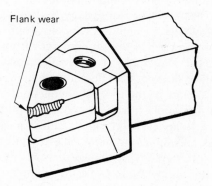

Flank wear

Figure 3.14. Flank wear on cutting tool.

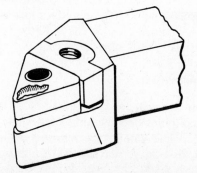

Figure 3.15. A crater is formed on the face of the tool by adhesive wear.

adjacent to it, because work hardening has not been retained. Therefore, the rate at which the asperities are pulled off diminishes.

Diffusion Wear. Diffusion wear involves the transfer of atomic particles from the tool to the workpiece and from the workpiece to the tool. Diffusion is accelerated by high temperatures caused by rapid movement of the work material near the tool surface.

Oxidation. At the elevated temperature range at which a cutting tool operates, oxidation can cause rapid wear. As oxidation takes place, it weakens the tool matrix and thus the cutting-edge strength. The oxides that form are easily carried away, leading to increased wear.

Breakage

Excessive pressures acting on the cutting edge of a tool may cause immediate failure by the loss of one or two large particles or several smaller particles. Breakage is usually attributed to mechanical shock, thermal shock, thermal cracks, fatigue, or excessive wear.

Mechanical shock, leading to breakage, is usually caused by severe interruptions of the cut on the workpiece surface or by hard inclusions in the workpiece. If this type of machining is necessary with carbide tools, tough rather than hard grades should be chosen.

Thermal shock breakage is usually due to sudden cooling of a very hot cutting edge. This often happens when a chip breaks off and the coolant strikes the tool or when the tool emerges from the cut.

Thermal cracks are caused by severe heat gradients that develop mainly on the face of the tool. The cutting edge gets hot first, then the heat penetrates into the interior of the tool by conduction. The surface is first put into compression and if it cools rapidly it contracts, putting it in tension and causing a thermal crack to appear. Repeated cycling causes more cracks, making the cutting edge weak and subject to breaking. Harder grades of carbide crack more readily than the softer grades. Although thermal cracks occur during dry machining, they are greatly accelerated by the interrupted application of cutting fluid.

Chipping. Chipping is a microscopic form of breakage due to the loss of many small particles from the cutting edge and is usually caused by unhoned carbide edges. Unhoned edges lack sufficient mechanical strength to withstand the mechanical forces encountered in cutting. Chipping may also be caused by excessive vibration and chatter.

Deformation

When a heavy load is applied close to the cutting edge of the tool, the surface at this point becomes indented while the adjacent face shows a corresponding bulge. The amount of deformation increases as the load is increased, until a critical strain is reached and fracture starts. A crack forms at the periphery of the indentation, away from the cutting edge, spreads to the edge, and then goes down the adjacent face until a small flake of carbide breaks away.

TOOL LIFE

Tool life may be defined as that period of time that the cutting tool performs efficiently. Since it is undesirable to replace a worn tool before it becomes necessary, various quantitative criteria have been suggested to determine the maximum permissible limit of tool wear, which is in essence tool life. The recommended flank wear land, shown in Fig. 3-16, is 0.030 in. (0.76 mm) for roughing cuts and 0.015 in. (0.38 mm) for finishing cuts.

Many factors can be considered as having an effect on life, such as the microstructure of the material being cut, the desired cubic-inch removal rate, the rigidity of the setup, and whether or not cutting fluids are used.

One of the best ways of showing the response of tool life to changes in cutting speeds is to calculate the various combinations according to Taylor's equation.

$$T = CV^{-n} \text{ or } \log T = C - (n \cdot \log V)$$

where: T = tool life
V = cutting speed
C = a constant representing one minute of tool life
n = a coefficient of the slope of the tool life curve on log-log paper

Significant changes in tool geometry, depth of cut, and feed will change the value of the constant C and cause slight changes in the slope of the curve, or the exponent n.

Tool life curves may be constructed from the results of laboratory tests or by actual on-the-job machining operations. Figure 3-17 compares the tool life

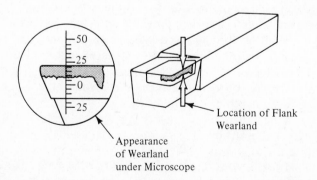

Figure 3.16. The flank wear of the tool used in determining tool life.

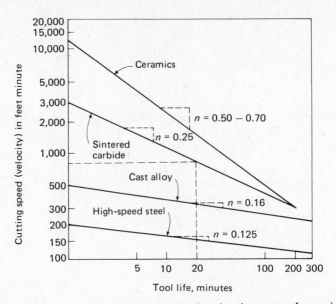

Figure 3.17. Tool-life curves as plotted on log-log paper for various tool materials.

curves as plotted for common cutting-tool materials. As an example, at a cutting speed of about 825 fpm, a carbide tool would last about 20 min. To get the same tool life for an HSS tool, the velocity (V) would have to be reduced to about 150 fpm. The values of n shown in the figure may be accepted as standard.

TOOL FORCES

The main forces acting on the cutting tool during a turning operation are tangential, longitudinal, and radial forces, as shown in Fig. 3-18(a). The tangential force F_t acts in a direction tangent to the revolving workpiece. This force is the highest of the three components. The horsepower for a turning operation is based largely on the tangential tool force and the cutting velocity.

The longitudinal force F_1 acts in a direction parallel to the axis of the rotating bar and is also called the feeding force. The power required to feed a turning tool is thus the product of the longitudinal tool force and the feeding velocity.

The radial tool force F_r is normal to the rotating work or is in a radial direction from the center of the workpiece. For turning, the radial force holds the tool at the desired depth of cut. In a turning cut, there is no velocity in the radial direction, hence there is no power required for the radial force. The longitudinal force is affected by the side rake of the tool, but it constitutes only a small percent of the total force. The radial force is much less, being about half of the longitudinal force. The three components of force may be resolved into three mutually perpendicular components, with the resultant force vector being as shown in Fig. 3-18(b).

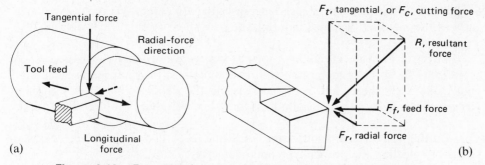

Figure 3.18. Forces acting on a cutting tool during a turning operation (a). The resultant force acting on the cutting tool (b).

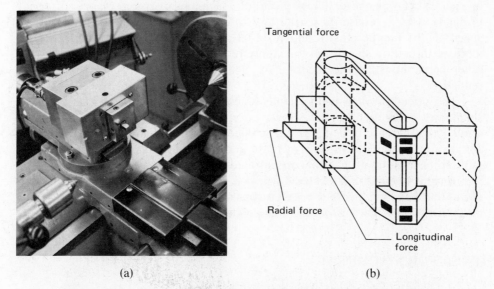

Figure 3.19. A three-component, strain gage type, metal-cutting dynamometer (a). Drawing showing location of strain gages (b).

Measurement of Cutting Forces. An indirect method of measuring cutting forces acting on the tool is with the aid of a wattmeter. A more exact method is with the aid of a tool dynamometer. Shown schematically in Fig. 3-19 is a commonly used three-force-component lathe tool dynamometer. Resistance-wire strain gages are mounted on an extended octagonal ring. The advantage of this design is that the instrument becomes independent of where the load is applied to the material.

Effect of Speed on Force. In general, as speeds increase, forces increase. However, negative-rake tools that show high forces at low cutting speeds may show more than 30% decrease in forces when the speed is increased from 50 to 300 sfpm. The increase in speed increases the temperature at the cutting edge, and as the material becomes more plastic, the shear strength is lowered.

Effect of Feed on Force. An increase in feed is usually considered an increase in efficiency of metal removal, but sometimes the increase in force does not make it the best choice. An increase in the depth of cut causes a direct increase in tangential and longitudinal force but not always in the radial force.

Cutting Forces and Tool Design. Cutting-force data is useful in designing tools or selecting the proper type and geometry.

Metal Removal Rate vs. Tool Life. The maximization of metal removal and tool life is an ever-present problem. Increased metal removal can be achieved by increasing any one or all of the three machining parameters; speed, depth of cut, and feed. These do not all cost the same in terms of tool life. Cutting speed has by far the most pronounced effect on tool life. For example, a 50% increase in cutting speed will commonly reduce tool life by about 90%. On the other hand, boosting the feedrate by 50% will result in about 60% tool-life reduction.

The effect of varying the depth of cut is more complex because it is a simultaneous function of feed. However, a depth of cut ten times the feedrate, a full 100% increase in depth of cut, reduces tool life by about 25%. Below the 10 : 1 ratio of depth to feed an increase in depth of cut has a greater impact on tool life because more heat is concentrated on the insert's nose radius.

A rule of thumb can be stated from the foregoing. Select the heaviest depth of cut and heaviest feedrate possible consistent with the part and then select a machining speed that will yield an economically acceptable tool life.

HORSEPOWER FOR MACHINING

Horsepower was first defined by James Watt as "the power exerted by a dray horse exerting a pull of 150 pounds while traveling at the rate of $2\frac{1}{2}$ miles per hour." This is equivalent to 33,000 ft-lb per min or 550 ft-lb per sec.

The horsepower required at the spindle (HP$_s$) is expressed as the summation of the individual force components and their corresponding velocities.

$$HP_s = \frac{F_t V_c}{33,000} + \frac{F_l \cdot F_r \cdot N}{33,000 \cdot 12}$$

where: F_t = tangential force
V_c = relative cutting velocity of the tool to the workpiece in feet per minute
F_l = longitudinal force
f_r = feed in inches/revolution
N = rpm

Often the longitudinal and radial forces are neglected since they are considered negligible; thus the formula becomes simply:

$$HP_s = \frac{F_t V}{33,000}$$

Unit Horsepower

In metal cutting, horsepower is most often calculated in terms of *unit horsepower*. Unit horsepower is defined as the power required to cut a material at the rate of one cubic inch per minute. Unit horsepower will vary with the cutting characteristics of each material. Thus unit horsepower is given for the different hardness levels of a material as shown in Table 3-3. A rule of thumb is that it takes 1 hp at the spindle for each 1 cu in. per min of metal removal for plain carbon steels using sharp tools and about 25% more if they are dull.

The horsepower requirement at the motor (HP_m) can be calculated using the unit power (P) from the equation:

$$HP_m = \frac{Q \times P}{E}$$

where: Q = the cubic inch removal rate
E = efficiency of spindle drive

Figure 3-20 illustrates the formulas used to determine Q for turning, drilling, and milling. Table 3-4 contains all the needed formulas for various shop power and time calculations for the operations listed. The difference between the horsepower required at the spindle of the machine and the horsepower at the motor is a measure of the efficiency of the machine, or how much power is lost in the gear train or belt drives before it is actually delivered at the spindle. The effi-

Table 3-3. Average unit horsepower values of energy per unit volume.

Material	BHN	Unit Power
Carbon steels	150–200	1.0
	200–250	1.4
	250–350	1.6
Leaded steels	150–175	0.7
Cast irons	125–190	0.5
	190–250	1.6
Stainless steels	135–275	1.5
Aluminum alloys	50–100	0.3
Magnesium alloys	40–90	0.2
Copper	125–140	0.7
Copper alloys	100–150	0.7

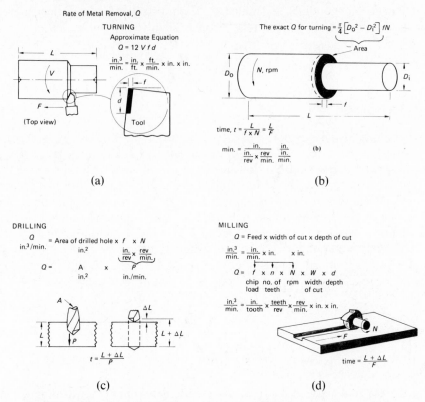

Figure 3.20. Q defined for various machining operations. Also shown is the formula used to calculate the time required to make cuts for each type of operation.

ciency factor applied is usually 0.9 for direct-belt drive and 0.7 to 0.8 for gear drives.

Figure 3-21 shows the useful relationships for quickly approximating metal removal rates and tangential force when the sfpm and axial feed rates are known. From the tangential force, horsepower can be determined. For example at 1000 sfpm and at 0.010 ipr feedrate, the cu in. removal rate would be approximately 18 cu in. per min and the tangential force would be about 425 lb.

METRIC POWER CALCULATIONS

Unit Power

Unit power (kW_u), which is similar to unit horsepower, is defined in terms of kilowatts required to remove one cubic centimeter of material per second.

$$kW_u = \frac{kW}{cm^3/s} \quad \text{or} \quad kW_u = \frac{kW}{1000\,mm^3/s}$$

Table 3-4. Shop formulas for turning, milling, drilling, and broaching.

	Turning	Milling	Drilling	Broaching
Cutting Speed, ft/min	$V_c = .262 \times D_t \times RPM$	$V_c = .262 \times D_m \times RPM$	$V_c = .262 \times D_d \times RPM$	V_c
Revolutions per Minute	$RPM = 3.82 \times \dfrac{V_c}{D_t}$	$RPM = 3.82 \times \dfrac{V_c}{D_m}$	$RPM = 3.82 \times \dfrac{V_c}{D_d}$	—
Feed rate, in./min	$f_m = f_r \times RPM$	$f_m = f_t \times n \times RPM$	$f_m = f_r \times RPM$	—
Feed per Tooth, inches	—	$f_t = \dfrac{f_m}{n \times RPM}$	—	f_t
Cutting Time, minute	$t = \dfrac{L}{f_m}$	$t = \dfrac{L}{f_m}$	$t = \dfrac{L}{f_m}$	$t = \dfrac{L}{12\,V_c}$
Rate of Metal Removal, $\dfrac{\text{cu in.}}{\text{min}}$	$Q = 12 \times d \times f_r \times V_c$	$Q = w \times d \times f_m$	$Q = \dfrac{\pi D_d^2}{4} \times f_m$	$Q = 12 \times w \times d \times V_c$
Horsepower Required at Spindle	$HP_s = Q \times P$	$HP_s = Q \times P$	$HP_s = Q \times P$	—
Horsepower Required at Motor	$HP_m = \dfrac{Q \times P}{E}$	$HP_m = \dfrac{Q \times P}{E}$	$HP_m = \dfrac{Q \times P}{E}$	$HP_m = \dfrac{Q \times P}{E}$
Torque at Spindle	$T_s = \dfrac{63030\,HP_s}{RPM}$	$T_s = \dfrac{63030\,HP_s}{RPM}$	$T_s = \dfrac{63030\,HP_s}{RPM}$	—

Symbols:

D_t = Diameter of workpiece in turning, inches.
D_m = Diameter of milling cutter, inches.
D_d = Diameter of drill, inches.
d = Depth of cut, inches.
d_t = Total depth per stroke in broaching, inches.
E = Efficiency of spindle drive.
f_m = Feedrate, inches per minute.
f_r = Feed, inches per revolution.
f_t = Feed, inches per tooth.
HP_m = Horsepower at motor, hp.
HP_s = Horsepower at spindle, hp.
L = Length of cut, inches.
n = Number of teeth in cutter.
P = Unit power, horsepower per cubic inch per minute.
Q = Rate of metal removed, cubic inches per minute.
RPM = Revolutions per minute of work or cutter.
T_s = Torque at spindle, inch pounds.
t = Cutting time, minutes.
V_c = Cutting speed, feet per minute.
w = Width of cut, inches.

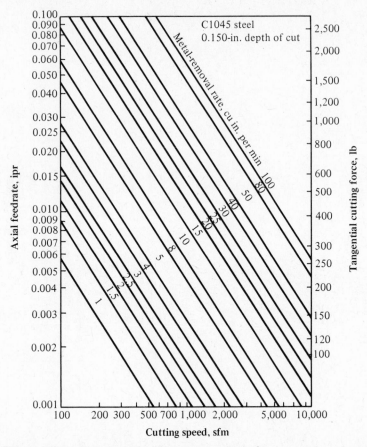

Figure 3.21. Relationships between speed, feed, metal removal rate, and tangential force plotted on log-log grid for constant cutting depth of 0.150 in. (3.81 mm) in C-1045 steel of 204 BHN. (*Courtesy* American Machinist.)

The unit of time in the metric system is the second (s) because the kilowatt is defined as the rate of work on one kilonewton-meter per second. The value of unit power can be calculated by the following equation:

$$kW_u = \frac{F'_t}{f'd'}$$

where: F'_t = tangential cutting force in kilonewtons
f' = feed rate in millimeters per revolution
d' = depth of cut in millimeters

Metric units of cutting conditions are given with the prime symbol so they can be distinguished from inch values of the same term. This practice is not necessarily universal.

Example: A shaft of AISI 1020 steel has a Brinell hardness of 165. The cutting speed is given as 90 m/min. The roughing depth of cut is given as 2.54 mm with a feedrate of 0.50 mm/rev. The lathe to be used is a geared-head type. Assume normal tool wear or a correction factor of 50%. What size motor will be required?

Solution:

$$kW_c = kW_u \ \frac{V'f'd'}{60}$$

$$= 1.4 \ \frac{0.90 \cdot 0.50 \cdot 2.54}{60}$$

$$= 2.667 \ kW$$

$$kW_m = \frac{kW_c}{e_m}$$

$$= \frac{2.667}{0.75}$$

$$= 3.556$$

Thus 3.55 kW are required at the motor for sharp cutting tool. Considering a tool wear factor of 50%:

$$kW_m = 3.556 \cdot 1.5$$
$$= 5.334$$

Converting kilowatts to horsepower (1 kW = 1.341 hp)

$$HP_m = 1.341 \cdot 5.334$$
$$= 7.153$$

Therefore, a 7.5-hp motor would be required.

MACHINABILITY

Machinability is an involved term with many ramifications, but briefly defined it is the relative ease with which the chip may be separated from the base material. The two main aspects of machinability are the properties of the material being cut and the machining conditions involved.

Machinability Properties of the Material. The material properties that affect machinability may be listed briefly as follows:

   a. Shear strength. The shear strength can be obtained by the ratio of the chip thickness to the depth of cut or feed, depending on the operation.
   b. Strain hardenability. The increase in strength and hardness with increasing plastic deformation.

c. Hardness. The characteristic of the material to resist indentation.
d. Abrasiveness of the microstructure. The abrasiveness and nature of inclusions and interfaces in the metallic matrix affect the machining qualities.
e. The coefficient of friction. The coefficient of friction varies with the type of material and the reaction of it to the tool material at the chip interface.
f. Thermal conductivity.

Methods of Evaluating Machinability.   Common methods of evaluating the machinability are by using one or more of the following criteria:

a. Tool life for a given surface-foot speed and tool geometry. This may also be expressed in terms of cubic-inch removal rate.
b. The power required, which is also related to cutting forces.
c. Surface texture. Surface-texture requirements vary widely; however, a material that produces a better surface texture under equal conditions is considered to possess a higher degree of machinability.

Items (a) and (b) have already been discussed, and item (c) is discussed in detail in Chapter 21.

Influence of Additives on Machinability of Steel.   Studies have been made that show a strong correlation between steel composition and machinability. Elements that tend to break up the soft ferrite structure, such as carbon, manganese, and sulfur, improve machinability. However, if these elements are used in excess, they may increase the hardness and abrasiveness of the metal.

Machinability Ratings.   Machinability ratings are based on AISI 1112, a resulfurized carbon steel. This steel is given a rating of 100% machinability. Table 3-5 lists various metals and their machinability ratings compared to 1112 steel. Factors that have a bearing on the rating are also shown as Brinell Hardness Number, cutting-speed range for a carbide tool, and a feedrate range. These are given only to have a comparable knowledge of the machining property of common materials and should be taken only as starting figures. Other factors that would have to be considered are size and type of machine, setup rigidity, use of coolant, tool life desired, power available, and required surface finish.

METAL-CUTTING ECONOMICS

The foregoing information on machinability is of little value unless it is used to obtain quality parts at minimum cost. The cutting speed and tool life that produce a minimum cost are called *economical cutting speeds* and *economical tool life,* respectively. Just what is the full significance of economical tool life? Taylor recognized some 80 years ago that a long tool life was not the goal but rather a minimum production time per workpiece.

The basic problem in determining the most economical cutting condition is that as cutting speeds increase, the efficiency of the machine tool increases but

Table 3-5. Machinability ratings of common metals (based on 1112 steel).

AISI No.	% Rating	BHN	Carbide Tool Roughing Cut	
			Sfpm	Feed/ipr
1212CD*	100	163	540–660	.010–.015
12L14†	160	140	800–940	.010–.015
1020CD	80	156	450–550	.015–.022
1045HR‡	50	217	400–450	.015–.020
Stainless Steel				
300 Series	50	—	275–330	.007–.015
Alloy Steel				
4145HR	55	217	400–450	.010–.020
Cast Iron				
Ferritic malleable	90	110–156	350–450	.011–.018
Nodular iron ferritic	110	140–200	525–600	.012–.025
Pearlitic iron	70	207–229	400–550	.012–.018
Soft gray 30	80	140–179	425–550	.015–.022
Cast Steel				
1040	65	190	400–600	.005–.015
New Ferrous Metals				
Cartridge brass	70	B-20	400–600	.007–.016
Naval brass	70	120	400–600	.007–.016
Beryllium copper	40	—	450–600	.003–.009
Free-cutting brass	140	120	800–1100	.006–.014
Aluminum bronze	55	140	250–340	.006–.013
Aluminum 2024–T4	150	120	1200–1800	.008–.020
Aluminum 6061–T6	95	170	800–1400	.008–.020

*CD: cold drawn.
†L: leaded.
‡HR: hot rolled.

the tool life goes down. Finding the theoretical optimum point at which there is a balance of three individual costs—machining costs, tool-changing costs, and setup costs—is the essence of metal-cutting economics. The relationships between these three factors are shown in Fig. 3-22. It can be seen that machining cost decreases with increasing cutting speed. Both the tool costs and the tool changing costs are observed to increase, since they are wearing out faster at the higher cutting speeds. The setup costs are independent of cutting speeds. Adding up each of the individual costs results in the total unit-cost curve which is observed to go through the minimum point.

If tooling is babied by using low cutting speeds, the unit cost goes up because the hourly burden rate of labor and overhead is spread over too few workpieces. On the other hand, if cutting speeds are set too high the unit cost will again go up because more tools will be consumed, but more importantly the cost of downtime for changing tools will be increased.

For any given job, with all other factors held constant, the relationship between tool life (T) and cutting speed (V) is expressed in Taylor's formula as:

$$VT^n = C$$

Using this relationship, it is possible to derive an equation to find the tool life that will result in the minimum unit cost (T_c, expressed in minutes) in terms of tool-changing time (TCT, in minutes), cost per cutting edge (C, in minutes), cost of labor and overhead (L, in dollars per minute), and the Taylor tool-life exponent (n). This equation is:

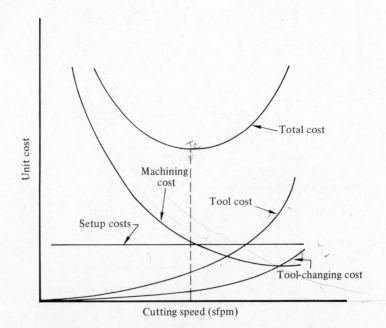

Figure 3.22.　The influence of cutting speed on costs.

$$T_c = \left(TCT + \frac{e}{L}\right)\left(\frac{1}{n} - 1\right)$$

A plot of the part production against cutting speed will pass through a maximum value (see Fig. 3-23). In the low-speed range it is obvious that a boost in speed will boost output. However, at the high-speed end a further boost causes the production rate to fall off due to the excessive tool-changing time. The tool life for maximum production (T_p) can be expressed as a function of tool change time and the Taylor exponent:

$$T_p = TCT\left(\frac{1}{n} - 1\right)$$

The range between the minimum unit cost on the lower speed and the maximum production on the higher speed was termed the *high-efficiency range* or Hi-E by William W. Gilbert of General Electric.

To select a machining speed within the high-efficiency range, therefore, it is only necessary to adjust the cutting speed so that the rpm falls between the paired values.

Insert Costs.   Insert costs are small in comparison to other machining costs. As an example, labor and overhead may come to $30.00 an hour and a one-

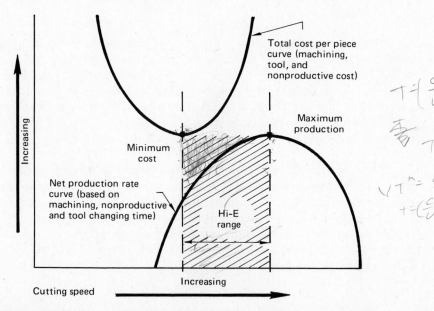

Figure 3.23.   The cutting-speed range between the low point on the cost curve and the high point on the productivity curve at the right is termed the "Hi-E" or high-efficiency range. Cutting speeds outside of this range sacrifice both cost and productivity.

minute tool index may cost 50¢, which is considerably higher than the cost per cutting edge of many inserts. Continuing progress in materials, design, and manufacture of inserts constantly adds to their economic effectiveness. It is the performance of the cutting tool that determines the ultimate efficiency of the machine tool.

The cost of inserts per cutting edge, labor and overhead, are easily available. Tool change can be quickly determined with a stopwatch. The Taylor exponent can be determined experimentally and will typically be in the range of 0.22 to 0.30 for tungsten carbide and 0.38 and up for ceramics.

HIGH-SPEED MACHINING

The ever-present question—"Why can't we do it faster?"—has led to considerable research in the topic of high-speed machining. By high-speed machining is meant the removal of metal at speeds in excess of 3000 sfpm for aluminum and 1,000–3,000 fpm (305–915 m/min) for steel and cast iron.

Detailed studies of the results in this area have shown that:

1. Aluminum oxide tools can be used successfully for high-speed machining. However, it is recommended there be a high effective negative rake, a high side-cutting edge angle, and a large nose radius. The insert must be evenly supported on a carbide pad and any tool deflection must be eliminated.
2. The machine must ensure rigidity with a minimum of deflection and vibration.

Ultrahigh-Speed Machining. Recent research has pointed toward ultrahigh-speed machining. Today's practical limit is about 50,000 sfpm (15240 M), but the goal is 250,000 sfpm (381,000 M). Experimental work at Lockheed in California was done on specially designed 25-hp 18,000-rpm milling machine with a feedrate of 300 ipm (7,620 mm). It is now used in productions to remove 30 cu in./min (762 cu mm) of aluminum with a 1.250-in. (31.75-mm) diameter cutter. The production rate of this machine is 3.4 times that of a conventional machining center.

It is expected that conventional electric motors will be replaced by turbine drives. Conventional bearings will be replaced by either liquid film or hydrostatic hydrodynamic gas bearings. Cutting tools will not have to be changed a great deal. Properly designed cutters will last considerably longer than off-the-shelf cutters due to the reduced cutter loads and temperatures. In high-speed machining, most of the heat is contained in the chip, which in some cases is just a glob of molten metal and/or metal vapor.

CUTTING FLUIDS

Cutting fluids are used in many metal-cutting operations, as well as in grinding, to maintain optimum production rates, minimize tool wear, improve surface finish, and inhibit rust. In most cases, the cost of the fluid, which is usually recirculated, is negligible compared to the benefits obtained.

Types of Cutting and Grinding Fluids

The main types of cutting fluids are cutting oils, emulsified oils (water miscible), and chemical and semichemical fluids (water miscible).

Cutting Oils.   Plain cutting oils are derived from petroleum, animal, marine, or vegetable base substances and are used straight or in combination. Oils may be used straight or in uncompounded form but are restricted to light-duty machining on metals of high machinability such as aluminum, magnesium, brass, and leaded steels.

Polar additives are blended with the mineral or fatty oils to wet and penetrate the chip–tool interface. This is accomplished by polarity or affinity of the polar additive for the metal substrate. The film formed on the metal provides lubrication by reducing the friction at the chip–tool interface. Since the melting point of the oil film is rather low, in some severe operations it must be further supplemented by the use of extreme pressure additives such as sulphur and chlorine.

Emulsified Oils (Water Miscible).   Water-miscible fluids form mixtures ranging from emulsions to solutions when mixed with water which, due to its high specific heat, high thermal conductivity, and high heat of vaporization, is one of the most effective cooling media known. Blended with water, usually in the ratio of one part oil to 15 to 20 parts water for cutting, and 40 to 60 parts water for grinding, the water-miscible fluids provide excellent cooling and lubrication for metal cutting at high speeds and light pressures.

Chemical and Semichemical Fluids (Water Miscible).   Chemical and semichemical solutions, when diluted with water, vary from translucent to completely clear. The emulsified oils are "milky" or opaque.

Although water has long been recognized as the most efficient cooling agent, it has a number of objectionable characteristics, such as a tendency to promote rust on the machine and workpiece and a lack of lubricating qualities. It may also contain nutrients for the growth of bacteria. Chemical coolants are modifiers of water to reduce its undesirable side effects. Where lubricity is required, soaps, wetting agents, and "chemical lubricants" are incorporated. Chemicals used are amines and nitrites—for rust prevention; nitrates—for nitrite stabilization; phosphates and borates—for water softening; compounds of phosphorus, chlorine, and sulphur—for chemical lubrication; soaps—as wetting agents and for lubrication; glycols—as blending agents and humectants; and germicides—to control bacterial growth.

Semichemical Fluids.   Semichemical fluids contain a small amount of mineral oil plus additives to further enhance lubrication properties. The semichemicals are gaining wide popularity in industry today because they incorporate the best qualities of both chemicals and normal water emulsions. Both chemical and semichemical coolants are now available that contain chlorine, sulphur, or phosphorous, additives that allow extreme pressure or boundary lubrication effects and can be used on some of the more difficult machining or grinding applications. Concentrations range from 2 to 10%.

Advantages of Chemical Fluids

1. Rapid heat dissipation.
2. A very light residual film that is easy to remove.
3. Excellent resistance to rancidity.
4. An easy concentration to control with no interference from tramp oils.
5. Good detergent properties that aid in the maintenance of open and free-cutting grinding wheels.

Disadvantages of Chemical Fluids

1. The high detergency may irritate sensitive hands over a long period of time.
2. In comparison to oils, there is less rust control, lubrication, and there is some tendency to foam.
3. The lack of lubrication "oiliness" may cause some sticking in the moving parts of machine tools.

Application of Cutting Fluids

The cutting fluid, to be effective, must be directed where cooling is required, right at the cutting edge of the tool. While a general flow over the workpiece does help cool the work, it is only by forcing the fluid into the cutting area that the heat can be removed as it is generated.

A good example of getting the fluid at the tool–work interface is in forcing the coolant through the wheel as in grinding.

Care must be taken not to have cyclic heating and cooling on carbide tools. Rapidly reversing heating and cooling of the cutting edge produces microcracks that grow until small pieces of the tool break away. Properly placed nozzles, or a mist apparatus, can give continuous gentle cooling. Each application of coolant should be studied to select the best approach for the specific job.

Problems

3-1. A new 1 in. wide, 6 in. dia., 8-tooth, face-milling cutter is given a performance test on a SAE 1020 CRS block (200 BHN), 4 in. square and 33 in. long. The cut is made over the long axis of the work. The following operating conditions were used:

Sfpm = 300
Chip load = 0.003 in.
Depth of cut = 0.125 in.

Idle horsepower = 0.5
Machine mech. efficiency = 0.9

Calculate the following:
(a) RPM
(b) Feedrate in ipm
(c) Time to make the cut
(d) Cubic-inch removal rate
(e) Horsepower required at the cutter
(f) Horsepower required at the motor

3-2. (a) Determine the spindle horsepower that is required to turn the surface of a 2-in. dia. soft cast iron bar under the following conditions:

> Sfpm = 200
> Depth of cut = 0.100 in.
> Feed = 0.050 ipr

(b) Is the increase in horsepower proportional to the increase in cutting speed if 300 sfpm were used?

3-3. (a) What causes a builtup edge to form on the tool when cutting metal? (b) How may it be eliminated? (c) Is there any disadvantage in having a builtup edge?

3-4. A 4-in. dia. gray cast iron bar 3 ft long is to be turned on a lathe. A roughing depth of cut of 0.100 in. is used and a 0.020 ipr feed. (a) Select a carbide grade and cutting speed. (b) Compare the machine time required to turn 10 bars with a WC tool and with a ceramic tool.

3-5. Make simple graphs to show the following relationships:

(a) Cutting force vs. continually increasing cutting speed.

(b) Cutting force vs. side-rake angle ($-20°$ to $+20°$)

(c) Cutting force vs. feed (0.005 to 0.100 in.)

3-6. Two pieces of aluminum—1100-O and 2024-T3—are machined orthogonally with the same cutting conditions: depth of cut—0.050 in., feed—0.020 ipr, back-rake angle—$+10°$.

The measured chip thickness for 1100-O aluminum was 0.080 in., and for the 2024-T3 aluminum it was 0.060 in.

(a) Compare the cutting ratios of the metals.

(b) Compare the relative shear angles of the two metals. Make a sketch labeling the shear angle for 1100–O aluminum as A and the angle for 2024–T3 aluminum as B. Also show: the tool, α, t_1, and t_2.

(c) Which material has the better machinability?

3-7. (a) What would be the recommended tool life, for the minimum cost per piece, when machining a 6-in. dia. SAE 1020 steel bar 30 in. long under the following conditions:

> Square-insert carbide tool with negative rake
> Feed = 0.020 ipr
> Length of cut = 25 in.
> Depth of cut = 0.250 in.

> Machine overhead rate = $6.00/hr
> Machine labor rate = $3.50/hr
> Carbide tool = $2.50/insert eq $2.50/8 = $0.31/cutting edge
> Tool changing time = 0.1 min
> Total handling time = 5 min/part

(b) What would the cutting speed be for this tool life if the tool-life curve, as shown in Fig. 3-17, is used?

(c) How long would the cut take under the conditions given?

3-8. What would the tool life be for maximum production using the information from Problem 3-7?

3-9. Using the information given in Problem 3-8, change the tool to ceramic. How much does this change the tool life for minimum cost?

3-10. (a) If the cutting speed for one minute of tool life with a HSS tool is 200 fpm and the assumption is made that a carbide tool can cut four times faster, estimate what the tool life will be for a carbide tool cutting at 300 fpm. Use log-log paper to construct the tool-life curves similar to those shown in Fig. 3-17.

(b) What would the cutting speed be for each tool material for each of the following times:

> 10 min   60 min   30 min

3-11. In machining a new steel alloy the data indicated that $V_{10} = 316$ fpm and $V_{100} = 200$ fpm. (a) What tool life would you expect if the cutting speed were 240 fpm? (b) What is the tool material that was likely used in this operation? Why?

3-12. In analyzing the economics of single-point turning, what are the four cost factors that contribute to the total unit cost? Make a chart as shown for your answer.

Cost Factors	Function of CS?	Included in Determining Max. Prod. Rate?

3-13. Use Taylor's equation to find the tool life in minutes with the following conditions given:

Carbide tool
sfpm = 206
C = 494

3-14. What horsepower would be required at the spindle if a 3-in. dia. bar of AISI 1045 steel were being machined with a carbide tool in a roughing cut at 0.150-in. depth of cut and 0.015-ipm feed? The BHN is 230.

3-15. (a) How long would it take to machine a 2-in. dia. free-cutting brass bar down to $1\frac{1}{2}$ in. for a length of 10 in.? Use roughing cuts (decide on the number) and a carbide tool. (b) What carbide tool should be used?

3-16. On 2 × 1 cycle log-log paper draw the following tool-life curves. Cutting parameters— 0.125 in. depth, 0.020 in. feed. Tool carbide, +15° SCEA, and $\frac{1}{32}$ NR.

Cast Iron sfpm	Tool Life min	SAE 1020 sfpm	Tool Life min
100	350	350	900
150	115	400	520
200	53	450	330
250	29	500	200
300	17.5	550	145
350	11.5	600	100

(a) What is the value of n and C for each of the materials?

(b) How does n compare to what is considered standard for carbide tools?

(c) What is sfpm for V_{60} for SAE 1020 steel?

(d) What may account for the slope of the curve being somewhat off?

3-17. Is the cutting speed for minimum cost the optimum point at which to be cutting? Why or why not?

3-18. What is meant by "Hi-E" machining?

3-19. Change the carbide tool designation from US to ISO for the following conditions. Also show suggested improvement.

Carbide Grade (US)	ISO	Material Cut	Results	ISO Improvement
C–1		Gray iron over 200 BHN	Tool chipping	
C–7		Steel casting	Excessive tool wear	
C–4		Very hard gray iron	Excessive tool wear	

3-20. Why is a coated carbide tool better than a straight carbide tool for machining alloy steel?

3-21. What is meant by a cermet tool and what are its properties?

3-22. How does a super HSS tool differ from a regular HSS tool?

3-23. If cutting speeds of over 2000 fpm can be had with ceramics and cermets, why use tungsten carbide tools at all?

3-24. How is coolant able to get to the chip–tool interface effectively?

Case Study 1

The part shown in Fig. CS3.1 was machined with conventional grade carbide tools on a numerically controlled turning center. The maximum depth of cut taken was 0.250 in. The part is turned from 8-in. dia. bar stock. One finishing cut is taken. The feed used for roughing was 0.020 in. and 0.005 in. for finishing.

The manufacturing engineer decided to try ceramic tools using the same feeds and depth of cut. He found the cost per cutting edge for the carbide tools was 60¢ and for ceramic tools 80¢.

The tool-changing time in either case was 1.5 min. The labor and overhead cost/hr = $25.00.

(a) What is the tool life for each tool?

(b) What is the sfpm for each tool?

(c) How many roughing cuts are required?

(d) What is the approximate time/part for each tool?

(e) What is the difference in cost of the parts per thousand for each tool?

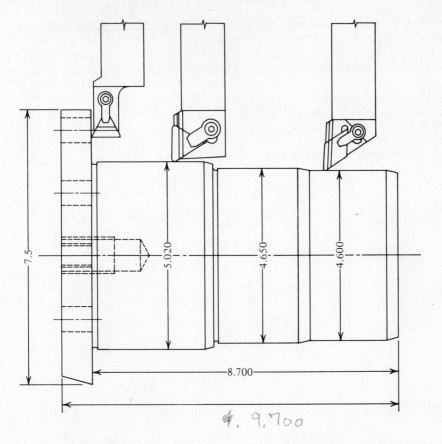

7.5

5.020

4.650

4.600

8.700

9.700

Figure CS3.1.

Case Study 2

Several years ago Lockheed was experimenting with high-speed milling. A four-axis machining center was purchased that had a 20-hp spindle capable of 18,000 rpm for high-speed contouring. A cutting speed of 5890 sfm was used along with a 1.25-in (31.75-mm) three-flute cutter at 0.005-ipt (0.13-mm) feed. What would the table feed be? The slots cut were $\frac{1}{2}$ in. (12.70 mm) deep. (a) What was the metal removal rate? (b) How does this compare with normal shop standards?

Bibliography

Adamas Carbide Corporation. *Carbide Technical Manual.* Kenilworth, N.J.

Gettelman, K. "Improve Productivity with Sound Basics." *Modern Machine Shop* 51 (December 1978):86–98.

Hatschek, R. L., ed. "Turning with Inserts."

American Machinist Special Report 707 122 (October 1978):119–134.

Kalish, H. S. "Carbide Grade Classifications— What They Mean." *Manufacturing Engineer* 76 (January 1976):49–52.

———. "Selecting the Optimum Cutting Tool

Material." *Cutting Tool Engineering* 28–29 (January–February 1977):6–9; 28–29 (March–April 1977):10–13.

Kennametal Inc. *Tool Application Handbook.* Latrobe, Pa., 1973.

King, R. I. "Ultra High-Speed Machining." *Production Engineering* (March 1978).

Machinability Data Center. *Machining Data Handbook.* Cincinnati, 1980.

Monarch Machine Tool Company. *Speeds and Feeds for Better Turning,* 5th ed. Sidney, Ohio, 1980.

Springborn, R. K., ed. *Cutting and Grinding Fluids: Selection and Application.* Dearborn, Mich.: American Society of Tool and Manufacturing Engineers, 1967.

Tipnis, V. "Finding Your Best Machining Conditions." *Modern Machine Shop* 49 (January 1976):68–76.

CHAPTER FOUR

TURNING MACHINES

Exactly when the machine age began is difficult to pinpoint; however, a need for boring cannon barrels precipitated the development of a vertical boring machine by a Swiss named Maritz in 1713. A horizontal boring machine was developed by a Dutch gun founder working with a Swiss engineer in 1758. Newcomen invented and built the first practical steam engine, which had cylinders of cast brass about 8 ft long with a bore of from 21 to 28 in. (533 to 685 mm). At that time a piston was considered a good fit if it came within $\frac{1}{8}$ in. of fitting the cylinder bore over its length of travel. Watt's steam engine, however, required a smaller, tighter fitting piston, and in 1775 John Wilkinson was able to bore the cylinders to the accuracy Watt required. This combined achievement is often credited as being the birth of both the machine and power age.

The constant search for improved methods of making wheels, shafts, and other cylindrical items led to the development of the lathe. Although the early lathes were used entirely for turning cylindrical items, they gradually became more versatile and many other tasks were assigned to them. Henry Maudsley, an Englishman, is credited with making the first screw-cutting engine lathe in about 1797.

The main types of lathes are engine, turret, screw machines, single-spindle automatics, and multispindle automatic. Figure 4-1 illustrates the engine lathe.

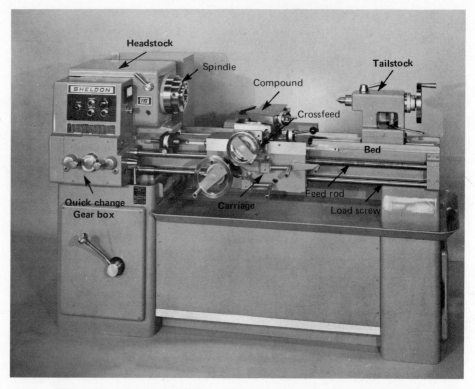

Figure 4.1. An engine lathe with five main components shown in bold type. (*Courtesy Sheldon Machine Co., Inc.*)

LATHE OPERATIONS

The engine lathe is very versatile and can be used for a wide variety of operations such as turning, facing, taper turning, threading, parting, knurling, drilling, reaming, and boring. With accessories additional operations can be performed, such as milling, grinding, and duplicating. Some of these operations are shown in Fig. 4-2.

Turning.   Turning is used to produce cylindrical parts. It is accomplished with the work rotating and the cutting tool feeding parallel to the axis of the workpiece (Fig. 4-3).

The three primary factors in any basic turning operation are *speed, feed,* and *depth of cut*. Other factors such as kind of material and type of tool are also important considerations.

Speed refers to the spindle speed or revolutions per minute (rpm) of the workpiece. An important figure for a particular turning operation is the *surface foot speed*, or the number of feet that pass the cutting tool per minute. This figure is simply the circumference (in feet) times the rpm (see Table 3-4).

Feed refers to how much the cutting tool advances per revolution of the

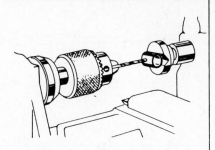

Drilling an oil hole in a bushing with crotch center in tailstock.

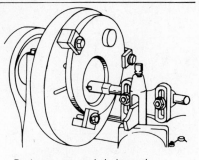

Boring an eccentric hole on the faceplate of the lathe.

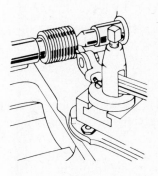

Cutting a screw thread with compound rest set at 29 deg.

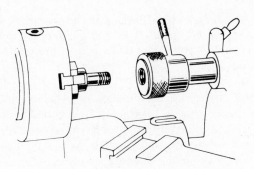

Die mounted in tailstock of lathe for threading studs.

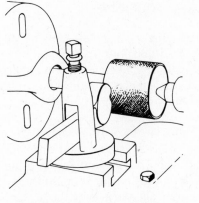

Knurling a steel piece in the lathe.

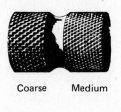

Coarse   Medium

Sample of knurling.

Figure 4.2. A variety of operations commonly performed on the engine lathe.

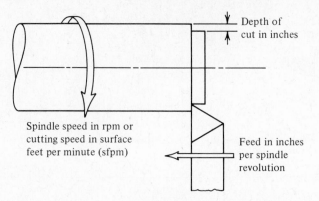

Depth of
cut in inches

Spindle speed in rpm or
cutting speed in surface
feet per minute (sfpm)

Feed in inches
per spindle
revolution

Figure 4.3. The basic turning operation. As in most machining operations speed, feed, and depth of cut must be controlled by the operator either manually or by a programmed input.

stock and is expressed in inches per revolution (ipr). Some machines have a hydraulic feed that is completely independent of the spindle rotation. In this case, the feed is given in inches per minute (ipm).

Depth of cut is made on the cross-feed diameter and is self-explanatory except to note that the diameter is reduced by twice the depth of cut.

Basic Structure of the Engine Lathe

The five main parts of the modern engine lathe are the headstock, tailstock, bed, carriage, and quick-change gearbox as was shown in Fig. 4-1.

Headstock.   The headstock contains the driving mechanism—either pulleys or gears—to turn the work. Provision is made so that 8 or 10 different speeds may be obtained on the smaller-sized lathes, and several times this number on larger ones. The headstock furnishes a means of support for the work, either with a center fitted into the spindle or by a chuck.

Tailstock.   The tailstock is the most common means of supporting the outer end of the stock. It may be positioned upon the bed at any point and locked securely in place. It contains the dead center which is held in place by a taper. Adjusting screws in the base of the tailstock are used to shift it laterally on the bed. This provides a means of maintaining accurate alignment with the headstock center. It also makes it possible to introduce a small amount of taper into the work (Fig. 4-4).

Bed.   The bed makes up the basic structure of the lathe, on which all other parts are mounted. It is usually made of aged gray cast iron in the smaller-sized lathes, and may be of welded-steel construction in the larger sizes. At the top of the bed are the ways (Fig. 4-5). They act as guides for accurate movement of the carriage and tailstock.

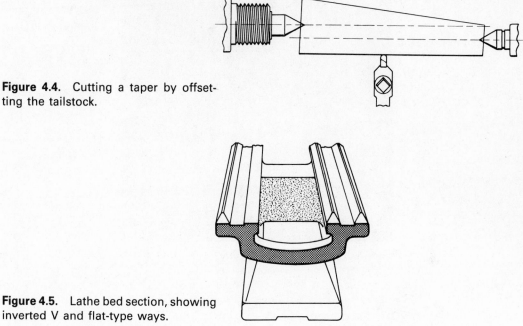

Figure 4.4. Cutting a taper by offsetting the tailstock.

Figure 4.5. Lathe bed section, showing inverted V and flat-type ways.

Much of the machine's accuracy is dependent on the preciseness of the ways. They are usually finished by a milling or planing operation followed by grinding and then are hand-scraped for further exactness of the mating parts. Some manufacturers prefer to harden and grind the ways and also make them replaceable.

Carriage. The carriage, as the name implies, is used to carry the cutting tool along the bed longitudinally (Fig. 4-6). A cross slide is mounted on top of the carriage. It moves the tool laterally across the bed. Another slide is mounted above the cross slide. This is referred to as the *compound rest*. The compound rest can be swiveled in a horizontal plane at any desired angle with respect to the work, thus furnishing another method of cutting tapers. Since this method is dependent on hand turning of the compound-rest knob, it is generally limited to short, steep tapers. The back side of the carriage is often equipped with a taper attachment that provides still another way of cutting tapers (Fig. 4-7). This is a more convenient method of cutting long tapers than by offsetting the tailstock, since it can be quickly engaged or disengaged so as not to interfere with straight turning.

Quick-Change Gearbox. The quick-change gearbox allows the operator to change the amount of longitudinal or cross-feed quickly. Feed refers to the amount the tool is made to advance per revolution of the work. The quick-change gearbox also provides settings for cutting all the commonly used threads.

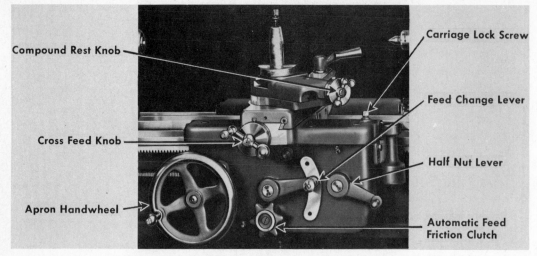

Compound Rest Knob

Cross Feed Knob

Apron Handwheel

Carriage Lock Screw

Feed Change Lever

Half Nut Lever

Automatic Feed
Friction Clutch

Figure 4.6. Lathe carriage. (*Courtesy South Bend Lathe, Inc.*)

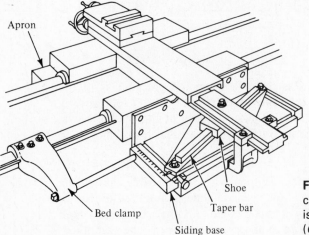

Apron

Bed clamp

Siding base

Taper bar

Shoe

Figure 4.7. The taper attachment, located on the back side of the lathe bed, is used for cutting long, shallow tapers. (*Courtesy* American Machinist.)

Lathe Size and Accuracy

Lathe Size. The size of a modern engine lathe is given by the maximum diameter of swing, in inches, and the maximum length of bar that can be turned between centers, in inches. *Swing* refers to twice the distance from the lathe center point to the top of the ways. An engine-lathe size specification may read: swing $12\frac{3}{4}$ in. over bed, $7\frac{5}{8}$ in. over cross slide; length 47 in.

Accuracy. Lathe dials are graduated in increments of 0.001 in. or in millimeters. Some are made to read directly; that is, for each 0.001-in. (0.03-mm)

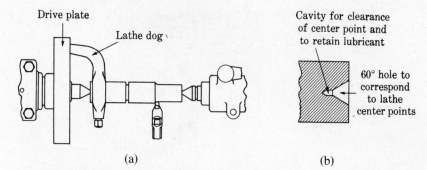

Drive plate

Lathe dog

Cavity for clearance of center point and to retain lubricant

60° hole to correspond to lathe center points

(a)

(b)

Figure 4.8. (a) Stock mounted between centers on a lathe. (b) Center holes drilled into each end of the stock.

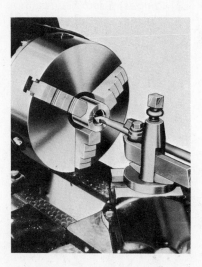

Figure 4.9. Round work held in a universal chuck. (*Courtesy South Bend Lathe, Inc.*)

depth of cut, a corresponding 0.001 in. will be removed from the stock. Other lathes are made so that the tool moves in the amount registered on the dial. Since the material is removed from both sides of the stock, the amount taken off will be twice the dial reading. Engine-lathe work can, with good equipment, be maintained within 0.001- or 0.002-in. (0.03–0.05-mm) accuracy.

Work-Holding Methods

Lathe Centers. The work to be turned is first cut to length, and then center holes are drilled into each end [Fig. 4-8(b)]. A dog is securely fastened to one end of the work, which is then placed between the lathe centers. The tail of the dog fits in a slot of the drive plate for positive rotation.

Chucks. Another common method of mounting work in a lathe is by means of a chuck (Fig. 4-9). The universal chuck shown in the figure is self-centering;

Figure 4.10. Centering work with a dial test indicator using an independent four-jaw chuck. (*Courtesy South Bend Lathe, Inc.*)

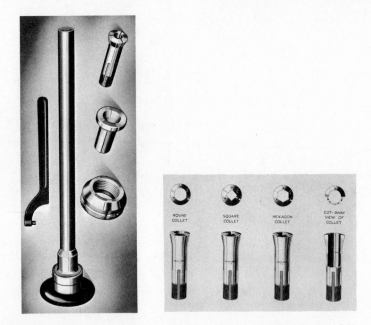

Figure 4.11. Various types of collet chucks and their uses. (*Courtesy South Bend Lathe, Inc.*)

all three jaws move toward the center on a scroll arrangement. This permits the operator to load and unload work quickly and easily.

A more versatile chuck is the four-jaw independent type (Fig. 4-10). Loading and unloading the work is slower in this chuck, since the jaws must be adjusted separately; however, it has more holding power and can be used for both on- and off-center work. It can also be used to advantage on odd-shaped pieces.

Figure 4.12. Rubber-flex collets. (*Courtesy The Jacobs Manufacturing Co.*)

Figure 4.13. Boring a bracket with an angle plate attached to the faceplate. (*Courtesy South Bend Lathe, Inc.*)

The collet chuck (Fig. 4-11) is the most accurate of all chucks. The collets have split tapers that are made to close by means of a draw-in handwheel that fits through the spindle of the headstock. The work mounted in the collet should not be more than a few thousandths of an inch larger or smaller than the size stamped on the collet.

A more versatile collet chuck is the rubber-flex type (Fig. 4-12). The parts to be chucked may vary as much as $\frac{1}{8}$ in. in diameter for each collet, except in the smallest size, $\frac{1}{16}$ in. The standard range of sizes for collets of this type is $\frac{1}{16}$ to $1\frac{3}{8}$ in.

Faceplate. The faceplate shown in Fig. 4-13 is made with bolt slots so that

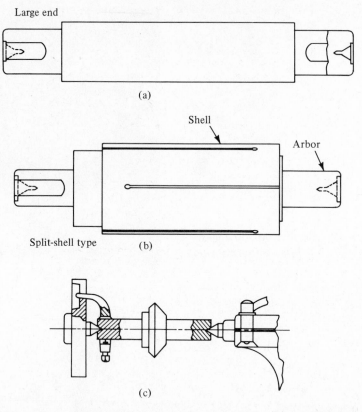

Large end

(a)

Shell

Arbor

Split-shell type   (b)

(c)

Figure 4.14. Solid (a) and expansion mandrels (b) used to hold parts for machining (c) so the outside surface can be machined concentric with the bore.

flat work may be clamped or bolted to it. More elaborate setups may call for angle plates, as shown, or for fixtures attached to the faceplate.

Mandrel.   It is often desirable to machine a surface concentric and true with the inside diameter. This can be accomplished with the aid of a mandrel. Two types of mandrels are shown in Fig. 4-14.

The solid mandrel (a) has an 0.008-in./ft taper. The part to be machined is placed on the small end of the mandrel and forced on with an arbor press until it is firmly in place.

Work-Support Equipment

Center Rest.   Stock extending from the chuck that is to be drilled or bored will not have the benefit of the tailstock support. If the stock extends from the chuck for a distance equivalent to more than 4 dia., a center rest (Fig. 4-15) is

Figure 4.15. Stock supported with a steady rest for a boring operation. (*Courtesy South Bend Lathe, Inc.*)

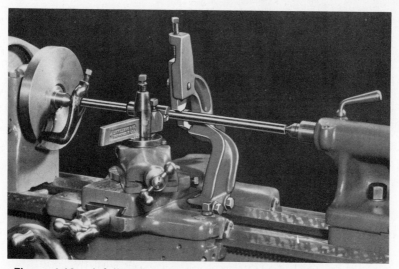

Figure 4.16. A follower rest used to support slender stock as it is being machined. (*Courtesy South Bend Lathe, Inc.*)

used. The steady rest may also be used to give support for long stock when it is machined between centers.

Follower Rest. Long, comparatively small-diameter work machined between centers on a lathe cannot be accurately cut. The stock tends to spring away from the cutting tool. A follower rest is used to overcome this difficulty (Fig. 4-16). The rest is bolted to the carriage and is adjusted to follow directly behind the cutting tool.

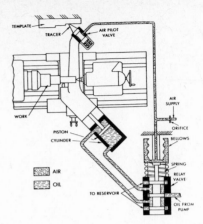

Figure 4.17. A schematic of an air–oil hydraulic tracer attachment used on an engine lathe. The tracer pilot valve controls the action of the hydraulic cylinder attached to the cutting tool. (*Courtesy The Monarch Machine Tool Company.*)

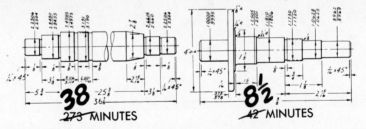

38 ~~273~~ MINUTES   **8½** ~~42~~ MINUTES

Figure 4.18. A lathe tracer arrangement with typical parts, and the time required compared to the same operation by a standard engine lathe. (*Courtesy Sidney Machine Tool Co.*)

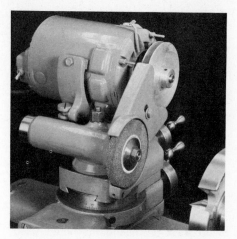

Figure 4.19. A toolpost grinding attachment. (*Courtesy South Bend Lathe, Inc.*)

Duplicating Lathe

The duplicating lathe is very much like an ordinary engine lathe except that it is equipped with a tracer attachment (Fig. 4-17). A template, which may be cut out on a metal-cutting band saw and then filed for accuracy, is attached to a stand so that a stylus can trace its contour. The stylus, in turn, actuates a pilot valve that is used to control the hydraulic cylinder attached to the cutting tool. Figure 4-18 shows a lathe with a tracer attachment along with typical time-saving operations.

Engine Lathe Accessories

Probably the most used engine lathe accessories are the taper and tracer attachments, already mentioned, a toolpost grinder, and a milling attachment.

The toolpost grinder consists basically of a small high-speed motor that drives a grinding wheel. The grinder is made so that it can conveniently fit on top of the compound of the lathe and be held by the toolpost. The grinder can be used to grind straight cylindrical work, tapers, or bevels by the use of the taper attachment or by use of the compound feed set at the desired angle (Fig. 4-19).

A milling attachment consists of a vertical vise that is used in place of the compound on the cross slide. A milling cutter is mounted in the headstock as shown in Fig. 4-20.

Turret Lathes

Turret lathes have a turret mounted on the ways in place of the tailstock. The turret generally has six faces on which tools can be mounted and can be indexed so that the appropriate tool is brought to bear on the work when required (Fig. 4-21). A square turret is mounted on the front side of the cross slide and a single tool is mounted at the rear. Once the tools are mounted for a given job, their

Figure 4.20. A milling attachment used to cut gears on a lathe. (*Courtesy South Bend Lathe, Inc.*)

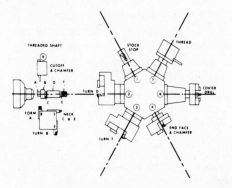

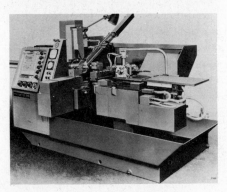

Figure 4.21. A manual or semiautomatic turret lathe usually has two turrets: one at the tailstock end of the bed and on the cross slide. A single tool is shown here at the rear cross-slide position. A fully automatic turret lathe has a tailstock turret with front, rear, vertical, and angular tool slides. (*Courtesy Herbert Machine Tools.*)

travel is controlled by prepositioned stops, thus drastically reducing the time required to produce a part.

Automatic turret lathes have the entire cycle required to produce a part programmed. This of course means longer setup times, but it also means higher production. The automatic controls are governed by hydraulic circuits or numerical control.

Figure 4-22 shows a more detailed view of some of the types of tooling that can be mounted on the tailstock turret.

Numerical Control Lathe. The newest lathe, called a *turning center,* is shown in Fig. 4-23. A major departure in the design of this lathe is that it has

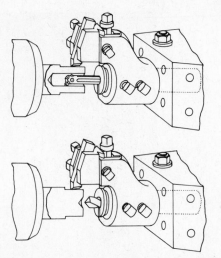

Figure 4.22. The adjustable knee tool can be used for both internal and external machining. The vertical external tool can be adjusted to turn different diameters.

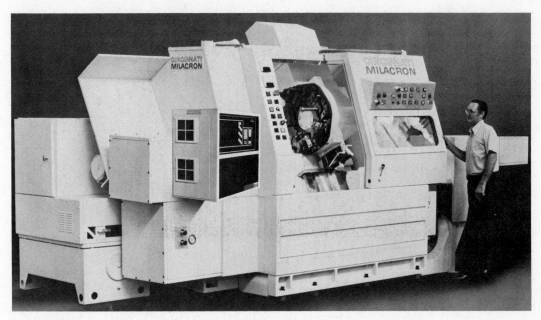

Figure 4.23. A turning center CNC lathe. (*Courtesy Cincinnati Milacron.*)

an overhead or crown turret with seven tools that can be brought into position when needed for external machining (Fig. 4-24). The turret is tilted so that it will not interfere with the chuck. A second turret, a heavy-duty disc-type turret [Fig. 4-25(a)], can be indexed into place when needed for both OD and ID operations. The OD tools are mounted in the outer portion of the tool block and ID tools are mounted on the inner portion of the block. A total of 12 or 14 tools

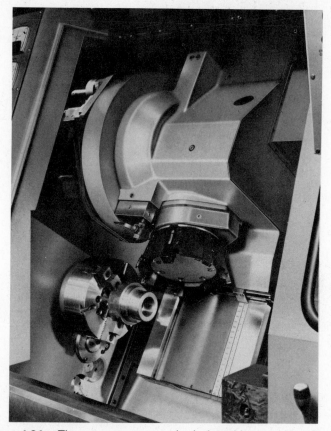

Figure 4.24. The crown turret can be indexed into position for any one of seven different turning tools. (*Courtesy Cincinnati Milacron.*)

can be held in this turret depending on the size of the machine. On long parts a tailstock support can be called up automatically by the NC program [Fig. 4-25(b)]. Figure 4-26 shows a one-piece touch-actuated impervious membrane control panel for the turning center. Three sets of three buttons each grouped to the right of the CRT screen allow the operator to call up various messages describing the steps that walk him through the correct procedures required for setup, operation, editing, and troubleshooting, termed the *menu*. The messages are displayed in simple English on the CRT screen. A 12-button numerical keypad to the right of the menu buttons provides for manual input of dimensions and other numerical values.

Division of the 480-character CRT screen into four display areas speeds communication. A vertical rectangular area on the right side is reserved for displaying the menu format in use. The major portion of the screen is used to display up to seven lines of current and command part program data. Diagnostic information is displayed below this space. During a program run, indicators of

(a)   (b)

Figure 4.25. The disc turret can be used for both ID and OD operations (a). The tailstock can be automatically programmed into position when needed to support long stock for OD turning operations (b). (*Courtesy Cincinnati Milacron.*)

Figure 4.26. The control for the turning center is one-piece touch-actuated impervious membrane operator panel. (*Courtesy Cincinnati Milacron.*)

actual and potential troubles—in the control, machine, or program—appear in this area. The area at the bottom of the screen is reserved for operator entry of data and for display of special instructions for the programmer to the operator. The turning center in the medium-size range can handle bar stock up to $2\frac{1}{2}$ in. dia. (63.5 mm) and up to 40 in. (1016 mm) long.

Vertical Turret Lathes (VTL)   *Single-station vertical lathes.*   Large castings, forgings, and other parts are difficult to mount and hold in chucks or between centers for machining on the horizontal turret lathe. The vertical turret lathe (Fig. 4-27) was developed to overcome some of the difficulties encountered in mounting and holding large workpieces. The tabletop is a combination three-jaw chuck and faceplate. The size of this lathe is given by the diameter of the table, in inches, which usually ranges from 30 to 46 in. The main structure of the machine, in addition to the table, consists of columns that support a crossrail and a side rail. On the horizontal crossrail are mounted one or two vertical slides, one with a turret toolholder, the other with a square toolholder. The side rail supports a square turret used for holding tools to cut the outside diameter of parts mounted on the table. As with the horizontal turret lathe, combined cuts and multiple tooling are possible.

Multistation vertical lathes.   The six- or eight-station vertical lathe or chucking machine is designed for high production. The six or eight independently driven spindles are mounted on a carrier around a stationary column. As the machining operations are completed, the tools retract and the spindle carrier indexes one station. Thus one part is completed with each indexing cycle. These machines have the advantage of requiring only the minimum amount of floor space (Fig. 4-28).

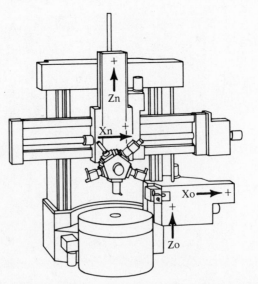

Figure 4.27.   A 4-axis vertical turret lathe with 5-position turret and a side head with a 4-position turret.

Figure 4.28. Multistation vertical turret lathe. (*Courtesy The Bullard Company.*)

HIGH-PRODUCTION LATHES

The borderline between medium- and high-production lathes is not well defined, but the distinction is made here to help the reader consider the machines in terms of relative quantity production. Generally, high-production machines run continuously, with very little operator attention. The amount of production depends on the difficulty of the job, the degree of automation, and the number of tooling stations available. Lathes in this category are automatic turret lathes, single-spindle, automatic-chucking-type lathes, and single- and multispindle automatic screw machines.

Tape-Controlled Automatic Turret Lathes. Most turret lathes are now equipped with automatic indexing devices for the hexagon turret, which minimize the machine- and work-handling responsibilities of the operator. Figure 4-29 shows one type of single-spindle automatic. These machines may be made for bar stock or for individual parts that are automatically loaded and unloaded in the chuck. In this case the machines are often referred to as a "chucker." The machine shown has CNC control (computer numerical control) so that the tooling on the turret can be quickly programmed or reprogrammed as described previously.

Swiss-Type Automatic Screw Machine. The Swiss-type screw machine was first developed at about the same time as the watch industry of Europe to provide rapid production of miniature pinions, screws, and studs. In all Swiss machines the work material is gripped in a collet contained in a sliding spindle that can be cam actuated longitudinally. Five tools are mounted radially around the stock (Fig. 4-30). By coordinating the forward movement of the stock with the cam-controlled radial tool, the desired contour can be generated. Accuracy can be maintained on small diameters to ± 0.0005 in. (0.013 mm).

Cold-finished steel, brass, and aluminum bar stock are traditional materials for screw machine work. Almost all plastic materials in both thermosetting and thermoplastic classes can be machined. Figure 4-31 shows a typical part made by a Swiss screw machine.

Figure 4.29. A single-spindle automatic lathe or chucker. (*Courtesy Lodge and Shipley.*)

Single-Spindle Automatic Screw Machine.   Figure 4-32 shows a single-spindle automatic screw machine. The bar stock, which may be round, square, or hexagonal, is fed through a collet chuck. The stock is automatically stopped at a predetermined distance of feedout from the chuck. The cutting tools are then actuated from the three hydraulically operated cross slides and the seven-station vertically mounted turret. The lower cross slide can be operated as a tracer tool to produce contour cuts.

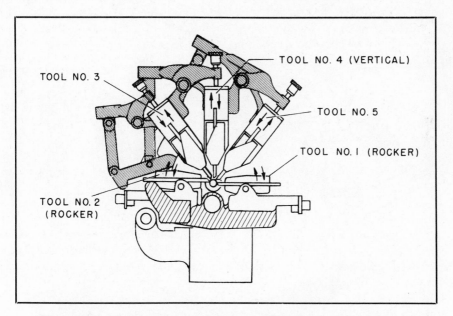

Figure 4.30. Swiss automatic screw machine. Cross-slide tools are grouped around sliding spindle.

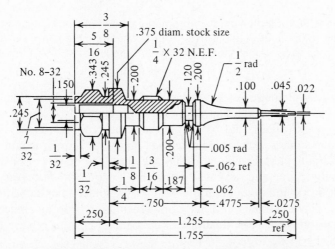

Figure 4.31. Typical part for a Swiss machine.

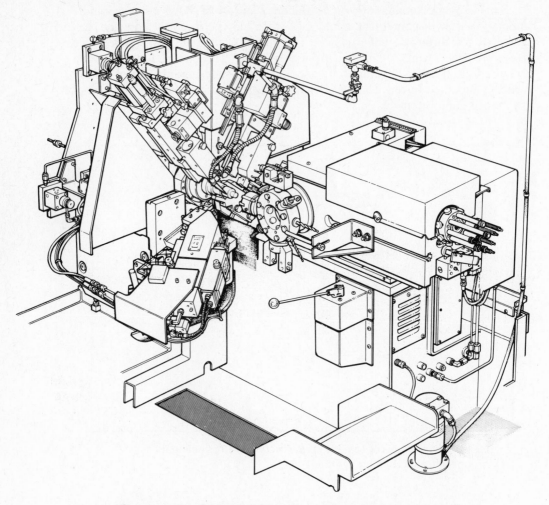

Figure 4.32. Single-spindle automatic screw machine with hydraulic radial slide tooling. The front horizontal slide can be used as a trace tool. (*Courtesy Brown and Sharpe.*)

Multispindle Automatic Screw Machines. The multispindle automatic lathes are built for very high production with four, six, eight, or more spindles with corresponding cross slides (Fig. 4-33). These lathes can be either bar-type or chucking-type machines. If bar-type machines, the stock moves up automatically by cam or hydraulic action. On chucking machines, the blanks or castings may be automatically loaded and unloaded in the chuck. After the necessary cuts are made at each work-holding station all the chucks rotate or index around to a new location. This continues until all workpieces have been machined and the bar stock is cut off, or if individual pieces the chuck opens to release the parts, being ready to be loaded again.

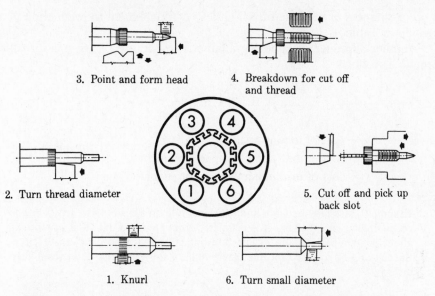

3. Point and form head

4. Breakdown for cut off and thread

2. Turn thread diameter

5. Cut off and pick up back slot

1. Knurl

6. Turn small diameter

Figure 4.33. Tooling for six-spindle automatic lathe used to complete a part in nine operations. (*Courtesy The National Acme Company.*)

NUMERICAL CONTROL (NC) FOR LATHES

Almost any lathe can be adapted for numerical control. In general appearance, most NC lathes look very much like engine lathes or turret lathes or tracer lathes with the exception that the familiar handwheels and levers are largely absent. They, in essence, have been moved to a control panel as was described for the NC turning center.

All NC lathes have access to a number of cutting tools during a machining cycle, and therefore they tend to be more versatile. For example, they can do boring, facing, turning, and backfacing all in one operation, using appropriate tools for each surface machined. Contouring, of course, can be included in any phase of the operation.

On NC lathes, all auxiliary functions (speed and feed changes, spindle direction, turret index, coolant on off) are controlled by the tape or in CNC lathes directly by the computer.

LATHE SELECTION

Because a given part can generally be produced on several lathes, the particular lathe used is selected on the basis of cost per part. Determining the type of lathe to use calls for careful analysis of the conditions such as number of workpieces, similarity of the parts, complexity of the machining (turn, bore, ream, chamfer, thread, etc.), size of the workpiece, and skill of the operators or knowledge of the programmers in the case of NC lathes. A rough guide to production volume is given in Table 4-1.

Another approach to choosing a lathe is by the use of a simple break-even formula as follows:

$$B.E. = \frac{C_t}{C_A - C_B}$$

where: C_t = special tooling cost

C_A = cost of machining per part on machine A

C_B = cost of machining per part on machine B

Example: Assume the labor and overhead rate on a turret lathe is $25/hr and on an engine lathe $15/hr. It is estimated from standard data that it will take about 30 min/part on the engine lathe and 3 min/part on the turret lathe. Special tooling required for the turret lathe for this particular job is estimated at $95.

$$B.E. = \frac{\$95}{(30/60 \times 15) - (3/60 \times 25)} = 15.2$$

Therefore, under this condition, if more than 15 parts are required, it would be more economical to use the turret lathe.

Table 4-1. Lathe selection based on quantity of parts.

Number of Parts	Machine Selection
1–3 parts nonrepetitive work, or 1–10 simple parts	Use a hand-operated engine lathe or turret lathe
1–50 simple* parts or 1–500 or 1000 parts of average complexity 1–on up—complex parts	Use an NC lathe
3–50 repetitive parts where idle time exceeds cutting time (substantially higher for small parts)	Use a hand-operated turret lathe: ram type for smaller parts, saddle type for larger parts
50–5000 parts	Use a single-spindle automatic
500–10,000 parts	Use a multispindle automatic, camless
5000–100,000 parts	Use a multispindle automatic

* Simple part is defined as one having work faces parallel or perpendicular to the spindle centerline, and only 90° angles with single radii at the corners.

Problems

4-1. (a) Figure P4.1 illustrates a pulley. It is sand cast and must be machined in the $\frac{11}{8}$-in. dia. bore and the 38° included V. The material is soft gray cast iron. If this part is made in quantities of a thousand several times a year, what lathe should be used?

  (b) What type of chuck should be used?

  (c) What rpm should be used for turning the OD (not the V) if a carbide tool is used? Assume a 7.5 in. dia.

  (d) What grade of carbide should be used?

  (e) If this were a cast aluminum pulley, what sfs could be used for turning with a HSS tool?

4-2. Could the part shown (Fig. P4.1) be completely machined on an engine lathe?

4-3. (a) Figure P4.2 illustrates a center with two tapers and small bevels. Assuming the part could be held in a chuck, what method would be used to cut the point taper?

  (b) What method would be used to cut the long taper?

  (c) What method would be used to cut the two bevels?

  (d) When the long taper is cut, would the tailstock center be needed for support?

  (e) After the tapered part is cut on the lathe and heat-treated to an R_c hardness of 58 could they be finished further on the lathe? Why or why not?

4-4. A given cylindrical part can be machined either on an engine lathe or on a turret lathe. Ten operations are required. The average time for each operation on the engine lathe is 5 min and on the turret lathe 0.3 min. The special tooling required for the turret lathe costs $160.00. The labor and overhead rate is $18.00/hr on the engine lathe and $25.00/hr on the turret lathe. It takes 30 min to set the job up on the turret lathe. Setup time per part on the engine lathe is included in each operation. How many parts would have to be made before it would pay to set up the turret lathe?

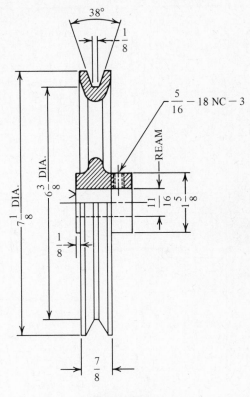

Figure P4.1.

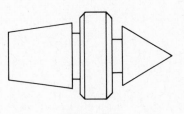

Figure P4.2.

Case Study Questions

1. The part shown in Fig. CS4.1 was machined on a single-spindle 40-hp chucking machine. This workpiece embodies several of the extremes encountered in tooling and programming lathe-turning operations. Some of the important considerations include:

  (a) The surface speeds range from 243 sfpm at bore H to 933 sfpm at periphery A.

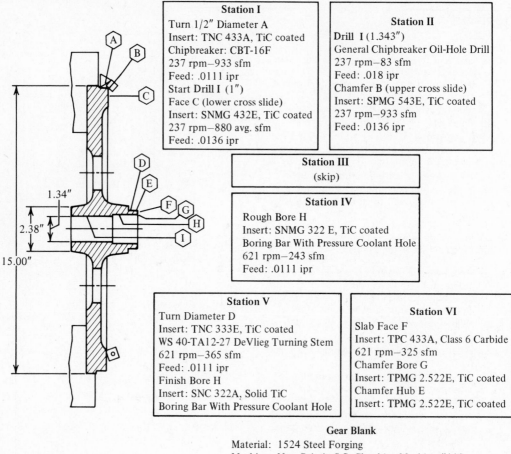

Station I

Turn 1/2″ Diameter A
Insert: TNC 433A, TiC coated
Chipbreaker: CBT-16F
237 rpm—933 sfm
Feed: .0111 ipr
Start Drill I (1″)
Face C (lower cross slide)
Insert: SNMG 432E, TiC coated
237 rpm—880 avg. sfm
Feed: .0136 ipr

Station II

Drill I (1.343″)
General Chipbreaker Oil-Hole Drill
237 rpm—83 sfm
Feed: .018 ipr
Chamfer B (upper cross slide)
Insert: SPMG 543E, TiC coated
237 rpm—933 sfm
Feed: .0136 ipr

Station III

(skip)

Station IV

Rough Bore H
Insert: SNMG 322 E, TiC coated
Boring Bar With Pressure Coolant Hole
621 rpm—243 sfm
Feed: .0111 ipr

Station V

Turn Diameter D
Insert: TNC 333E, TiC coated
WS 40-TA12-27 DeVlieg Turning Stem
621 rpm—365 sfm
Feed: .0111 ipr
Finish Bore H
Insert: SNC 322A, Solid TiC
Boring Bar With Pressure Coolant Hole

Station VI

Slab Face F
Insert: TPC 433A, Class 6 Carbide
621 rpm—325 sfm
Chamfer Bore G
Insert: TPMG 2.522E, TiC coated
Chamfer Hub E
Insert: TPMG 2.522E, TiC coated

Gear Blank

Material: 1524 Steel Forging
Machine: New Britain S.S. Chucking Machine #118
               40 hp
               18″ Buck Chuck
               9-pound/jaw (3)
               Centrifugal Force: 1100 lb/jaw @ 621 rpm
                                          1800 lb/jaw @ 805 rpm
               (10,000 pound drawbar force @ 40 psi air)

Figure CS4.1.

(b) A maximum of 621 rpm is used for small boring and turning. Although a maximum of 805 rpm is available from the machine tool, this speed was discontinued because it caused varying imbalances in the part that created vibrations and a questionable safety condition.

(c) TiC-coated inserts and solid titanium inserts are used for the lighter cuts and offer a better performance at low surface speeds because of the lower affinity for the builtup edge.

A. Determine which of the six stations of the turret operations requires the longest time. Add the following dimensions: Face *C* = 0.875 in. Total bore length = 3.750 in. Rough bore *H* = 1 in. length. *D* = 0.875 in. Slab facing and chamfering operations do not involve tool-feed motion.

B. Could the longest operation be

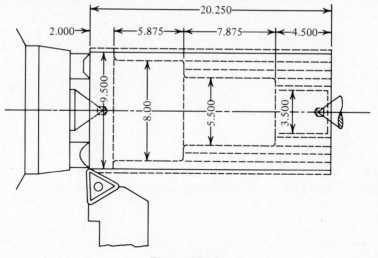

Figure CS4.2.

speeded up? Tell why or why not.

C. What other lathes could be used for this part, assuming low, medium, high, and very high production?

2. Figure CS4.2 shows a shaft that has to be machined over its entire length. Previously a chuck had been used to drive the part on the large end. This required two operations since the part had to be rechucked on the small end so that a finishing cut could be made on the large end. A solution to the problem was to use face-driving centers so that the part can be machined over the entire length. Feeds are limited by the diameter of the driver and the amount of tailstock pressure that can be applied. Speeds are usually not limited to any significant extent.

The company wanted to turn this part out in the least amount of time possible. They decided to use ceramic inserts and run at 3000 sfpm. The maximum depth of cut was $\frac{1}{4}$ in. Roughing feeds were 0.0111 ipr and a final finish cut is made over the entire surface at 0.005 ipr.

A. What is the machining time required for rough and finish machining? Note that rpms should be rounded off to the nearest 100. Use the average rpm for the finish cut over the entire surface. The original stock dia. was 10 in.

B. What horsepower would have to be available at the spindle for the first roughing cut if the material machined is carbon steel with a BHN of 240?

C. If the cuts were limited to 0.125 depth, what horsepower would be required?

D. What lathe would be used for (a) low production and (b) medium production?

Bibliography

Bittence, John C. "Understanding Machine Tools." *Machine Design* 46 (August 22, 1974): 82–87.

Gettleman, Ken. "Non-NC Should Be Programmed Too." *Modern Machine Shop* (October 1976).

CHAPTER FIVE

Hole Making and Sizing Operations

The bow drill (Fig. 5-1), one of man's most ancient tools, was developed about 3800 B.C. It shows the early use of rotary motion. The upper cap or socket was supported by one hand, which left the other hand free to work the bow. The bow was used not only for drilling but also for starting fires.

The standard drill is the most common method of originating holes in metal; however, other processes such as punching, electric discharge machining, and piercing with an oxyacetylene torch and laser are also used. If accuracy is critical, the drilled hole is often followed by boring and reaming.

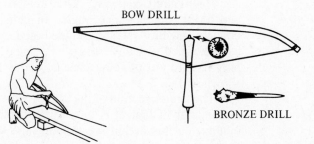

BOW DRILL

BRONZE DRILL

Figure 5.1. The bow drill was used in ancient times.

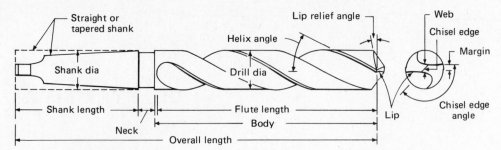

Figure 5.2. The design and nomenclature of a standard twist drill.

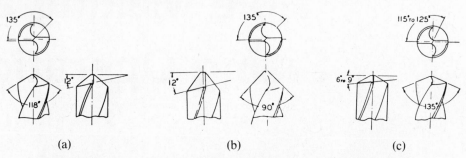

(a) (b) (c)

Figure 5.3. Drill-point geometry for various materials: (a) standard, (b) for soft materials, (c) for hard materials.

DRILLING

There are now a variety of drills that have been developed to serve special purposes in the metal-cutting field. These are twist drill, half-round, spade, indexible insert, gun drill, and trepanning.

Twist Drill. The twist drill (Fig. 5-2) is manufactured in standard sizes from 0.0059 in. to $3\frac{1}{2}$ in. and is sized according to four commonly used systems: fractional inch, number (or wire gage), letter, and metric. Common metric size drills range from 0.4 mm (0.0157 in.) to 69 mm (2.716 in.).

Drill-Point Geometry. The standard drill point has a 118° included lip angle and a $12\frac{1}{2}°$ lip-clearance angle. In general, soft material can be drilled more efficiently with a smaller included angle; that is, bakelite, wood, or hard rubber can be drilled with a 90° drill point. The opposite is also true. Very hard materials may require a drill-point angle of about 135° and 6 to 9° lip-clearance angle, as shown in Fig. 5-3.

The center section of the drill or web does not produce a good cutting action; as a matter of fact, it is more an extruding action than cutting. Two means of minimizing this action are thinning the web and drilling a lead hole. Thinning the web (Fig. 5-4) may be done freehand by tool-grinding personnel or, if many drills are to be ground, with the aid of a fixture to provide accurate centering.

Figure 5.4. A standard drill point showing a thinned web.

Standard chisel point after web thinning

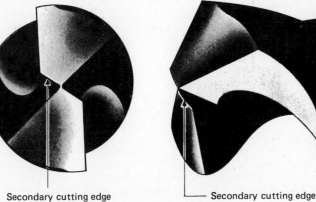

Secondary cutting edge — Secondary cutting edge

Figure 5.5. The split-point drill virtually eliminates the web effect at the cutting edge.

A lead hole is a small-diameter hole drilled before the desired-size drill is used. It eliminates the section that would otherwise be subjected to the extruding action of the web.

Modifications in drill-point geometry have been developed. Shown in Fig. 5-5 is a split-point drill. This point creates a secondary cutting edge that virtually eliminates the web effect at the point. The split point was developed for drilling high-strength materials with heavy-web drills. Because the splitting cuts are rotated toward the main cutting edges, the intersection of the two cutting edges is stronger. The geometry also gives an improved centering action.

Shown in Fig. 5-6 is the action of a conventional drill as it breaks through the bottom side of the material as compared to the radial-lip grind. The radial lip produces a smooth breakthrough as compared to the shock of a conventional point. Proponents state this drill point has four to ten times longer life and that the holes produced are smooth, round, on size, and burr-free. The point is

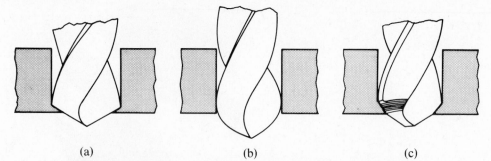

(a)   (b)   (c)

Figure 5.6. The cutting action of a conventional drill point (a), and a radial lip (b), as it breaks through the material, is shown schematically. The double angle drill (c) is often used on cast iron.

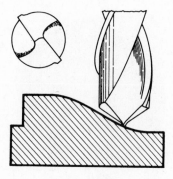

Figure 5.7. The spiral-point drill is used to accurately start a hole, without previous preparation.

patented by the Radial Lip Machine Corporation and is based upon grinding the drill cutting lips to the shape that most evenly distributes wear over the entire cutting area. A drill, similar to radial lip, is a double-angle point, (c), which is often used in cast iron to compensate for premature corner wear.

Another drill point ground on special drill-pointing machines is the "spiral point." The special S-shaped curve (Fig. 5-7) provides a high point at the center that eliminates wandering (moving off from center) upon starting without a guide bushing. The drill is conventional in other respects.

Half-Round Drill. The half-round drill is a hole-making tool of relatively simple geometry (Fig. 5-8). These drills are often used on screw machines but may be used to good advantage in other applications. They are available in sizes ranging from 0.003 in. or smaller up to 1 in. The half-round drills have no chisel edge, but rather a conical point on dead center. Thus they will accurately start in position without a guide bushing and they have no tendency to *walk* or *skate*. That is, the drills tend to start and run true to center. They exceed the capability of two-lipped drills on diameter tolerances and internal finish.

What is lacking is that the drill does not have a means of bringing the chips out of the hole; therefore these drills are held below a diameter-to-depth ratio

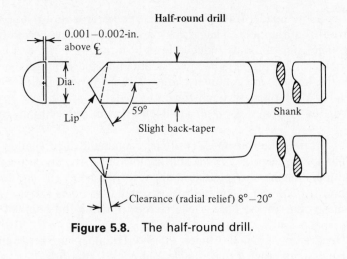

Figure 5.8.   The half-round drill.

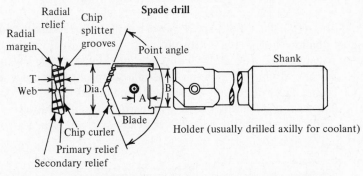

Figure 5.9.   The spade drill. (*Courtesy* American Machinist.)

of less than 10 : 1 in the horizontal position and 2 : 1 or 3 : 1 in the vertical position.

Spade Drills.   The spade drill is made by inserting a spade-shaped blade into a slot in a bar and fastening it securely with a screw, as shown in Fig. 5-9. The use of these drills has risen sharply in the past decade. The major reasons are their efficiency in making holes in the 1.00 to 10 in. (25.40 to 254.0 mm) dia. range and the lower cost. Over 3.5 in. (88.90 mm) in diameter, they are the only drills made as stock items. A standard 2-in. (50.80-mm) diameter twist drill costs about $220, whereas a 2 in.-diameter (50.80-mm) spade drill, including the shank and blade, is about $125. The shank can be used with more than one size tip, so it may cut the cost of a 2-in.-spade drill to about $50. Standard blade material is M2 or M3 HSS and carbides. Some advantages of spade drills are: (1) rigidity— they are capable of taking heavy feeds; (2) relatively low cost—only the insert need be replaced; (3) capability of deep-hole drilling (up to the length of the blade holder) with low-pressure coolant; and (4) chips are broken up at the cutting edge.

It is essential in spade drilling that the chips be broken up so they can be easily removed from the hole. That is the reason for the chip-breaking features of the point. The chips must be reduced in size so that they can exit around the head of the holder. This problem becomes acute in deep drilling smaller diameter holes where the holders are relatively large. Except in the shallowest holes, a generous supply of coolant through the tool holder is mandatory to flush out chips. In deep holes, compressed air at 40–60 psi is introduced into the coolant line.

The use of drill-guide bushings is often recommended in spade-drill applications to enhance the accuracy of the starting location. An alternative procedure is to start with a short-holder spade drill and after the hole is 1 dia. deep switch to a longer holder. The use of centerdrills or small lead holes should be avoided unless the sharp edge of the hole is chamfered to match the point of the point angle on the spade drill.

Heavy feedrates—0.015–0.050 ipr (0.38–1.27 mm)—are recommended in order to produce chips that do not tangle and pack in the hole.

Indexible-Insert Drills.   Industry's newest drill is the indexible-insert drill for producing shallow holes (Fig. 5-10). In addition to providing the economic benefits of throwaway inserts, along with easy indexing for new cutting edges, the use of these new drills provides the capability of producing holes at carbide speeds. This is especially important in the use of numerically controlled machines where the tooling turrets generally bristle with carbide-tooling stations. Currently the indexible drills are available in diameters ranging from approximately $\frac{5}{8}$ in.

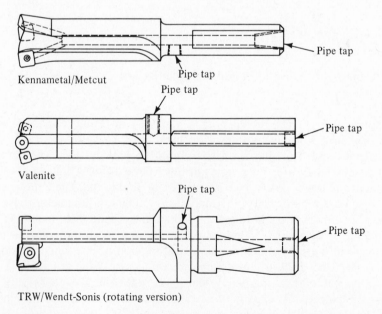

Kennametal/Metcut — Pipe tap

Valenite — Pipe tap

TRW/Wendt-Sonis (rotating version) — Pipe tap

(a)

Figure 5.10.   Indexible-insert drills, both straight and helical-flute types. (*Courtesy Valenite.*)

(b)

(c)

Figure 5.10. (continued).

to 3 in. (15.87–76.20 mm). Usually the depth to diameter ratio is limited to 2 : 1 to 3 : 1.

Various insert shapes are used in making the drills by different manufacturers. However, the drills are generally two-fluted with two or three inserts fixed to the cutting end. One insert cuts to the center, one cuts the ID of the hole, and if a third insert is used (usually on the larger sizes), it serves to fill the cutting gap between the first two.

A unique feature of the indexible drills is the capability to be used in boring operations. The surface quality of holes produced by the indexible drills is somewhat rough but it is possible on an NC lathe with a cross-sliding turret, for example, to step the tool radially outward and make a second pass—boring the hole to a final diameter and providing a much-improved ID finish. In fact one manufacturer states the indexible drill is a multifunction tool that can be used on an NC lathe for drilling, boring, ID contouring, facing, and OD turning and contouring. In the latter case, the tools are made 0.030 in. (0.76 mm) smaller than the nominal size to allow for secondary boring.

Typical speeds for starting conditions are 300–500 sfm in free-machining steels and cast iron, 150–300 fpm in alloy steels and austenitic stainless steels,

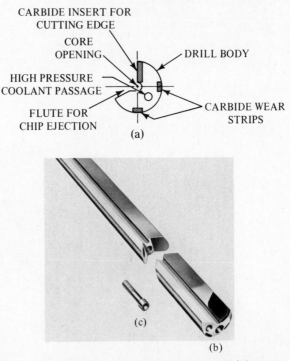

Figure 5.11. Cross section of a typical gun drill (a), and a quick detachable end (b). A bushing is used for accurate starting of the drill (c). (*Courtesy National Twist Drill Inc., Division of Lear Siegler, Inc.*)

and 500–1000 sfm or more in aluminum. Feedrates are about the same as those used on twist drills.

Gun Drills. Gun drills (Fig. 5-11) differ from conventional drills in that they are usually made with a single flute. A hole provides a passageway for the pressurized coolant, which serves as a means of both keeping the cutting edge cool and flushing out the chips. Gun drills may be used to shallow depths, as shown in Fig. 5-12, or to depths of up to 130 dia. Even though the drill can be made to advance rapidly (from 15 to 60 in./min) (381.0–1524 mm) good tolerance and finish can be maintained. This is due in part to the high (60 to 1500 psi) cutting fluid pressure. Some parts that commonly require deep-hole drilling are gun barrels, camshafts, crankshafts, machine tool spindles, and connecting rods. Standard gun drills are made in diameters from 0.078 in. (2 mm) to 2 in. (50.80 mm) or more.

Tolerances that can be held in gun drilling under $\frac{1}{2}$ in. are 0.0005 in. (0.013 mm) total, and over $\frac{1}{2}$ in. (12.70 mm) the tolerance should not exceed 0.001 in. (0.03 mm). Because of the burnishing effects of the guide pads (hardened strips in drill body), hole finishes can be produced to 5–12 microinches under favorable conditions. The hole deviation should not exceed about 0.002 in. (0.05 mm) in a 4-in. (101.60-mm) depth at any diameter and can in most cases be held to 0.001 in. (0.03 mm) per ft.

The basic setup for gun drilling requires a drill bushing close to the work

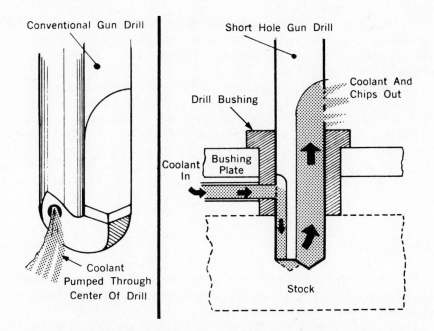

Figure 5.12. The gun drill as used for short holes.

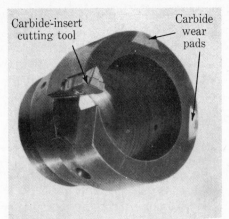

Carbide-insert cutting tool

Carbide wear pads

Figure 5.13. A trepanning head used with inserted wear pads. It is threaded on the end for inserting into a drive tube. (*Courtesy* The Tool and Manufacturing Engineer.)

surface as shown in Fig. 5-11(c). With carbide cutting edges, cutting speeds are high. An rpm of 10,000 is not unheard of for gun drilling small holes in easily machined alloys.

Trepanning.   Someone in the drilling business, who was producing holes $1\frac{3}{4}$ in. (44.45 mm) and larger, wondered why all this material should be reduced to chips? So the trepanning drill was invented. In trepanning, only an annular ring is reduced to chips (Fig. 5-13). Some very impressive holes can be produced by this method with diameters well above 24 in. (609.6 mm) and a depth-to-diameter ratio exceeding 100 : 1.

A problem with trepanning is the core remains if the hole is not cut all the way through. One solution is a special core-cropping tool that incorporates a pivoting cutoff blade that is fed across the base of the core to produce a groove like that of a lathe cutoff tool.

On large holes, trepanning offers considerable advantages. A 6-in.-dia. (152.40-mm) hole 10 in. (254 mm) deep requires the removal of 282.6 cu in. (7050 mm cu) of metal in chips by a conventional drill. A trepanning drill using a $\frac{3}{4}$-in. (190.50-mm) wide cutter would remove only 126.6 cu in. (3165 mm cu) of metal chips.

Typical penetration rates are about $2\frac{1}{2}$ ipm (63.50 mm) for a $1\frac{3}{4}$-in. (44.45-mm) hole and about $\frac{3}{4}$ ipm (19.0 mm) for a 12-in.-dia. (304.57-mm) hole in mild steel.

DRILL MATERIALS AND COST

Materials.   Comparatively few metal-cutting drills are now made from low-capacity carbon steel. Production demands have made high-speed steel, carbide-tipped, and solid-carbide drills imperative.

The high-speed steels M1, M2, M8, M10, and T1 are most commonly used. These steels combine toughness, edge strength, and wear resistance with good red hardness. When these prove inadequate, as in drilling stainless steel, special high-speed steels containing more vanadium and carbon may be used.

Carbide-tipped drills give excellent results in drilling cast iron, highly abrasive materials, and certain nonferrous alloys.

Carbide-tipped and solid-carbide drills do not give good performance unless the setup is extremely rigid, the spindle is precisely aligned, and there is a minimum of tool overhang.

Cost Comparison.   An example of the cost comparison for three different $\frac{3}{16}$-in.-dia. drills: HSS, 48¢; carbide-tipped, $4.10; solid-carbide, $11.70. Although initial cost of the three drill materials shows a wide disparity, the higher cost carbide drill may be the right one to lower hole costs on many applications.

Horsepower Requirements

Horsepower required for drilling is made up of torque and thrust.

One might expect that the drilling torque or moment would be proportional to the amount of metal removed per revolution, which would be $(\pi/4)fd^2$, where: f = feed (ipr) and d = drill diameter. However, it is not. It has been found to be closely related to the empirical relationship of $(\pi/4)f^{0.8}d^{1.8}$. This means that doubling the feed per revolution will increase the torque by a factor of 1.75 instead of 2.

Thrust forces are important when determining drill, jig, and machine rigidity but are only 2% or less of the total power requirement and are often disregarded.

Horsepower requirements for drilling operations can be determined even though torque measurements are not available by the use of unit power data given in Table 3-3 and the formula (Table 3-4).

Drilling Speeds and Feeds

Due to the drill geometry, the cutting speed for drilling should be reduced to about 80% of that used in turning. It is difficult to provide efficient cooling for the tip of the drill except in the case of gun drills or oil-hole drills. Oil-hole drills provide a passage for coolant from the shank to the tip, as shown in Fig. 5-14.

Feeds are dependent on cutting conditions and what the drill will take without deflection. A rule of thumb for feed is to make it $\frac{1}{100}$ of the drill diameter. Thus a 0.5-in. drill would have a recommended feed of 0.005 ipr. Shown in Tables 5-1 and 5-2 are speeds and feeds used for various materials with HSS drills; Appendixes B and C include speeds for both HSS and carbide tools. These figures should be reduced to 80% for drilling.

Drilling Equipment

Drilling equipment ranges from the portable hand drill to large multispindle machines capable of driving hundreds of drills at one time. The main types are sensitive, radial, turret, multispindle, and individual drilling units.

Sensitive and Upright Drill Press.   The most common type of drill press is a hand-feed type (Fig. 5-15). The operator is actually sensitive to how fast the drill is cutting and can control it according to the conditions of the moment.

Table 5-1. Speeds for hss twist drills. (*Courtesy Cleveland Twist Drill.*)

Material	Speed, sfm
Aluminum alloys	200–300
Brass, bronze	150–300
High-tensile bronze	70–150
Zinc-base diecastings	300–400
High-temp alloys, solution-treated & aged	7–20
Cast iron, soft	75–125
medium hard	50–100
hard chilled	10–20
malleable	80–90
Magnesium alloys	250–400
Monel or high-Ni steel	30–50
Bakelite and similar	100–300
Steel, 0.2%–0.3% C	80–100
0.4%–0.5% C	70–80
tool, 1.2% C	50–60
forgings	40–50
alloy, 300–400 Bhn	20–30
High-tensile steel, heat treated to 35–40 Rc	30–40
heat treated to 40–45 Rc	25–35
heat treated to 45–50 Rc	15–25
heat treated to 50–55 Rc	7–15
Maraging steel, heat treated	7–20
annealed	40–55
Stainless steel, free machining	30–100
Cr-Ni, nonhardenable	20–60
Straight-Cr, martensitic	10–30
Titanium, commercially pure	50–60
6AI-4V, annealed	25–35
6AI-4V, solution-trt & aged	15–20
Wood	300–400

Table 5-2. Feeds for hss twist drills in mild steel. (*Courtesy Cleveland Twist Drill.*)

Diameter, in.	Feed, ipr
Under $\frac{1}{8}$	0.001–0.002
$\frac{1}{8}$–$\frac{1}{4}$	0.002–0.004
$\frac{1}{4}$–$\frac{1}{2}$	0.004–0.007
$\frac{1}{2}$–1	0.007–0.015
1 and over	0.015–0.025

Figure 5.14. An oil-hole drill provides a passageway for coolant to the tip, where it can be used effectively. (*Courtesy Morse Cutting Tools.*)

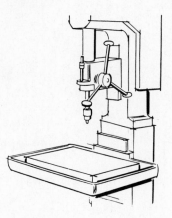

Figure 5.15. The sensitive-type drill press is characterized by hand feed.

Upright drill presses are very similar to the sensitive type except they are equipped with a power feed. This machine is usually of heavier construction and is suited to a wide range of work. It may be equipped with a universal table that allows the table to be accurately positioned both longitudinally and laterally.

Radial Drill Press. Large, heavy work pieces are difficult to reposition for each succeeding hole. The radial drill press (Fig. 5-16) makes it easy to reposition the drilling unit, which is independently powered. The size of a radial drill is given by the radius, in feet, of the largest plate in which a center hole can be drilled and the size of the supporting column in inches.

Turret-Type Drill Presses. When a number of drill sizes or other tools such as reamers and taps are required to complete a part, a turret-type drill is very useful. Any one of six, eight, or ten tools can be quickly indexed into place (Fig. 5-17). Many drill presses of this type are now numerically controlled.

Multispindle Drilling Heads. A multispindle drilling head may be attached to a single-spindle machine to provide a means of drilling many holes at one

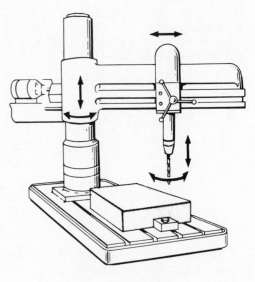

Figure 5.16. The radial drill press.

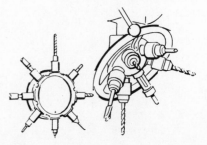

Figure 5.17. The turret-type drilling machines have six, eight, or ten tools ready to use.

time. These heads are of two main types: adjustable, for intermediate production; and fixed, for long, high-production runs.

The adjustable head [Fig. 5-18(a)] is driven by a gear from the spindle through a universal-joint linkage to each drill. Another variation is the oscillator head [Fig. 5-18(b)], which allows drill placement anywhere within the head area and as close as the sum of the two drill-holder diameters.

For really high-volume production of complex components requiring many holes on several faces of each part, more specialized machines are frequently justified. These may be single-station machines with multispindle heads advancing into the workpiece from different sides, or multistation machines in which the part is transferred in either linear or rotary fashion for single or multiple operations (Fig. 5-19).

Headchanger Machine. A newer type of machine that has just begun to emerge in recent years is often used for drilling operations in automated machines. This machine has individual heads, generally multispindle, for hole-making applications (Fig. 5-20). The heads are changed automatically for sequential operations on workpieces at a single station. The headchanging machines are a logical cross between NC machining centers and moderately high production transfer machines.

Self-Contained Drilling Units. Self-contained drilling units consist of a power source, a feeding arrangement, and a chuck. These units come in a wide range of sizes, from small air-powered units, as shown in Fig. 5-21, to large

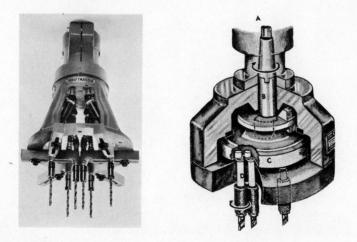

Figure 5.18. Two types of adjustable, multiple-spindle drilling heads. The spindle at (a) has adjustable arms driven by universal joints from gears in the head. The spindle at (b) transmits power through the drill sleeve (A) which drives the crank (B) and oscillator (C). The oscillator turns the individual spindles (D) in the same direction and speed as the drive crank.

Figure 5.19. Specials are often used in large-volume manufacturing industries. This multispindle, multihead Drillmation machine is for drilling tractor differential cases. (*Courtesy Wickes Albion Machine.*)

Figure 5.20. Shown is a 28-spindle drillhead used to completely machine a family of disc-brake housings on an Automatic Toolhead Changer Machining Center. This type of machine is much faster and more accurate than single-spindle machines because of the fixed relationship between tools that is built into multispindle toolheads. (*Courtesy Cincinnati Milacron.*)

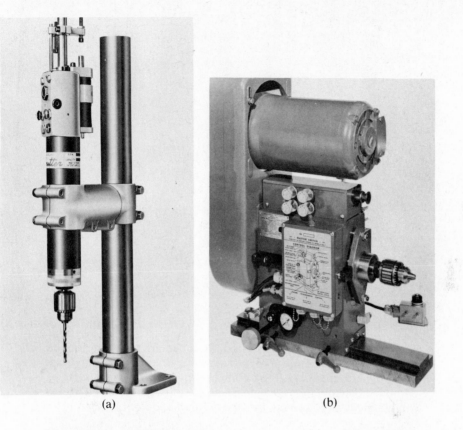

(a)   (b)

Figure 5.21. Individual drill units range from small air-powered units (a), to large electric-motor types (b). (*Courtesy Desoutter Corp. and Universal-Automatic Corp.*)

electric-drive motor types. These units are assembled into specialized automation equipment, as shown in Fig. 5-22.

Drill Jigs

Large quantity production requires quick and accurate hole location. One method of accomplishing this is with the aid of a drill jig (Fig. 5-23). The part to be drilled is placed in the jig against a number of locating pins. Clamping holds the work tightly against these pins to maintain accuracy of the part. A well-designed drill jig will provide for chip clearance and removal. It also provides for quick and easy loading and unloading and foolproofing. The hardened and ground drill bushings guide the drill to the precise location.

Drill jigs used for NC machines have become simpler. Standardized components such as tooling plates and blocks are made available. The plates and blocks are made to fit on an NC drill press table and are made with T-slots (for

Figure 5.22. Individually powered drill units are mounted around a four-station indexing table for automatic drilling, counterboring, tapping, and reaming. Programmed-sequence selector switches permit single-spindle operation on any two, three, or four spindles simultaneously. (*Courtesy Universal-Automatic Corp.*)

Figure 5.23. The drill jig provides for accurate hole location.

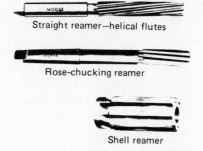

Straight reamer—helical flutes

Rose-chucking reamer

Figure 5.24. Hand, machine, and shell reamers. (*Courtesy Morse Cutting Tools.*)

Shell reamer

easy bolt location) and numerous tapped holes to make fastening workpieces very much simpler. Also with numerical control short, stocky tools may be used for accurate hole starting followed automatically by tools of the desired size and length.

Reamers

Reamers are used to size a hole and provide a good finish. Hand, machine, and shell reamers are shown in Fig. 5-24. Hand reamers are recognized by their square shank, whereas machine (or chucking) reamers have a round or tapered shank to fit into the machine spindle. The shell or hollow reamers are used on arbors that have driving lugs. Other common types of reamers are tapered, tapered-pin, and adjustable.

To work efficiently, a reamer must have all teeth cutting. A proper stock allowance is very important. For holes up to 0.5 in. (12.7 mm) in diameter, 0.015 in. (0.38 mm) is allowed for semifinish reaming on most metals. For holes over 0.5 in. in diameter 0.030 in. (0.76 mm) is recommended. For finish reaming, the allowance is 0.005 in. (0.13 mm) for holes less than 0.5 in. (12.7 mm) and 0.015 in. (0.38 mm) for holes greater than 0.5 in. in diameter. The tolerance expected is normally ±0.0005 in. (0.013 mm).

Drilling Accuracy

Drilling accuracy may be considered with regard to hole location, size, and geometry. The location will be determined by starting conditions. A center punch mark often serves as the drill location. This of course is subject to the skill of the layout man and the drill press operator. In production work, where drill bushings guide the drill, a high degree of accuracy is obtained. The drill bushing should be no more than 0.0005 in. (0.013 mm) larger in diameter than the drill. Excessive clearance permits the drill to wander until a hole is established. This results in oversized, tapered, or bell-mouthed holes. A drill that has unequal lip lengths will also produce oversize holes. Short-hole gun drills as previously shown can produce holes with a location tolerance of ±0.0005 in. (0.013 mm) and a size tolerance of ±0.0002 in. (0.005 mm).

BORING

Up until about 1775 there was not a great demand for accuracy in bored holes. Cannons were using cannon balls in the as-cast condition, so the carefully cored holes, after being cleaned out, proved satisfactory. However, the coming of the steam engine put much higher requirements on machining capabilities. Fortunately, Wilkinson, Watt, Smeaton, and others accepted this challenge and began designing and building machines of ever-increasing accuracy. Today's boring machines are able to work to within a few ten-thousandths of an inch on a production basis.

Boring is the process of using a single-point tool to enlarge and locate a previously made hole. Drills tend to wander or drift, thus, where greater accuracy is required, drilling is followed by boring and reaming. Shown in Fig. 5-25 is a sketch of how the centerline of a hole may be shifted by means of a boring operation. The single-point boring tools are usually made out of HSS or have carbide inserts, as shown in Fig. 5-26. This bar has an adjustable dial that allows it to be set to within 0.0001 in.

Boring Machines.   The main types of boring machines are the horizontal and the vertical, as shown in Fig. 5-27. The horizontal machine is well adapted to *in-line boring,* boring two or more holes on the same axis, as shown at Fig. 5-27(a). In this case, the boring bar is supported at the outboard end. Machines of this type also are used for milling and drilling operations, as shown at Fig. 5-27(b). The vertical boring mill is used to machine large, heavy parts that are difficult to mount in any other position. Vertical boring mills range in size from about 54 in. (1371.6 mm) to 40 ft (12 M) in diameter (table diameter).

Jig-boring machines (Fig. 5-28) get their name from the fact that they are so often used to drill and bore the holes needed for drill jigs. Outside of grinding,

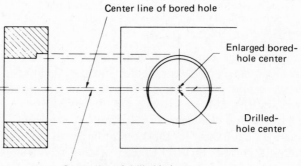

Figure 5.25.   A boring tool is used to enlarge and correct the hole location.

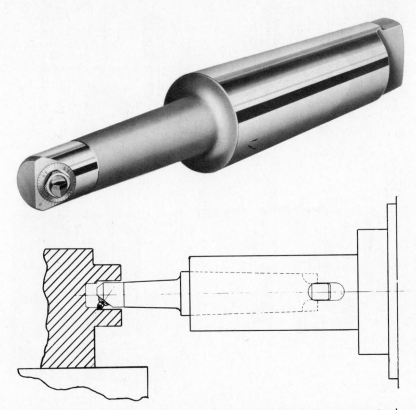

Figure 5.26. Boring bars are often equipped with accurate tool-adjustment settings as shown. Boring tools may also be used for undercuts or for internal threading.

they are considered to be the most accurate machines of the machine shop or toolroom. They can be used to locate holes to an accuracy of ±0.0001 in. (0.003 mm).

HOLE PUNCHING

Hole punching is, in many instances, much more economical than drilling. This is especially true because of the heavy-duty hydraulic punches that have been developed. Punches of the type shown in the sketches of Fig. 5-29 may be either portable or stationary and are capable of punching 1-in. diameter holes in steel ranging from 12 gauge up through $1\frac{1}{16}$ in. thick (26.30 mm). A wide variety of hole configurations can be made.

The biggest savings in utilizing hole-punching equipment is in the time saved per hole. As an example, a 1-in. diameter hole can be punched through a 1-in.

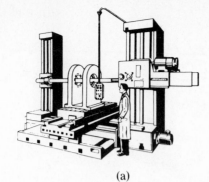
(a)

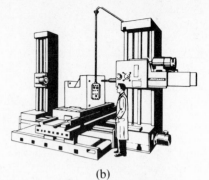
(b)

(c)

(d)

Figure 5.27. Horizontal and vertical boring machine. Basic operations performed on horizontal boring machine are boring (a), drilling (b), and milling (c). (*Courtesy Giddings and Lewis Machine Tool Co.*)

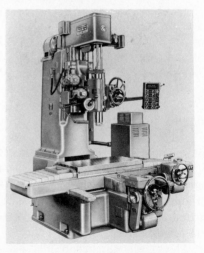

Figure 5.28. A jig-boring machine. (*Courtesy Pratt and Whitney, Division of Colt Industries.*)

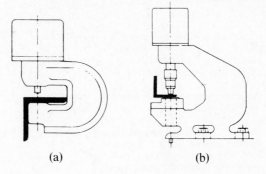

Figure 5.29. Two types of hydraulic punching units. Shown at (a) is a portable unit and at (b) is a stationary type bolted to rails or a base.

(a)   (b)

thick (25.40-mm) steel plate in approximately 2 sec, compared to about 46.8 sec required for drilling.

Oxyacetylene Flame-Cut Holes.   A flame-cut hole may be of any desired size. An oxyacetylene torch can be used to cut holes quite accurately, $\frac{1}{32}$ in. tolerance, in thin sheet and within $\frac{1}{16}$ in. in 2-in. plate. For most purposes the flame-cut hole need not be resurfaced or resized. Flame cutting is, however, restricted to iron-base metals with alloy constituents totaling less than 10%. Thin sheets can be cut at the rate of 28 ipm but 24-in. thick plate requires 2 ipm. Steel plates thicker than 4–6 in. are often pierced with an oxygen lance to obtain a starting hole.

Laser Hole Cutting.   The laser creates a hole by vaporizing a small amount of workpiece material. As the laser beam strikes the surface of the material, localized heating occurs at rates up to 10^{10} degrees per second. As the surface material vaporizes, the beam penetrates further and the pressure within the hole, due to the vaporization, rises to 10^2 to 10^3 atmospheres. A jet of vaporized material then flows out of the hole at supersonic velocity causing further erosion of the hole.

Industrial YAG and ruby lasers are primarily suited to drilling holes 0.005 in. to 0.050 in. (0.13 to 1.27 mm) in dia. in metallic parts up to $\frac{3}{8}$ in. (9.52 mm) thick within a depth to dia. range of 20 : 1.

Electric Discharge Machining (EDM) Hole Making.   When introduced about 20 years ago, EDM was the answer for complex holes or shapes in hard-to-machine material. For an explanation of the process see Chapter 20. EDM is no longer limited to "impossible" jobs. It is now used on the production floor for cutting intricate designs and contours, narrow slots, and blind cavities that are difficult or impossible with conventional machines. A typical application is the production of a plastic extruding die, where 384 holes, 0.093 in. (2.36 mm) in dia. are cut in $\frac{3}{16}$-in. (4.75-mm) carbide plate in 11 hr.

Problems

5-1. Would an upright drill press equipped with a 2-hp motor be large enough to operate a $\frac{3}{4}$-in. dia. drill to make holes in low-carbon steel if the feedrate is 0.010 ipr?

5-2. What torque would be required at the spindle for Problem 5-1?

5-3. How long would it take to pay for a multi-spindle drilling head to drill five $\frac{1}{2}$-in. holes in a $\frac{3}{4}$-in. thick soft-cast iron pipe flanges under the following conditions:

Individual layout − center punching and drill alignment = 4 min/hole
Drilling time/hole = $\frac{1}{2}$ min

A multispindle head can be used to drill all five holes at once. How many parts would be needed if the multispindle head cost $2500? The total time required for positioning and drilling all five holes is 2 min. The labor and overhead rate is $25/hour.

5-4. (a) What horsepower would be required at the spindle for the multispindle head mentioned in Problem 5-3?

(b) What size motor would be required? Assume the drills will not always be sharp.

5-5. Compare the drill rpm for 2-in. dia. HSS drill and indexible carbide drill for cutting alloy steel, 300–400 BHN.

5-6. Approximately how much faster would a spade drill be for drilling a 2-in. dia. hole in mild steel than an HSS drill?

5-7. (a) Which of the following drills could be used to drill a hole $\frac{3}{4}$ in. in dia. 5 in. deep: HSS twist drill, half-round drill, spade drill, indexible insert gun drill, or trepanning drill? The drilling is done in the vertical position.

(b) State why each of the drills not listed in answer (a) are not able to be used.

5-8. Compare the penetration rate of 2-in. dia. carbide-tipped spade drill and a carbide-tipped trepanning drill in mild steel.

5-9. Fifty small-cast iron parts approximately 3 × 4 × 6 in. must have a total of 30 holes drilled in five of the surfaces. There are about 6 holes per side ranging in size from $\frac{1}{8}$ in. to $\frac{1}{2}$ in. State what equipment would be best suited for this job. State why you chose the equipment you did.

5-10. How does the cutting speed in sfm compare in drilling a piece of high-tensile-strength steel that is heat-treated to 35–40 R_c as to the same steel heat-treated to 50–55 R_c?

5-11. What would the advantages and disadvantages of drilling the 50–55 R_c steel be compared to 35–40 R_c?

5-12. Why is the recommended feedrate for small-diameter drills relatively low?

Case Study 1

Company A had traditionally been engaged in rebuilding machine tools. The shop was moderately equipped with standard drill presses, lathes, milling machines, etc. An opportunity became available to fabricate hydraulic general purpose presses. The structural members consisted of heavy plate stock, I-beams, and channel. Many holes were required to be drilled in this material.

The manufacturing engineer wanted to have the company invest in heavy hydraulic punching equipment. A preliminary investigation showed that to be efficient two heavy-duty C-type portable hydraulic punches would be needed at a cost of $3000 each. Also a hydraulic pumping unit and heavy-duty lines and connections would cost an additional $10,000.

The sales forecast showed it would be reasonable to expect to build 20 presses a year for several years. Each press requires sixteen $\frac{3}{4}$-in. dia. holes in $\frac{1}{2}$-in. thick steel and thirty-two 1-in. dia. holes in $\frac{3}{4}$-in. steel plate. The material to be used was a medium carbon steel A1S1 1045.

The required holes could be made on the existing plant facilities. A time study on a prototype press revealed that the time required to locate, center punch, and get the part into position for drilling averaged about

5 min a hole. Clamping was not required for punching and with simple stops the holes could be located on the average of 15 sec each. Labor and overhead can be calculated at $25/hr.

Case Study 2

A method of evaluating drilling operations is to do it on the basis of multiplying the number of holes drilled per year by the depth of the hole, giving the total depth of hole per year.

An example of this type of evaluation by Company X is shown below:

> Part No. 12456
> 6-in. pipe flange
> ASTM-A105 forging
> Drill 12 holes, 1-in. dia.
> by $1\frac{1}{2}$ in. through
> Yearly production: 5000 pieces
> Total yearly hole penetration:
> $12 \times 1\frac{1}{2} \times 5000 = 90{,}000$ in.
>
> Yearly penetration time:
>
> $$\frac{90{,}000}{3.6 \times 60} = 416.7 \text{ hr}$$
>
> Machine: 6-ft. radial drill
> Existing plant standard for hole penetration only, with HSS drills and flood coolant:
> $180 \text{ rpm} \times 0.020 \text{ ipr} = 3.60 \text{ ipm}$

Would the manufacturing engineer have a hard time selling his plan to management or not? Discuss.

Using the same basis for evaluation Company X decided to find out what the yearly penetration rate would be with the same number of holes using an oil-hole drill under the following conditions:

> RPM = 500
> Feed = 0.024 ipm

a. How much were they able to change the yearly penetration time?

b. They decided to make a further investigation to see what effect carbide-tipped drills would have under the following conditions:

> Cutting speed = 100 fpm
> Feedrate = 0.020 ipr

c. If the labor and overhead rate is $25/hr, what is the direct savings in the cost of floor-to-floor time using the oil-hole drill over the HSS drill? What are some factors not taken into consideration in this cost savings?

Bibliography

Baker, Alan. "The Fundamentals of Twist Drills." *Cutting Tool Engineering* 30 (November/December 1978): 3–5.

Baxter, Donald F., ed. "The Twist Drill." *Metal Progress* 114 (December 1978): 34–39.

Burant, R. O., and Michael J. McGinty. "Cutting Tools/Drills." *Manufacturing Engineer* (a series of 9 articles from March 1979 to November 1979).

Clement, George. "Nineteen Tips for Better Holes." *Modern Machine Shop* 51 (May 1978): 92–101.

Conn, Harry. "A Realistic Look at Fabricating Costs." *Metal Fabricating Institute Technical Paper 550.* Rockford, Ill., 1970.

Hatschek, R. L., ed. "Fundamentals of Drilling." *American Machinist Special Report 709* 123 (February 1979): 107–130.

Hegland, D. R., ed. "Trends in Holemaking." *Automation* 21 (July 1974): 38–44.

Wagner, William. "Drill Selection Care and Use." *Modern Machine Shop* 49 (November 1976): 88–95.

Wick, Charles, ed. "Advances in Coolant-fed Tooling." *Manufacturing Engineering* 81 (November 1978): 44–51.

CHAPTER SIX

Straight and Contour Cutting

Most machines are normally associated with the main type of work they perform. For example, the lathe is normally associated with cylindrical work although it can be used for a wide variety of operations. In this chapter, the machines normally associated with both straight and contour cutting will be discussed. Some machines are associated with both contour and flat surfaces.

STRAIGHT OR PLANE SURFACE MACHINING

Plane surfaces are normally associated with shapers and planers.

Shapers. Shapers are made to cut by means of reciprocating single-point tools. The tool is held in a toolpost on the end of a ram. Typical surfaces that can be cut with a shaper are shown in Fig. 6-1.

A shaper is usually not considered a production machine; however, it is widely used in machine shops and toolrooms, since it is easy to set up and operate. The cutters are low in cost and are easily sharpened. Because the amount of metal removed at one time is relatively small in area, little pressure is imposed upon the work, and elaborate holding fixtures are not needed. Al-

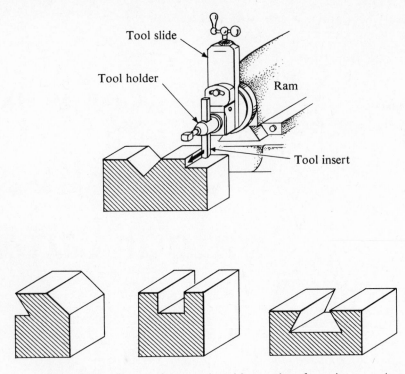

Figure 6.1. Types of external and internal surfaces best produced by a shaper.

though many shaper operations can be more rapidly performed by other machining processes, usually more costly tooling and setups are involved. These alternate methods are economical where the number of parts to be machined is large enough to justify a greater initial investment.

Planers.   Planers are used for much the same type of surface cutting as shapers are, but on a larger scale. On the planer the tool remains stationary, and the work reciprocates back and forth. At the end of each cutting stroke, the tools are fed a small distance across the work in order to give a fresh bite in the work. The work is fastened to the table which, in turn, rides over the bed of the machine (Fig. 6-2).

Three or four tooling stations are available, depending on the type of planer. Two toolheads are mounted on the overhead crossrail and one on each of the columns. When more than one surface is to be cut, three or four cutting stations can be engaged at one time (Fig. 6-3).

Planer Action.   The single-point tool used in the planer imposes relatively little pressure on the work, and only a small amount of heat is generated. This is important when producing large, accurate, distortion-free surfaces. Modern planers are able to move carbide-cutting tools in semisteel at the rate of 400 fpm.

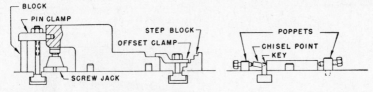

Figure 6.2. Various methods used in clamping work to the planer table.

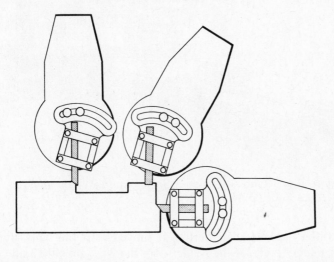

Figure 6.3. Three or four tooling stations may be used at one time on a planer.

Large heavy castings usually receive their first cuts on a planer. A roughing cut used to true the casting surface will be approximately $\frac{3}{4}$ in. deep, with a feed of $\frac{1}{16}$ to $\frac{1}{8}$ in. per cut. Two or three finishing cuts are taken to obtain accuracy on large surfaces. As little as a 0.001- or 0.002-in. depth may be left for the last finishing cut. Although planers are most often associated with the machining of large parts, such as machine tables, beds, rams, and columns, they may be used economically on many similar parts. The practice of setting up duplicate pieces in tandem is known as *string planing*.

Planer Classification.   Planers are classified according to their main structures—*openside, double-housing, pit,* and *edge* or *plate* type. The double-housing type shown in Fig. 6-4 is of extremely solid construction. The support furnished the tool by the two columns and heavy crossrail allows heavy cuts to be taken even far above the table.

The openside planer is made with only one supporting column. Its main advantage is that work may be extended out over the bed. It is also somewhat easier to get at for setup work.

The pit-type planer is more massive in construction than the double-housing

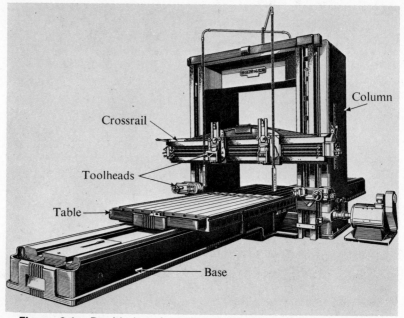

Figure 6.4. Double-housing planer. (*Courtesy Giddings & Lewis Machine Tool Co.*)

planer. It differs also in that the tool is moved over the work by means of a gantry-type superstructure which rides over the ways on either side of the table. The clapper boxes, which are of the double-block type, allow planing in both directions. Since the pit-type planer has its bed recessed in the floor, loading and unloading are facilitated.

Another more specialized planer is the *plate planer,* used for squaring or beveling the edges of heavy plate stock.

Work-Clamping Methods.   The planer table is made with evenly spaced T-slots and holes. Use of the T-slot is the most common method of holding work against the surface of a machine. Blocking materials are used to support the heel of the clamp at the proper height. These may be any scrap materials or, more conveniently, step blocks, parallels, and jack screws. Flat work is held on the table by means of toe dogs, chisel points, and poppets.

Other methods of holding work to the planer table include electromagnetism and hydraulic and vacuum clamping. Owing to large inertial forces and intermittent cutting action, these methods usually must be aided by mechanical blocking methods at each end.

Speeds and Feeds.   It is difficult to make definite speed and feed recommendations for planers, since cutting conditions vary widely. Some of the considerations involved in setting speeds and feeds are the size of the workpiece, clamping facilities, ability of the workpiece to withstand the pressure of the cut, and the type and number of tools used at one time.

STRAIGHT AND CONTOUR-CUTTING MACHINES

Machines normally associated with contour-type surfaces are also used for plane surface machining. These are: milling, broaching, and sawing machines.

MILLING-MACHINE CLASSIFICATION

Many machine-tool builders are engaged in the manufacture of milling machines. Each machine has a special purpose, requirement, or condition. As a result, we have a wide variety of milling machine types. These can be classified, depending on their use, in three general groups—low-production, high-production, and special types. In the first category are the column-and-knee machines. Fixed-bed machines comprise the second group. Special types include duplicating or profiling machines, pantograph mills, rotary-table mills, and others especially adapted for certain jobs.

Column-and-Knee Milling Machines. The column-and-knee mills are so named because of their two main structural elements (Fig. 6-5). The column-

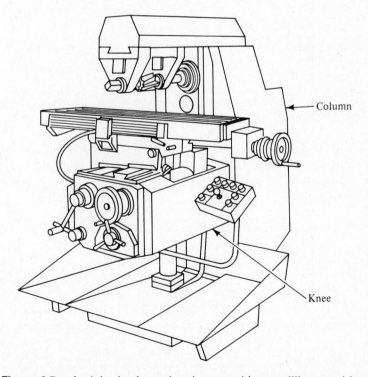

— Column

Knee

Figure 6.5. A plain, horizontal, column-and-knee milling machine. (*Courtesy* American Machinist.)

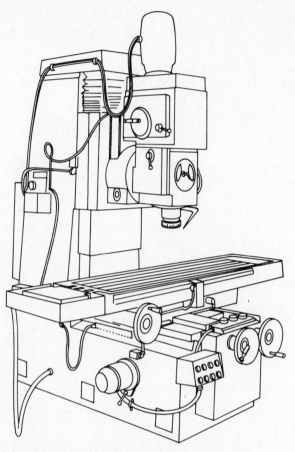

Figure 6.6. A bed-type vertical milling machine. (*Courtesy* American Machinist.)

and-knee mill is classified as plain or universal, depending on whether or not the table can be swiveled in a horizontal plane. The table on the universal machine can be swiveled up to 45° to the right or left, making possible angular and helical milling.

Horizontal and Vertical Mills.   The table of a column-and-knee milling machine has a wide range of feeds that can be applied in all three directions—vertically, longitudinally, and laterally. The table can also be quickly positioned by means of a rapid traverse mechanism, and some incorporate automatic table movements. Graduated dials allow the operator to set the table to within 0.001 in. accuracy. The rigid overarm and bearing support used on horizontal machines also increase accuracy.

The vertical milling machine is so called because the spindle extends in a vertical position (Fig. 6-6). Some vertical spindles are adjustable only vertically; others can swivel through 360° across the face of the machine.

A variation is a combination vertical and horizontal machine (Fig. 6-7).

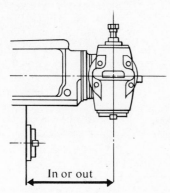

In or out

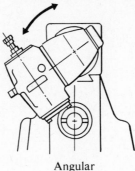

Angular
adjustment

Figure 6.7. A combination vertical and horizontal column-and-knee milling machine. The Universal ram-mounted head adds range and versatility to the machine. (*Courtesy Cincinnati Milling Machine Co.*)

Some machines of this type have an independently driven head mounted on the overarm ram. This head is of the universal type and can be swiveled for cuts at any angle between the cutter and a horizontal plane.

. To gain a better understanding of the use of both the vertical and the horizontal milling machines, study the types of work done by each as shown in Fig. 6-8. You may see that there is some overlapping in the type of work performed; however, each kind of machine has its own advantages.

Planer Mill. A planer mill is very similar in structure to the double-housing planer discussed previously. On the planer mill, the single-point tools are replaced by rotary milling heads (Fig. 6-9).

Special Types of Milling Machines

Many types of special milling machines are made to accomplish specific kinds of work more easily than the standard types. In this category are the duplicating mills, profiling machines, and pantographs. Most of these machines are vertical

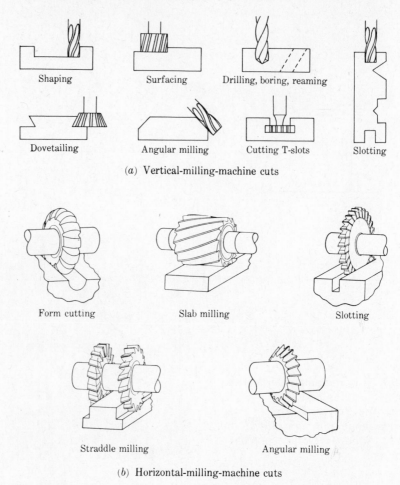

Shaping Surfacing Drilling, boring, reaming

Dovetailing Angular milling Cutting T-slots Slotting

(*a*) Vertical-milling-machine cuts

Form cutting Slab milling Slotting

Straddle milling Angular milling

(*b*) Horizontal-milling-machine cuts

Figure 6.8. Applications of vertical and horizontal milling machines.

mills that have been adapted to reproduce accurately, by means of a tracer, the forms or contours from a master pattern (Fig. 6-10).

THE MILLING PROCESS

Milling-Cutter Rotation. Milling cutters may be mounted on the arbor so that the cutting action will bring the cutting forces down into the work (called *climb milling*), or the forces may be directed up as in *up*, or *conventional*, *milling* (Fig. 6-11). The advantage of climb milling is that the downward action of the cutter helps hold the material in place. Thus the fixture can be less rugged and of simpler design. This is especially helpful in milling thin materials that are hard to clamp. Higher feedrates may be used, with a corresponding increase in pro-

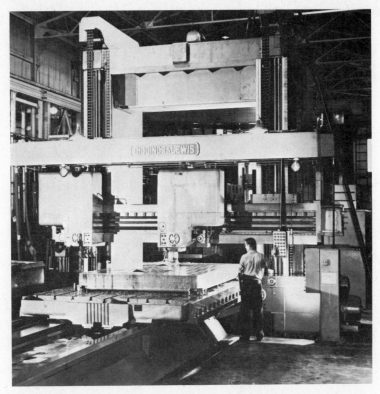

Figure 6.9. Planer-type mill. (*Courtesy Giddings & Lewis Machine Tool Co.*)

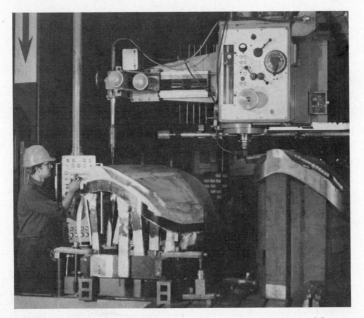

Figure 6.10. A duplicating mill tracing a plaster pattern. (*Courtesy Quality Machines Inc.*)

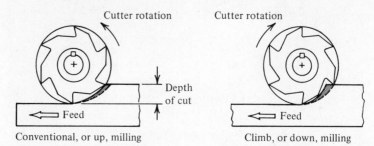

Figure 6.11.   Cutter relationship to work in conventional and climb milling.

duction, increased cutter life, and lower horsepower per cubic inch of metal removed. However, as the name implies, there is a tendency for the cutter to climb over the work. This imposes undue strain on both cutter and arbor. The climbing action may be eliminated if the work can be made to feed into the cutter at a steady pace. On production machines, this is taken care of by a hydraulic backlash eliminator.

Up milling tends to pull the work up away from the fixture. However, where no provision has been made to eliminate the lead-screw backlash, this method is used exclusively.

MILLING METHODS

The methods used to mill a part or a number of parts may vary widely. Wherever possible, ways of increasing production are given prime consideration. The practice of placing more than one cutter on the arbor at one time or more than one workpiece on the milling-machine table may be the deciding factor as to whether or not a job will prove profitable.

Straddle Milling.   When two cutters are mounted on an arbor and are spaced so as to cut on each side of the material (Fig. 6-12), the term applied is straddle milling. The cutters are spaced very accurately and are held rigidly. Therefore, the parts produced are uniform.

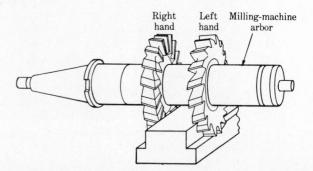

Figure 6.12.   A stradle-milling setup.

Figure 6.13. Gang-milling setups used to machine a combination of surfaces simultaneously. (*Courtesy Ingersoll Milling Machine Co.*)

Gang Milling.   Many cutters may be placed on the same arbor to make an entire surface or contour change in one pass (Fig. 6-13). A combination of gang- and straddle-milling cutter arrangement is also shown.

PLANNING FOR PRODUCTION MILLING

After a preliminary study of the part prints, it is necessary to specify the most suitable type of milling-machine fixture, tools, cutters, speeds, and feeds to produce the part at the lowest unit cost. The following outline may aid in accomplishing this.

(1) Determine the type of machine to be used. Keep in mind, when selecting a machine, that the column-and-knee machines are easily set up and operated. They can be changed quickly and are therefore ideal for short-run jobs. The bed-type machines require more setup time, since feeds and speeds are set by change gears; therefore, longer runs are desirable. Specialized machines are considered only for parts not conveniently handled on a regular machine and where the quantity is large enough to warrant the extra cost.

(2) Determine the type of milling to be done—whether straddle, contour, etc. Also, decide on the method of mounting the work—whether several parts can be placed on the table for abreast milling, string milling, or reciprocal milling, or if indexing or other fixtures are to be used.

(3) Check the Brinell hardness of the part in several places to obtain a basis for cutting speed.

(4) Choose the type of cutters needed, including the style, size, and material. Milling-cutter dimensions are given by outside diameter, width, and inside diameter. The cutter diameter is based on the depth of cut plus the clearance between the arbor collars and the work. Too large a cutter adds to the time of the job and to the horsepower requirements. Heavy cutting requires rigid cutter mounting. A 2-in.-dia. arbor is 16 times as rigid as

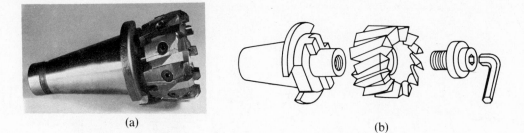

(a)

(b)

Figure 6.14. A blade-type face mill mounted on a stub arbor (a), and conventional shell end mill (b). (*Courtesy Kearney & Trecker Corp.*)

a 1-in. arbor. Face mills mounted on stub arbors (Fig. 6-14) have minimum overhang and can be counted on for heavy roughing cuts as well as for accuracy in the finished cut. The cutting-tool material, whether high-speed steel, carbide, or ceramic, will be governed by the rigidity of the machine and the setup. Carbides and ceramics demand rigid setup of both the cutter and the work, or chipping of the tool will occur.

(5) Determine proper speeds and feeds. The speed at which the cutter rotates is based on the formula:

$$\text{RPM} = 3.82 \times \frac{V_c}{D_m} \text{ (see Table 3-4)}$$

Speeds. The cutting speed is always given in terms of the number of surface feet per minute (sfpm), rather than rpm. It is related to the physical properties of the work material and the tool material. Suggested starting cutting speeds for carbide tools on steels of various hardness are shown in Fig. 6-15. It may be necessary to increase or decrease these speeds to suit actual job conditions. As mentioned previously, a builtup edge forms when machining carbon and low-alloy steels at too low a speed. By careful observation, the speed at which the BUE is no longer formed can be found. This may be termed the *critical speed* (CS). The CS will be affected by hardness, chip thickness, and depth of cut. The general recommendation for use of carbide cutters is to set a speed that is 50 to 100% higher than the CS for HSS cutters.

The feed, in inches per minute of table travel, will be based on the machine horsepower, the finish desired, and the chip clearance of the cutter. The feedrate will be calculated from the formula:

$$f_m = f_t \times n \times \text{RPM} \text{ (see Table 3-4)}$$

Recommended feeds per tooth for carbide-milling cutters are given in Table 6-1. As a general guideline, at least two or three teeth should be in the cut at one time to reduce vibration, cutter deflection, and insert chipping.

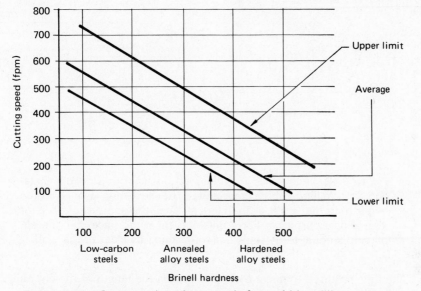

Figure 6.15. Suggested cutting speeds for carbide-milling cutters.

Table 6-1. Recommended feeds per tooth for milling steel with carbide and HSS cutters.

Type of Milling	Feed per Tooth	
	Carbides	HSS
Face	.008–.015	.010
Side or straddle	.008–.012	.006
Slab	.008–.012	.008
Slotting	.006–.010	.006
Slitting saw	.003–.006	.003

Horsepower. Horsepower required for milling can be determined on the basis of the cubic-inch removal rate:

$$Q = w \times d \times f_m \text{ (see Fig. 3-4)}$$

The horsepower at the spindle:

$$HP_s = Q \times P$$

The horsepower at the motor:

$$HP_m = \frac{Q \times P}{E}$$

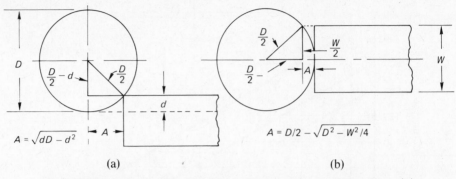

Figure 6.16. Allowance for approach on plain or slot milling (a) and face milling (b). If the diameter of the cutter is only a little larger than the width of the work surface, the approach is $D/2$ for all practical purposes. For a finishing cut with a face mill, the approach is equal to D, since the cutter must be clear of the work.

The time required to make a cut is based on the length of the cut plus cutter approach and overtravel divided by the feedrate:

$$\text{Time} = \frac{L + \Delta L}{f_m}$$

where: L = length of cut
ΔL = approach of cutter to the work (Fig. 6-16). (An approximation of one-half the cutter diameter is sometimes used for ΔL.)
f_m = feedrate in inches per minute

MILLING-CUTTER GEOMETRY

Three main types of insert-tooth-type milling cutters are now available: double negative, double positive, and shear angle. The names given refer to the radial and axial rake, as shown in Fig. 6-17. The double-negative rake tool is especially good for interrupted cutting or heavy roughing cuts and the removal of scale. It tends to put more stress on the workpiece and is therefore not recommended for parts that may distort during machining. It is also not the type of cutter to use on materials that strain-harden readily. The horsepower requirements are higher than for a conventional cutter. It has the advantage of being able to utilize all edges of indexible insert-type tooling.

A double-positive tool requires less horsepower and exerts the minimum of force on the workpiece. There is little tendency to chatter. As can be seen from the sketch, it is not possible to reverse the tool, so on a square insert only four cutting edges are available instead of eight. Also, the clearance angles make the cutting edges weaker.

The shear-angle cutter is of relatively new design. It combines a negative

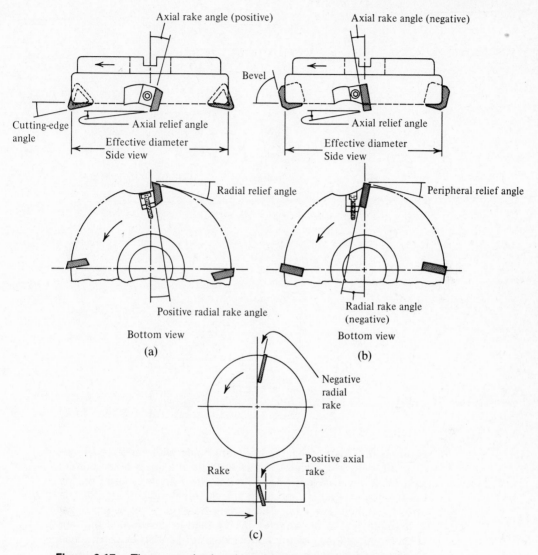

Figure 6.17. Three standard tool geometries for insert-type milling cutters: double-positive rake (a), double-negative rake (b), and a combination negative radial rake and positive axial rake (c).

radial rake with a positive axial rake. This design causes the chips to flow away from the cutter so that higher feedrates can be used. It also produces a fine surface finish.

Figure 6-18 shows an enlarged view of the edge of the cutter. Shear-angle cutters are usually made with high lead angle, whereas the double-positive angle may vary from 2 to 30°. The standard lead angle for the double-negative cutter is 10°. As with single-point tools, the lead angle serves to protect the cutting

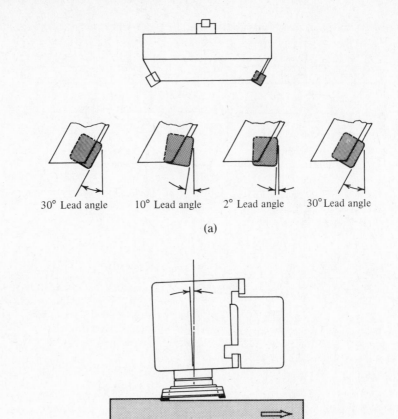

30° Lead angle   10° Lead angle   2° Lead angle   30° Lead angle

(a)

(b)

Figure 6.18. Insert-milling cutters are made with various lead angles, 30° with a chamfered flat is used on the shear-angle cutters to provide an excellent finish. The 10° lead angle is used on the double-negative cutter and from 2 to 30° lead angle is used on the double-positive cutter (a). The cutting head is offset slightly, just a degree or so, in the direction of the feed to ensure that the rear (noncutting portion) of the cutter does not scratch or carry small chips around on the cut surface (b).

point and thin out the chip. This allows higher feedrates without overloading the insert. A 2° lead angle permits cutting up to a shoulder.

BROACHING

The broach is used equally well on straight or irregular surfaces, either externally or internally, and it can perform many of the operations that are done more laboriously on milling, drilling, boring, shaping, planing, or keyway-cutting machines.

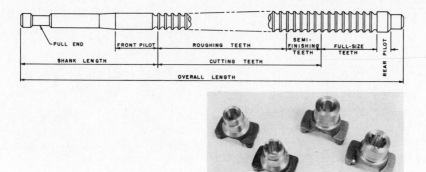

Figure 6.19.   An internal broach with sample workpieces. (*Courtesy National Broach Division, Lear Siegler, Inc.*)

Internal Broaching Tools.   Internal broaching tools are designed to enlarge and cut various contours in holes already made by drilling, punching, casting, forging, etc. The tool shown in Fig. 6-19 has the essential elements found in most broaches. The first teeth are designed to do the heaviest cutting and are called the *roughing teeth*. The next portion has the *semifinishing teeth*, followed by the *finishing teeth* with progressively lighter cuts. Some broaches are made so that the last few teeth do no cutting at all but have rounded edges for a burnishing action.

The amount of stock that each tooth can remove varies with the type of operation and the material. A general average is 0.002 or 0.004 in. per tooth for high-speed-steel broaches. The space between the teeth must be sufficient to provide ample chip room, since the chip is carried by each tooth until it clears the stock. On ductile materials, it is necessary to have staggered grooves in the cutting teeth to act as chip breakers.

The internal forms shown in Fig. 6-20 represent only a few of the shapes that have become standard. Special broach shapes are built by manufacturers for odd and difficult contours.

Spline and Gear Broaching.   The involute type of spline, shown in the center sample of Fig. 6-21, is an improvement over the regular straight-sided spline in that it is much stronger and tends to centralize under a turning load. When there is an endwise movement under load, the spline acts as a gear, transferring the load uniformly from tooth to tooth.

The broaching of internal gears is similar to that of internal splines. Usually, however, involute form, tooth spacing, runout, and finish must be held to closer

Figure 6.20.   A variety of forms produced by internal broaches. (*Courtesy Colonial Broach Co.*).

Figure 6.21.   Internal gears and splines cut with internal broaches, some to extremely close tolerances. (*Courtesy Colonial Broach Co.*)

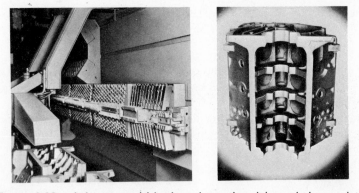

Figure 6.22.   A large, carbide-tipped, sectional broach is used to machine several surfaces of an engine block in one pass. (*Courtesy Ex-cell-o Corporation.*)

limits. Broaching of internal gears may be applied to either straight or helical gears. If the helix angle is less than 15°, the broach will turn as it cuts its own path. Angles greater than 15° require a spiral-lead-type drive head to turn the broach as it is pulled through the work.

External Broaching Tools.   External surface broaching competes with milling, shaping, planing, and similar operations. It offers a combination of a high degree of accuracy and excellent surface finishes, combined with high output rates and low downtime.

Automobile manufacturers have replaced many milling operations with surface broaching because of the combined speed and accuracy. An example of this kind of work is shown in Fig. 6-22. A maximum of $\frac{3}{16}$ in. of stock is removed in a single pass. The broach consists of carbide-tipped inserts made up of five sections in combinations of half-round and facing broaches. As many as 22,000 cast-iron engine blocks can be run before the tool needs resharpening. Some broaches of this type are made on the same principle as the single-point lathe tool with throwaway inserts. As the carbide tips become dull, they can be unclamped and indexed 90°; they will then be ready to cut again. By incorporating

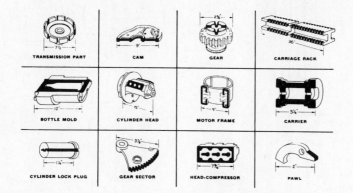

Figure 6.23. A variety of surfaces prepared by broaching. (*Courtesy* Cincinnati Milling Report.)

negative rake into the toolholder, six or eight cutting edges can be utilized before the insert is discarded.

Broaching speeds have increased tremendously with this kind of tooling. Formerly, 20 to 40 fpm was considered average; now, speeds in excess of 200 fpm are used on cast iron. A further time saving is made on large surface broaches by mounting them in pairs. With this arrangement, cutting can be done on both the forward and the return strokes. The top broach on a horizontal machine may be cutting on the forward stroke and the lower broach on the return stroke.

Some idea of the wide variety of surface broaching operations can be gained by referring to Fig. 6-23.

BROACHING MACHINES

The main types of broaching machines are vertical, including pullup, pulldown, and ram; and horizontal, both plain and the continuous-chain-type surface broach. These operate primarily as a means of either pushing or pulling the tools through or over the work.

Vertical Broaches. The various vertical machines are similar in appearance. In fact, one broach builder makes a convertible machine that can perform all four of the basic operations—push down, surface, pull down, and pull up (Fig. 6-24).

Horizontal Broaches. *Plain Machines.* Horizontal broaching machines (Fig. 6-25) are used primarily for broaching keyways, splines, slots, round holes, and other internal shapes or contours. They have the disadvantage of taking more floor space than do the vertical machines. However, long broaches and heavy workpieces are easily handled on the horizontal machine.

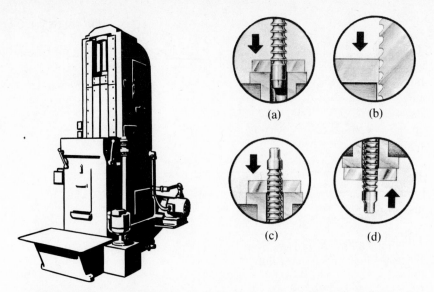

Figure 6.24. Vertical broaching machine convertible to four basic operations (a) Push down, (b) surface, (c) pull down, (d) pull up. (*Courtesy Sundstrand Machine Tool Co.*)

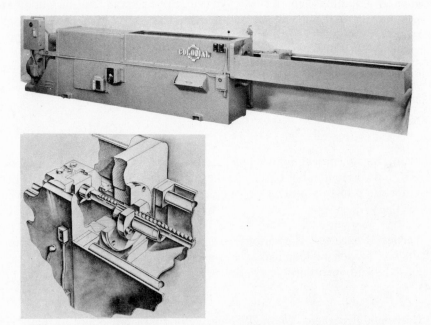

Figure 6.25. Horizontal broaching machine and operation. (*Courtesy Colonial Broach Co.*)

Continuous-chain-type Surface Broach.   The continuous-chain-type surface broach (Fig. 6-26) is used where extremely high production is desired. This machine consists mainly of a base, driving unit, and several work-holding fixtures mounted on an endless chain. The work is pulled through the tunnel where it passes under the broach. The operator has only to place the work in the fixture as it passes the loading station. Work is automatically clamped, machined, and unloaded.

Broaching Machine Sizes.   The size of a broaching machine is expressed by the total ram travel, in inches, and the maximum pressure exerted on the ram, in tons.

Finish and Accuracy.   A 30-microinch finish can be consistently held when broaching steel of uniform microstructure. Better finishes are available, but costs are higher. Irregular and intricate shapes can be broached to tolerances of ±0.001 in. from a location established on the part. Tolerances of ±0.0005 in. can be held between surfaces cut simultaneously.

Broach Length and Machining Time.   The length of a broach will determine the cutting cycle. Also, the length of the broach must be suited to the machine on which it is to be used. By way of example, if the average amount of stock removed per tooth is 0.003 in., the pitch (or tooth spacing) is $\frac{1}{2}$ in., and the material to be removed is $\frac{1}{8}$ in., the *effective length* of the broach can be calculated. By effective length is meant the area containing the broach teeth.

$$L_e = \frac{C_d}{C_t} \times p \text{ or } L_e = \frac{0.125}{0.003} \times 0.500 = 20.8 \text{ in.}$$

where: L_e = effective length of broach;
 C_d = depth of cut to be made;

Figure 6.26.   Continuous-chain surface-broaching machine.

C_t = cut per tooth (average); and

p = pitch.

The length of the broach will change with the cut per tooth and the pitch. On internal operations, the amount of metal to be removed from the diameter will be divided in half.

The pitch of the broach varies with the length of the cut to be made, for example, a cut only $\frac{3}{32}$ in. long would require a $\frac{3}{64}$-in. pitch, whereas a cut 1 in. long should have a pitch of $\frac{3}{8}$ in.

Broach Horsepower Requirement. The horsepower required at the motor may be found as follows:

$$HP_m = \frac{Q \times HP_u}{E}$$

where: $Q = 12 \times w \times d \times v_c$ (cutting velocity in fpm) (Table 3-5)

E = efficiency of the machine

Economic Considerations. Broaching tools are usually more costly than milling cutters. However, this initial outlay can be offset by other factors. The machine time for broaching is generally less, and the piece part is often completed in one pass. Further time savings per piece are realized when more than one part is broached at a time, by stacking. Stack broaching generally is best adapted to internal operations.

The relatively low speed and small cut per tooth give broaches a long life. Contour broaches can be resharpened by grinding on the face of the teeth. Generally, special fixtures are needed for sharpening the broach, making it a rather expensive operation. Surface broaches are often made in sections so that, as the teeth are reground and become shorter, they can be moved forward. The last section will be all that has to be replaced. The cost of broaching tools runs high because of the care needed in forging and grinding. Generally, a high quantity of production is necessary to justify the cost of the tool.

SAWING

Many machining operations start with the sawing of the stock to length or size. Improvements in the speed, accuracy, and the variety of operations that can be performed on sawing machines give this process an important place in the field of manufacturing.

Basic Types of Sawing Equipment. Sawing equipment may be classified by the motion used for the cutting action. There are reciprocating saws, represented by the power hacksaw; band saws, including cutoff and contour (traditional types and other applications); and circular saws, such as the cold saw and those with friction discs or abrasive discs (Fig. 6-27).

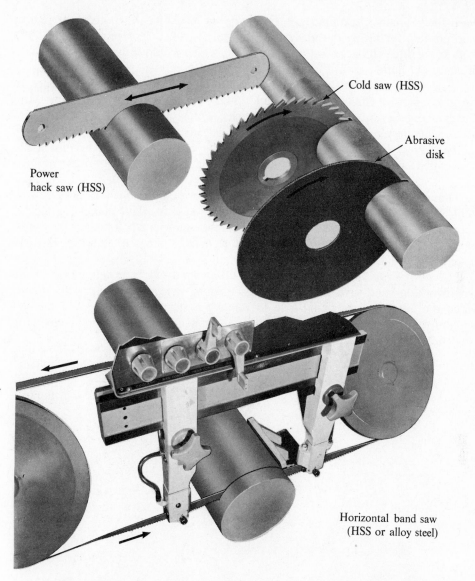

Cold saw (HSS)

Abrasive disk

Power hack saw (HSS)

Horizontal band saw (HSS or alloy steel)

Figure 6.27.   Four types of cutoff machine action and cutters. (*Courtesy DoAll Company.*)

Reciprocating Saws.   The reciprocating motion in sawing is familiar to all who have used the hand hacksaw. Although it is hardly a production process, its usefulness is appreciated in many toolroom and maintenance situations.

Power Hacksawing.   The power hacksaw is probably the simplest of the metal-cutting machines. It consists of a vise for clamping the work and a means of reciprocating the saw frame. Hydraulic pressure is used to regulate the downward feed force on the blade. A hydraulic or mechanical arrangement is also incorporated for lifting the blade on the return stroke. This prevents the teeth from being dragged backward over the work, which would cause them to become prematurely dull.

The stock to be cut is held between the clamping jaws. Several pieces of bar stock can be clamped together and cut at the same time. Both square and angular cuts can be made. Some of the larger heavy-duty hacksaws have hydraulic feeding and clamping devices which automatically move the bar forward to the correct length, and clamp it. After the cut, the stock is automatically unclamped, and the cycle repeats as long as the stock lasts (Fig. 6-28).

Band Saws.   *Cutoff Type.*   Band saws have a continuous cutting action in contrast to the intermittent action of the power hacksaw. This makes for shorter cutting time—up to 10 sq in./min on some steels. Another advantage of the cutoff band saw is the narrow *kerf,* or cut, that it makes. Saws of this type usually have a 16-in.-dia. cutting capacity.

Contour Type.   The metal-cutting contour band saw is an outgrowth of the woodworking band saw. One of the first models was produced in 1935. It provided the advantage of being able to remove large pieces of unwanted metal without reducing them to chips, and thus an immediate market for this machine (Fig. 6-29) was created.

Hydraulic clamping
and bar feed unit

Figure 6.28.   Power saw, showing hydraulic bar feed. (*Courtesy DoAll Company.*)

Many developments followed the early models. First among these improvements was a variable-speed control. A wide choice of speeds resulted in greater accuracy, faster cutting, and increased tool life. Other refinements were added gradually, such as the flash butt resistance welder used to join the saw bands right on the machine for internal cutting. A grinder mounted on the column of the machine is used to smooth the weld joints, permitting them to pass through the saw guides.

Numerous attachments added to the versatility of the machine. Among these are the circle-cutting attachment, ripping fence, and magnifying lens. Power feeds and coolant systems have also been added. On some of the larger machines, the work can be bolted directly to the power-fed table. This reduces operator fatigue and increases accuracy.

Other operations performed on the regular contour band saw are filing and polishing.

Friction Contour-Sawing Machines. Special contour-cutting band saws are capable of performing operations beyond the scope of the regular machine. For example, friction sawing is a comparatively new term when applied to contour-

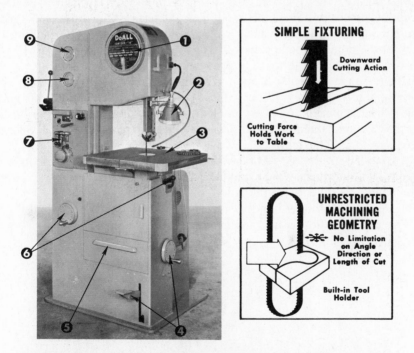

Figure 6.29. A standard metal-cutting band saw is normally used for straight, angular, and contour cutting but may also be used for filing and polishing. (1) Job selector dial; (2) reflector flood lamp; (3) tilt table—45° right, 10° left; (4) work feed controls; (5) chip drawer; (6) speed control and indicator; (7) saw blade welder; (8) band tension indicator; (9) blade sfs indicator. (*Courtesy DoAll Company.*)

sawing machines. The process consists of a fast-moving blade that produces enough friction to heat the material to the softening point. The heat of friction is confined to a small area just ahead, and a little to each side, of the blade. When the material becomes soft and loses its strength, it is removed by the saw teeth.

The right blade for friction sawing is more critical than in ordinary sawing. Too few teeth for the material thickness will tend to remove more material than is softened, which will ruin the blade. Too many teeth, on the other hand, will tend to clog and will not remove the material at all. Specially made blades are required for most friction-sawing operations. Saw velocities range from 6000 fpm on $\frac{1}{4}$-in.-thick carbon steel to 13,500 fpm on 1-in.-thick armor plate. The latter can be cut at the rate of 5 in./min.

The chief advantage of friction sawing is that it can cut many times faster than conventional methods on hard materials (Fig. 6-30).

Circular Saws. Circular saws cut by means of a revolving disc. The disc may have rather large teeth, as in the case of cold saws, or almost no teeth, as in the friction disc.

Cold Sawing. Cold sawing would simply imply that no heat is generated in the cutting operation. Since this is not true, the term is used only to distinguish it from friction-disc cutting, which heats the metal to nearly the melting temperature at the point of metal removal. Also the blades are usually large, which allows cooling time for the tooth after leaving the cut until reentering.

Friction Saws. Circular friction saws operate on the same principle as the friction band saw. However, since the saw friction heat is high and natural cooling is insufficient, an auxiliary coolant must be forced against the blade to keep it from becoming red hot. The saws may be as large as 6 ft in dia., operating at speeds up to 25,000 sfpm.

Abrasive-Disc Cutting. Abrasive-disc cutting, frequently used in cutoff operations, is not a true sawing technique, but, since it serves the same purpose, it is frequently classified with the cutoff sawing operations.

Thin resinoid or rubber-bonded wheels rotating at 12,000 to 14,000 sfpm are used. The cutting action is fast and accurate. For example, ordinary 2-in.-dia.

Figure 6.30. This fast-action photo stops the saw blade traveling 9000 fpm in a high-speed milling cutter. Heat penetration is at the immediate saw point. (*Courtesy DoAll Company.*)

pipe can be cut in 8 sec. For most work, the cut surface need not be refinished (Fig. 6-31).

Machines for abrasive-disc cutting are fairly simple and usually feature a powerful drive motor with a belt-driven wheel head. Most of the machines use a swing frame and are fed into the work manually.

To speed cutting action and handle diameters from 2 to 12 in., an oscillating wheel is used (Fig. 6-32). The wheel is power fed and made to travel back and forth across the material. Cutting capacities are high. Wheels range from 16 to 34 in. in dia., with corresponding motors of $7\frac{1}{2}$ to 30 hp.

Figure 6.31. The abrasive cutoff wheel can be used to cut both soft and hardened materials. In most cases no further machining needs to be done on the cut ends. (*Courtesy Wallace Supplies Mfg. Co.*)

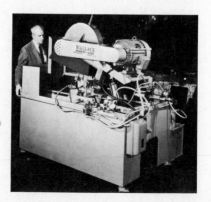

Figure 6.32. A swing-frame-type cutoff machine that is designed to move in and out as well as up and down. Units of this type are available that can travel up to 12 ft for cutting off plate and other special applications. (*Courtesy Wallace Supplies Mfg. Co.*)

Problems

6-1. A new 6-in. dia., 8-tooth, face-milling cutter is given a performance test on a SAE 1020 CRS block (150 BHN), 4 in. square and 33 in. long. The cut is made over the long axis of the work. The following operating conditions were used:

> Sfpm = 300
> Chip load = 0.003 in.
> Depth of cut = 0.125 in.
> Idle horsepower = 0.5
> Machine mech. efficiency = 0.9

Calculate the following:

(a) RPM

(b) Feedrate in ipm

(c) Time to make the cut

(d) Cubic-inch removal rate

(e) Horsepower required at the cutter

(f) Horsepower required at the motor

6-2. A large, medium-hard, cast-iron casting is machined on a planer. It is 20 ft. long and 48 in. wide. Two roughing and two finishing cuts are required. Find the machine time needed if carbide tools are used at 250 fpm. Both of the crossrail tools are used at the same time in tandem, set at different depths with the cutting edges 8 in. apart. The roughing feed is 0.060 in. and the finish feed is 0.020 in. The return stroke is made at two times the speed of the cutting stroke.

6-3. What HP_c would be required for the roughing cut (both tools engaged) if the depth of cut/tool = 0.075 in.?

6-4. A hole in a 1-in.-thick casting is $\frac{7}{8}$ in. in dia. and is to have a $\frac{3}{16}$-in.-deep, $\frac{1}{8}$-in.-wide keyway broached in it. (a) If the average amount of material removed by each tooth is 0.003 in. and the pitch of the broach teeth is $\frac{1}{2}$ in., how long will the cutting portion of the broach have to be? (b) How long will it take to broach 1000 parts if the broaching speed is 30 fpm and a total of 1 min is allowed for the return stroke, handling each part, and placing the broach in position for the cut? Allow 2 in. for approach and overtravel at each stroke.

6-5. A cast iron cylinder head 8 in. wide and 20 in. long with $\frac{1}{4}$ in. of material to be removed can be surface broached or milled. Milling is done with a 10-in. dia. face mill in one pass. The carbide mill has 20 teeth. The cutting speed is 300 fpm. No return of table is required since milling can be done in both directions. The carbide broach has an average chip load of 0.005 in. per tooth with a pitch of $\frac{3}{8}$ in. and moves over the surface at 150 fpm. One foot of travel is required before the broach is in full cut. Which method will be faster? If labor and overhead at $25.00/hr are the same for each machine, how long will it take to pay for the broach, which costs $300 more than the milling cutter?

6-6. What would be the horsepower required at the motor for the milling operation as given in Problem 6-5 if the material being milled is soft gray cast iron?

6-7. (a) What are the three types of insert milling cutters in regard to radial and axial rake?

(b) What are the advantages and disadvantages of each?

6-8. Why is the cutting head of a vertical milling machine tilted slightly in the direction of travel?

6-9. What is the principal difference between a shaper and a planer in how workpieces are cut?

6-10. (a) What are the advantages of climb milling?

(b) What is the main disadvantage of climb milling?

6-11. Two $\frac{1}{2}$ in. wide by $\frac{1}{2}$ in. deep cuts are to be made longitudinally on a block of steel 2 × 6 × 10 in. A horizontal mill will be used. To obtain maximum production state the following:

(a) milling method

(b) type of milling cutters

(c) what would the machining time be per piece if the high-speed steel cutters are 4 in. in dia. with 10 teeth? The recommended surface feet per minute is 80.

(d) Horsepower at the cutter?

Bibliography

Jablonowski, J., ed. "Fundamentals of Milling." *American Machinist Special Report 701* 122 (February 1978):SR1–SR-24.

Loy, J. D. "Indexable Versus Grind-Type Cutters." *The Carbide Journal* 1 (April–May 1973):3–6.

Metal Cutting Principles, 2nd ed. Rockford, Ill.: The Ingersoll Milling Machine Company Cutting Tool Division, 1979.

Milling Handbook of High-Efficiency Metal Cutting. Detroit: General Electric Company Carboloy Systems Department, 1980.

Oxford, C. "Get to Know Your Form Milling Cutters." *Machine and Tool Blue Book* 70 (May 1975):74–79.

Singh, K. "Chip Thickness—How It Affects Milling Efficiency." *Cutting Tool Engineering* 30–31 (September–October 1979):12–15.

———. "Factors Affecting the Choice of Climb or Conventional Milling." *Cutting Tool Engineering* 30–31 (July–August 1979): 3–6.

Smith, Darrel. "Tips on How to Do a Better Job of Finish Milling." *Cutting Tool Engineering,* Part I 32–33 (January–February 1980):19–24; Part II 32–33 (March–April 1980):29–24.

Tipnis, V. A. "High Rate Copy Milling Carves New Niche." *Modern Machine Shop* 52 (February 1979):105–117.

CHAPTER SEVEN

Grinding and Related Abrasive-Finishing Processes

Abrasives have been largely responsible for ushering in the precision capabilities of the twentieth century. Modern abrasive machines have made tolerances in the ten-thousandth range commonplace and surface finishes are measured in millionths of an inch. In recent years extensive research has changed grinding from a strictly metal-finishing operation to a competitive metal-removal method. Exacting tolerances are not achieved by grinding alone, but with the related processes of lapping, honing, and superfinishing.

Abrasives are also used to improve product appearance. These processes are barrel finishing, vibratory finishing, polishing, and buffing.

ABRASIVES

Abrasives are of two main types—natural and synthetic.

Natural Abrasives

Abrasives found in nature are emery, sandstone, corundum, and diamonds.

Both Turkish and American emery are natural mixtures of aluminum oxide (Al_2O_3) and magnetite (FeO_4). Owing to the presence of some softer accessory

249

minerals, both types have a milder cutting action than synthetic abrasives. Emery is often preferred for use as an abrasive on coated cloth and paper as well as in many buffing compositions. It is no longer used in making grinding wheels; thus the use of the term "emery wheel" is incorrect.

Diamonds are another type of natural abrasive, having the highest known hardness of any substance. Because of their nature, diamonds require a distinctive bond which is more specialized than that of conventional grinding wheels.

Synthetic Abrasives

Silicon Carbide.   The search for a method of synthesizing precious stones led Edward Acheson into the discovery of silicon carbide (SiC) in 1891. He created a small carbon-arc furnace by wrapping some wire around a pail and putting a carbon rod in the center of it. Around the carbon rod he had placed a clay mixture containing white quartz sand. After this makeshift furnace had cooled, he noted the clear crystals that adhered to the carbon rod. They were in fact silicon carbide crystals. However, Acheson termed the experiment the discovery of carborundum. Although this experiment was relatively simple, it could not have been exploited at any earlier date due to the prodigious amount of power required to make it feasible. Thus the Carborundum Company was established at Niagara Falls to be near the large hydroelectric plant.

Aluminum Oxide.   About the time silicon carbide was synthesized, another American, Charles B. Jacobs, succeeded in producing aluminum oxide by fusing small quantities of coke and iron borings with aluminum ore (bauxite). The Norton Company acquired the rights to the process and also located at Niagara Falls because of the need for large quantities of power.

Aluminum oxide is softer than silicon carbide, making it tougher and more resistant to fracture. Thus Al_2O_3 wheels are not used in grinding very hard materials such as tungsten carbide because the grains would get dull prior to fracture. Aluminum oxide performs best on carbon and alloy steels, annealed malleable iron, hard bronzes, and similar metals.

Cubic Boron Nitride (CBN).   This is a newer (1968) man-made abrasive that is harder than either Al_2O_3 or SiC, which have ratings of 2100 and 2480, respectively, on the Knoop scale. Boron nitride (rating of 4700) is the second hardest substance ever developed by man or nature. Diamonds developed by the General Electric Company have a hardness rating of 7000 (Fig. 7-1).

CBN is a tight network of interlocking and alternating boron and nitrogen atoms. It is especially good for grinding hard and tough tool steels. Why CBN or Borazon (the General Electric tradename) is a more effective abrasive on these hardened materials is not fully understood. It is presently theorized that the alloying elements found in many steels have a peculiar affinity for carbon and thus prematurely erode the diamond abrasive through a chemical process. Also, diamonds revert to plain carbon at about 1500°F (816°C), while CBN can withstand temperatures above 2500°F (1371°C).

HARDNESS KNOOP VALUES

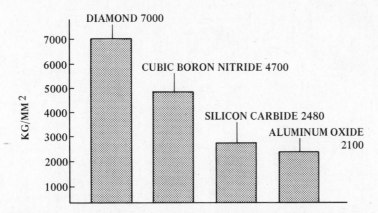

Figure 7.1.   A comparison of the Knoop hardness values of various grinding abrasives. (*Courtesy* Cutting Tool Engineering *and Excello Corporation.*)

Diamonds.   Diamonds may be classified as both natural and synthetic abrasive materials. Natural stones which are unsuitable for gems are termed *bort* and are crushed down into a series of sizes for abrasive use.

The General Electric Company was the first to manufacture diamonds on a commercial scale. In 1955 they announced the production of a commercial diamond capable of sustaining pressures up to 470,000 psi for long periods of time at very high temperatures. Other manufacturers are now using high temperatures and explosive shock waves to convert graphite into polycrystalline diamonds. Due to its polycrystalline structure, the synthetic diamond has many more cutting points than the natural diamond. Also, the random orientation of the crystals, 30 to 300 Å in size, means there are no large cleavage planes that often cause a natural diamond to split.

Commercial diamonds are now manufactured in high-, medium-, and low-impact strengths. Figure 7-2 shows the relative performance of each type of stone on ceramics and metals. The letters refer to the type of bond used to hold the abrasive grains: (MBS)—metal bond saws; (MBG 11)—metal bond grinding; and (RVG)—resinoid and vitrified grinding. Metal bonding of the abrasive grains requires a much tougher diamond than that of the vitrified clay or resinoid material. The effectiveness of the various diamond types is directly related to the modulus of resilience (MOR) of the workpiece, which is defined as the integral under the straight-line portion of the stress–strain diagram and has units of inch-pounds per cubic inch or centimeter-kilograms per cubic centimeter.

The RVG-W diamond is a modern, low-impact strength, irregular, polycrystalline structure coated with metal. The metal coating strengthens the crystal and enhances the retention of the diamond in the wheel, which increases wheel life and reduces cost. It is prepared specifically for wet grinding operations and

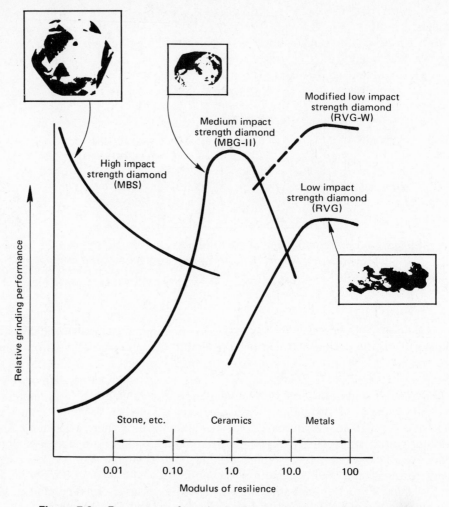

Figure 7.2.   Four types of synthetic diamonds used in grinding various materials. The modified, low-impact diamond (not shown) is similar to the medium-impact type but is metal-coated. (*Courtesy General Electric, Plastics Division.*)

has provided significantly improved performance in the grinding of hard materials such as tungsten carbide.

Bond Materials

The grinding wheel is made up of two materials: the abrasive grains and the bonding material. New bonding materials are being developed all the time. The most commonly used bonding materials in today's grinding wheels are vitrified materials (ceramics), resinoid materials (plastics), rubber (both natural and synthetic), shellac, and metal (sintered powdered metals).

Vitrified Materials.   There are two forms of ceramic bond used in grinding wheels: the glass type used with Al_2O_3 abrasives and the porcelaneous type used with SiC abrasives. Approximately 50% of all grinding wheels are made with the vitrified bond. The porosity and rigidity that this bond provides make it possible to obtain excellent stock-removal rates; yet it is also well suited to precision grinding.

Resinoid Materials.   Resinoid bonding materials consist of thermosetting plastics. Because of their high strength and rigidity, wheels made of this material may be operated at speeds up to 12,000 sfpm. The resilient action of the resin bond enables it to produce an unusually fine finish even when a coarse grit is used. Boron nitride crystals are given a metallic coating which increases the ability of the resinous bond to hold the crystals and to dissipate the heat generated in grinding.

Rubber.   The resiliency of the softer grades of rubber makes them excellent materials for polishing wheels. The harder types provide a degree of resiliency plus water-resisting qualities for safe, cool cutoff wheels, used where burr and burn must be held to a minimum.

Shellac.   Shellac-bonded wheels are especially good for producing the high finish required in roll, camshaft, and cutlery grinding.

Metal.   Metal bonding is used only in the manufacture of diamond wheels. The diamonds are firmly held in a shock-resistant bond and are not easily removed until fully used.

Grain Size

Grain-size numbers used in grinding wheels are derived from the number of grains of a given size which, when laid end to end, would equal one inch. Thus, in interpreting the grit-size markings on wheels, simply remember that the larger the base number the smaller the grain. The American National Standards Institute has now established sieve sizes, but the old numbers are still retained. Some idea of grain size can be obtained from the fact that approximately 3.5 million grains of the 240-grain size are contained in a single gram.

Finding the correct grit size to use goes deeper than the old generalization: "Use coarse grits for soft materials and rough finish; use fine grits for hard materials and fine finish." It is true that the fine grits are generally used on hard materials. The reason for this is that they have what is termed a *higher impulse ratio;* that is, they permit a large number of grains to contact the work per unit of time. Thus, though the grains cannot penetrate as deeply into the hard material, a greater number of cuts are being made and production rates are maintained.

It is also true that a fine-grit wheel, with a high impulse ratio, will generate more heat than a 46-grit wheel. Using a 46-grit wheel as a base, the 60-grit wheel will produce 1.8 times more heat. A 90-grit wheel has an impulse ratio of 12:3.6, a 120-grit wheel of 40:1.

Figure 7.3. The grinding wheel may be compared to a milling cutter having hundreds of tiny teeth. The voids in the wheel provide the chip room. (*Courtesy The Carborundum Co.*)

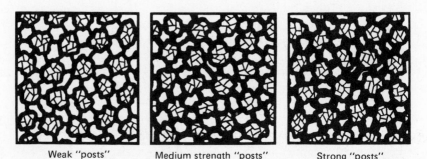

Weak "posts"   Medium strength "posts"   Strong "posts"

Figure 7.4. The grade of a grinding wheel is based on the strength of the bond posts. (*Courtesy The Carborundum Co.*)

Wheel Structure

Structure relates to the spacing of the abrasive grain. Close grain spacings are identified by low-structure numbers and are referred to as being dense. High-structure numbers indicate open grain spacing. A grinding wheel may be compared to miniature teeth on a milling cutter (Fig. 7-3). The spacing of the grains allows chip room. If this space is too small for the type of material being ground, the wheel will tend to "load up." A loaded wheel heats up and is inefficient in its cutting action. Of course, too large a space is also inefficient since there are not enough cutting edges.

Wheel Grades

The wheel grade refers to the strength of the binding posts (Fig. 7-4). As the binding posts become thicker and stronger, the wheel tends to act harder, and vice versa. Grades are designated with increasing hardness by letters from C to Z.

Grinding-Wheel Designation

Grinding-wheel codes are easily understood since they give in logical order the complete information needed to determine the wheel's characteristics. Listed in order are the abrasive type, abrasive size, grade (hardness), and bond type, as shown in Fig. 7-5. This is a standardized code used throughout the grinding wheel industry.

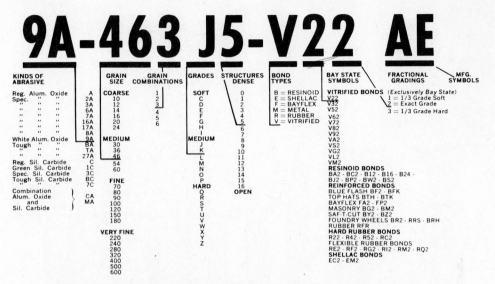

Figure 7.5. The standard marking system for grinding wheels. (*Courtesy Bay State Abrasives Division, Dresser Industries, Inc.*)

Balancing and Dressing the Grinding Wheel

Balancing. When a new grinding wheel is used it should first be checked for proper balance. Large wheels are placed on arbors and set in balancing stands. Weights are shifted in the wheel flange until the wheel is in balance. Smaller wheels, 10-in. (25.4-cm) dia. or less, can be balanced on the machine spindle by shifting movable weights in the flange. Wheels that are operated without proper balancing produce undue strains on the machine and a poor finish.

Dressing. After use, a grinding wheel becomes dull; that is, the sharp abrasive grains become rounded over. The wheel may also become *loaded,* a condition where the metal, or whatever is being ground, becomes imbedded in the wheel face. To restore the grinding efficiency of the wheel, it is necessary to cut away a small portion of the face to expose sharp abrasive and an open structure. The dressing can be done with a variety of tools, as shown in Fig. 7-6.

Grinding-Wheel Selection Principles

With the extreme range of abrasive requirements and the multiplicity of wheels from which to choose, the question arises of how to select the proper grinding wheel for a specific job. The demands on grinding wheels are diverse and exacting. They may range from snagging wheels that remove as much as 60 lb (27.21 kg) of steel per hour to precision wheels that remove very little material but produce finishes as fine as 1 or 2 microinches. Often wheels must be capable of maintaining dimensional tolerances as close as 25 millionths of an inch.

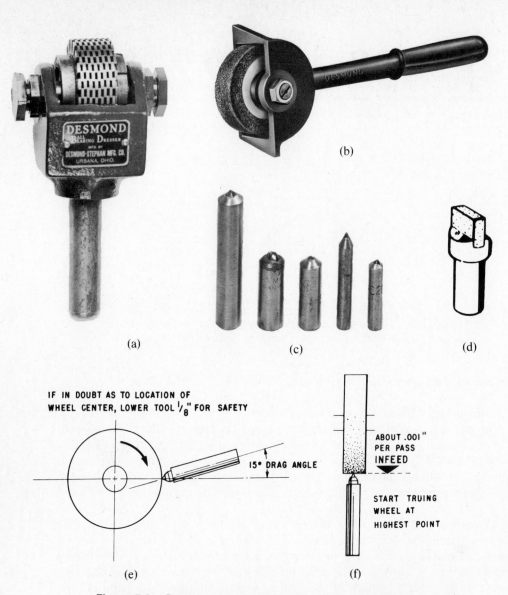

(b)

(a)

(c)

(d)

IF IN DOUBT AS TO LOCATION OF
WHEEL CENTER, LOWER TOOL $\frac{1}{8}$" FOR SAFETY

15° DRAG ANGLE

(e)

ABOUT .001"
PER PASS
INFEED

START TRUING
WHEEL AT
HIGHEST POINT

(f)

Figure 7.6. Dressing and truing tools. (a) A multiple-disc type, (b) an abrasive wheel, (c) straight-shank diamond nibs, (d) diamond cluster, (e) diamond position, (f) recommended infeed per pass. (*Courtesy The Desmond Stephan Manufacturing Co.*)

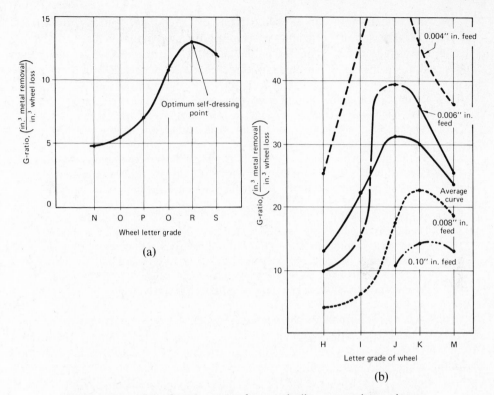

Figure 7.7. (a) A G-ratio curve for a grinding operation using re-senoid-bonded wheel in a snagging operation. (b) The G-ratio obtained using various infeed rates while grinding SAE 3145 steel on a centerless grinder.

To meet this range of performance specifications, all the factors discussed previously concerning abrasive types, abrasive grit sizes, hardness of the wheel, structure or grain spacing, and bond types must be considered in making a wheel selection for a given workpiece or a production run.

Experience will help narrow the selection field, but to pinpoint it to an optimum wheel may be difficult. Even an expert may have to test several varieties. A technique that has been developed and is now used in analyzing grinding-wheel efficiency is referred to as finding the optimum G-ratio.

G-Ratio

G-ratio refers to the ratio of cubic inches of stock removed to the cubic inches of grinding wheel worn away. Figure 7-7 shows the plot of a G-ratio curve. The G-ratio is plotted as a function of wheel grade or hardness. The ratio of metal removal to wheel loss in a unit of time is also a measure of the production rate and the amount of work that a wheel is capable of performing during its useful life.

A soft wheel shows up as being relatively inefficient, but as the wheel hardness is increased, the efficiency increases until a maximum is reached. Then, despite the fact that the wheel becomes harder, efficiency drops off. Thus the curve as shown in Fig. 7-7 represents three phases of wheel action:

1. From the soft-grade, low-efficiency wheel, to where the curve begins to peak, the wheel is constantly breaking down, but at a rate that is too rapid for the most economical use. On this part of the curve, the wheel does not load appreciably, nor does it produce excessive burn on the workpiece.
2. At the peak of the curve, the wheel is continuously breaking down at a rate that makes the most effective use of the abrasive. The abrasive grains have not become dull to the extent that the cutting rate is retarded, but they are continuously released by the bond and new sharp grains are exposed.
3. The harder wheel, as represented by J and K, show the efficiency dropping off due to a discontinuous wheel breakdown. This consists of wheel loading and unloading in cycles, accompanied by alternate pressure buildups and releases. An increase in temperature results, the wheel burns the work, the cutting rate is decreased, and more power is consumed.

The wheel performance, as indicated by the G-ratio curve shown in Fig. 7-7(a), is not restricted as to bond, grit size, or any specific grinding operation. It may be used for heavy cutting, as in snag grinding, or equally well in precision grinding where the metal-removal rate as well as the wheel loss is measured in grams. Figure 7-7(b) shows a family of G-ratio curves based on a 60-grit wheel using increasing *infeeds* from 0.004 to 0.010 in. By infeed is meant the amount the wheel is advanced for each cut. In vertical-spindle surface grinding the cut per pass is referred to as *downfeed*. The effect of infeed on wheel performance changes the optimum grade of wheel to the harder side when heavier cuts are taken. This is logical since, as the infeed is increased, the grinding pressure becomes greater and a harder wheel is required to maintain wheel life and optimum performance.

G-ratio studies are also useful in measuring the efficiency of items related to the grinding process, such as grinding fluids, and the resistance to abrasion of various metals and nonmetallic materials.

Rubbing, Plowing, and Cutting

Researchers have delineated precisely when a group of abrasive grains begins to cut. As the force increases, three distinct processes take place: rubbing, plowing, and cutting. In rubbing, plastic deformation of the metal occurs, although essentially no metal is removed. In plowing, there is a plastic flow of material which may be likened to extruding the material under the grain with only small stock removal. In cutting, fracture occurs just ahead of the grain causing chip formation and there is a fairly rapid stock-removal rate. The three distinct processes are shown in Fig. 7-8 where wheel depth of cut is plotted against force.

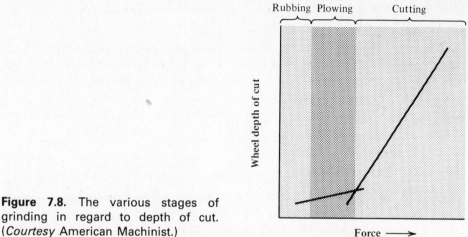

Figure 7.8. The various stages of grinding in regard to depth of cut. (*Courtesy* American Machinist.)

The cutting mode is most often used but the other modes are also important. The plowing mode is a description of what happens during *sparkout*. Sparkout means that the sparks continue to show for a pass or two even though no more infeed is added to the cut. Sparkout ensures that there is no cutting pressure left in the machine and it is also useful in refining the finish at the end of the grinding cycle.

Grinding Speed and Feed

Traditionally, vitrified wheels have been made to operate at 5000 to 6500 sfpm. Now, however, manufacturers are looking critically at all factors involved in order to increase production. If adequate power is available, an increase in grinding wheel speed will generally increase productivity, and also abrasive costs will decrease because of a decrease in the force involved. Both of these principles can be shown mathematically.

$$\text{Production ratio} = \frac{v \times b \times d}{V}$$

where: v = work speed in sfpm
b = width of grind (in.)
d = depth of grind (in.)
V = speed of grinding wheel in sfpm

An increase in productivity ($v \times b \times d$) and a corresponding increase in wheel speed (V) will allow the abrasive cost to remain fixed even though the production rate is higher. Thus, production can be increased by increasing any one of the three factors (v, b, d) if a corresponding increase is made in V.

Grinding Force

In cylindrical grinding, the workpiece is usually the most compliant member of the machine's structural system. Thus the level of grinding force is often limited by the workpiece geometry. A diagram of the forces involved is shown in Fig. 7-9. The tangential force (F_t) or power component is limited by the power available from the wheel drive motor.

$$F_t = \frac{33,000}{V} \times HP$$

From this equation it is obvious why higher wheel speeds (V) are attractive, particularly for cylindrical grinding. If the wheel speed is doubled, twice as much power can be used without increasing the grinding force.

The normal grinding force (F_n), which tends to push the work away from the wheel, is related to F_t but is dependent on the wheel specifications, the work material, coolant, type and condition of wheel dressing, etc. In general, F_n is the highest when fine-grain wheels are used on hard materials and lowest with coarse-grained wheels on soft materials. For materials such as cast iron and steel, the F_n ranges from two to five times the F_t. As an example, a grinding machine operating at 6000 sfpm with a 10-hp motor would have resultant forces (F_r = vector sum of F_n and F_t) on the work ranging from a low of 120 lb up to nearly 300 lb. A wheel running at 12,000 sfpm would cut these forces in half.

High-Efficiency Grinding

As stated previously, the traditional speed for vitrified grinding wheels has been 6500 sfpm. Now industry is adapting to speeds of up to 16,000 sfpm and higher. Infeed rates are also comparably higher since the force on the work is reduced, as shown in the previous example.

However, high-speed grinding is not just a matter of increasing the wheel

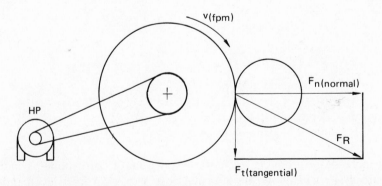

Figure 7.9. Tangential force is limited by the power available for the wheel but is not changed if horsepower and speed are increased proportionately.

speed and/or the infeed rate on conventional machines. The interaction of the various process variables must be considered and balanced with the operating variables such as machine condition, quality of surface finish required, coolant application, wheel structure, and the hardness of the material.

The beneficial effects of high wheel speeds are mainly the result of a change in the thickness of the uncut chip (t), as shown in Fig. 7-10. If the work speed is doubled for a given depth of cut and wheel speed, the cutting force will also be doubled, as shown in Fig. 7-10. Thus there are two operating conditions to consider: high work speed that produces high cutting forces and large metal removal, or high wheel speed with low cutting forces and no change in metal removal. A marriage of these two conditions is illustrated in Fig. 7-10, where the ability to remove metal is doubled, grain loading is reduced to normal values, and the only disadvantage is a very high heat input.

Thus, to increase the metal removal rate, the infeed rate must be increased proportionately with the wheel speed to present a full layer of material for each revolution. In theory this cuts the grinding time in half. However, production is not necessarily doubled since loading and unloading times must also be considered.

Grinding Operations

The wide variety of grinding operations can be classified by the main types of surfaces to be ground. These include cylindrical, surface, and internal surfaces.

Cylindrical grinding

As can be seen in the schematic drawing of Fig. 7-11, there are two distinct types of outside cylindrical grinders; those that hold the workpiece between centers or in a chuck, and those in which the work is rotated between two wheels, or *centerless grinding*.

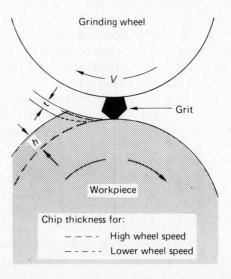

Figure 7.10. The effect of wheel speed on the undeformed chip thickness. (*Courtesy Society of Manufacturing Engineers.*)

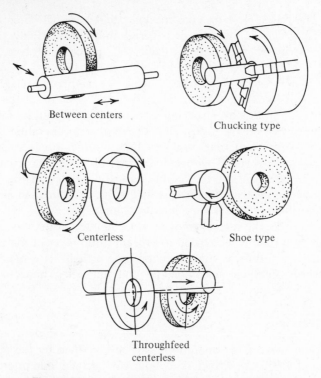

Between centers

Chucking type

Centerless

Shoe type

Throughfeed
centerless

Figure 7.11. Cylindrical grinding operations.

As the name implies, cylindrical grinding is used to grind cylindrical parts that may be straight, stepped, or tapered. The workpiece is driven by an independent motor and rotates the workpiece opposite to the wheel rotation at the point of contact.

Plunge cutting may be done by feeding the rotating workpiece straight into the wheel. Sometimes after the wheel has been plunged into the workpiece and the diameter is almost down to size, the feed is programmed to traverse longitudinally still cutting at the plunge depth. This is known as *peel grinding*. The action peels a layer of stock away from the workpiece cutting primarily on the flat face of the wheel.

Multiple-wheel grinders are often used when a workpiece has multiple diameters. In this case, the multiple wheels are fed into a preset depth but no traverse movement is used. A typical application of this type of machine is in grinding the bearing surfaces of a crankshaft or camshaft.

Centerless Grinding.   In centerless grinding, the workpieces are held between the grinding wheel, a regulating wheel, and workrest blade (not shown). By tilting the rotational axis of the regulating wheel with respect to the grinding wheel, the workpiece is given a longitudinal movement or *throughfeed*. Once in contact with the grinding wheel the workpiece automatically feeds to the far side of the grinding wheel. *Infeed* grinding is much the same as plunge grinding and is used on workpieces that have a shoulder larger than the ground diameter. The work is inserted and either the grinding wheel or the regulating wheel is fed in a preset distance to obtain the proper diameter. The wheels are then separated and the piece is either removed by hand or by an automatic ejector.

Shoe-Type Grinding.   Shoe-type grinding is characteristically used for such parts as bearing rings and parts of similar shape.

Magnetic adhesion keeps the workpieces in place but allows for sliding play for self-centering in the grinding process. This type of grinding avoids clamping the rings on the OD and the risk of deforming the workpiece circularity. The parts rest on a cemented-carbide shoe and turn with an opposite rotation to that of the grinding wheel, which is the same motion as that of centerless grinding.

Surface grinding

All grinding can be construed to be surface grinding. However, the term here refers to flat surfaces. Many types of machines are available for grinding flat surfaces depending upon the workpiece size and shape. The principal types are shown schematically in Fig. 7-12. The machines can be broadly divided into horizontal spindle and vertical spindle and further subdivided as reciprocating table and rotary table.

The most common surface grinder is the horizontal type with a high-speed reciprocating table. In this type the work is typically held on an electromagnetic chuck.

A variation of the reciprocating table type of machine is the rotating table which also has an electromagnetic holding surface. Whereas reciprocating table machines generally incorporate head-tilt capabilities, rotary table designs feature table tilt. This enables a flat-faced grinding wheel to cut dish-shaped workpieces to flat, concave, or convex cross sections.

Vertical-spindle machines are made so that the flat face of the wheel is in contact with the workpiece rather than only with the periphery of the wheel as in the horizontal spindle. With greater wheel contact area these machines are made so that the table can be tilted, which allows only the leading edge of the wheel to be in contact with the metal. This results in a greater unit force on the cutting abrasive and causes two grinding conditions to change: the wheel acts softer and power consumption is reduced. A softer-acting wheel allows it to break down before glazing (dulling condition) occurs. With reduced power consumption, higher feedrates can be maintained.

A particular advantage of the vertical surface grinder is that the work, whether multiple pieces or a single part, revolves in a continuous motion. There is no "end" to the cutting operation and therefore all parts in the same setup receive equal treatment.

Wheels used on vertical-spindle machines may be made in segments, held in a special adapter, or made in the form of a shallow cylinder. Some of them have wire or steel strap windings around the periphery.

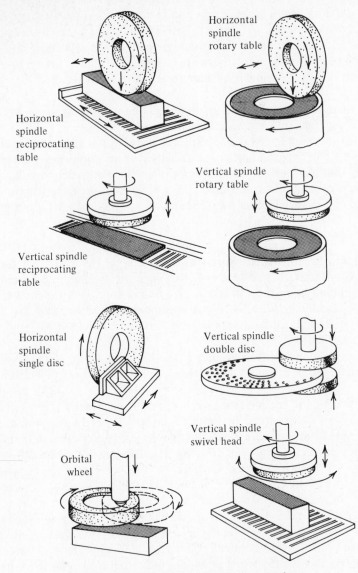

Horizontal spindle rotary table

Horizontal spindle reciprocating table

Vertical spindle rotary table

Vertical spindle reciprocating table

Horizontal spindle single disc

Vertical spindle double disc

Vertical spindle swivel head

Orbital wheel

Figure 7.12. Surface grinding operations.

The double-disc grinder is used extensively for grinding small parts in which flatness and parallelism must be closely monitored on a high-volume basis.

Internal grinding

Internal grinding is used to obtain very accurate size and good finish on holes made by other operations. Machines are made that are able to finish holes from a few thousandths of an inch in diameter to 5 feet or more. The basic types of

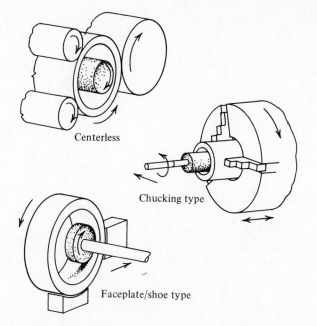

Centerless

Chucking type

Faceplate/shoe type

Figure 7.13. The basic types of internal grinders.

internal grinders are shown in Fig. 7-13; they are the chucking and centerless types. A variation of the centerless, faceplate shoe is also shown.

Chucking Type. The chucking-type machine is the most common type for internal grinding. Chucks may include four jaw chucks, collet chucks, faceplates, and magnetic chucks. One problem often associated with internal grinding is the pressure of the chuck jaws in causing the part to distort.

Centerless Internal. One of the advantages of centerless internal grinding is that the part need not be chucked, thus avoiding the distortion mentioned previously. Perfect concentricity can be maintained between the ID and the OD.

A slightly different variation of centerless internal grinding is shoe-type grinding, which is similar to external shoe-type grinding. The tubular part is held against a magnetic faceplate and rests on hard-faced shoe supports. An advantage of this system is the ease with which the workpieces can be loaded and unloaded. This method is especially useful for parts that require only a short cycle.

Most internal grinders are automatic. The operator loads the workpiece and starts the machine. Traverse to the work, cutting feeds, and dressing the wheel are all preset and done without operator assistance.

Internal Grinding Wheel Size. As with other grinding operations, it is desirable to maintain a cutting speed for the wheel of about 6000 fpm. For a $\frac{1}{2}$-in. dia. wheel this would mean an rpm of 46,000.

$$\text{RPM} = \frac{12 \times 6000}{\pi \times 0.5 \text{ in.}} = 46,000$$

Ultrahigh-speed grinding heads use high-frequency direct-drive motors or air turbines to reach speeds in the 200,000 rpm range. The relation between spindle speed and cutting speed is shown by the formula:

$$\text{sfm} = \text{rpm} \times \pi \times D/12$$

where D is the diameter of the grinding wheel in inches.

Abrasive Machining and Grinding Economics

Abrasive Machining.   Abrasive machining is a term used to denote those grinding operations in which the metal-removal rate is the main consideration, and surface finish and accuracy are secondary. It includes all operations where cost is a major factor in determining whether to grind or to use other machining methods.

There are a number of factors that make abrasive machining competitive such as: (a) rough and finish cuts can be done in one operation; (b) fixtures can be largely eliminated; (c) often cutting tools and their maintenance are carried as overhead costs whereas the abrasive wheel is charged as a direct expense to the job; fixtures can be eliminated or made much simpler; and the handling time is greatly reduced.

Another consideration that does not show up in direct comparison of machining methods is that less stock needs to be allowed for abrasives than for cutters. Thus there is also an initial savings.

Crush-Form Abrasive Machining.   Abrasive machining is most often associated with flat surfaces; however, advances in *crush-form grinding* have made it competitive to other machining methods. Crush-form grinding involves a cylindrical or centerless grinder in which the wheel is formed by a hardened steel or carbide roller of the desired profile (Fig. 7-14). The dressing wheel is driven by the grinding wheel, which rotates at approximately 300 sfpm. The infeed rate is a few thousandths of an inch per revolution. Forming the wheel takes only a few minutes, and redressing it only a few seconds. Figure 7-15 is an automotive front-wheel spindle that is abrasive machined in a single *plunge cut* to finish tolerances of 0.0007 in. on a crush-true grinder. Plunge cut refers to feeding the wheel straight into the work with no table traverse movement. In the case of the automotive spindle, the total abrasive machining time was 42 sec compared to 15 min by conventional machining.

Grinding Economics.   A detailed analysis of grinding costs is not within the scope of this text. The analysis can become somewhat cumbersome because there are more variables to be considered in grinding than in any other metal-

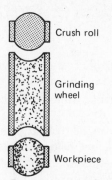

Figure 7.14. The grinding wheel may be crush-formed with a hardened steel roller with almost perfect transfer of form.

Figure 7.15. An automotive spindle abrasive-machined from a rough casting to a finished dimensional tolerance of 0.0007 in. in 42 sec. (*Courtesy Bendix Automation and Measurement Division.*)

working process. In addition to all the variables associated with a grinding wheel, such items as setup, load and unload, and dressing times, which in some cases may be a significant portion of the grinding cost, must be included. However, these costs are normally independent factors and can be calculated by a separate analysis. What remains is the specific grinding cost, which depends on the parameters used with a particular wheel on a specific workpiece and is associated with removing a definite amount of material from the workpiece, as measured in in.^{3}/min.

A commonly used equation for calculating the specific grinding costs is

$$C = \frac{C_a}{G} + \frac{L}{tF}$$

where: C = specific cost of removing a cubic inch of material
C_a = cost of abrasive in $/in.^{3}
G = grinding ratio (volume of material removed/volume of wheel used)
L = labor and overhead charge in $/hr
F = machine feed rate in in.^{3}/hr
t = fraction of time the wheel is in contact with the workpiece

Examining each term at the right of this equation we have:

$$\text{Specific wheel cost, } \frac{C_a}{G}$$

The wheel cost in removing a cubic inch of material is a function of G. An aluminum oxide wheel, for example, may have an abrasive cost of $.30/in.^{3}. Thus, with a grinding ratio of 10 on a particular workpiece, the wheel cost would be $.03/in.^{3} of material removed. G may be easily determined by weighing the wheel and the work before and after grinding.

$$\text{Fixed labor and overhead cost, } \frac{L}{tF}$$

This term shows the labor and overhead costs, assuming the wheel acts as a perfect cutter and the feedrate equals the material removal rate. The factor t, the fraction of time in which the wheel is in contact with the work, acts to increase cost as it becomes smaller by taking into consideration such items as override and approach. (This factor need not be considered in a vertical-spindle surface grinder since the wheel is in constant contact with the workpiece.)

Example: What is the specific wheel cost of grinding 0.030 in. off of both sides of a gray cast iron plate 1" × 10" × 10" using an aluminum oxide wheel and with the following conditions: labor and overhead = $20/hr, length of stroke = 12 in., grinding ratio = 0.5, feedrate is adjusted to 6 in.^{3}/hr, wheel cost = $.50/in.^{3}, fraction of time of wheel contact = $\frac{10}{12}$ = 0.83%.

Solution:

$$C = \frac{(.50)}{(0.5)} + \frac{(20)}{(.83)(6)} = 5.00$$

Therefore the specific grinding cost of removing 0.030 in. from both sides of the plate would be $5.00 × 2 = $10.00.

From the foregoing discussion some generalizations can be made:

1. As G becomes larger, the cost of using the wheel decreases.

2. As the feedrate F increases, costs become less.
3. If the labor and overhead cost L increases, the cost of using a high-wearing wheel increases.

Abrasive-Belt Machining

In the past, cloth-coated abrasives have been regarded as being for polishing or finishing operations. In more recent years the picture has changed, so that now they are considered competitive with grinding. Some abrasive-belt machines are now built with 150-hp motors and can be used in plunge grinding operations.

Abrasives.   Five types of abrasive minerals are used, both natural and synthetic: flint, emery, garnet (natural), aluminum oxide, and silicon carbide (synthetic).

The grit size varies from a 12-mesh number, which is the coarsest, to a 600-mesh number, which is the finest. Simplified markings are also used, such as very fine, fine, medium, coarse, and very coarse. Just as in grinding, the metal-removal rate and surface finish will be determined by the coarseness of the abrasive, the surface foot speed of the belt, and the spacing of the abrasive grains. Closed and open coats are shown in Fig. 7-16. The open coat is used for faster cutting and on softer materials that tend to load the closed coat.

Two methods are used in applying the abrasives to the cloth backing: the gravity process and electrostatic process. In the gravity process, the mineral grains are dropped from an overhead hopper onto the adhesive-coated cloth. In the electrostatic process, the mineral grains pass through an electrically charged field. As the mineral and the backing pass simultaneously through the electrostatic field, the mineral grains are propelled upward and are imbedded in the adhesive on the cloth backing. This process results in having the sharpest edge of the mineral grains exposed for the best cutting action.

Process Parameters.   There are many factors to be considered in the use of abrasive belts for metal removal. For example: grit size, surface finish, belt speed, contact wheel, lubrication, and pressure. Each of these factors will be discussed briefly.

Grit Size.   The rate of stock removal will vary inversely with grit size. A grit size of 220 will be extremely slow. As the grit sizes grow the rate of metal removal increases almost linearly until a grit size of 36. Above 36 the curve rises exponentially.

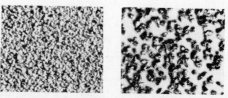

Figure 7.16.   Closed and open coat patterns of abrasive cloths. (*Courtesy 3M.*)

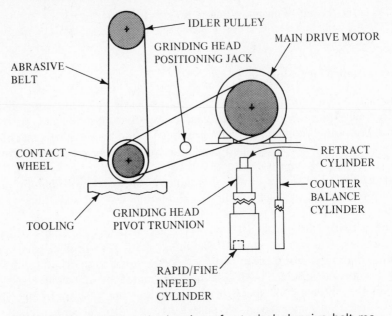

Figure 7.17. A schematic drawing of a typical abrasive belt machine. The most important part (after the belt) is the contact wheel which controls metal removal rate and finish. (*Courtesy* Manufacturing Engineering.)

Surface Finish.   The surface finish for a given grit changes as the belt is used. Finish varies less in the last 70% of use than it does in the first 30% of the belt life.

Belt Speed.   This parameter governs the rate of metal removal and the surface finish. The rule is simple; the faster the belt speed, the greater the metal removal rate and the finer the finish.

The Contact Wheel.   The contact wheel, shown in Fig. 7-17, controls many parameters. The diameter of the wheel affects efficiency. The smaller the wheel, the greater the rate of stock removal—and therefore the greater the efficiency. Usually the larger the wheel the better the finish.

The harder the contact wheel, the better the penetration of the abrasive particles and therefore the better the rate of stock removal. Generally, the softer the wheel the finer the finish.

Lubrication.   Both the stock-removal rate and the finish are affected by lubrication. A film of oil produces the best overall results. Too much oil or grease reduces the stock-removal rate. Cutting dry or with a water emulsion produces initally high metal-removal rates but drops off rapidly. The thicker the film of lubricant, the better the surface finish.

Table 7-1. Material removal rate for abrasive belt machining. (*Courtesy* Manufacturing Engineering.)

Material	Hardness	Feedrate (ips)
2024 Aluminum	131 Bhn	0.8
Cast Iron (Gray, Malleable, Ductile)	187 Bhn	0.4
416 Stainless Steel	94 R_B	0.3
1018 Mild Carbon Steel	155 Bhn	0.17
Carbon Steel Weldment	200 Bhn	0.15
1080 Carbon Steel	289 Bhn	0.15
304 Stainless Steel	40 R_C	0.125
1023 Cast Steel	25 R_C	0.125
Ni-Hard (Elvarite)	375 Bhn	0.125
420 Stainless Steel	52 R_C	0.08
Inconel 1100 Series	45 R_C	0.08
Rene 95 Series	45 R_C	0.08
Ni-Hard (Elvarite)	65 R_C	0.08
T15 Tool Steel	70 R_C	0.08

Pressure. Pressure is not really a process parameter but rather a process variable that changes from low to high as the belt progressively wears. It is desirable to keep a constant cut. The work of gradually increasing the pressure is called *pressure programming*. Because of the many variables involved, actual values for programming must be obtained on the job.

Metal-Removal Rate. Many variables concerning the metal-removal rate have been discussed. The 3M Company has developed a standard test for the metal-removal rate that provides insight into the capabilities of the process. In this test a 1-in. square (25.4 × 25.4 mm) bar approximately a foot and a half in length (457.0 mm) is fed into a coated abrasive wheel. Table 7-1 lists the feedrate for a variety of materials in inches per second. These figures can easily be converted to cubic-inch removal rate.

Lapping, Honing, and Superfinishing

Designers are aware of the ever-increasing demands for machines that can run faster, last longer, and operate more precisely than ever. This requires bearings, shafts, and seals that are dimensionally and geometrically accurate. Even grinding cannot meet the accuracy and surface finish required on a production basis.

As industry seeks to approach perfection, interest has focused on a family of precision-finishing processes: lapping, honing, and superfinishing. These processes are often referred to as *microfinishing*.

Each of these processes is designed to generate a particular surface finish and to correct specific irregularities in geometry. Briefly stated, microfinishing is used on parts:

1. Dimensioned with tolerances in the fourth or fifth decimal place.
2. With surface finish requirements in the 1-microinch to 8-microinch range.
3. Where an accurate geometric shape must be maintained.
4. Where surface integrity is critical.

Lapping.   Lapping is an abrading process done either with loose-grain abrasive or with bonded-abrasive wheels or discs.

In the loose-grain process, a fine-grain grit of silicon carbide, diamond, or aluminum oxide is mixed with a carrier (usually oil) and placed on the *master*. The master is usually a soft porous metal such as gray cast iron which holds the abrasive temporarily as it rubs or abrades against the surface to be cut. For example, gear teeth are lapped by running the newly cut gear with a master gear that is supplied with abrasive compound. Valves are often lapped to valve seats by rotating the two surfaces together with a lapping compound between them. The type of abrasive and carrier are dependent on the materials to be lapped. Soft materials are lapped with aluminum oxide and hard materials with diamond or silicon carbide grit. As contrasted with other abrasives, diamonds do not break up and become smaller during the lapping process. The permanence of the diamond particle, which cuts at a constant rate, means relatively consistent lapping times.

It is especially important in using diamond-lapping compounds that the grit be uniformly graded. If not, the lapping load will fall on the larger particles, which will cause scratches (Fig. 7-18). For diamonds, an oil-base vehicle is no longer used. A diamond compound has been developed that consists of graded diamond particles held in a permanent uniform suspension. A visual grade identification of the diamond compound is attained by adding a liquid color to the vehicle.

The most popular way of dispensing the diamond compound is by a plastic syringe. The syringe enables the operator to deposit metered amounts of the compound on the lapping surface without exposing the diamond compound to contamination.

There is some difference of opinion as to whether the process can be called lapping if the abrasive is entirely free, that is, with no imbedding in the master. In this case, it may be referred to as *free-abrasive machining*.

Figure 7-19 shows a lapping machine used to produce flat surfaces. The parts to be lapped are confined to cages that impart a rotary and gyratory motion at the same time, covering the entire surface of the lapping table. Parallelism is maintained by having a stationary lapping plate on top of the workpieces. Cylindrical parts are lapped by a cage arrangement as shown in Fig. 7-20.

Figure 7.18. The diamond grit must be uniformly graded to be effective. Large grit particles will do all the work and also cause scratches.

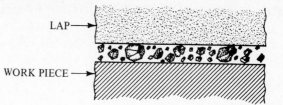

LAP →

WORK PIECE →

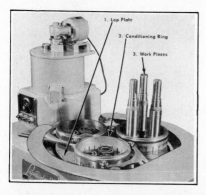

1. Lap Plate
2. Conditioning Ring
3. Work Pieces

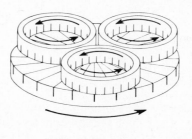

Examples of workpieces with lapped surfaces

Figure 7.19. Work to be lapped is placed within the conditioning rings, which are held in place but are free to rotate. The work tends to abrade the lap plate but the rotating action of the conditioning rings causes the lap plate to wear evenly, maintaining a flat surface. Standard machines handle parts from ⅛ in. to 32 in. in cross section. Steel, tool steel, bronze, cast iron, stainless steel, aluminum, magnesium, brass, quartz, ceramics, plastics, and glass can be lapped on the same lap plate. (*Courtesy Crane Packing Co.*)

273

Figure 7.20. A cage arrangement used for cylindrical lapping. (*Courtesy Warner & Swasey Co.*)

Bonded-abrasive lapping is similar in appearance to the machine shown in Fig. 7-20. The upper section contains the bonded abrasive.

Advantages and Limitations. Extreme accuracy in both tolerance and geometry are the hallmarks of lapping. As an example, gage blocks used as a measurement standard are lapped to flatness of ±0.000003 in. and a parallelism of ±0.000002 in. One of the reasons this can be accomplished is that very little heat is generated in the lapping process.

Since lapping is essentially a finishing operation, parts should not be far from the expected size and geometry before lapping.

The residue of compound left on the surface of the part after lapping must be removed. This is not true of bonded lapping. Although lapping appears to be a simple process, it is considered somewhat of an art to achieve the extremely close tolerances expected.

Honing. Honing is a controlled, low-velocity, abrasive-machining process that is most often applied to inside diameters or bores. One or more abrasive stones and one or more nonabrasive shoes (to equalize force throughout the workpiece) are mounted on an expanding (or contracting) mandrel (Fig. 7-21). The tool is inserted into the bore and adjusted to bear against the walls. A combination of motions gives the stones a figure-eight travel path. This motion causes the forces acting on the stones to be continually changing their direction, equalizing wear.

Advantages and Limitations. As with lapping, very little heat is generated, hence there is no submicroscopic damage to the workpiece surface. Tolerances can be easily maintained to within 0.0001 in. on a production basis. The surface finish produced can be very smooth (32 to 44 microinches) and part geometry can be corrected as shown in Fig. 7-22. The amount of stock removal is usually slight (0.001 to 0.0025 in.), so the cycle times are relatively short, usually 40–50 parts per hour. Materials that may be honed include soft materials like silver and brass as well as very hard materials, including the hardest alloys, carbides, and ceramics.

More recently single-stroke bore finishing has been introduced for high pro-

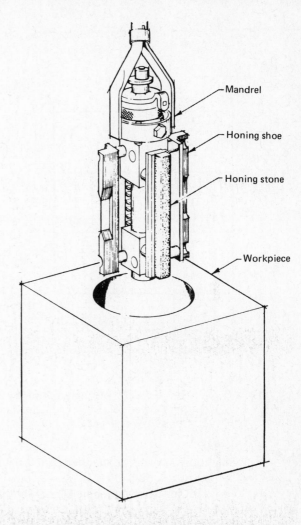

Labels on figure: Mandrel, Honing shoe, Honing stone, Workpiece

Figure 7.21. The shoes and stones are uniformly spaced on the internal hone to produce even cutting forces. Honing does not distort the workpiece and produces a nearly perfect diameter. (*Courtesy* Machine Design.)

duction. The process utilizes what are sometimes referred to as superabrasives—diamonds and CBN. The hone makes only one complete stroke, in and out of the bore. Typical tolerances that can be maintained are 0.0002 in. (0.005 mm) for the diameter, and 0.00005 in. (0.013 mm) for roundness and straightness. Stock of up to 0.001 in. (0.03 mm) is removed from the bore in one stroke.

Superfinishing. Superfinishing is somewhat similar to honing but is applied primarily to outside surfaces. The process is particularly useful in finishing bearing surfaces. In ordinary use, the oil film between mating members is in danger of being punctured by any sharp points left by grinding. When the sharp points penetrate the oil film, the concentration of weight and friction causes the two metal surfaces to momentarily weld and tear apart, causing rapid-wear

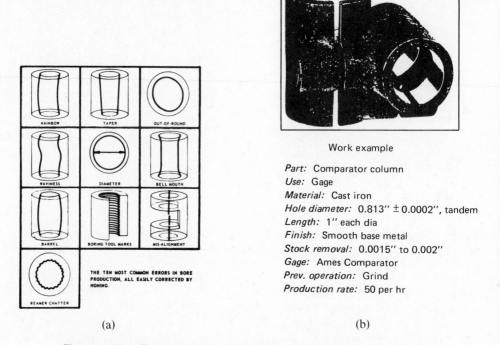

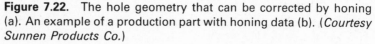

Work example

Part: Comparator column
Use: Gage
Material: Cast iron
Hole diameter: 0.813″ ± 0.0002″, tandem
Length: 1″ each dia
Finish: Smooth base metal
Stock removal: 0.0015″ to 0.002″
Gage: Ames Comparator
Prev. operation: Grind
Production rate: 50 per hr

(a)   (b)

Figure 7.22.   The hole geometry that can be corrected by honing
(a). An example of a production part with honing data (b). (*Courtesy
Sunnen Products Co.*)

conditions. Also, the minute particles sloughed off by this process further score
the bearing surfaces.

The superfinishing process, as shown in Fig. 7-23(a), provides a gentle pressure over a wide contact area. The surface irregularities are reduced to a "near
perfect" surface finish. The amount of material removed is usually small, about
0.0001 to 0.0004 in. The surface finish produced ranges from less than 1 to 80
microinches, with the average being around 3 microinches. When a surface
smoother than 3 microinches is desired, the time needed to produce it goes up
rapidly. As an example, a part that can be brought to a 3-microinch finish in
1 min may require 3 min more to bring it to a 2-microinch finish or better.

Flat surfaces may be superfinished as shown in Fig. 7-23(b). The upper
spindle is a springloaded quill on which the stone is mounted. The lower spindle
carries the work. Both spindles are parallel so that, when they rotate, the result
is a very flat, smooth surface.

Advantages and Limitations.   Superfinishing produces excellent surface
finish in a minimum time after grinding, honing, or lapping. The type of surface
finish may be preselected, ranging anywhere from 1 to 80 microinches. The
process is easily adapted to high-volume production.

Superfinishing cannot be used to correct feature location such as concentricity or out of squareness.

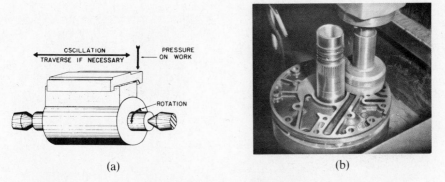

Figure 7.23. A schematic of the superfinishing process for round surfaces (a), for flat surfaces (b).

Deburring Processes

One of the most troublesome spots in the production of metal parts is the burr that remains after the cutting operation is finished. In most cases it must be removed before another operation can be performed. Formerly, much of deburring was a hand operation accomplished by the use of a file. Now the manufacturing engineer has a wide choice of processes that can be used for abrasive cleaning and deburring. Some of the more common methods used are barrel finishing, vibratory finishing, abrasive flow machining, thermal deburring, and electropolishing.

Barrel Finishing. Barrel finishing or tumbling consists of putting a number of parts in a steel drum or barrel that is usually six- or eight-sided or has shelves inside. The abrasive medium used and the proportion in relation to the drum capacity is very important. Some operations are done dry, but the results are usually enhanced when done wet.

There are more than 25 kinds of abrasive media used in tumbling and vibratory finishing.

The two most commonly used for today's high-speed, mass-production requirements are plastic and ceramic of various shapes, as shown in Fig. 7-24.

Bonded Media. Plastic-bonded preformed media are now used extensively because they are lightweight and tough (break-resistant). Finishes can be produced to as low as 2 microinches. The plastic or synthetic bond mixed with a finely grained silica-flour abrasive produces a "plater's polish," yet is relatively fast.

A faster-cutting medium is a 220-grit size, fused-aluminum oxide abrasive. This aggressive media is especially good for nonferrous metals where machine lines or milling marks must be removed. It also leaves a good surface for paint adherence.

Preformed silicon carbide media are used mostly on exotic metals such as zirconium, titanium, zircoloy, and stainless steel.

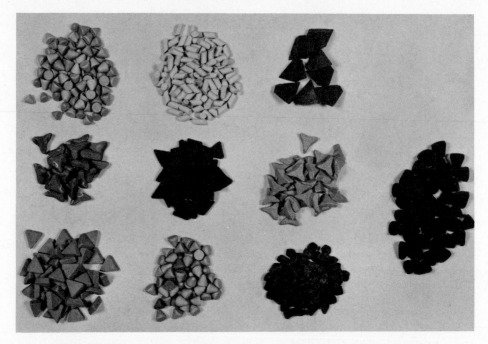

Figure 7.24. Various media as used in tumbling operations. Plastic media are shown in the top row, ceramics in the middle row, and various aluminum oxide abrasive chips in the bottom row. (*Courtesy Almco, Queen Products Div., King-Seely Thermos Co.*)

High-luster finishes can be obtained by the use of abrasive impregnated walnut shells or peach pits.

Compounds. Commercial compounds are composed of synthetic wetting agents, water conditioners, and various abrasives such as aluminum oxide, quartz, silicon carbide, garnet, silica flour, and cleaning chemicals. The purpose of the compound, which may be in either liquid or powder form, is to keep parts and media clean, inhibit corrosion, lubricate, cut, cushion parts against damage, and suspend tiny, loose, abrasive particles. Chemical grinding compounds, in addition to maintaining a clean tub, keep the abrasive media from becoming glazed. Burnishing compounds are designed to impart high-luster finishes on all metal parts.

Descaling chemical compounds are used to remove heat-treat scale, smut, oil, grease, and rust, and to "bleach" the metals. A positive step toward pollution control has been the development by various suppliers of a line of biodegradable compounds.

Machine Features.   The two primary styles of vibrating finishing machines are the tub-type and the round bowl unit (Fig. 7-25). Vibratory motion is typically induced in the tub or bowl-shaped chamber by an eccentric weight system

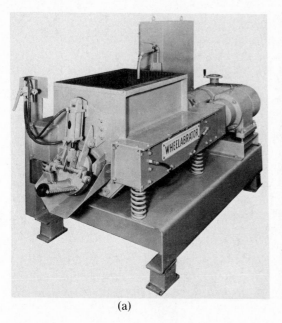

(a)

(b)

Figure 7.25. Two primary types of vibratory finishing machines are the tub-type (a) and the round bowl (b). The action of the tub-type shown at (c) consists of two shafts mounted eccentrically, driven by a motor to produce an aggressive scrubbing action. (*Courtesy Wheelabrator Corporation and Almco Division, King-Seeley Thermos Co.*)

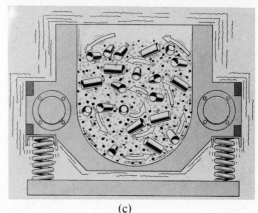

(c)

mounted on a drive mechanism. The chamber is usually mounted on coil springs to isolate the machine base from the vibrating chamber. Adjusting the degree of eccentricity and/or drive speed induces a rolling motion in the media/parts mass resulting in a constant scrubbing action between media and parts.

The tub-type machine can run both small and high production parts particularly when set up on a continuous basis. Some long, continuous tubs range up to 20 ft (6.1 m) in length. Modular type machines have been built to handle parts over 100 ft (30.5 m) long.

Round bowl or toroidal machines have a doughnut-shaped chamber positioned horizontally. Parts and media flow in a toroidal pattern around the bowl. At the end of a cycle, parts and media are deflected up onto a screen and the medium drops back into the bowl. The parts are automatically carried out across the screen deck. Other features frequently included are magnetic separators and secondary screen decks for separating parts with critical finish requirements.

Abrasive-Flow Machining (AFM).   Abrasive-flow machining is a relatively new process that involves the controlled extrusion of a specially formulated abrasive-laden media.

Fixtures are used to hold the part in relation to the machine and to direct the flow of media across the area requiring finishing.

The medium consists of abrasive grains suspended in a semisolid matrix. Included with the abrasive polymer are lubricants that reduce the friction of the material as it extrudes through the workpiece. The puttylike medium flows along a given path carrying the abrasive grain with it. When the medium reaches a point of restriction (caused either by a burr or intentionally by tool design), the outermost grains in the flow stream are forced against the workpiece surface and a gentle grinding action occurs.

One example of an application of AFM is removing burrs from the inside of oxyacetylene welding tips. The tip hole 0.026 in. (0.66 mm) in dia. showed a superior quality of flame over the unhoned tip. Over 1000 tips per hour are honed using a 20-cavity fixture.

Thermal Deburring.   Thermal deburring is also a relatively new process that utilizes heated gases. The part to be deburred is placed on a pallet and pushed into the thermal chamber. A gas mixture of 62% hydrogen and 38% oxygen is forced into the chamber. An igniter, similar to that of a spark plug, detonates the gas mixture. Combusion takes place in about 2 milliseconds and the temperature climbs to over 6000°F (4947°C). The extreme temperature and the pressure, in excess of 6000 psi, vaporizes and oxidizes small objectionable burrs, but does not harm the parent metal.

Electropolishing and Buffing

Electropolishing.   Electropolishing is the reverse of electroplating, that is, the work is the anode instead of the cathode and metal is removed rather than added. Many products can be efficiently and economically electropolished.

The process is especially good on thin sheet-metal parts, intricately formed wire products, and deeply recessed complex parts. The metal loss is low, from about 0.0002 in. to 0.001 in., depending upon the original surface. Brightness and regularity of finish are dependent on uniform metal removal; therefore a smooth, dense surface, free from scratches and tool marks, is necessary.

Generally a good, bright finish may be obtained in 2 to 10 min, depending upon the original surface and range of current densities used. One ampere per square inch will remove 0.001 in. in a 10-minute period. Premixed acids may be obtained for electropolishing from E. I. du Pont de Nemours & Co., Inc.

Buffing.   Many types of contact wheels have been developed for polishing and buffing surfaces. Finishes can be varied from that of textured satin to a high luster. Harder wheels are made out of abrasive-impregnated rubber with plain, serrated, or cogtooth surfaces.

Buffing is used to produce a high luster or bright appearance. To achieve this, the abrading action is reduced to a minimum. A lubricant blended with the abrasive promotes a flowing rather than a gouging action.

Buffing is usually divided into two operations: (1) cutting down, and (2) coloring. The first operation is used to change a relatively rough surface into a smoother one. The second produces a high luster.

The abrasives used are extremely fine powders of aluminum oxide, tripoli (a porous form of silica), crushed flint or quartz, silicon carbide, and red rouge (iron oxide). The degree of ductility of the metal determines the type of abrasive to use. Soft metals, for example, do not require a cutting abrasive but a flow or blending as achieved by red rouge.

Buffing wheels are made from a variety of soft materials. The most widely used is muslin. Wheels of flannel, canvas, sisal, and heavy paper are used for special applications. A newer type made of nonwoven nylon-web material has been developed to produce a satin finish without sacrificing accuracy (Fig. 7-26).

Muslin wheels are made in many forms (Fig. 7-27) to provide a wide range of properties. Stitched wheels are usually used for cut-down buffing. The wheels

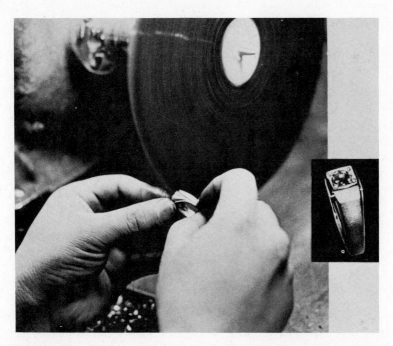

Figure 7.26.   Nonwoven nylon-web buffing wheels used to produce a satin finish without sacrificing accuracy. (*Courtesy American Buff International, Inc.*)

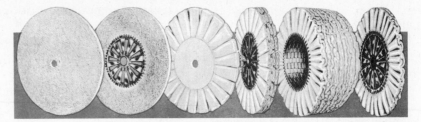

Figure 7.27. A variety of buffing wheels ranging from the hard stitched to the softer pocketed type. (*Courtesy American Buff International, Inc.*)

are made 0.25 in. thick and two or more are mounted on the arbor to provide the desired thickness.

Pocketed buffing wheels are used for cut-down operations on irregular contours.

Loose buffing wheels are stitched only at the hub and are used for color buffing.

The abrasive may be applied in liquid form by letting it slowly drip on the wheel or by holding a stick of the compound up to the rotating wheel. The heat of the contact is sufficient to transfer the abrasive to the wheel. Some of the sticks are made with various oils and greases and others with glues or cements. The advantage of the latter is that degreasing is not required before lacquering.

Problems

7-1. Why are Al_2O_3 wheels not used for grinding WC materials?

7-2. What abrasive is recommended for hard, tough tool steels? Why?

7-3. (a) Which diamond classification is recommended for grinding WC?
(b) How does it compare in impact with the regular RVG diamond?

7-4. (a) What are the four main factors to consider in selecting the right grinding wheel for a given job?
(b) After making the grinding-wheel selection, you notice it has a tendency to "load up." What change should be made?

7-5. What is meant by G-ratio in grinding?

7-6. (a) Fifteen 2-in. rectangular flange heads are machined on a turret lathe, one at a time. The setup time was 1.09 hr. The machine time per piece is 0.296 hr. As a compar-

ison, the same parts were abrasive machined on a vertical grinder, where the setup time was 0.4 hr. The total amount removed from the flange was 0.030 in. The grinding chuck can accommodate eight flanges per run. The downfeed rate was 0.015 ipm. Compare the time per lot by each method.
(b) What is the time saved by grinding, given in percent?

7-7. (a) A new 10 in. dia., 2 in. wide wheel was placed on the grinder. A dressing operation was performed to true the wheel to the spindle. Four passes, each of 0.001 in., were made. After grinding a cast iron plate 5 in. × 15 in. from a thickness of 1.5 in. to 1.484 in., the wheel diameter measured was 9.980 in. What was the grinding ratio?
(b) Comment on this grinding ratio.
(c) What is the actual wheel cost for this operation if the lost abrasive is valued at 90 cents/in.^{3}?

7-8. Compare the time for milling and grinding both sides of a 40 in. × 40 in. plate. Assume the following milling conditions: a 10 in. dia. face mill is used that has 14 indexible carbide inserts. Allow an overlap of an inch width per pass. The chip load per tooth is set at 0.008 in. The loading and unloading time is 8 min. Assume the loading and unloading time for grinding is 3 min per part. The longitudinal speed is 50 fpm. The crossfeed is 1 in. per pass, and approach and overtravel are 1 in. each. Wheel width = 20 in.

7-9. (a) If a cylindrical grinder with a 10-hp motor has the capability of increasing the wheel speed from 6000 spfm to 15,000 sfpm and the infeed used was kept constant, what would be the change in tangential force on the work material?
(b) What would the approximate normal force be if the operation (a) were done with a fine-grain wheel on hard material for each wheel velocity?

7-10. Can the wheel infeed be increased proportionately with the increase in wheel velocity? Why?

7-11. A hardened steel plate 1 in. × 8 in. × 16 in. is to be ground on both 8 in. × 16 in. sides with the following conditions:
Wheel diameter after dressing = 9.88 in.
Wheel width = 0.750 in.
Wheel cost = $0.90/in.^{3}
Labor and overhead charge = $10/hr
Traverse feedrate = 0.5 in./pass
Material removed from each side = 0.1020 in.
Wheel diameter after grinding = 9.850 in.
Length of stroke = 18 in.
Longitudinal feed = 1440 in./hr, 2 cuts/side
What is the specific grinding cost for this operation?

7-12. Distinguish the difference in the action of the abrasive between the three grinding modes.

7-13. If work speed, width of grinding wheel, and depth of cut are increased will there be any significant increase in production? Explain why or why not.

7-14. What is the main limiting factor on the force used in cylindrical grinding?

7-15. How can the tangential force be kept down even though with a deeper cut the tangential force would normally increase?

7-16. Why must the infeed rate of a grinding wheel be proportionate to the wheel speed?

7-17. In plunge-cut grinding can the wheel be moved longitudinally along the work axis after the plunge? Why or why not?

7-18. What type of grinder is well suited to grind bearing rings?

7-19. What is the effect of tilting the wheel slightly on a vertical-spindle surface grinder?

7-20. What is a problem associated with chucking work for internal grinding?

7-21. What is the relative hardness of CBN compared to diamonds and silicon carbide?

7-22. At what rpm would a $\frac{1}{4}$-in. dia. internal grinding wheel have to be run to maintain a surface foot speed of 6000 fpm?

7-23. How does abrasive machining differ from surface grinding?

7-24. Why is it that the grinding wheel does not grind the crushing roller?

7-25. How long would the actual machining time take to remove a 0.020-in. layer of 416 stainless steel from one side of a 3 ft × 6 ft plate by abrasive-belt machining?

7-26. What is the main difference in using diamond grit for lapping as compared to other abrasives?

7-27. What finish can be produced by tumbling or vibrating finishing?

7-28. What is the main advantage of electropolishing?

7-29. What are the two main kinds of buffing?

7-30. What are some of the abrasives used in buffing?

7-31. About how much force would be exerted on the workpiece if a grinding wheel is operating at 6000 sfpm?

7-32. When a grinding wheel is new for a given surface grinder it is 10 in. in dia. The spindle rpm is 2500. How far could this wheel wear down before the surface foot speed would be below 5000 fpm?

Bibliography

Fundamentals of Grinding. Westborough, Me.: Bay State Abrasives, Division of Dresser Industries, Inc., 1979.

Brown, E. L.; F. L. Shierloh; and A. R. McMillan. Dearborn, Mich.: "High-Speed Plunge Grinding." *SME Tech Paper MR 79-952.* Society of Manufacturing Engineers, 1979.

Carlson, G. A., Jr. "Advances in Abrasive Finishing." *Manufacturing Engineer* 82 (February 1979):59–62.

Centerless Grinding. Cincinnati: Cincinnati Milacron, 1975.

Cosler, Ronald I. "High Speed Abrasive Machining." *Automation* 22 (October 1975):52–56.

Dallas, Daniel B. "What You Should Know About Abrasive Belt Machining." *Manufacturing Engineering* 98 (June 1977):59–64.

Jablonowski, Joseph, ed. "Fundamentals of Grinding." *American Machinist Special Report No. 684* 120 (February 1976):65–88.

Lindberg, Charles E. "Double Disc Grinding—A Three Part Series." *Cutting Tool Engineering* 30 (May–June):12–14; (July–August):12–14; (September–October 1978):8–12.

Spiotta, Raymond H. "Abrasive Flow Machining." *Machine and Tool Blue Book* (February 1976).

Stauffer, Robert N. "Abrasive Belt Machining Equals High-Speed Machining." *Manufacturing Engineering* 83 (December 1979):69–70.

Stauffer, Robert N. "What You Should Know About Vibratory Finishing." *Manufacturing Engineering* 83 (July 1979):48–54.

Williams, Robert N. "Keeping Grinding Costs Down." *Welding Design and Fabrication* 51 (March 1978):128–132.

CHAPTER EIGHT

Metal Forming

In recent years there has been an unprecedented interest in metal-forming technology. This is due in part to the continuing demands on industry to produce lighter weight, yet stronger, more rigid components. Ever higher strength-to-weight ratios are being used in forming parts for missiles, aircraft, and more recently for automobiles. As new metals are tried, such as aluminum alloys and high-strength, low-alloy steels, they are found to be less forgiving than mild steels. Higher forming forces are required, which often result in more fractures.

In the past the quality of a stamping depended on the skill of the artisan, who depended on knowledge gained by trial and error. Now, with new materials and greater demands, industry cannot rely on lengthy trial-and-error methods. Therefore considerable work has been done by both metallurgists and process engineers to provide a method of predicting formability of metals.

FORMABILITY

A simple definition of formability is the ease with which metal can be forced into a permanent change of shape. Although formability is easy to define, it is an elusive quality to measure. No single index can reliably predict the formability of a specific material for production conditions. A material that is readily formable for one stamping design may break when used in a die of a different configuration. This necessitates die tryout and modification which can be time-consuming and costly. Since there is no simple solution to this problem, it is necessary to understand the attempts at predicting sheet-metal forming.

Predicting Sheet-Metal Formability

In the past there have been two main methods of determining sheet-metal formability, testing the mechanical properties and testing the forming properties by simulating forming operations. The mechanical properties most often tested are hardness and tensile properties.

Hardness Testing.   The most popular test for the mechanical properties of sheet metal is the Rockwell hardness test. The number obtained is a measure of the metal's resistance to penetration by an indentor. The test is quick and easy to perform, which accounts for its popularity. However, there are problems associated with the test and using the numbers obtained as an index of formability. Readings are sensitive to surface conditions, flatness of the specimen, and test procedures. The thickness of the sheet must also be taken into account. For example, sheet metal 0.035 in. (0.89 mm) thick and softer than R_b44 will have an erroneous reading because the indentor will penetrate too far into the metal and the hardness of the anvil supporting the metal will influence the reading.

Hardness values are strongly affected by the amount of cold-working near the surface of the material as in temper-rolling. This makes a soft material appear harder than it is.

Keeping in mind the errors that may be introduced, only a generalization can be made that steels with high Rockwell hardness readings are more difficult to form. Hardness tests are useful, however, in making comparisons between several shipments of identical grades of steel. The tests serve to indicate whether all the steel was processed in the same way.

Uniaxial Tensile Test.   The tensile test, described in Chapter 2, will yield more accurate information about the fundamental mechanical properties of the sheet metal than a hardness test. A stress–strain curve of a temper-passed mild steel is shown in Fig. 8-1. Temper-passed means that the sheet metal has been lightly cold-rolled to improve flatness, minimize stretcher-strains, and obtain the desired texture and mechanical properties. Stress–strain curves of a temper-rolled and an annealed sheet are shown in Fig. 8-2. The annealed sheet has a yield-point elongation. It is the yield-point elongation that produces the stretcher-strains in sheet-metal forming. The stretcher-strains, often referred to as Luder's lines, are irregular surface patterns or ridges and valleys that develop during the forming of annealed or aged sheet steel (Fig. 8-3). Aging is a term applied to changes that take place in the mechanical properties of a temper-rolled sheet with the passing of time. Sheet products become less ductile partly because of the dissolved nitrogen in the metal. Aluminum-killed steels have very little dissolved nitrogen, the nitrogen being combined with the aluminum in the form of stable nitrides, and thus they have practically no susceptibility to strain-aging.

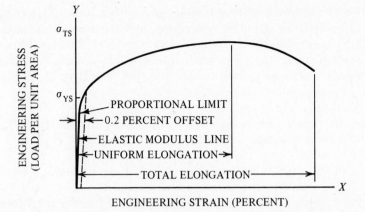

Figure 8.1. A stress–strain curve of a temper-passed mild steel.

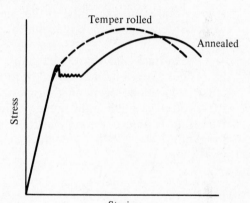

Figure 8.2. Stress–strain curve for both temper-rolled and annealed mild steel sheet.

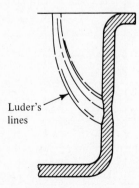

Figure 8.3. Luder's lines develop during sheet-metal forming operations because of yield-point elongation.

When a material has been temper-passed sufficiently, the yield-point elongation is eliminated and a smooth stress–strain curve will develop, as was shown in Fig. 8-1.

Several properties associated with the stress–strain curve, such as yield stress, ultimate tensile strength, and total elongation, have been used to evaluate formability. Yield stress has the disadvantage of being sensitive to speed of deformation. It increases with increasing speed. Tensile tests are run at speeds lower than production stamping operations, and therefore the measured yield stress will be lower than that encountered in stamping. The method of specimen preparation also affects the yield stress. Mechanical properties such as ultimate tensile strength and total elongation are also very sensitive to test conditions, surface scratches, and specimen geometry.

Strain-Hardening.   Forming limits relates primarily to the exhaustion of ductility. To achieve better formability a material must be capable of straining uniformly.

Research has shown that two properties strongly influence formability: the strain-hardening coefficient, or n value; and the anistropy coefficient; or r value. The n value determines the ability of the material to be uniformly stretched. The r value represents the thickness-to-width strain.

Figure 8-4 illustrates a true stress–strain curve plotted on log-log paper. The n value represents the slope of the line a/b. The n value also shows the rate of work-hardening that took place in the metal as it was being strained.

Strain-Hardening Coefficient.   Strains describe the actual dimensional changes in a sheet. Strain-hardening refers to the fact that as a metal deforms in some area, dislocations occur in the microstructure. As these dislocations pile up, they tend to strengthen the metal against further deformations in that area. Thus the strain is spread throughout the sheet. However, at some point in the deformations, the strain suddenly localizes and necking, or localized thinning,

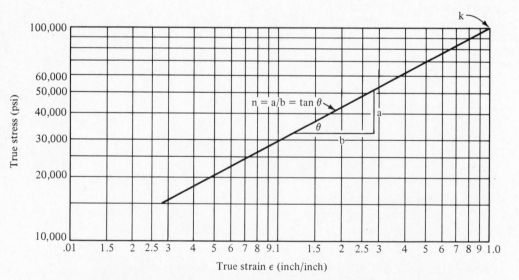

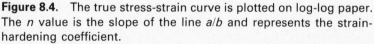

Figure 8.4.   The true stress-strain curve is plotted on log-log paper. The *n* value is the slope of the line *a/b* and represents the strain-hardening coefficient.

begins. When this occurs, little further overall deformation of the sheet can be obtained without its fracturing in the necked region.

Thus the strain-hardening coefficient reflects how well the metal distributes the strain throughout the sheet, avoiding or delaying localized necking. The higher the strain-hardening coefficient, the more the material will harden as it is being stretched and the greater will be the resistance to localized necking. Necks in the metal may affect structural integrity or harm surface appearance.

For most practical stamping operations, stretching of the metal is the critical factor and is dependent on the strain-hardening coefficient. Thus stampings that entail much drawing should be made with metals having high average strain-hardening coefficients.

The Normal Anisotropy Coefficients or r Value. The anisotropy coefficient is derived from the ratio of the plastic width strain ε_w to the thickness strain ε_t, as measured in a uniaxial tensile test. A material that has a high plastic anisotropy also has a greater "thinning resistance." In general, the higher the anisotropy coefficient the better the metal deforms in drawing operations.

Tensile specimens oriented at 90°, 45°, and parallel with the direction of rolling are shown in Fig. 8-5. These specimens, when pulled, will yield an average $\bar{r}$ value. This orientation is necessary since rolling the sheet produces mechanical anisotropy. The equation for $\bar{r}$ is:

$$\bar{r} = \frac{r_L + 2r_{45} + r_T}{4}$$

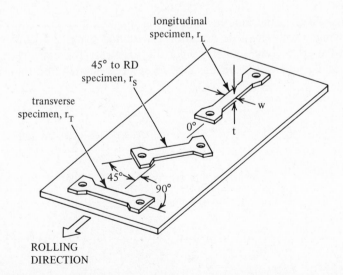

Figure 8.5.   The tensile specimens are oriented 90°, 45°, and parallel to the direction of rolling for determining the average anisotropy coefficient.

Table 8-1. Typical $\bar{r}$ values for several common metals used in forming.

Normalized steel	1.0
Rimmed steel	1.0–1.35
Aluminum-killed steel	1.35–2.0
Copper, brass	0.8–1.0
Lead	0.2
Hexagonal, close-packed metals	3–6+

If $\bar{r}$ is greater than unity, the material is characterized as having resistance to thinning and has a good thickness strength throughout. Typical $\bar{r}$ values for several common metals are shown in Table 8-1.

Simulating Forming Conditions

Many individual operations may be required to produce a complex stamping. "Forming" is used to cover all the operations needed to form a flat sheet into a part. The operations may include deep-drawing, stretching, bending, buckling, etc.

Deep-Drawing.   In deep-drawing, also called cup or radial drawing, a parallel-walled cup is created from a flat blank (Fig. 8-6). The blank may be circular, rectangular, or of a more complex outline. The blank is drawn down into the die cavity by action of the punch. Deformation is restricted to the flange and draw radius. No deformation occurs under the bottom of the punch. As the punch forms the cup, the amount of material in the flange decreases.

In drawing, the limit of deformation is reached when the load required to deform the flange becomes greater than the load-carrying capacity of the cup wall. The failure site will be in the unworked area near the bottom of the cup

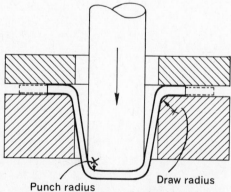

Punch radius
should be 6 to 10t
to prevent tearing.

Draw radius
should be 4 to 5t.

Figure 8.6. A cup or deep-drawing operation.

wall. The hold-down pressure on top of the blank must be sufficient to keep the material from wrinkling as it goes into the die. The deformation limit (limiting drawing ratio or LDR) is the maximum blank diameter that can be drawn into a cup without failure. The percentage reduction can be calculated as follows:

$$\% \text{ reduction} = \frac{D - d}{D} \times 100$$

where: D = blank diameter
d = punch diameter

The calculation is usually made with the cup-wall inside diameter, which would also be the draw-punch diameter. A drawing ratio or percentage reduction of 48% is considered excellent on the first draw. Succeeding draws must be reduced. That is, the percentage reduction that can be expected, under favorable conditions, without annealing, for a second draw would be about 30% and about 20% for a third draw.

Unwanted deformation such as buckling or wrinkling of the flange in deep-drawing is caused by excess metal in the flange. This can be minimized by decreasing the required yield stress of the metal by raising the hold-down pressure, reducing the clearance between the punch and die, increasing the thickness of the blank, or by reducing the blank size.

Stretch-Forming.   In stretch-forming the blank is securely clamped between the blank holder and the die (Fig. 8-7). Deformation is restricted to the area initially within the die. The stretching limit is the onset of failure. Location of the failure is dependent on the material and the forming conditions. Forming operations that consist of only deep-drawing or pure stretching are rare. Most often deformation is a combination of deep-drawing and stretching.

Forming Tests

Tests that subject sheet metal to some type of deformation are used to evaluate formability. Two of the most common stretch-forming tests are the Erichsen and

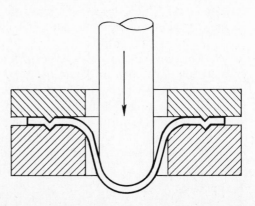

Figure 8.7. Stretch-forming; the flat blank is securely clamped between the blank holder and the die.

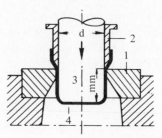

Figure 8.8. The second draw of the Erichsen deep-drawing cup test.

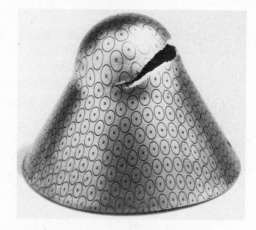

Figure 8.9. A Fukui cup test used to predict the formability of sheet metal. No blank holder is needed, and wrinkles can be avoided by proper attention to the dimension ratios in the test disc. (*Courtesy Institut de Recherches de la Siderurgie Francaise.*)

Olsen tests in which small punches are pushed into a large sheet of clamped metal. In some cases large punches 4, 8, and even 15 in. (101.6, 203.2, and 381.0 mm) in diameter are used. A sketch of the Erichsen deep-drawing cup test is shown in Fig. 8-8.

Combined stretching and drawing are simulated in the Fukui conical cup test. This test utilizes a hemispherical, smoothly polished punch. No blank holder is required, and therefore the influence of friction on metals of differing metallurgical qualities can be neglected. In each test, a drawing ratio that will result in a broken cup (Fig. 8-9) is used. Wrinkles are avoided by using a fixed ratio between the thickness of the sheet, the size of the blank, and punch and die diameters. Under these conditions, the test produces a known amount of stretching, drawing, and bending under tension.

Forming-Limit Curve.   The forming-limit curve is one of the more recent methods developed to determine the formability of sheet metal. Essentially, it consists of drawing a curve that shows a boundary line between acceptable strain levels in forming and those that may cause failure.

The curve is based on relating major and minor strains that are perpendicular to the plane of the sheet. To determine the major and minor strains, a grid of circles is printed on the sheet metal. This may be done by an electrolytic stencil-etching process or by using a photosensitive resist method as produced by the

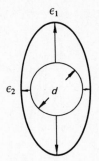

Figure 8.10. The relationship of major, ε_1, and minor, ε_2, strains is established by measurement after forming.

Eastman Kodak Co. After the metal is deformed, the circles are measured to obtain the major strain ε_1 and the minor strain ε_2, as shown in Fig. 8-10. Typically, about ten data points are obtained from a test specimen in the region of fracture. Ellipses lying both in the failed region and just outside of it are measured. The forming-limit curve is then drawn to fall below the strains in the necked and fractured zones, and above the strains found just outside these zones (Fig. 8-11).

The major strain, ε_1, is calculated from the formula:

$$\% \text{ strain } = 100 \, (l_a - d_o)/d_o$$

where: l_a = length of the axis of the ellipse
                    (major strain)
          d_o = diameter of the initial circle

The minor strain, ε_2, is calculated similarly.

Controlled variation in specimen size and lubrication permit the plotting of an entire forming-limit curve from one test setup. A reasonably accurate forming-limit curve may be obtained with four specimens, or 40 data points. A precision curve may be obtained with eight specimens.

Credit for recognizing the major–minor strain relationship goes to Dr. Stuart P. Keeler of the National Steel Corporation. He found the forming-limit curve to be immensely helpful in diagnosing press-shop failures, even though hundreds of production stampings were needed to produce the data for plotting one forming-limit curve. Through further study of this method at General Motors Research Laboratories (GMR), a simple procedure was developed for obtaining this curve for any metal. It was also found that "local" ductility varies for different metals, so no universal forming-limit curve could be developed. For example, two candidate metals may have peak local ductilities of 20% and 50% at a given minor strain. The metal with the 20% local ductility (high strain-hardening coefficient) may turn out to be the best choice because the strain will then have a better distribution throughout, allowing the entire sheet to be stretched 20%. If the other sheet showed little strain-hardening, it might stretch by 50% in a local area, but leave the rest of the sheet relatively unstrained.

Through the use of formability-prediction techniques, designers and fabri-

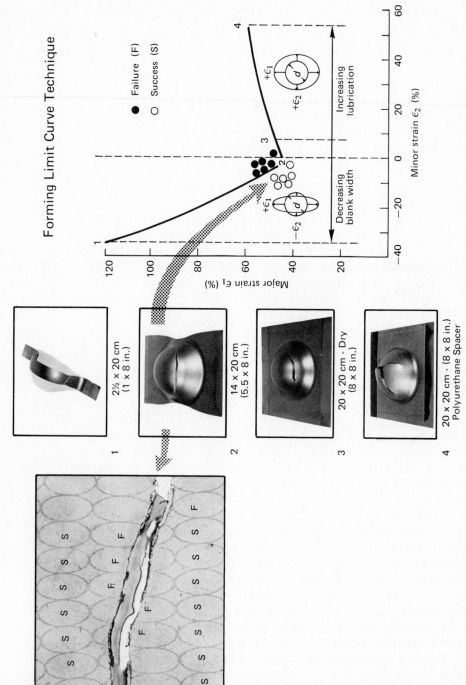

Figure 8.11. A forming-limit curve may be obtained directly from four test specimens. The narrower specimens (1) and (2) provide data for negative minor strains (where original circles become "squeezed"). Data for positive minor strains (where the original circles are stretched in both directions) are obtained through controlled increase in lubrication. Detail for specimen (2) shows source of individual data points. These are obtained directly from measurements of major and minor axes of deformed ellipses in both successful (S) and failed (F) regions.

cators are able to make a wiser choice of metals and obtain data quickly on newer metals. The essential data can be obtained before the die is designed. Also, metal suppliers will be able to establish whether a material possesses required formability before it is shipped from the plant.

   Bend Tests.   The bending ratio r/t (ratio of radius to plate thickness) is also a measure of formability. For example, a $\frac{1}{2}$-in. mild steel plate is bent with a $\frac{1}{2}$-in. radius ($r/t = 1$) and splits result in the ends of the plate at the bend. When the bend radius is increased to $\frac{3}{4}$ in. ($r/t = 1.5$) no breakage occurs. The strain becomes more uniformly distributed, thus reducing peak strains. A rule of thumb states that a material that has an $r/t = 1$ has good forming qualities.

Formability, Temperature, and Recrystallization

   Temperature.   Forming operations are classified as cold or hot. Cold-working is usually associated with operations done at room temperature or below the recrystallization temperature. The properties of yield strength, strain-hardening rate, and ductility are all very much temperature-dependent. With increasing temperature, it is generally true that the yield strength and rate of strain-hardening will progressively reduce and ductility will increase. Hot-working may be defined as metalworking at a temperature above which no strain hardening takes place.

   The product designer should be well aware of the benefits of cold-working, such as higher strength which permits the replacement of many components made from alloy steel with materials having a lower alloy content. Other savings are in overall weight, heat-treating costs, and reduced machining or finishing operations.

   Hot-working, on the other hand, allows forming operations to be done with less energy input. Not only does the metal deform more easily, but it can also accept a very large amount of deformation before cracking. As an example, billets are sheared off prior to forging. Even in moderately alloyed steels the sheared face frequently contains cracks that run into the billet and may result in forging defects. Heating the billet prior to shearing, even in the low range of 300 to 600°F (149–316°C), increases the ductility to the point where simple shear occurs and no cracks propagate into the billet.

   Recrystallization.   As discussed in Chapter 2, the strain-hardening that takes place during cold-working may be relieved by annealing. In the process, the grains recrystallize or new grains are formed that are strain-free. The amount of recrystallization is a function of time, temperature, and the degree of prior cold-working. The whole relationship of hardness and strength, as the result of cold-working, to annealing temperatures is shown schematically in Fig. 8-12. A pictorial sketch showing the effect of both cold-working and hot-working on the microstructure is shown in Fig. 8-13.

   A newer term that has been introduced in metal forming is *warm-forming*. In this process, the temperature is closely controlled so that no microstructural changes take place in the metal. As an example, AISI 5140 H steel was heated

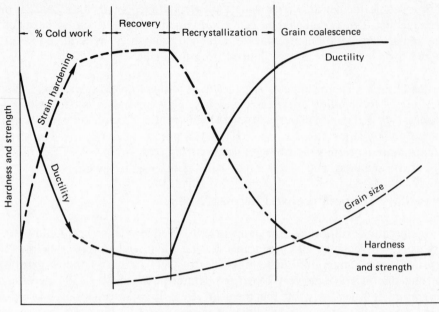

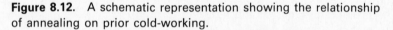

Figure 8.12. A schematic representation showing the relationship of annealing on prior cold-working.

to 1255°F (655°C) to make possible roll-forming the gears for an automatic transmission. This process avoided the nearly 50% material waste inherent by the machining method. The 1225°F temperature selected provided optimum formability under the critical point of 1300°F (723°C) and a combination of the advantages of both hot- and cold-forming were realized (surface scale was minimized, unit loads on tooling were decreased, and dimensional control comparable to that of cold-forming was possible). Induction heating maintained the operation within ±25°F (±14°C) of the set temperature.

The force–energy curve (Fig. 8-14) shows there is little benefit to be gained by heating SAE 5147 steel up to 400°F. From 400 to 1100°F (204–593°C) there is a proportionate reduction in force and stress. Between 1100 and 1500°F (593–816°C) the curve levels off. Above 1500°F (816°C) increases in temperature once again result in proportionate reduction in force and stress.

FORMING PROCESSES

The enormous variety of plastic deformation processes for metals makes it desirable to classify them into general groups and subgroups. Traditionally, metal-forming operations have been classified as those done by "hot-working" pro-

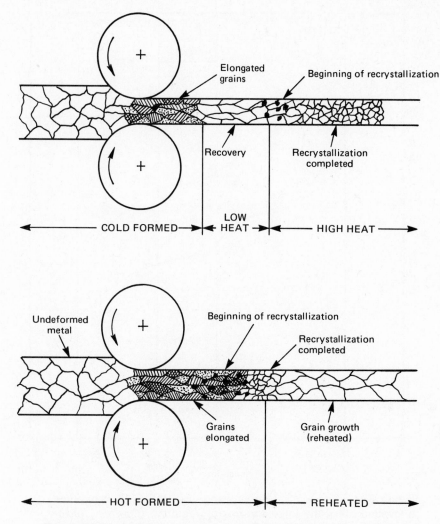

Figure 8.13. A sketch showing the effect of both cold-working and hot-working on the microstructure of cast metals.

cesses or "cold-working" processes. This classification no longer seems adequate since many forming operations would fit in either classification and now there is also warm-forming.

A more appropriate method of classifying metal-forming operations is to divide them into processes that intentionally alter the cross-sectional area and those that do not. With one or two exceptions, most of the metal-deformation processes can be classified as forging (either hot, cold, or warm) or as a process that merely alters the profile, as does sheet-metal forming.

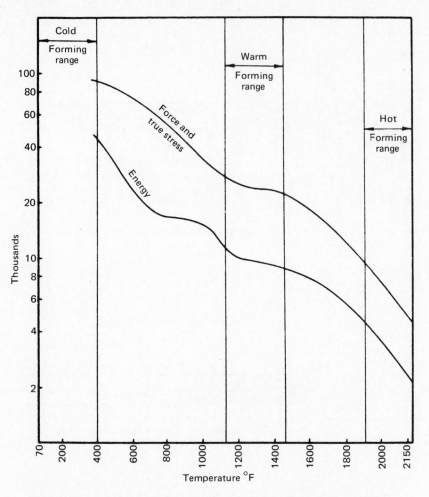

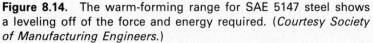

Figure 8.14. The warm-forming range for SAE 5147 steel shows a leveling off of the force and energy required. (*Courtesy Society of Manufacturing Engineers.*)

Sheet-Metal–Forming Operations

Sheet-metal–forming operations are usually associated with conventional press-working, which includes cutting the blank and then bending or drawing it to the desired shape.

Cutting operations

Common sheet-metal–cutting operations are shown in Fig. 8-15. Blanking, piercing, and notching are usually done with punches and dies that are mounted in a standard-type die set, as shown in Fig. 8-16. Die sets facilitate exchanging one setup for another, since the punch and die can be kept intact. This of course

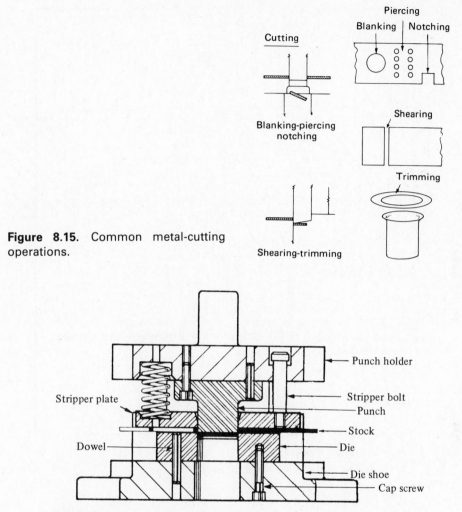

Figure 8.15. Common metal-cutting operations.

Figure 8.16. A standard die set with a punch and die mounted in place.

necessitates a die set for each different job; however, the time saved in getting the punch, die, stripper, etc., mounted and in alignment each time makes it economically feasible.

Bending

Bending operations involve stretching the outside fibers and compressing the inside fibers. The type of press commonly used in making bends is a press brake. The press and the type of dies used are shown in Fig. 8-17.

Bending Force. The approximate force required to bend a piece of metal in a V-die, Fig. 8-17(a), can be found by the formula:

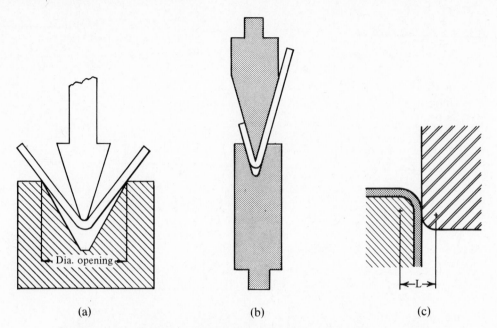

Figure 8.17. Press-brake bending of metal. Air bending is shown at (a) and die bending at (b). Shown at (c) is the distance *L* as used in the formula for wiping dies.

$$F = 1.33 \frac{S_u W t^2}{L}$$

where: S_u = ultimate strength
         W = width of metal at bend, in inches
         t = metal thickness, in inches
         L = length of span across V-die opening, in inches

The same formula can be used for wiping dies, as shown in Fig. 8-17(c). In this case the constant is changed from 1.33 to 0.33 and *L* becomes the distance between the punch and die radii as shown.

For mild steel up to 0.5 in. thick, the recommended die opening is eight times the metal thickness. For mild steel plates thicker than 0.5 in., the recommended die opening is 10 to 12*t*.

Springback.   One of the problems encountered in forming metal, particularly in bending, is the tendency after the force has been removed to spring back or establish internal equilibrium. As shown in Fig. 8-18, when a bending load is applied, tension and compression forces are set up about the neutral axis, but when the load is relieved, the outer surface shortens and the inner surface lengthens. The amount of springback is in proportion to the *sum* of the tension and compression recovery forces.

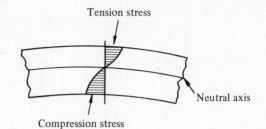

Figure 8.18. Springback is caused in bending by tension and compression forces on each side of the neutral axis of the metal.

Tension stress

Neutral axis

Compression stress

Forms any angle

Forms different shapes

Figure 8.19. Polyurethane makes a bottom press-brake die that can be used for a wide variety of shapes.

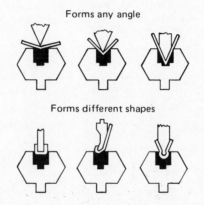

In press-brake operations, as was shown in Fig. 8-17, air bending (a) is much preferred over die bending (b) for metals that produce springback problems. Minimum bend radii are particularly hard to achieve on metals of very high yield strength due to substantial springback.

Polyurethane inserts are sometimes used, as in the lower die in Fig. 8-19. These inserts save making special forms since they can assume any shape. The polyurethane forces the sheet to hug the punch. Therefore a 90° punch can be used to make a bend up to 90° by regulating the amount of punch penetration into the "rubber." Where a lot of springback is encountered, polyurethane dies tend to reduce springback, and the upper die can be modified so that overbending occurs to offset the springback.

One method of compensating for springback is to overbend the metal slightly. As an example, if a 90° bend is desired, a bend of 87° can be made that will springback to 90°. It will be necessary to experiment because the amount of springback will increase with the following variables:

1. Metal hardness—harder metals have a higher elastic limit.
2. Larger bend radius—a larger radius has a larger elastic zone.
3. Thicker metals—thicker metals have more plastic deformation.
4. High yield stress—high-yield-stress metals have higher elastic limits.

Bend Allowance. It is sometimes necessary for production engineers and designers to know the exact width of a strip that will be contained in various bends. A standard bend allowance formula is:

$$B.A. = \frac{\theta}{360} 2\pi (r_i + kt)$$

where: r_i = inside radius, in inches
k = 0.33 when r_i is less than or equal to $2t$ and 0.50
     when r_i is greater than $2t$
t = metal thickness

Example: The width of a $\frac{1}{8}$-in. thick strip required for a channel 3 in. wide and 2 in. deep with $1t$ bend radius would appear as shown in Fig. 8-20: $A + B + C + 2(BA)$

A = 2 in. $-$ (bend radius + t), B = 3 in. $-$ 2 (bend radii + t), C = same as A
A = 2.0 $-$ (0.125 + 0.125) = 1.750 in.
B = 3.0 $-$ (0.250 + 0.250) = 2.5 in.
C = 1.750 in.
$BA = \dfrac{90}{360} 2\pi\,(0.125 + 0.33 \times 0.125) = 0.260$ in.

total width = 1.750 in. + 2.50 in. + 1.750 + 2(0.260)
         = 6.52 in., the strip width before bending

When $r_i \approx t$, the strip width can be found by adding the inside dimensions. For example, a channel that is $\frac{1}{8}$ in. thick, 3 in. wide, and 2 in. in depth would be 2(1.875) + 2.750 = 6.5 in. wide before bending.

Drawing Operations

In sheet metal, *drawing* is a process of forming flat sheet metal into hollow shapes by means of a punch that causes the metal to flow into the die cavity. If the depth is one or more times the diameter, the process is called *deep-drawing*. The forming of shallow shapes is sometimes referred to as *stamping,* but the distinction is not clear since stamping is also used to describe cutting flat figures or patterns.

Most drawing operations start with a sheet-metal blank whose surface area is only slightly larger than the finished product, allowing just enough material to hold the blank during the drawing operations and for trimming after drawing.

There should be no appreciable change in the thickness of the material between the blank and the finished part. When it is desirable to reduce the metal thickness, it may be done in secondary operations as in ironing. Schematic drawings of deep-drawing, reverse drawing with a cushion, and ironing are shown in Fig. 8-21.

The die cushion, which is underneath the press bed, serves to hold the blank against the punch in the initial stage of operation. Then as the punch descends, the cushion also moves down maintaining contact with the blank. When the stroke is completed, the die cushion returns to its original position, flush with the top of the punch. This action serves to strip the part off the punch. Presses are not ordinarily equipped with a die cushion but this attachment greatly extends

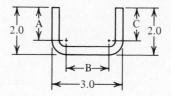

Figure 8.20. An example of how strip width is calculated before bending.

Drawing operations

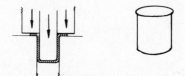

Deep double-action drawing

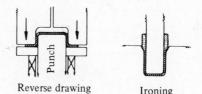

Figure 8.21. Sheet-metal drawing operations.

Reverse drawing
with cushion

Ironing

the versatility of the press. The cushion may be spring, air, or hydraulically actuated.

Figure 8-22 consists of drawings made in three stages to show in detail the action of the metal as it is being drawn into the die. As the metal starts to move into the die, bending occurs. Greater depth is obtained by drawing or pulling in metal from the flange region. The bottom of the cup is completed in the first stage. Very large, compressive forces occur in the flange and die radius area. Normally wrinkles would start to form as the metal is pulled over the die radius. Fortunately, the proper application of a blank holder can keep the wrinkles from forming. However, if the blank-holding pressure is too great, the metal will tear.

The final stage is a continuation of the two previous stages of bending, compressing, overcoming friction, and straightening.

Important variables of the drawing operation are friction, squeezing force, and formability of the metal.

The amount of friction encountered will be dependent on the surface finish of the metal and die components and the blank-holding force. Friction is usually reduced by the use of the proper drawing compounds.

The squeezing force that takes place in the flange and draw radius is related to the percentage reduction of area, ductility of the metal, yield strength, blank thickness, temperature, and the depth of draw. How well the metal forms will be governed by the strain-hardening exponent n and average strain ratio $\bar{r}$ as discussed previously.

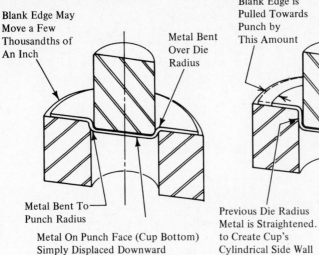

Blank Edge May
Move a Few
Thousandths of
An Inch

Metal Bent
Over Die
Radius

Metal Bent To
Punch Radius

Metal On Punch Face (Cup Bottom)
Simply Displaced Downward

(a)

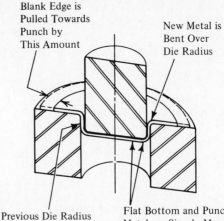

Blank Edge is
Pulled Towards
Punch by
This Amount

New Metal is
Bent Over
Die Radius

Previous Die Radius
Metal is Straightened.
to Create Cup's
Cylindrical Side Wall

Flat Bottom and Punch Radius
Metal are Simply Moved Downward

(b)

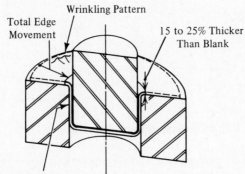

Total Edge
Movement

Wrinkling Pattern

15 to 25% Thicker
Than Blank

No Further Squeezing Action
After Metal Flows Over Die Radius

All Conditions Continue:

Die radius bending
Straightening of die radius
Compressing flange metal
Overcoming dynamic friction
Displacing bottom and punch radius

(c)

Figure 8.22. The progressive stages of deep-drawing a cup. At (a) the metal deformation is mainly that of bending. At (b) the deformation is bending, straightening and radial compression. At (c) is a continuation of all the steps of (b). (*Courtesy Society of Manufacturing Engineers.*)

Embossing, Coining, and Stamping. Embossing, coining, and stamping are three closely related techniques for making shallow impressions and patterns on sheet metal or extruded shapes.

Embossing. Embossing is a severe stretching of small areas of metal to create a desired shape. The process as it takes place between closed matching male and female dies is shown in Fig. 8-23. This operation generally requires high pressure to produce clear-cut outlines. The thickness of stock remains consistent throughout the section, which is the identifying characteristic of the embossing technique.

Coining. Coining consists of imbedding images or characters that are on a set of dies into the plane surface of metal blank under great pressure. Top and bottom dies are frequently different. The metal is made to flow between the punch and die to assume the negative contour of the two dies. The most common example of this process is used, as the name implies, in making coins.

Stamping. In this process, cut lines of letters, figures, and patterns are pressed onto a smooth surface under the impact of a stamp that has sharp projecting outlines. Impressions are made only on one side of the metal, generally between the depths of 0.020 to 0.040 in. (0.51 to 1.02 mm).

Compound, Progressive, and Transfer Dies

Compound Dies. Frequently, in the interest of saving time, two distinctly separate operations such as blanking and forming are done on the same die. A compound blank and draw die is shown in Fig. 8-24. As can be seen in the sketch, the metal enters as a flat sheet, is cut to the right length, and is then

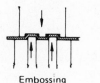
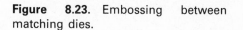

Embossing

Figure 8.23. Embossing between matching dies.

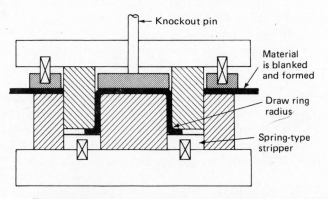

Figure 8.24. A compound blank and draw die.

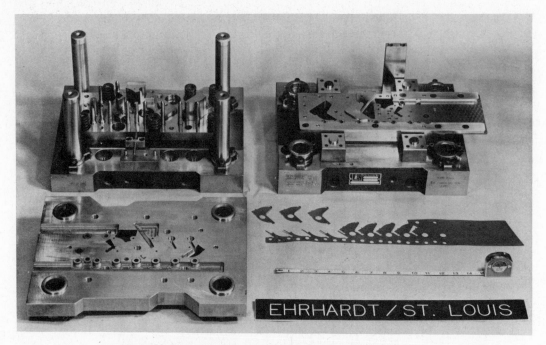

Figure 8.25. A progressive die shown in the open position with punch (upper left), die (upper right), stripper plate (lower left), and strip with stamped out parts (lower right). (*Courtesy Ehrhardt Tool and Machine Co., Inc.*)

formed over a reverse-type punch. Spring action on a pressure plate strips the part off the punch. The knockout pin, shown at the top, pushes the part out of the upper die if it stays on that side when the die opens.

Progressive Dies. Progressive dies are made to cut and form a part in successive "stages" or "stations" of the die. The parts are held together by the strip skeleton until the last station or cutoff. Force for the movement comes from the strip feeder attached to the press. A progressive die in the open position with the punch at the top and the die below it is shown in Fig. 8-25. The blanked-out parts that have been made progressively are also shown.

The cost of progressive dies is high and therefore they are usually limited to high-production operations.

Transfer Dies. Transfer dies are used when the size and complexity of the part require that they be moved from one die to another. An example of the type of part that would require a transfer die is shown in Fig. 8-26. As the part is formed in one die, fingers of the transfer mechanism pick the part up and move it to the next die, and so on, until the part is completed.

Fluid-Activated Diaphragm Forming. Drawing of sheet-metal parts may be accomplished by the use of fluids and a diaphragm, as shown in Fig. 8-27. The

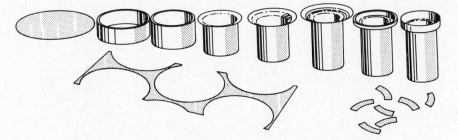

Figure 8.26. A deep-drawn part formed in a transfer die.

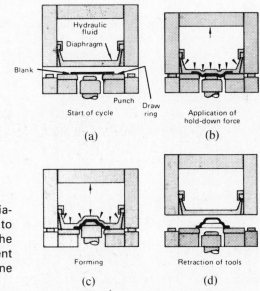

Figure 8.27. The fluid-activated diaphragm applies a hold-down force to the blank, with the major portion of the forming done by the upward movement of the punch. (*Courtesy* Machine Design.)

process has the advantage of closer control of the drawing operation so that parts that usually require two drawing operations can be done in one.

Multiple-Action Dies. A newer approach to deep-drawing is by the use of multiple-action dies. In this setup (Fig. 8-28) the entire sheet is held between tooling plates at all times. Unsupported metal does not flow with the punch as it does in conventional tooling. The multiple, matched, male-female dies are made so that segments can be timed to move for an optimum forming sequence. The tools are mounted in a conventional hydraulic press and the part is made in one stroke. The sequential power is supplied by the press, which is programmed to activate independent, external hydraulic circuits.

Although both the fluid-activated diaphragm and the multiple-action dies have good control over metal flow, which is necessary to avoid excessive thinning and buckling, it appears that the newer multiple-action die provides more versatility.

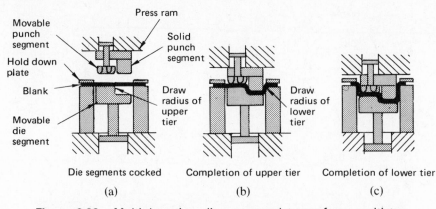

Figure 8.28. Multiple-action dies are used to perform multistep forming in one operation. The punch and die segments are operated in a sequence that is coordinated with the press stroke. (*Courtesy* Machine Design.)

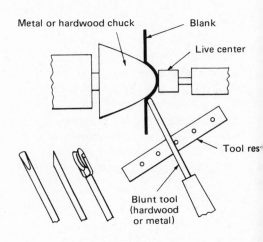

Figure 8.29. The spinning process is used to form metal over a revolving chuck or punch.

Metal spinning

Production quantities often are such that it is not economically feasible to make the dies required to form the part. One approach to this problem is to make only one of the forms, usually the punch, and then form the metal over it by a spinning operation, as shown in Fig. 8-29.

The desired form, either hardwood or steel, is mounted on the lathe headstock. A live center is brought up to hold the metal blank in place. The forming operation is accomplished by pressing the metal with wood or steel rods as it rotates. Skill and experience are required to cause the metal to flow at the proper rate, avoiding wrinkles and tears.

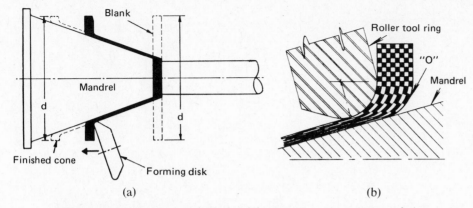

Figure 8.30. Power-roll forming (a), and an enlargement of the extruding process (b).

Displacement Spinning or Power-Roll Forming. Power-roll forming is similar to hand spinning except that the part is shaped to the mandrel by rollers that produce a progressive extruding action (Fig. 8-30).

The control over workpiece contours and thickness gives the process significant advantages over drawing. Springback encountered in both press-working and conventional spinning is eliminated. The increase in tensile strength of the finished product may be as much as 100% or more over that of the original metal. Most parts are formed from flat blanks, but some cylinders require preformed blanks so that a sufficient volume of metal can be maintained at the desired places, such as the bottom or top flange.

Continuous-roll forming

Roll forming may be defined as a process used to change the shape of coiled stock into desired contours without altering the cross-sectional area. The process utilizes a series of rolls to gradually change the shape of the metal as it passes between them (Fig. 8-31).

The intricacy of the shape, the size of the section, the thickness, and the type of material will determine the number of rolls required. A simple angle or channel with straight web and flanges can usually be made with three or four pairs of rolls, whereas complicated shapes may require ten or twelve roll passes. Usually some additional rolls are needed for idle stations and straightening.

The main advantage over other production methods is speed. For example, an item that needs no piercing or punching can be run at 300 fpm (9000 cm/min). Everyday production speeds, however, are more likely to be between 800 and 1200 fph (24,000–36,000 cm/hr) for heavy-gage steel and 1800 to 2200 fph (54,000–66,000 cm/hr) for light-gage steel.

The process has become more attractive in recent years since it can handle prepainted or electroplated surfaces without damage. As may be surmised, roll

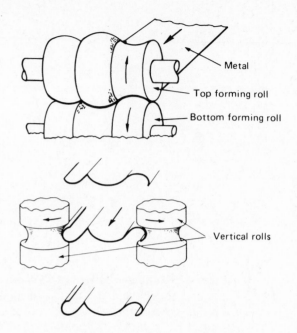

Metal

Top forming roll

Bottom forming roll

Vertical rolls

Figure 8.31. An example of continuous-roll forming.

forming is for quantity production. As a rule of thumb, at least 50,000 ft (15,240 m) should be required annually to amortize the cost of the tooling. Tolerances on normal production items can usually be held within 0.005 in. (0.13 mm) and on some light-gage steel within 0.001 in. (0.039 mm).

Stretch-forming

Stretch-forming of metal consists of placing sheet material in a tensile load over a form block. The material is stressed beyond its elastic limit and into the plastic range causing it to take a permanent set. In stretch-forming, all the fibers are stretched but those on the inside of the radius of bend are stretched less than those on the outside (Fig. 8-32). Since the metal is placed under one type of load (tensile), there is no tendency to springback. However, allowances must be made for dimensional changes that occur. That is, during stretching the length increases and the width decreases.

There are two types of stretch-forming. One is termed *stretch-forming* and the other *stretch-wrap forming,* as shown in Fig. 8-33.

Stretch-forming is used to produce compound curves in sheet stock. The ends of the stock are gripped in hydraulically operated serrated jaws. Stretching is accomplished by one or more hydraulic pistons moving up under the die. Forming of this type is usually restricted to parts that do not have sharp edges, which tend to produce accelerated stretch in localized areas. The forming block or punch is usually highly polished and lubricated. The process is very useful in making prototype models of aircraft and automotive parts. It is sometimes used on a production basis for truck and trailer bodies.

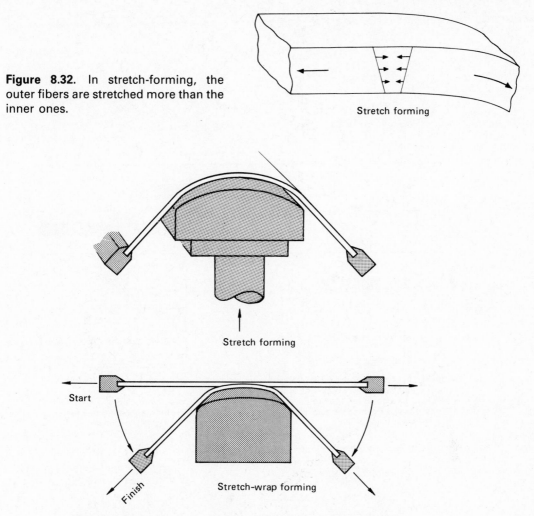

Figure 8.32. In stretch-forming, the outer fibers are stretched more than the inner ones.

Stretch forming

Stretch forming

Start

Finish   Stretch-wrap forming

Figure 8.33. The two methods of stretch-forming of metal.

Stretch-wrapping consists of first stretching the metal beyond its yield point while it is straight and then wrapping it around a form block. It is particularly suited for long sweeping bends on tubes and extruded shapes.

A big advantage of this type of forming is that only one die is needed. Also, the die may be made out of inexpensive material.

Thinning and strain hardening are inherent in the process. It is important to know the elongation values for the metal being used. Metals having a so-called infinite gage length elongation are well suited to this process. The thickness reduction should not exceed 5% of the original thickness.

The production or manufacturing engineer is not only concerned with how much strain-hardening takes place during a given deformation or what is the best

BEADS ON LARGE PANELS

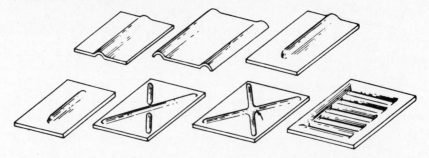

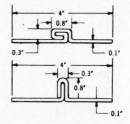

COMPARATIVE
STRENGTHS

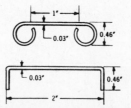

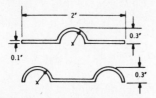

Curled edges are stronger than flanged and present a smooth, burr-free edge. Production, however, may require one more operation since the curl is usually started as a flange.

Although locking ability is sacrificed, vertical standing seams are 3 times as strong as flattened seams.

Ribs are even more efficient than flanges. Dual-rib design yields 56.5 percent more strength for 10.8 percent more material.

Corrugated sheets are common examples of ribs, used as a continuous form.

Figure 8.34. Methods of increasing strength and rigidity on sheet-metal parts. (*Courtesy* Machine Design.)

process to use to produce a given item, but he is equally concerned that the design lends itself to production. A fundamental question is: Can any changes be made that will make it easier to produce, yet not adversely affect the quality of the product?

Sheet-Metal Product Design Considerations

Strength and rigidity

Many design and fabrication techniques are available that add strength and rigidity to simple parts fabricated in sheet metal. For example, strength can be incorporated into the structure by means of flanges, ribs, corrugations, beads, etc. (Fig. 8-34).

Although ribs, beads, and flanges are most often thought of in connection with flat stock, they can be equally effective in supporting cylindrical shapes. Sometimes beads serve the dual purpose of adding strength and rigidity and eliminating *dishpanning*. Dishpanning, as shown in Fig. 8-35, is an oval or bulged bottom. To restore it to flatness, the excess metal is incorporated into beads.

Press-size requirements

Sheet-metal parts are commonly formed on presses such as those shown in Fig. 8-36.

The press tonnage required for blanking is based on the perimeter of the cut, the thickness, and the shear strength of the material, as shown in the formula:

$$F = \frac{(LtS_s)}{2000} \times F_s$$

where: F = total blanking force in tons
          L = length of sheared edge

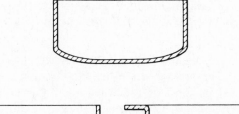

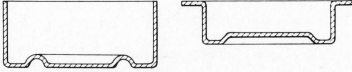

Figure 8.35. Beads at the bottom of a container eliminate dish-panning and add rigidity and strength.

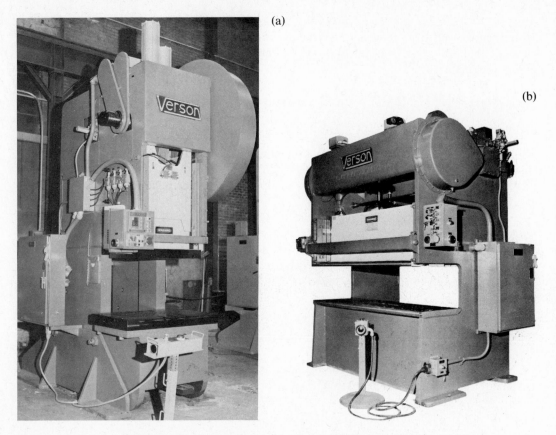

(a)

(b)

Figure 8.36. An open-back inclinable (OBI) press (a), and a 60-ton gap press (b). (*Courtesy Verson Allsteel Press Company.*)

t = thickness of the material being sheared
S_s = shear strength of material in psi
F_s = safety factor (usually 2x)

The press size is designated both by the tonnage capacity that can be delivered by the ram to the press bed and by the bed area. As an example, the press shown at the left in Fig. 8-36 is termed a C-frame press or an OBI press since it has an open back and is inclinable. That is, it can be tilted backward. A standard size designation may be given as:

$$OBI - 22 - 12\tfrac{3}{4} \times 17\tfrac{1}{2}$$

The letters stand for the type of press, 22 is the tonnage rating, $12\tfrac{3}{4}$ and $17\tfrac{1}{2}$ are bed size measurements, left to right and front to back, respectively.

Press energy

The press-tonnage rating is usually the most significant factor in selecting a press but the energy capacity is also critical when it is used for continuous operation. The energy of a press is the product of force times distance (working stroke), usually expressed in inch-tons. As the energy is expended in forming, the flywheel slows down.

If there is not enough energy stored in the flywheel for a given press stroke, the press will slow down on each successive stroke, causing belt slippage, wear, and/or overloading of the main drive motor. There is also the possibility of the press sticking at the bottom of the stroke. If this happens, considerable work is required to release the jammed condition.

A general rule-of-thumb is that a press will not have more work capacity than its tonnage rating times the distance from the bottom of the stroke at which it is rated. For example, a 100-ton press rated at $\frac{1}{4}$ in. from the stroke bottom would usually be capable of providing only 25 inch-tons of energy to work on a continuous basis. For presses using multistation dies, the energy requirements at all stations must be added.

Since blanking, piercing, and similar operations require the application of force over a short distance, less energy is needed. Frequency of load application is also an important consideration since higher press speeds reduce the time allowed for the flywheel to recover speed. As the press speed approaches the area of 250 to 300 spm, the energy may be absorbed so rapidly that the flywheel is of limited value as an energy source.

Deep-drawing and similar operations require the force to be exerted over a considerable distance and more energy is needed. Because slower draw presses (usually geared) often have sufficient time between strokes for the flywheel to regain its speed, relatively small motors can be used.

Numerical Control Punching

The first time flat metal was punched by numerical control (NC) was 1955. That first punching exhibit had far-reaching repercussions and triggered major changes in machinery for punching holes and contoured cuts in sheet metal, plate, and structural members. In recent years it has spawned hybrid machines that not only punch but also cut by plasma arc or laser beam and even perform such functions as milling. A plasma-arc torch fixed to a punch press is shown in Fig. 8-37.

In 1972 computerized numerical control (CNC) entered the metal-punching field and drastically reduced NC programming time and effort, as well as production cycles. CNC manipulates the punch-controlling numbers electronically, performing such tasks as optimization of punching instructions and other kinds of number juggling that an eight-track punched tape could not do.

The NC or CNC punching machinery differs from conventional mechanical, hydraulic, or pneumatic presses in that the workpiece is automatically positioned under a basic punch that is gripped by a turret bushing or adapter in a single-

Figure 8.37. The plasma-arc torch is fixed to the punching head of a variable type NC punch press. The torch is fixed and the workpiece moves by NC control. (*Courtesy W. A. Whitney Corp., an Esterline Company.*)

station type of press. A 22-ton CNC press capable of more than 200 strokes per minute on moves shorter than 5 in. (127.0 mm) is shown in Fig. 8-38. A detail of one tooling station of the turret is shown in Fig. 8-39.

The workpiece is automatically shifted from one position to another according to the NC or CNC program. An accuracy of positioning under the punch is ±0.005 in. (0.10 mm) and a ±0.001-in. (0.03-mm) repeatability.

Programming. The NC tape is prepared from a part drawing, from which dimensions, configurations, and other information is entered in an in-house or remote computer or simply with a calculator on the programmer's desk. The programs required are simple enough that a new programmer can learn the job in a few hours. The programmer lists the hole locations on the coordinate axis as well as the patterns and contours desired, the tooling to be used, and the quantity of parts required.

With the software in place, the tape-preparation computer figures out the most efficient way for the press to operate. If changes in the part program are

Figure 8.38. A CNC turret punch press capable of more than 200 spm on moves shorter than 5 in. (127.0 mm). (*Courtesy Wiedemann Division of Warner & Swasey Co.*)

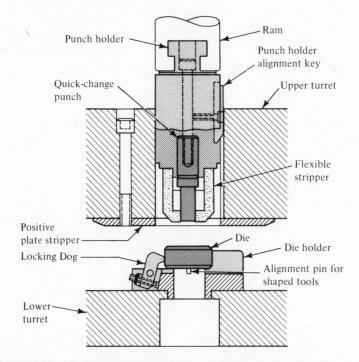

Figure 8.39. A detailed view of the quick-change punch, die, and stripper of a turret-type punch. (*Courtesy* American Machinist.)

required after the first part is produced, the programming can be checked on either a cathode-ray-tube (CRT) screen or a plotter and fast and easy revisions can be made.

The press operator keeps the press supplied with stock and removes the finished parts, although even these tasks are automatic in some installations. The press operator ensures that the press runs properly and corrects most problems as they occur. He performs maintenance work after consulting the control panel's diagnostic readout. On many CNC punch presses, the computer flashes numbers to coded remedies in a troubleshooting manual or displays messages in English on the CRT.

Programs are prepared in advance, often on magnetic tape cassettes. The cassettes can be inserted into the machine control at any time to reproduce specific parts exactly regardless of the skill and experience level of the operator. All of this points up the fact that the responsibility for making parts has shifted from the operator to the programmer.

Forging

Whenever the word "forging" is used, the impression of strength and toughness immediately comes to mind. Usually, however, only a vague idea exists of why forgings have these properties. Frequently, castings and weldments compete for

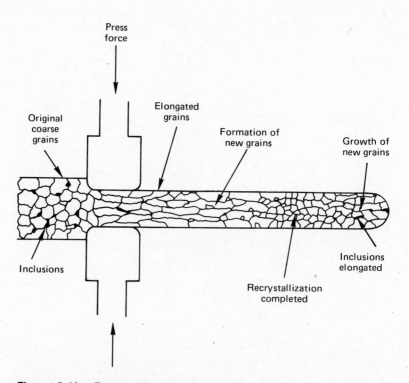

Figure 8.40. Recrystallization and grain refinement take place as a result of the forging process.

the same use as forgings. As more and more is understood of metal properties and newer manufacturing processes, a wiser, more economical selection can be made as to which is the best process. Generally, if stress loads are high, forgings are preferred even though the unit cost is higher.

As the hot metal is forged, recrystallization takes place (Fig. 8-40) and the grain flow closely follows the outline of the component (Fig. 8-41). Continuous-flow lines decrease the susceptibility to fatigue and corrosion failures. Internal flaws are largely eliminated, resulting in low inspection costs and consistent results in both heat treatment and machining.

Forging processes can be divided into several groups as follows: hammer and press forging, upsetting, and extrusion forging. Some special techniques include heated-die forging, high-energy-rate forging, and swaging.

Hammer and Press Forging. Hammer forging is essentially the same process it has been since its early development. The quality of forgings produced under the drop hammer depends to a large degree on the skill of the operator. All hammer-forging operations, from bar to final shape, are done by repeated blows between two platens (open-die forging) or in several impressions in a die set (closed-die forging) (Fig. 8-42).

Preforming operations used to prepare the metal for the forging dies are often done by using the blocking and edging portions of a standard closed die, or on separate machine tools. The use of roll forging to make the preforms is increasing. Another approach is to use pieces cut from an extrusion (Fig. 8-43).

Press forging, although of later development than hammer forming, is becoming a major forging method. Press forging utilizes a slow squeezing action

Figure 8.41. The grain flow of the metal closely follows the part contour, contributing to the strength of the forging.

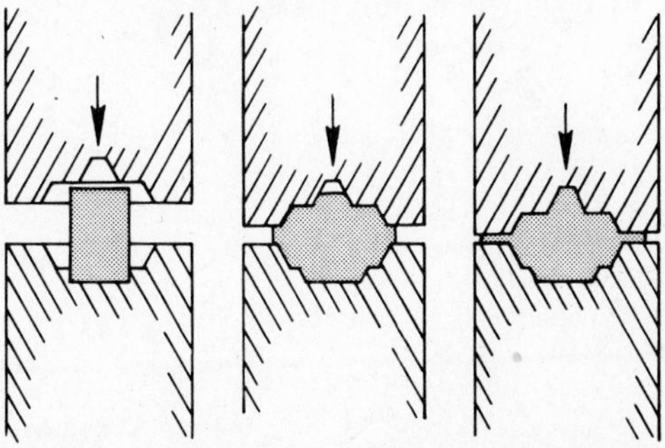

Figure 8.42. A closed-die forging sequence.

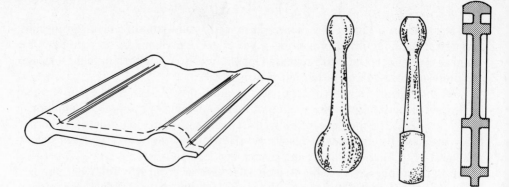

Figure 8.43. An extruded preform shape that can be sliced up and forged into connecting rods. (*Courtesy* Machine Design.)

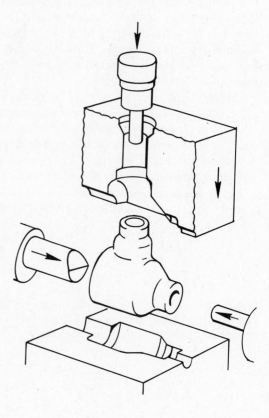

Figure 8.44. Top- and side-acting auxiliary rams are used in some forging presses to produce hollow parts and undercuts.

and relies less on operator skill and more on the proper design of the preform and the forging dies. At first, press forming was limited to axially symmetrical components, but now almost any shape can be forged by this method. Newer developments make it possible to produce bevel gears with straight and helical teeth. Rotation of the die during penetration will press bevel gears with spiral teeth. Also, hollow parts with undercut shapes can be press-forged by hydraulically operated, side-acting, auxiliary rams (Fig. 8-44).

Flashless Forging.   Continuous work is being done in forging design to eliminate the *flash,* or excess metal that squeezes out between the dies as they close. One approach has been to force the excess metal into a compensating cavity, in the center of the forging, that will be pierced later. Flashless forgings call for close tolerance on billet weight. Some gears are now being made by this process with significant savings in material, energy, labor, and die life.

Heated-Die Forging.   Heating the forging dies to 1600°F (871°C) improves tolerances and makes it possible to achieve thinner sections with fewer die sequences. As an example, aircraft forgings of AISI 4340 steel were made 81% deeper than that attained on conventional dies by using heated dies at 300 to 800°F (149–427°C). The process is not extensively used due to shortened die life.

Roll Forging.   Originally developed to forge simple shapes, roll forging has expanded its field of application to many finished products. In its most simple form, a preheated billet passes between a pair of rolls that deform it along its length (Fig. 8-45).

Cross Rolling.   Cross rolling is a process of gradually deforming a shaft in the transverse direction as it passes between the rolls [Fig. 8-46(a)]. U-shaped tools are fastened to the rollers for unilateral or bilateral displacement of the

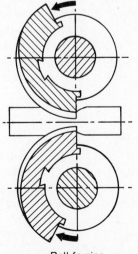

Figure 8.45. Roll-forging is often used in preforming for drop-forging or pressing. It may also be used to produce finished forgings.

Roll forging

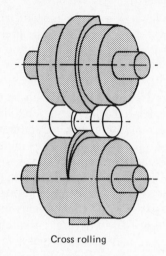

Cross rolling

(a)

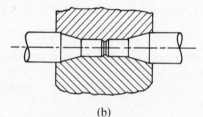

Figure 8.46. Cross rolling showing U-shaped tooling (a); a typical part (b).

(b)

metal, which may be hot, warm, or cold. The part is brought to its final dimensions in one rotation. A part such as that shown at Fig. 8-46(b) may be made by hot cross rolling from 1⅜-in. hot bar stock in a total of 3.5 sec. Forming proper takes 1.5 sec, while 0.2 sec is required for separation of the part from the bar and 1.8 sec for bar feeding.

The cross-rolling process is capable of showing considerable savings in material and production time over conventional closed-die forging for smaller parts. The accuracy [0.004 in. (0.10 mm) on diameters up to 1.2 in. (3.04 cm) and 0.008 in. (0.20 cm) on diameters up to 2 in. (5.08 cm)] allows the parts to be used without further machining. Where precision is called for, parts may be finished by grinding.

Upset Forging.   Upsetting is accomplished by inserting a blank of a specific length into a stationary die (Fig. 8-47). The bore of the die is a few thousandths of an inch larger than the outside diameter of the blank. A punch moves toward the die blank to upset that portion of the blank protruding from the die. The maximum length that can be upset in one stroke is 2.25 to 2.50 dia. of the stock. If this length is not enough to form the part, two or more blows may be used. As a general rule, 4.5 dia. can be upset in two blows. When a sliding punch is used to support part of the blank during upsetting, up to about 6.5 dia. can be

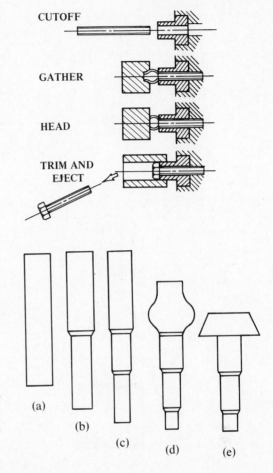

CUTOFF

GATHER

HEAD

TRIM AND
EJECT

Figure 8.47. Upset-forging or cold-heading utilizes three dies and four stages to produce a hexagon bolt blank. (*Courtesy Republic Steel.*)

Figure 8.48. A combination of extrusion and upsetting is used to make this stem-pinion blank. Bar stock cut to exact required length (a), forward extrusion (b), a second forward extrusion (c), combined forward extrusion and upsetting (d), and heading (e). (*Courtesy* Machine Design.)

(a)

(b)

(c)

(d)

(e)

formed. A rule of thumb used to estimate impact values required for upsetting is that the impact value must be seven times the yield strength of the material.

Multiple-die machines often combine upsetting with extrusion to form a large head on small shank parts. Contrary to upsetting, extrusion reduces the diameter of the initial stock while increasing its length (Fig. 8-48). The reduction, expressed in percentage reduction in area, may be 30 to 40% or more. The advantage of the combined process is that a typical part may have six to ten or more shank diameters.

Extrusion Forging.   Extrusion forging is often called a cold-working process. Three principal types of metal displacement by plastic flow are involved: backward, forward, and a combination of both backward and forward (Fig. 8-49). Impact extrusion consists of hitting a slug held in the die with a punch. The metal plastically deforms when the yield point is exceeded but extrusion does not start until the pressure becomes seven to fifteen times the initial yield strength of the alloy.

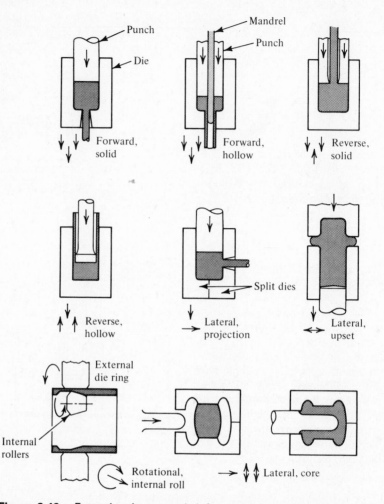

Figure 8.49. Extrusion is a metal deformation process with many variations as shown. In most forging operations one or more of these occur. (*Courtesy* American Machinist.)

Advantages and Limitations.   Metals particularly suited to impact extrusion are softer and more ductile such as aluminum, copper, and brass. However, low- and medium-carbon steels are also used even though the impact factor is high.

Generally, steel impacts are limited to 2.5 times the punch diameter on reverse extrusions. Approximate limits for aluminum extrusions are: 14 in. (35.56 cm) in diameter and 60 in. (152.4 cm) in length. Hydraulic presses are used for loads over 2000 tons (1800 metric tons) because they have greater variation in stroke lengths, speed, and economic advantages.

Tolerances vary with materials and design, but production runs calling for 0.002- to 0.005-in. (0.05–0.13-mm) tolerance are regularly made.

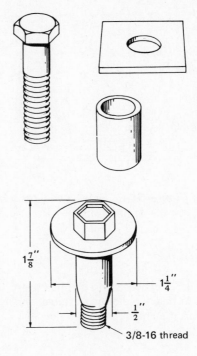

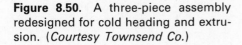

Figure 8.50. A three-piece assembly redesigned for cold heading and extrusion. (*Courtesy Townsend Co.*)

Applications. Impact extrusions compete with other press-working operations and with metal machining. Common products are aerosol cans, cocktail shakers, lipstick cases, flashlight cases, and vacuum bottles. Secondary operations, such as beading, thread rolling, dimpling, and machining are sometimes needed to make the complete item. Figure 8-50 shows a three-piece assembly that represents a considerable saving in time and materials when manufactured as one part by cold-heading and impact extrusion.

Parting Line. The parting line, where the die halves meet, helps determine grain flow and affects production rates, die costs, and die life. For maximum economy, the parting line should be kept in one plane. This makes die sinking, forging, and subsequent trimming operations simpler and therefore less costly. Also, the parting line will have a pronounced effect on the metal-flow lines, as shown in Fig. 8-51.

To avoid mismatch of the forging dies, every effort should be made to balance the forces, especially on nonsymmetrical parts. Side thrust increases as the parting line inclines away from parallel to the forging plane (Fig. 8-52).

Corner and Fillet Radii. Small corner and fillet radii can be incorporated into modern forging designs and can be of considerable cost saving when they are used to eliminate machining operations. When this is not necessary, more generous corner and fillet radii should be used to reduce the costs and prolong die life. Also, connecting sweeps between pockets and recesses should be as generous as feasible to obtain good metal flow.

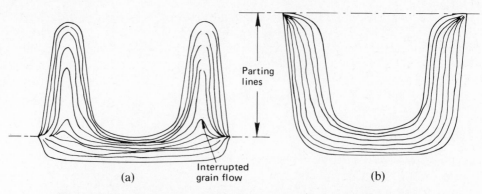

Figure 8.51. Parting-line placement will have a significant effect on the grain structure of a forging as contrasted by (a) and (b).

Figure 8.52. The parting line has been placed to balance die-thrust forces and avoid mismatch.

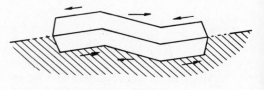

Draft Angles. Draftless forgings are now being produced, particularly for aircraft and missile applications. Generally the draft angle on small forgings is about 5°.

Holes. Forging dies can be designed with integral plugs. The plugs form recesses that can be pierced during the normal trimming operation, often at considerable savings in subsequent machining operations.

As-Forged Surfaces. The as-forged surface can often be used in the finished part. However, it must be determined that the forging tolerances that apply to the finish (normally 500 microinches) are acceptable.

Finished Size. The designer knows what the finished product's dimensions and tolerances should be. The forger knows how much excess metal (the machining envelope) will be required to achieve this finished size. The designer knows excess metal is necessary but wants it kept to a minimum, hence a problem develops. There are many variations on this theme that arise to defeat the application of forging tolerances, such as die closure, mismatch, warpage, random surface deviations, die wear, concentricity, and out-of-roundness. Two obvious approaches to the problem are to add enough material to take care of all contingencies or insist on tighter forging tolerances. Both approaches will add considerably to the cost. A more realistic approach is for the designer to establish tooling points.

Tooling points are set by the designer and are based on the fact that any part can be located and dimensioned within a framework of three planes located at 90° to each other. The principle is that of 3–2–1. Three locating points are used on the main or largest surface, two on the next longest surface, and one on the shortest. The machined surface of the tooling points will be provided by the manufacturer when shown on the design. When this is done, many manufacturers are willing to guarantee the machining tolerances.

Equipment

The most widely used type of forging equipment is the hammer, which can deform the stock with one or more sudden blows of substantial force.

Closed-die forging hammers are classified by the force that drives the weighted ram downward: either gravity or a controlled combination of gravity and compressed air. (Steam, hot air, was originally used, but such hammers are currently losing favor.)

A type of gravity hammer still in wide use, particularly for producing forgings weighing only a few pounds, is the board drop hammer. The ram is keyed to one or more wooden boards, which are then propelled upward by powered friction rolls (Fig. 8-53). A trip on the ram engages a roll-release lever that moves the friction rolls away from the boards, allowing them to drop. The height of fall, adjustable mechanically, determines the force of the blow. Board hammers are rated in pounds of falling weight, the range being from 400 to 10,000 lb. Air-lift gravity hammers in the same (light) range have approximately the same force.

A completely automatic counterblow hammer is a horizontal type shown in Fig. 8-54. A moving tong advances hot billets through the dies and turns them 90° if required. The press is air-operated and is built in sizes ranging from about 3000 to 70,000 ft lb of striking energy. All functions are programmed from a central control, which can include stock heating, by resistance-type heaters, and advancement through all the impressions of the forging dies.

The largest forging presses in the world are hydraulic; two in the United States, each of which is rated at 50,000 tons, and one in France with a squeeze capacity of up to 66,000 tons. Hydraulic-powered forging presses efficiently form large high-strength aircraft and aerospace components from necessarily high-cost alloys with high precision and minimal machining waste. No other type of metal-forming machine can do such work.

More recently NC has been made available for hydraulic forging presses so that any sequence can be programmed, such as rapid approach of the ram, and then decelerate as contact is made, all the while maintaining just the right velocity to make the metal flow at the desired rate. After a controlled dwell period, the ram is driven up quickly.

High-energy-rate forming (HERF) presses are also used in forging as discussed under high-velocity forming at the end of this chapter.

Swaging

Swaging is a process of reducing the cross-sectional area of rods or tubes. The process is shown schematically in Fig. 8-55. In actual practice, a ring of hammers

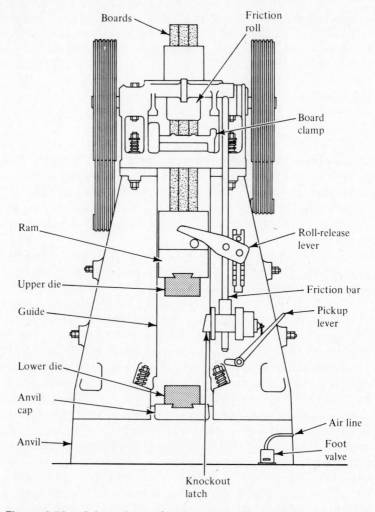

Figure 8.53. A board-type forging press. The height to which the boards are lifted determines the striking force of the gravity hammer. (*Courtesy* American Machinist.)

is rotated around the workpiece at a speed of about 200 rpm, delivering 1000 spm (strokes per minute). The outer surface is always circular in cross section but the inner surface of a swaged tube can be any shape from which a forming mandrel can be removed. Swaging may be started or stopped at any point along the length of the stock and is often used for pointing the ends of tubes or bars or for producing stepped diameters.

Drawing and extrusion

Bars or tubes that have been hot-rolled are often given a cold-finishing operation to reduce the size, increase the strength, improve the finish, change the shape, or provide better accuracy. The drawing process is shown in Fig 8-56.

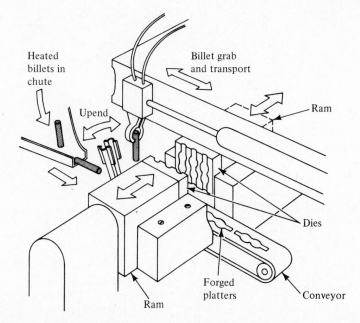

Figure 8.54. A horizontal, automatic, counterblow-type forging press. (*Courtesy* American Machinist.)

Figure 8.55. Swaging is used to reduce the cross-sectional area of rods or tubes.

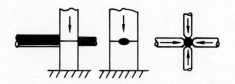

Figure 8.56. Essential features of the cold-drawing process. (*Courtesy* Steelways, *published by the American Iron and Steel Institute.*)

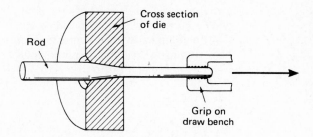

Drawing is confined largely to the metal manufacturers and is not generally considered a fabricating process. Extrusions are also made by the metal manufacturer but have such a wide application that the engineer should be aware of their properties, uses, and relative economy. A sketch of the basic extrusion process is shown in Fig. 8-57.

Extrusion was previously mentioned as a forging process, whereas here it is presented as a principal method of producing long structural shapes.

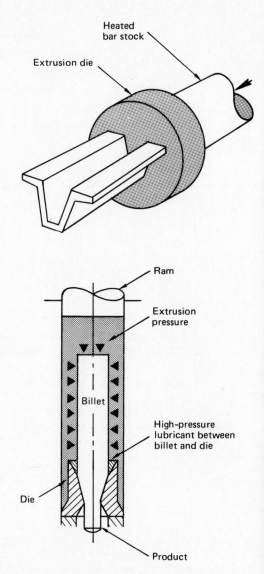

Figure 8.57. Basic concept of the extrusion process.

Figure 8.58. Hydrostatic extrusion uses high-pressure fluid to force a billet through a die orifice. The fluid completely surrounds the billet, which virtually eliminates all billet-container friction, reducing energy requirements. (*Courtesy* Machine Design.)

Hydrostatic Extrusion A more recent development is hydrostatic extrusion, which makes it possible to cold extrude many difficult-to-form materials such as the high-strength super alloys, arc-cast tungsten, and molybdenum. In this process, the metal billet is inserted into the chamber behind the die where it is surrounded by a suitable liquid. The billet is then extruded by applying pressure to the liquid. The included angle in front of the die has a pronounced effect on the total force needed to form the metal, on the flow patterns, and on the soundness of the extrusion (Fig. 8-58). The extruding stress changes continuously with the die angle and reaches a minimum at the optimal angle.

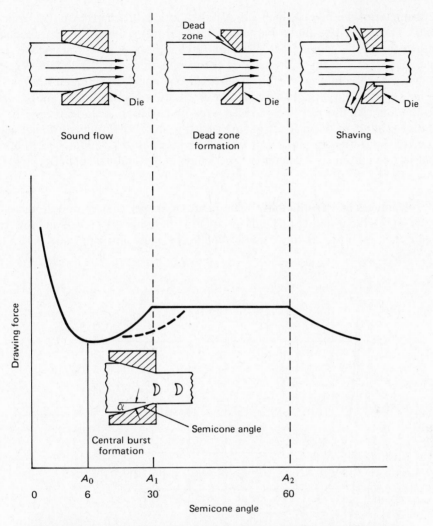

Figure 8.59. The effect of semicone angles and mode of flow on drawing or extruding force. Prepared by Evans and Avitzur. (*Courtesy Society of Manufacturing Engineers.*)

Extrusion Die Angles.   Excessive die angles cause the metal to shear within itself and form its own cone angle that does not conform to the contour of the die as shown in Fig. 8-59. The new angle that forms is known as the dead-zone cone angle. For any semicone angle between A_1 and A_2, the dead-zone angle will form. As the semicone angle goes beyond A_2, the die acts like a cutting tool and shaves the billet off.

Three interacting parameters—ideal deformation, shear resistance, and friction losses—vary with the die semicone angle to produce a minimum resistance curve. The ideal deformation factor involves only the power required to reduce the diameter of the billet, which varies only a little with changes in the semicone angle. The shear-resistance parameter is determined by the force needed to produce a specific amount of distortion and increases as the semicone angle becomes larger. In either drawing or extrusion, friction is dependent upon the contact with the die wall. With a small semicone angle, the contact between the die and billet is large and therefore the friction losses are large. As the angle becomes larger, the die-billet contact area and friction loss decrease. As the curves for all three of these effects are added to make the total resistance curve, there is found, for any one set of friction and reduction conditions, a minimum point at one particular semicone angle, which is the optimal semicone angle, as shown in Fig. 8-60.

Advantages and Limitations.   The range of shapes that can be produced by extrusion is almost infinite. The dies required are relatively simple and low cost. Simple dies may run as low as $100, while the more intricate may cost $1500

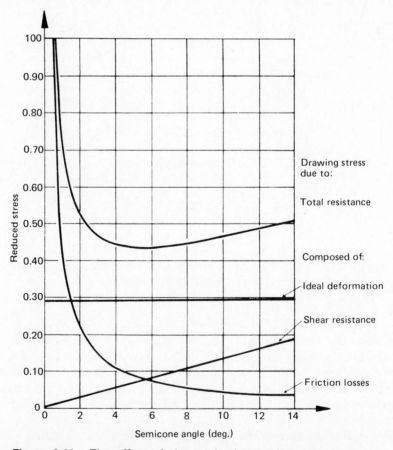

Figure 8.60. The effect of change in the semicone angle on extruding stress. (*Courtesy Society of Manufacturing Engineers.*)

Figure 8.61. A multiple-hole extrusion die provides high production and decreased die wear. (*Courtesy Elox Corp. of Michigan.*)

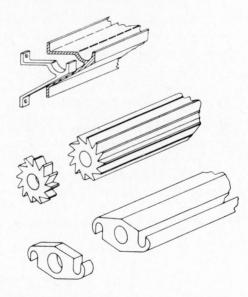

Figure 8.62. Extrusions are made in bar form and can be sectioned off. (*Courtesy Kaiser Aluminum and Chemical Sales, Inc.*)

or more. A multiple-hole extrusion die that can be used to produce four parts simultaneously, resulting in less die wear and higher production is shown in Fig. 8-61. Sections may be designed to reduce the number of parts needed in an assembly or to reduce machining costs. Extrusions made in "bar" form can be sectioned for special purposes, as shown in Fig. 8-62.

An example of the part extrusions play in building a prototype aluminum boat is shown in Fig. 8-63.

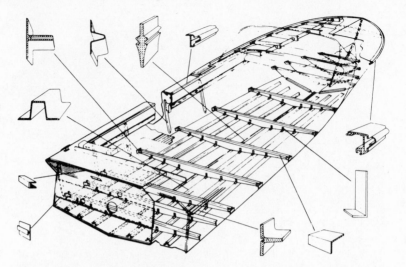

Figure 8.63.   A wide variety of extrusions are used to construct this prototype aluminum boat. (*Courtesy Kaiser Aluminum and Chemical Sales, Inc.*)

Extrusions are usually limited to parts that can be circumscribed by a 17-in. (43.18-cm) circle for aluminum and a 6-in. (15.24-cm) circle for steel. Extremely thin sections should be avoided, as should extreme thicknesses. Hollow shapes having unsymmetrical voids are not recommended.

High-Energy-Rate Forming (HERF)

High-energy-rate forming is a term that has been widely used to describe a method of efficiently applying more energy to the workpiece per unit of time than has been applied by previous techniques. HERF can be applied to both light- and heavy-gage materials; in the former it is usually thought of as another forming method and in the latter as a forging technique.

Theory of high-energy metal forming

The plastic flow of metal is both time and temperature dependent.

Time Dependence.   Basic strain-rate curves for various materials are shown in Fig. 8-64. True stress increases with strain rate. In general, the flow stress is raised considerably with increased strain rates, whereas true strain to fracture decreases a small amount.

The toughness of the material can be thought of as the area under the true stress–strain curve and is a measure of the energy-absorption ability of the metal. Replotting the curve as toughness versus strain rate gives the graph as shown in Fig. 8-65.

This graph shows that the toughness of all metals increases to some critical

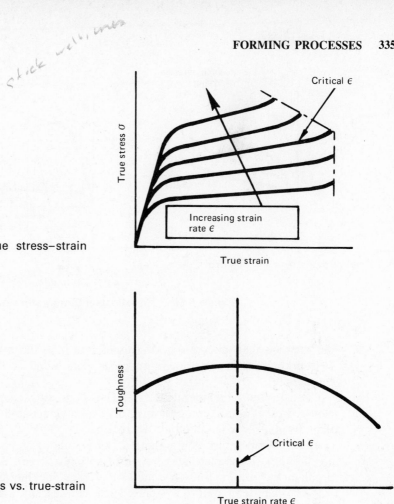

shock velocities

Figure 8.64. Basic true stress–strain curves.

Figure 8.65. Toughness vs. true-strain rate.

strain rate and then diminishes. The toughness is the ability of the metal to plastically deform under shock without fracture. Thus it can be seen that all high-velocity methods have decided limitations on the materials that can be formed. Metals that have a high toughness and strain rate should be selected.

Temperature Dependence. As shown previously, increased temperatures lower the flow stress of a material. An even greater benefit is the greatly increased true strain to fracture, which can be tremendous in some metals. As an example, tungsten increases from a strain of less than 1% at room temperature to 55% at 1950°F (1066°C). Annealed titanium 6A1-4V increases from 12% at room temperature to 52% at 1200°F (649°C). Of course, not all of this increased true strain is usable in forming because of necking.

Temperature also increases toughness. The increase in toughness does not benefit slow forming, such as hot-fluid forming, but it does benefit high-velocity–forming systems where both slip and twining occur in rapid succession.

A third benefit of increased-temperature forming is *creep,* or the plastic

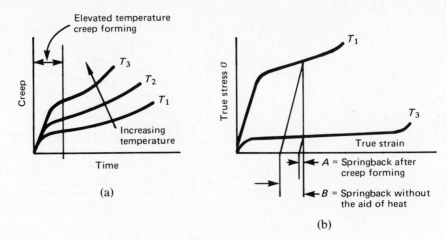

Figure 8.66.   The effect of temperature on creep.

deformation that takes place at stresses less than those required to cause slip. It is primarily a high-temperature phenomenon and is shown schematically in Fig. 8-66(a).

Increasing the temperature from T_1 to T_3 greatly accelerates creep. Practical elevated-temperature creep forming is generally accomplished in a period of less than 10 min for a temperature approximately 500°F (277°C) above the service-operating temperature of the metal. The results of creep forming [Fig. 8-66(b)] are the almost complete elimination of springback and the elimination of other problems in forming thin-gage metals, such as buckling and distortion.

The high-temperature systems, such as integrally heated dies and hot-fluid forming, result in deformation that is almost totally slip. Some twinning may occur in the hexagonal metals where slip is impeded by the small number of slip planes. Slip deformation in these systems has been termed elevated-temperature creep forming and is also dependent on time and the number of dislocations in the metal.

Hydrostatic Pressures and HVF.   Bridgman (see bibliography at end of chapter) has shown that high transverse pressure acting on a tensile specimen during pulling will exhibit a considerable increase in a total true strain to fracture (Fig. 8-67). The true stress–strain curves show that flow stress does not change appreciably but that the true strain to fracture increases with hydrostatic pressure. Also, the mode of fracture changes from a combination of ductile-brittle failure below 170,000 psi to a totally ductile failure above this pressure, as shown by the tensile specimens.

Investigators of HVF have attempted to correlate hydrostatic pressures to formability. Transverse pressure does not benefit forming for the large-mass systems unless the pressure exceeds the compressive flow stress of the material. When this happens, as it often does in HVF, the shear stress τ greatly increases, allowing a considerable increase in plastic flow.

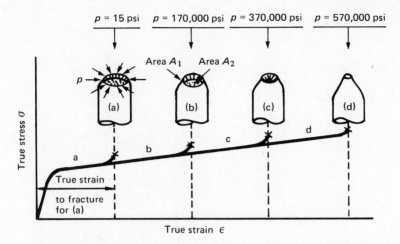

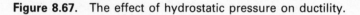

p = applied transverse hydrostatic pressure
Area A_1 = ductile fracture = slip
Area A_2 = brittle fracture = separation

Figure 8.67.   The effect of hydrostatic pressure on ductility.

Mass vs. velocity in HERF

Energy can be increased by increasing either the mass or the velocity as shown in the kinetic energy equation:

$$KE = \tfrac{1}{2}mv^2$$

A much greater effect can be obtained by increasing the velocity since the energy is a function of the velocity squared.

A system that makes use of the velocity factor is that of the floating piston, shown schematically in Fig. 8-68.

A comparison of the kinetic energy of the two systems can be made briefly as follows.

Mass System:
  Assume a 10,000-lb punch and holder and 20 fps velocity.

$$KE = \tfrac{1}{2}mv^2 = \tfrac{1}{2}\frac{(w)}{(g)}v^2$$

$$= \tfrac{1}{2}\frac{(10,000)}{(32.2)}(20)^2 = \tfrac{1}{2}(311)(400)$$

$$= 62,100 \text{ ft lb}$$

Velocity System:
  Assume a 325-lb mass and an impact velocity of 100 fps.

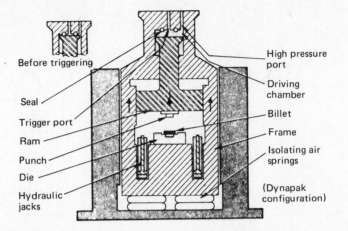

Before triggering

High pressure port

Seal

Driving chamber

Trigger port

Billet

Ram

Frame

Punch

Isolating air springs

Die

Hydraulic jacks

(Dynapak configuration)

Figure 8.68. This pneumatic-mechanical system uses the rapid release of highly compressed gas to accelerate a ram to high velocity. The ram is raised against the seal by hydraulic jacks. After the work has been positioned, the jacks are lowered and a small volume of high-pressure gas is admitted through the trigger port. This forces the ram down enough to expose the top of it to the high-pressure gas, accelerating its downward movement. At the same time, the upper end of the driving chamber is exposed to the gas and the frame moves upward. (*Courtesy* Machine Design.)

$$KE = \tfrac{1}{2} \frac{(w)}{(g)} v^2 = \tfrac{1}{2} \frac{(325)}{(32.2)} (100)^2 = \tfrac{1}{2}(10)(10,000)$$

$$= 50,500 \text{ ft lb}$$

By this simple example, we see that with an assumed velocity five times greater and a mass 30 times smaller, the total energy developed is of the same order of magnitude as with the lower-velocity, larger-mass systems. The velocity factor contributes 1000 times as much as the mass in the total energy, hence the newer term high-velocity metalworking (HVM) or simply high-velocity forming (HVF).

High-Velocity Metal Forming. In the high-velocity metal-forming systems such as explosive-forming, electrohydraulic, and electromagnetic processes (shown schematically in Fig. 8-69), the mass contribution toward total energy is negligible compared to velocity. Explosive forming can be used as an example.

Explosive Forming [Under water similar to Fig. 8-69(a)]:
Assume 6 lb for weight of moving water front (*w*) and an impact velocity (*v*) of 700 fps.

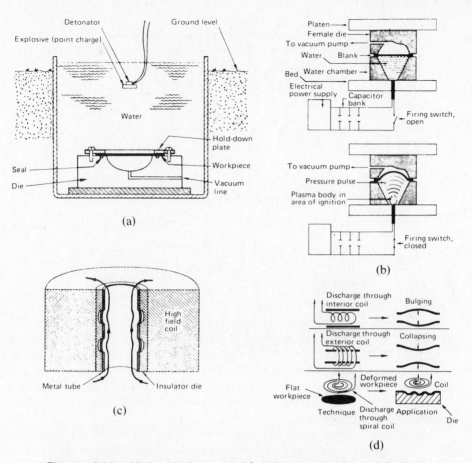

Figure 8.69. High-velocity metal-forming methods: explosive forming (a), electrohydraulic forming (b), and electromagnetic forming (c). In the expansion ring or bulging concept of electromagnetic forming, the metal takes the form of the die when the field outside is removed and the unbalanced forces cause the metal to expand. The three main concepts of electromagnetic forming are shown schematically at (d). (*Courtesy* Machine Design.)

$$\mathrm{KE} = \tfrac{1}{2}\frac{(w)}{(g)}\,v^2 = \tfrac{1}{2}\frac{(6)}{(32.2)}\,(700)^2 \approx \tfrac{1}{2}(0.2)(5 \times 10^5)$$

$$\approx 46{,}000 \text{ ft lb}$$

The 46,000 ft lb of energy in the example is only a small portion of the actual energy generated by the explosion as shown by the following simple calculation:

E = specific energy content of explosive = 3×10^6 ft lb/lb (assumed)
w = weight of average charge = 0.1 lb (assumed)

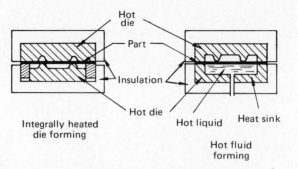

Figure 8.70. Two basic systems that use heat to aid high-velocity forming. (*Courtesy Society of Manufacturing Engineers.*)

$$v = Ew = (3 \times 10^6)(0.1) = 300,000 \text{ ft lb}$$

$$\text{Efficiency} = \frac{\text{KE}}{v} = \frac{46,000}{300,000} = 15\%$$

Most explosive-forming operations have an efficiency of 15 to 20%. The bulk of the energy, as in the example, is used in blowing the water into the air.

High-Temperature Metal Forming. Two basic systems that use heat for high-energy forming are shown schematically in Fig. 8-70. At the left is an integrally heated die with heat supplied by cartridge-type heaters. Insulation surrounds the die completely.

In hot-fluid forming, both heat and pressure are supplied by a hot, liquid-metal alloy, giving considerable advantage by lowering the flow stress of the metal. As an example, stainless steel has a flow stress of approximately 100,000 psi at room temperature, but reduces to 5000 psi at 1200°F (649°C). This reduces the mechanical work by a ratio of 20:1.

Some advantages and limitations of HVF

Advantages.   A wide variety of materials can be formed or forged, including exotic and refractory metals, stainless steels, nonferrous alloys, and high-strength materials, some of which are not usually forgeable.

Draft allowances are reduced and in some cases eliminated.

Complex parts can be formed in one blow.

Tolerances and surface finish are improved over those obtained with conventional forging techniques. (The surface finish usually lies in the range of 20 to 60 microinches.)

Because strength and fatigue resistance are improved, parts can be made smaller.

Repeatability is excellent.

Usually the more expensive the material, the greater the savings.

Large (20 ft or more in diameter) one of a kind domes and cylinders have been explosively formed using ice as a die.

Limitations. The main limitation of HVF is in tooling materials and capacity (except for explosive forming). Tooling materials present a serious limitation. There is a need for less expensive, long-life materials that can withstand the heavy loads involved. The tooling materials also set limitations on the production rate since the dies can operate only so long without being allowed to cool.

Part configuration is usually limited to one-piece dies. If two-piece or split dies are necessary, the economics of the process becomes debatable.

If several blows are required, as is sometimes done in the pneumatic-mechanical press shown in Fig. 8-68, the process again becomes economically questionable. Usually two blows are acceptable but three blows become borderline.

Problems

8-1. A 10-in. dia. aluminum blank, $\frac{1}{8}$ in. thick, is to be drawn into a 4-in. dia. cup.
   (a) Is this possible in one draw? Why or why not?
   (b) What function does the draw radius serve? How large should it be for this application?
   (c) What punch radius should be used? Why?

8-2. For drawing, what is the significance of the strain-hardening coefficient?

8-3. Why isn't a metal that has a large yield elongation good for forming?

8-4. What are some of the advantages of hot-working metal?

8-5. (a) What are some ways to compensate for springback?
   (b) What should the die opening be for a high yield-strength material $\frac{1}{8}$ in. thick? Make a sketch to show a sectional view of the die with the material in place and the die-opening dimension.

8-6. What is the force required to bend a piece of low-carbon steel plate $\frac{3}{16}$ in. thick and 8 in. wide? Assume the die opening is $8t$.

8-7. A bolt blank is made by upsetting. The material used is SAE 1018 steel. The bar stock used is $\frac{1}{2}$ in. in diameter. The desired hexagonal-head size is $\frac{3}{4}$ in. in diameter and 0.333 in. thick.

   (a) How many strokes will be required?
   (b) Estimate what impact will be required.

8-8. (a) What is the main advantage of forging over a cast or welded structure?
   (b) What is a disadvantage of using the forging process?

8-9. (a) Compare the impact obtained per blow from a conventional drop-hammer forging press with that from a floating-piston press under the following conditions.
   Conventional drop-hammer forging press:
      Drop hammer = 1200 lb.
      Velocity of hammer = 16 fps
   Floating-piston press:
      Floating piston mass = 400 lb.
      Velocity of piston = 150 fps
   (b) What are the ratios of the mass and velocity to the total energy delivered between the two presses?

8-10. What is the relationship between the flow stress of a part with increasing strain rate and true strain to fracture?

8-11. (a) What is meant by the toughness of a metal?
   (b) Should tough metals be used for HERF? Why or why not?
   (c) Would Silly Putty be a good material to use to illustrate the requirements for HERF? Why or why not?

8-12. The advantages of elevated-temperature forming are quite obvious, but what may be some disadvantages?

8-13. (a) What is the effect of hydrostatic pressure on true strain?
   (b) How does hydrostatic pressure affect fracture?

8-14. When are hydrostatic pressures especially beneficial for metal forming?

8-15. What metallurgical changes make HVF possible?

8-16. Why must the die angle for forward extrusions be carefully controlled?

8-17. State the special purpose of each of the following dies:

(a) compound,

(b) progressive,

(c) transfer.

8-18. (a) Assume the forming-limit curve technique is used. The original circle diameters on the metal are 1 in. After forming $\varepsilon_1 = 1.5$ in. and $\varepsilon_2 = 0.5$ in. how would you determine if this is satisfactory or not?

(b) Would it be possible to show a negative ε_2 strain? Explain.

8-19. (a) What is the main objection to the Erichsen cup test in determining the drawability of metal?

(b) How does the Fukui test overcome the objection of the Erichsen test?

8-20. What material could be used for a die to form a dome 15 ft in diameter to a depth of 10 in. at the center? The material is mild steel $\frac{1}{4}$ in. thick. Only one is required.

8-21. Why should yield strength be low for a material that is used for drawing?

8-22. What may result if a material has a large yield elongation and is subjected to deep drawing?

8-23. What is the effect of temperature on the percent reduction of area that can be achieved?

8-24. When does hydrostatic pressure benefit forming?

8-25. What semicone angle produces the least total resistance in extrusion?

8-26. How wide a strip of metal would be required to form a channel of $\frac{1}{8}$ in. thick steel 2 in. high and 4 in. wide if the bend radius is $2t$?

8-27. Why is a high n value desirable for a metal that is to be formed?

8-28. What are two reasons the Rockwell B hardness test is not a good indicator of the formability of metal?

8-29. What happens to steel plate in the strain-aging process?

8-30. What purpose does temper-passing serve for sheet metal?

8-31. Which of these two metals would tend to have the better formability and why? Rimmed steel or copper?

8-32. Define hot-working of metal.

8-33. Why might the recrystallization temperature differ for two identical metals?

8-34. A 90° $2t$ bend was made in a strip of 0.125 in. thick strip steel and in $\frac{1}{4}$ in. thick steel. Which of the two would have less springback and why?

8-35. What purpose does a die cushion serve?

8-36. What size press would be required to blank a 3-in. dia. hole in 0.125-in. mild steel plate?

8-37. How much work energy in tons would a 100-ton press have on a continuous basis?

Bibliography

Blickwede, D. J. "Sheet Steel-Micrometallurgy by the Millions." *ASME Transactions* 61 (1968): 633–679.

Bridgman, P. W. "The Effect of Hydrostatic Pressure on Ductility." *Journal of Applied Physics* 17 (1947).

Dallas, Daniel B. "Presswork: The Punching Machines Have Arrived." *Manufacturing Engineering and Management* 70 (February 1973): 18–25.

Dwyer, John J. Jr., ed. "Sheet Steel Today."

American Machinist Special Report 687 120 (May 1976): 65–88.

Eary, D. F. "Analysis of Cup Drawing Variables." *Society of Manufacturing Engineers Report MF69-518* (April 1969).

Evans, W. M., and B. Avitzur. "Toward Better Extrusion Die Design." *ASTM Paper No. MF67-582* (1967).

Keeler, Stuart P. "Understanding Sheet Metal Formability." *Machinery*, six parts (February 1968 through July 1968).

Winship, John T., ed. "Fundamentals of Forging." *American Machinist Special Report 705* 122 (July 1978): 99–122.

Winship, John T., ed. *"A New Era in NC Punching." American Machinist Special Report 717* 123 (November 1979): 139–158.

Winship, John T., ed. "Pressworking Equipment." *American Machinist Special Report 696* (April 1977): 89–108.

CHAPTER NINE

Metal-Casting Processes

Most metal castings are made by pouring molten metal into a prepared cavity and allowing it to solidify. The process dates from antiquity.

The largest bronze statue in existence today is the great Sun Buddha in Nara, Japan. Cast in the eighth century, it weighs 551 tons (496 metric tons) and is more than 71 ft (21 m) high.

Artisans of the Shang Dynasty in China (1766–1222 B.C.) created art works of bronze with delicate filigree as sophisticated as anything that is designed and produced today. In the Cauca Valley in Columbia, aboriginal Quimbaya goldsmiths created remarkable hollow gold castings long before the time of Columbus.

There are many casting processes available today and selecting the best one to produce a particular part depends on several basic factors such as cost, size, production rate, finish, tolerance, section thickness, physical–mechanical properties, intricacy of design, and weldability.

Casting processes may be classified into six groups: sand casting, shell molding, plaster molding, investment casting, permanent-mold casting, die casting, and continuous casting.

SAND CASTING

Green-Sand Molding Process

Green-sand molding is the most frequently used casting process and therefore will be discussed in more detail than some of the other processes.

Green sand refers to moist sand (from 2 to 8% water). This method of molding is the most popular and widely used process in the foundry industry. The process is well suited to a wide variety of miscellaneous casting, in sizes of less than a pound to as large as 3 to 4 tons. This versatile process is applicable to both ferrous and nonferrous materials.

Green sand can be used to produce intricate molds since it provides for rapid collapsibility, that is to say the mold is much less resistant to the contraction of the casting as it solidifies than are other molding processes. This results in less stress and strain in the casting.

Dry-Sand Casting.   Most large and very heavy castings are made in dry-sand molds. The mold surfaces are given a refractory coating and are dried before the mold is closed for pouring. This hardens the mold and provides the necessary strength to resist large amounts of metal, but it increases the manufacturing time.

Molds that are hardened by a carbon dioxide process, explained later, can also be considered in the dry-sand class. The process of sand casting begins with the design, which in turn is made into a three-dimensional pattern. The pattern can be made of wood, metal, plaster, styrofoam, or other materials. It is made larger than the desired casting to provide for metal shrinkage, distortion, and machining. In hand molding the pattern is placed on a *molding board,* which is a little larger than the open box or *flask* in which the sand mold is to be made. The molding board is placed under the top half of the two-part flask and sand is poured around the pattern. The sand next to the pattern is always sifted or riddled to provide a better detail and is termed *facing sand.*

The sand is rammed or compacted around the pattern by a variety of methods including: hand or pneumatic-tool ramming, jolting (abrupt mechanical shaking), squeezing (compressing the top and bottom mold surfaces), and driving the sand into the mold at high velocities (sand slinging). Sand slingers are usually reserved for use in making very large castings where great volumes of sand are handled (Fig. 9-1). The jolt and squeeze method of filling the flask, usually a two-part box containing the pattern, is by far the most used. First the drag, or lower half of the flask, is filled and jolted, then the flask is turned over, the cope, or top half of the flask, is filled, and the two-part mold, with the pattern and molding board sandwiched in between is squeezed.

Patterns.   Patterns of simple design, with one or more flat surfaces, can be molded in one piece, provided that they can be withdrawn without disturbing the compacted sand. Other patterns may be split into two or more parts to facilitate their removal from the sand when using two-part flasks. The pattern

Figure 9.1. A sand slinger can be used to fill large floor molds or a number of smaller molds in a relatively short time.

Sandslinger — medium and large castings.

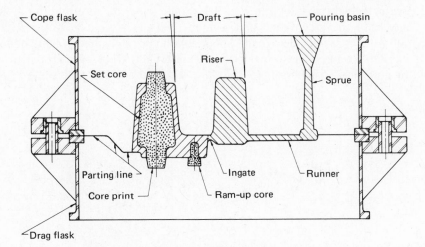
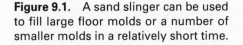

Figure 9.2. Sectional view of a casting mold.

must be tapered to permit easy removal from the sand. The taper is referred to as *draft*. When a part does not have some natural draft, it must be added. A more recent innovation in patterns for sand casting has been to make them out of foamed polystyrene that is vaporized by the molten metal. This type of casting, known as the *full-mold* process, does not require pattern draft. This process will be discussed later.

Sprues, Runners, and Gates. Access to the mold cavity for entry of the molten metal is provided by *sprues, runners,* and *gates,* as shown in Fig. 9-2. A pouring basin can be carved in the sand at the top of the sprue, or a *pour box,* which provides a large opening, may be laid over the sprue to facilitate pouring. After the metal is poured, it cools most rapidly in the thin sections and along the outer surfaces where it gives up most of its heat to the sand mold. Thus the outer surface forms a shell that permits the still molten metal near the center to flow toward it. As a result the last portion of the casting to freeze will be deficient in metal and, in the absence of a supplemental metal-feed source, will result in some form of shrinkage. This shrinkage may take the form of gross shrinkage (large cavities) or the more subtle microshrinkage (finely dispersed porosity). These porous spots can be avoided by the use of risers, as shown in Fig. 9-2, which provide molten metal to make up for shrinkage losses. The science of providing the proper gate and runners to ensure sound castings is discussed later.

Cores.   Cores are placed in molds wherever it is necessary to preserve the space it occupies in the mold as a void in the resulting castings. As shown in Fig. 9-2, the core will be put in place after the pattern is removed. To ensure its proper location, the pattern has extensions known as *core prints* which leave cavities in the mold into which the core is seated. An example of a core box and core placement is shown in Fig. 9-3. Sometimes the core may be molded integrally with the green sand and is then referred to as a *green-sand core.* Generally, the core is made of sand bonded with core oil, some organic bonding materials, and water. These materials are thoroughly blended and placed in a mold or core box. After forming, they are removed and baked at 350 to 450°F

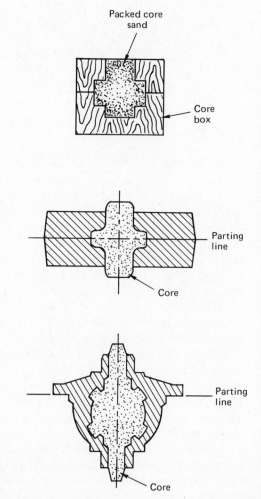

Figure 9.3. Examples of a core box and core placement.

(177–232°C). Cores that consist of two or more parts are pasted together after baking.

CO_2 Cores.   A newer method of curing cores is with CO_2. Carbon dioxide reacts with the sodium silicate constituent of the binder to produce silicon dioxide, as mentioned under CO_2-setting molds. This method has the advantage of being able to use the cores as soon as they are removed from the core box. Also, less draft is required since high strength is developed before removal from the core box. However, a major problem with the CO_2-set cores is that they do not burn out and are quite difficult to remove since they do not collapse and may even cause hot cracks to occur in the casting.

Shell-Mold Cores.   Cores may also be made by the shell-molding process, discussed in more detail later. Only a brief description will be given here. A core box is heated and sand that contains 3 to 6% resin is blown into it. After a dwell period, which establishes the thickness of the shell, the unheated interior sand may be drained out. The shell wall is usually $\frac{1}{4}$ to $\frac{1}{2}$ in. (6.35–12.70 mm) thick and can be easily stripped from the box and handled directly. No further baking is required. The cores may be placed directly in the molds. Although it is possible to recover the sand from the sand–resin mixture, the procedure is questionable unless the quantity is large. Recovery consists of burning out the resin at high temperatures.

Cores should be used only when they can produce real economies by reducing later machining operations. Cores are relatively expensive to make, require expensive tooling, are a nuisance to set (place in the mold), produce scheduling problems, are often difficult to remove from the casting, and in general complicate the casting process. In some cases the cost associated with cores can constitute the greatest single cost factor in producing a casting.

Parting-Line Placement.   The parting line is the line along which a pattern is divided for molding or along which sections of the mold separate. This surface and the pattern are sprinkled with parting dust so that the cope and drag will separate without rupturing the sand. The selection of the parting line can play an important part in the economics of production. Complex parting lines require expensive pattern equipment and are much harder to mold. On the other hand, ingenuity in the selection of a parting line can often take advantage of the natural draft found on the part, eliminate cores, allow more parts to be squeezed onto a match plate, and even eliminate some risers. It is obvious that the point where the metal enters the mold cavity, *the gate,* often lies on the parting line. While the parting line may, with sufficient justification, be allowed to wander up and down, it is most economical when it is restricted to a flat plane, thereby allowing the pattern halves to be affixed to a flat pattern board.

Generally, the flask is removed before the metal is poured to prevent it from becoming damaged and to speed up its reuse (Fig. 9-4). A pouring jacket may be put in its place if there is danger of the molten metal breaking out.

Mechanized Molding

The preparation of molds entirely by hand is slow and costly. In most modern foundries, molds of small and medium sizes are prepared by machines. For mass production, *match-plate patterns* are normally used (Fig. 9-5). The match-plate pattern is made with the cope part of the pattern on one side of a plate, usually aluminum, and the drag part of the pattern on the other side. The plate fits between the cope and the drag. After each half of the mold has been filled with sand and compacted, the match plate is removed, leaving a nearly completed mold since runners, gates, and risers are included on the match plate. A provision for the proper location of sprue is also made.

A variation of this process is called the *cope-and-drag pattern*. This allows both the cope and drag to be "rammed up" separately but at the same time. Separate pattern plates require accurate alignment of the two halves by means of a guide and locating pins.

Molding-Sand Preparation and Control

The selection and mixing of molding sand constitutes one of the main factors in controlling the quality of the castings. Deposits of natural molding sand are characterized by a clay content of 10 to 30%, with about 10% of the deposits being in the 15 to 20% range. Natural sands are sometimes altered by additions to improve certain characteristics. Such sands are referred to as semisynthetic sands. The most common alteration is the addition of more clay to increase the green strength. The reconditioning is also done to produce certain other desirable properties as follows:

1. The sand mold should present a smooth surface. This surface is governed by grain size, grain-size distribution, and clay and fines content.

The clay content is determined by washing a sand sample through several cycles in a solution of sodium hydroxide. This will remove the clay as well as the fines and the solubles. After drying, the sand is weighed and the percentage of clay removed is defined as the American Foundry Society (AFS) clay content. It should be appreciated that the AFS clay content includes not only the active clay but also all other materials washed off, in particular the entire fines content. When casting metals of high melting temperature, such as steel, fire clays are required. A mixture of 7 to 8% fire clay and 1 to 2% Western bentonite may be used. Bentonite is a weathered volcanic ash.

2. The sand, after being moistened with water, should possess good moldability, produce maximum mold hardness, and be *permeable*. Permeability refers to the ability of the sand to vent steam. The moisture may be measured by weighing a sample of the moist sand and then drawing off the moisture with hot (300°F, 140°C) air and reweighing it. Perhaps a more popular method is that of "feel" by an experienced foundryman. A handful of sand is squeezed and upon releasing it the sharp edges formed between the fingers should stand up without crumbling. If they fall down the sand is too dry. If, on the other hand, it breaks at the edges and rounds off, then it is too wet. The break should be sharp and clean.

Figure 9.4. Metal flask and mold. (*Courtesy Hines Flask Co.*)

Figure 9.5. A match-plate pattern. (*Courtesy Central Foundry Division, G.M.*)

Figure 9.6. A universal sand-strength testing machine showing a sand sample being tested for shear strength. (*Courtesy Harry W. Dietert Co.*)

The hardness of the rammed-up mold sand may be checked with a hardness tester that has a spring-loaded steel ball. The maximum hardness that can be expected from a hand-rammed mold is about 82–85. If no penetration occurs, as on a machine-rammed mold, the hardness reading will be 100. As the mold hardness increases, it produces a casting with a better surface finish and greater dimensional accuracy. It also helps eliminate loose-sand inclusions and mold erosion.

Both the green compressive strength and the green shear strengths of the molding sands can be checked by the use of a standard AFS specimen in a universal testing machine (Fig. 9-6). However, in practice the easily taken mold hardness is usually monitored for a given sand system. This characteristic of the sand can be closely correlated to the green-sand compressive and shear strength.

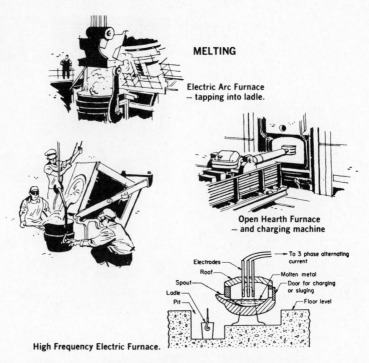

Figure 9.7.   Furnaces used to melt steel or other metals.

Sand Reconditioning.   In most foundries, after the casting has been shaken out of the mold, the sand is recirculated. In mechanized foundries the sand is automatically returned by belt conveyor for reconditioning. It is aerated, lumps are broken up, magnetic particles are removed by an electromagnet, new clay is added to make up for the "burnout" or deactivated portion, and it is brought back to proper "temper" by the addition of a suitable amount of water. In addition, the sand is passed through a muller that mechanically fluffs it up and makes it suitable for remolding.

Melting and Pouring the Metal

The metal for casting may be melted in one of several types of furnaces, as shown in Fig. 9-7. The choice of furnace is based on various factors given briefly as follows:

1. Economy, including the cost of fuel per pound of melted metal and the initial cost of the equipment plus installation.
2. The availability of the desired energy. As an example, if electric power is to be used it may require several years to make all the necessary arrangements needed for the heavy demands of a large foundry.
3. Temperatures required.
4. Quantity of metal required per hour or shift.

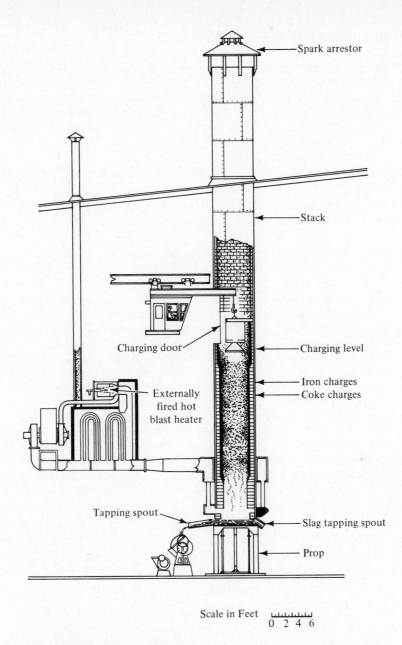

Scale in Feet
0 2 4 6

Figure 9.8. Sectional view of cupola construction. Cupolas are now equipped with dust collectors and other antipollution equipment. (*Courtesy Whitting Corporation.*)

5. The compatability of the process to the metal, e.g., cupolas are good for cast iron but not for aluminum.
6. The compatibility of operation to environmental considerations. For example, in recent years foundrymen have experienced that cleaning up a cupola operation can double the cost of the whole foundry. A cupola furnace is shown in Fig. 9-8.

Crucible, or Pot-Tilting, Furnaces.   Crucible furnaces, as the name indicates, consist simply of a crucible to hold the metal while it is being melted by a gas or oil flame. It is so arranged that it can be easily tilted when the metal is ready for pouring (Fig. 9-9). The capacity is generally limited to 1000 lb.

Cupolas.   Cupolas are designed more specifically for producing a cast-iron melt. The cupola, as shown schematically in Fig. 9-8, has several advantages:

1. Like a blast furnace used in making steel, it can be tapped at regular intervals as required under conditions of production.
2. The operating efficiency of the cupola is higher than any other foundry melting technique due to the countermovement of the heating gases in respect to the charge. As mentioned previously, the economic advantage of the cupola has now been somewhat mitigated by the complex problem of cleaning the inherently dirty stack effluent as well as by the increased shortage of "metallurgical grade" coke (low sulfur content).
3. The chemical composition of the melt can be controlled even under conditions of continuous melting. Since the molten metal comes in contact with the carbonaceous coke, the carbon content of the melt tends to be as high as in the range of cast irons. If steels are desired, the melt must be transferred to a different type of furnace, such as an *open-hearth,* an *electric-arc,* or a *basic-oxygen* furnace, where the carbon content is lowered and the alloy content adjusted.

Pouring the Metal.   Several types of containers are used to move the molten metal from the furnace to the pouring area.

Large castings of the floor-and-pit type are poured with a ladle that has a

Figure 9.9.   A crucible or pot-tilting furnace. (*Courtesy Randall Foundry Corp.*)

plug in the bottom, or as it is called, a bottom-pouring ladle. It is also employed in mechanized operations where the molds are moved along a line and each is poured as it is momentarily stopped beneath the large bottom-pour ladle.

Ladles used for pouring ferrous metals are lined with a high alumina-content refractory. After long use and oxidation, it can be broken out and replaced. Ladles used in handling ferrous metals must be preheated with gas flames to approximately 2600–2700°F (1427–1482°C) before filling. Once the ladle is filled, it is used constantly until it has been emptied.

For nonferrous metals, simple clay-graphite crucibles are used. While they are quite susceptible to breakage, they are very resistant to the metal and will hold up a long time under normal conditions. They usually do not require preheating, although care must be taken to avoid moisture pickup. For this reason they are sometimes baked out to assure dryness.

The pouring process must be carefully controlled since the temperature of the melt greatly affects the degree of liquid contraction before solidification, the rate of solidification which in turn affects the amount of columnar growth present at the mold wall, the extent and nature of the dendritic growth, the degree of alloy burnout, and the feeding characteristics of the risering system. These points are discussed in more detail later.

A newer high-speed automatic, continuous pouring system is shown in Fig. 9–10. The system is capable of pouring 120 aluminum castings per hour. This is the equivalent to three manual lines. The molten metal can be moved from the furnace to the pour in 6 sec. The exact amount required for each mold, within 2%, is regulated automatically. This keeps gate weight variances to within plus or minus 2%, reducing variances of normal pouring by 65%.

Finishing Operations

After the castings have solidified and cooled somewhat, they are placed on a shakeout table or grating on which the sand mold is broken up, leaving the casting free to be picked out. The casting is then taken to the finishing room where the gates and risers are removed. Small gates and risers may be broken off with a hammer if the material is brittle. Larger ones require sawing, cutting with a torch, or shearing. Unwanted metal protrusions such as fins, bosses, and small portions of gates and risers need to be smoothed off to blend with the surface. Most of this work is done with a heavy-duty grinder and the process is known as *snagging* or *snag grinding*. On large castings it is easier to move the grinder than the work, so swing-type grinders are used (Fig. 9-11). Smaller castings are brought to stand- or bench-type grinders. Hand and pneumatic chisels are also used to trim castings. A more recent method of removing excess metal from ferrous castings is with a carbon-air torch. This consists of a carbon rod and high-amperage current with a stream of compressed air blowing at the base of it. This oxidizes and removes the metal as soon as it is molten. In many foundries this method has replaced nearly all chipping and grinding operations.

Checking and Repairing.   After the castings are cleaned, they are inspected for flaws. Surface defects in steel castings may be repaired by welding.

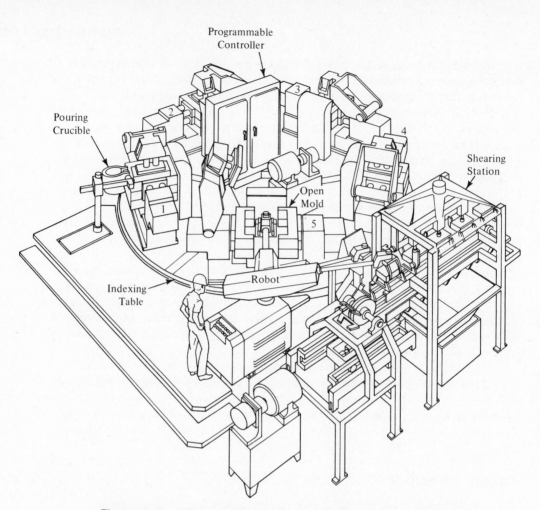

Programmable
Controller

Pouring
Crucible

3

2

4

Shearing
Station

Open
Mold

1

5

Robot

Indexing
Table

Figure 9.10. The indexing turntable shown has five permanent mold-type machines that can pour up to 120 aluminum castings per hour. A programmable controller can easily be changed to offer unlimited flexibility in the pouring operations. A robot is used to automatically remove castings from the mold and position them on the shear station where gating systems are removed. (*Courtesy Cast Industrial Products.*)

Figure 9.11. A battery of swing grinders being used to smooth the casting surfaces where excess metal has been removed. (*Courtesy Lebanon Steel Foundry.*)

356

Castings are tested both destructively and nondestructively. One destructive test consists of sawing the casting in sections to inspect for possible porosity. Other destructive tests are standard impact and tensile tests. A wide variety of nondestructive tests may be used, such as X rays, gamma rays, magnaflux, and ultrasonics. Typically, however, much of the testing is done at the time the casting is first being run, in order to prove out the gating, runnering, and risering systems. Once all the variables have been pinned down, the testing program can be drastically reduced. Usually after the initial defects have been determined and corrected, inspection is done merely on a visual basis at the time of cleanup. Of course critical parts will require more thorough testing.

Heat Treatment of Castings.   Castings are sometimes heat-treated to develop greater strength, refine the grain structure, produce more isotropic properties, or produce a more homogeneous structure. Castings are typically harder and stronger in their thinner, more rapidly cooled sections. In order to improve machinability, ductility, and/or elongation, heat treatment may in some cases be incorporated into the casting process. A separate heat treatment adds considerably to the cost and should not be specified unless it is essential to the finished product.

Advantages.   Sand casting provides a great deal of flexibility. There are very few limitations on casting size or shape. Inserts can form part of the mold. Fast production rates can be achieved for small castings. Sand casting provides one of the quickest ways to get from the drawing board to mass production.

Disadvantages.   Sand casting is not used where the highest dimensional accuracy is desired. Surfaces usually need finishing operations.

Cement Molding.   A variation of sand casting is a special sand–cement mixture that is mulled as in green-sand molding. Once mixed the mold is hurriedly used.

Advantages.   Molds may be air-dried, without costly equipment. Large high-strength molds can be made such as for ship propellers.

Disadvantages.   Cement molding is more costly than green-sand molding. Materials must be used rapidly and there is some difficulty in shakeout.

Sand-Casting Process Review.   The sand-casting process can become quite involved, especially for an initial presentation. Therefore a brief summary of the important steps are reviewed with a series of line drawings (Fig. 9-12).

Shell-Mold Casting

Shell molding is a special form of sand casting that uses conventional but finer foundry sands coated with a thermosetting (heat-hardening) phenolic resin. The pattern, usually cast iron, is carefully machined and polished.

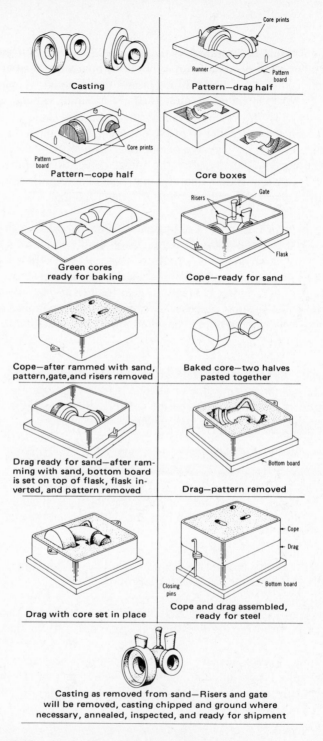

Casting

Pattern—drag half

Pattern—cope half

Core boxes

Green cores
ready for baking

Cope—ready for sand

Cope—after rammed with sand,
pattern, gate, and risers removed

Baked core—two halves
pasted together

Drag ready for sand—after ramming with sand, bottom board
is set on top of flask, flask inverted, and pattern removed

Drag—pattern removed

Drag with core set in place

Cope and drag assembled,
ready for steel

Casting as removed from sand—Risers and gate
will be removed, casting chipped and ground where
necessary, annealed, inspected, and ready for shipment

Figure 9.12. A summary of the steps involved in producing a steel casting. (*Courtesy Adirondack Steel Casting Co., Inc.*)

Jolting, squeezing, and ramming of the mold are all eliminated in this process. Instead a very fine sand is mixed with a thermosetting plastic (usually phenolic). This mixture is blown on a hot, 450°F (232°C) metal pattern until it builds up to a "shell" $\frac{1}{4}$ to $\frac{3}{8}$ in. (6.10–9.55 mm) thick. The shell is made in two parts. After the shell has cured, about 30 to 40 sec, it is removed from the pattern, usually by an automatic system of ejection pins.

Mold Assembly.   The shell halves are usually glued together with resin paste, which sets under pressure and the residual heat of the shells. A tongue-and-groove joint at the seam helps prevent the adhesive from entering the mold cavity and minimizes *fins* (small, thin, metal protrusions at the parting line).

Pouring.   When there is likely to be considerable pressure on the inside of the mold, it should be supported with backup material such as metal shot, gravel, or molding sand.

Advantages.   Shell molding is substantially more accurate than conventional sand molding. High-speed shell-molding machines are available that can handle from 30 to 200 full molds per hour depending upon the pattern-plate size and the number of stations available.

The transmission drum shown in Fig. 9-13 was changed from a green-sand mold to a shell-mold casting. The results were longer tool life in machining and less metal removal required. Balance and drill operations were also reduced. The savings in machining alone were enough to make the shell process feasible.

The shell-molding process is excellent for thin-wall sections with intricate holes and pockets and deep fins.

Disadvantages.   The shell-molding process requires an expensive, machined-metal pattern that makes it uneconomical unless production quantities are relatively high. Size limitations on shell-molding machines also restrict it to relatively small components.

The process is usually limited to castings under 100 lb (45 Kg). Precise

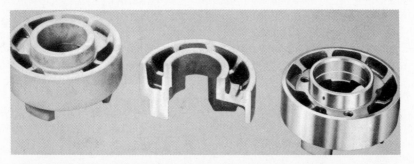

Figure 9.13.   The transmission drum is a good example of the use of shell molding because of the deep pockets that can be made with only a small amount of draft.

(a)   (b)

Figure 9.14. Pouring plaster into a core box to make cores for a tire-mold tread ring (a), assembly of plaster cores into mold (b). (*Courtesy Aluminum Co. of America.*)

control is required at all times. There is an objectionable odor in the molding area.

Plaster-Mold Casting

Plaster-mold casting is somewhat similar to sand casting in that only one casting is made and then the mold is destroyed. In this case the mold is made out of a specially formulated plaster, 70 to 80% gypsum and 20 to 30% fibrous strengthener. Water is added to make a creamy slurry.

The slurry is poured into a flask over the pattern, which is mounted on a match plate. When the plaster reaches its semiset state, the match plate is removed and the mold is put in an oven for drying. Cores, if needed, are made in the same way. The core boxes are usually made from brass, plastics, or aluminum. The process of pouring a plaster core and the assembly of its parts are shown in Fig. 9-14.

Patterns are made out of lightly lacquered plaster, sealed wood, or polished metal. Flexible molded-runner compositions and plastic compositions may also be used when undercuts are encountered. The patterns are covered with a thin film of soap, lard oil, or commercially available waxes rubbed on with a soft cloth.

Expansion plasters are available that expand as they set. The amount of expansion can be controlled by the plaster-to-water ratio. Values of $\frac{1}{16}$ to $\frac{1}{4}$ in./ft can be obtained. Thus the plaster can be made to automatically compensate for the shrink allowance of the cast metal. As an example, if a new pattern is to be made from an existing part, it can be cast right over it. When it sets, it will be larger than the part; but when the metal is cast, it will shrink and be the correct size.

Advantages and Limitations. Plaster-mold castings provide a finish that is superior to that of the sand cast. A surface finish of 125 microinches is typical; however, 50 to 60 microinches is attainable. Excellent reproducibility of dimensions is also an advantage of the process. When working to critical tolerances

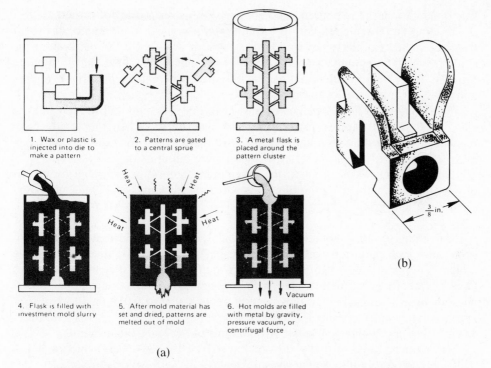

1. Wax or plastic is injected into die to make a pattern

2. Patterns are gated to a central sprue

3. A metal flask is placed around the pattern cluster

4. Flask is filled with investment mold slurry

5. After mold material has set and dried, patterns are melted out of mold

6. Hot molds are filled with metal by gravity, pressure vacuum, or centrifugal force

(a)

(b)

Figure 9.15. A schematic representation of the investment-casting process (a), an example of an investment casting (b). (*Courtesy* Product Engineering *and* Casting Engineers.)

such as ± 0.005 in./in. (0.005 mm/mm) shrinkage, allowance must be estimated closely. Because of the low chilling rate of plaster molds, walls as thin as 0.060 in. (1.52 mm) can be cast in small parts. A draft angle of 2° is usually required, but zero draft can often be achieved.

The major disadvantage of the plaster-casting method is that chemical decomposition of the gypsum begins at 1000°F (538°C) and thus limits the process to nonferrous, low-melting alloys, mainly aluminum and aluminum-zinc alloy. It is also used for rubber, plastic vacuum, and blow-form molds discussed in the chapter on plastics.

Investment Casting

Casting processes in which the pattern is used only once are variously referred to as "lost-wax" or "precision-casting" processes. In any case they involve making a pattern of the desired form out of wax or plastics (usually polystyrene). Formerly, frozen mercury was also used but now OSHA safety regulations have ruled it out. The expendable pattern may be made by pressing the wax into a split mold or by the use of an injection-molding machine. The patterns may be gated together so that several parts can be made at once, as shown schematically in Fig. 9-15. A metal flask is placed around the assembled patterns and a refractory mold slurry is poured in to support the patterns and form the cavities. A vibrating table equipped with a vacuum pump is used to eliminate all the air from the mold. Formerly the standard procedure was to dip the patterns in the slurry several times until a coat was built up. This was called the *investment* process. After the mold material has set and dried, the pattern material is melted and allowed to run out of the mold.

The completed flasks are heated slowly to dry the mold and to melt out the wax, plastic, or whatever pattern material was used. When the molds have reached a temperature of 1000°F (538°C) they are ready for pouring. Vacuum may be applied to the flasks to ensure complete filling of the mold cavities.

When the metal has cooled, the investment material is removed by means of vibrating hammers or by tumbling. As with other castings, the gates and risers are cut off and ground down.

Advantages and Limitations. Investment casting is particularly advantageous for small precision parts of intricate design that can be made in multiple molds. Thin walls, down to 0.030 in. (0.76 mm), can be cast readily. Surfaces have a smooth matte appearance with a roughness in the range of 60 to 90 microinches. The machining allowance is about 0.010 to 0.015 in. (0.25 to 0.38 mm). Tolerances of ± 0.005 in./in. (0.005 mm/mm) are normal; closer tolerances can be obtained without an excessive amount of secondary operations.

Investment molds usually do not exceed 24 in. (60.96 cm) maximum dimension. Preferably the casting should weigh 10 lb (4.53 kg) or less and be under 12 in. (30.48 cm) in maximum dimension. Sections thicker than 0.5 in. (1.27 cm) are not generally cast. The process has a relatively low rate of solidification. Grain growth will be more pronounced in larger sections, which may limit the toughness and fatigue life of the part.

Metals used in investment casting are aluminum, copper, nickel, cobalt, carbon and alloy steels, stainless steels, and tool steels.

Ceramic Process

The Ceramic or Shaw process, developed in England by Clifford and Noel Shaw, is somewhat similar to the investment-casting process in that a creamy ceramic slurry is poured over a pattern. In this case, however, the pattern, made out of plastic, plaster, wood, metal, or rubber, is reusable. The slurry hardens on the pattern almost immediately and becomes a strong green ceramic of the consistency of vulcanized rubber. It is lifted off the pattern while it is still in the rubberlike phase. The mold is ignited with a torch to burn off the volatile portion of the mix. It is then put in a furnace and baked at 1800°F (982°C), resulting in a rigid refractory mold. The mold can be poured while still hot.

Advantages and Limitations. The Shaw process finds its biggest advantage in being able to mold steels to very precise dimensions. Steel castings weighing over 100 lb (45.35 kg) have been made by this process. On small castings, tolerances may be held to within 0.001 in./in., but on larger parts the tolerance is generally about 0.010 in./in. The surface finish varies from 80 to 120 mi-

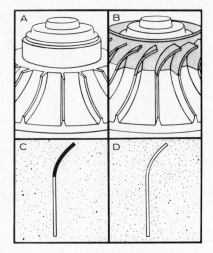

Figure 9.16. A combination of the Shaw process and lost-wax process were used to produce this impellor hub. The parts shown are metal pattern (a), wax pattern added to metal pattern (b), cross-sectional view after investment with metal pattern removed (c), cross section after wax has been melted out (d). (*Courtesy Lebanon Steel Foundry.*)

croinches. Pattern to finish casting can be done in 2 hr. The major tonnage of Shaw cast tooling goes into drop-forging dies, but the process is also used in making plastic molds, die-casting molds, glass molds, stamping dies, and extrusion dies.

Although the Shaw process is generally quite fast, it may take as long as 5 hr to "bake out" a large ceramic mold. Parting lines may be seen on the casting where the mold halves have been put back together. The molds are quite flexible before baking but some pattern contours do not lend themselves to stripping.

A casting that was made by both the Shaw process and the lost-wax method is shown in Fig. 9-16. All but the curved portion of the impellor hub was made by the Shaw process; then lost wax was used to make the curved portion of the blade.

Full-Mold Casting

Full-mold casting may be considered a cross between conventional sand casting and the investment technique of using lost wax. In this case, instead of a conventional pattern of wood, metals, or plaster, a polystyrene foam or Styrofoam* is used. The pattern is left in the mold and is vaporized by the molten metal as it rises in the mold during pouring (Fig. 9-17). Before molding, the pattern is usually coated with a zirconite wash in an alcohol vehicle. The wash produces a relatively tough skin separating the metal from the sand during pouring and cooling. Conventional foundry sand is used in backing up the mold.

Advantages and Limitations.   Because the pattern does not have to be withdrawn from the mold, undercuts are permitted, pattern draft is no longer required, coring is simplified, and loose pieces are done away with. The patterns can be

* Tradename of the Dow Chemical Company.

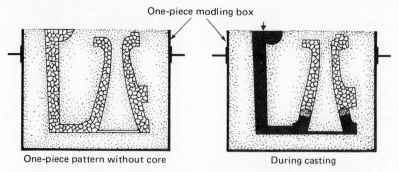

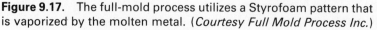

Figure 9.17. The full-mold process utilizes a Styrofoam pattern that is vaporized by the molten metal. (*Courtesy Full Mold Process Inc.*)

made in a relatively short time. Thus it can be used for individualized machinery repair. Where production is high, it is possible to produce the patterns by steam expansion of the polystyrene beads in a metal mold, as used for the steam iron core shown in Fig. 9-18. Large patterns can be cut from Styrofoam sheets with wood-cutting tools or a hot wire. Inserts such as lubrication lines, pipes, heating elements, wear strips, cutting edges, etc., can be cast in place in the pattern or put in the pattern when it is built.

The repeatability of the full-mold process may not be as great as that where the same pattern is used over and over. This of course is dependent on the skill of those who cut and glue the patterns. Substantial savings can be obtained in making multiple patterns by the use of templates, fixtures, special patterns, tools, and jigs.

NONEXPENDABLE MOLD-CASTING PROCESSES

Permanent-Mold Casting

One distinct advantage of the sand-casting process is that each time a casting is made the mold is destroyed. In the Middle Ages, iron molds were used to produce pewterware such as cups, pitchers, and other utensils. Later, tin soldiers were made by pouring metal into molds hinged together. After a time, the bulk of the still-liquid metal was poured out. The result was a thin-wall casting produced by what is now called *slush molding*.

Reusable metal molds (usually gray iron), or refractory materials, are used for nonferrous metals and cast irons. Machined graphite molds are used to a limited extent for steel castings. The molds can be used for up to about 3000 castings before they require redressing. The hot mold surface is coated with a refractory wash or acetylene soot prior to casting to protect the surface, to facilitate release of the casting, and to control cooling rates. Both metal and sand cores are used to form cavities in the cast parts. A typical permanent mold complete with cores is shown in Fig. 9-19.

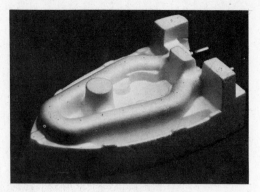

Figure 9.18. Styrafoam cores, as used in a steam iron, are made in metal molds. (*Courtesy Full Mold Process Inc.*)

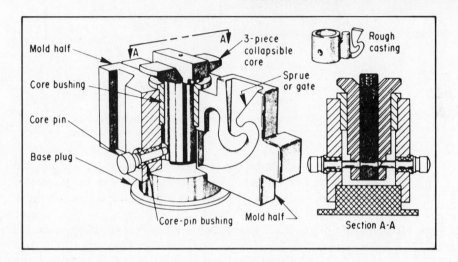

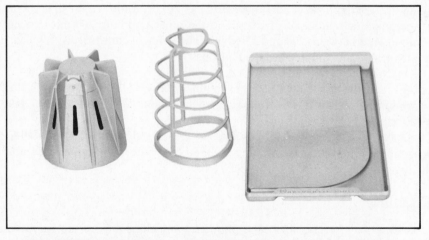

Figure 9.19. A typical permanent-mold arrangement for nonferrous castings; some examples of permanent-mold cast products. (*Courtesy Vacuum Die Casting Corporation.*)

Figure 9.20. A turntable used to help automate the permanent-mold casting process. (*Courtesy Eaton Corp.*)

A turntable used in automating the pouring of permanent molds is shown in Fig. 9-20. On this type of equipment a complete cycle can be made in 2 to 6 min, during which time the metal is poured, cooled and ejected.

Advantages and Limitations.   Permanent-mold castings generally have better grain structure than sand castings due to the faster cooling rate. As a rule-of-thumb, parts cast in metal molds show an increase of 20% in tensile strength and elongation increase of 30% over sand casting. The surface finish is controlled by the mold coating but it is typically in the range of 150 to 200 microinches for aluminum and between 200 and 350 microinches for ferrous metals. The castings are also less contaminated with inclusions.

Due to the high production rates (60–400 parts/hr), the high quality of the castings, and the small amount of machining required, the process is used in the production of automobile components such as aluminum pistons and also for forging dies.

The main disadvantage of permanent-mold casting is the high initial cost of the mold. The size of the casting is limited by the mold-making equipment, which is usually not over 200 lb (90.71 kg).

Die Casting

Die casting may be classified as a permanent-mold casting system; however, it differs from the process just described in that the molten metal is forced into the mold or die under high (1000 to 30,000 psi; 6.89 to 206.8 MPa) pressure. The metal solidifies rapidly (within a fraction of a second) because the die is water-cooled. Upon solidification, the die is opened and ejector pins automatically knock the casting out of the die. Most of the castings will have flash where the

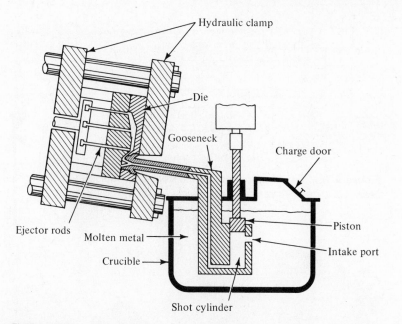

Figure 9.21. The hot-chamber die-casting machine. (*Courtesy* American Machinist.)

two die halves come together. This is usually removed in a trimming die but may be done with the aid of abrasive belts or wheels or by tumbling. If the parts are small, several of them may be made at one time in what is termed a *multicavity die*.

There are two main types of machines used: the *hot-chamber* and the *cold-chamber* types.

Hot-Chamber Die Casting. The hot-chamber machine is shown in Fig. 9-21. The metal is kept in a heated holding pot. As the plunger descends, the required amount of alloy is automatically forced into the die. As the piston retracts, the cylinder is again filled with the right amount of molten metal.

An integral feeding system makes the hot-chamber die-casting machines easy to automate. Recent improvements in design make the machines particularly advantageous for magnesium parts.

Cold-Chamber Die Casting. This process gets its name from the fact that the metal is ladled into the cold chamber for each shot, as shown in Fig. 9-22. This procedure is necessary to keep the molten-metal contact time with the steel cylinder to a minimum. Iron pickup is prevented, as is freezing of the plunger in the cylinder.

Advantages and Limitations. Die-casting machines can produce large quantities of parts with close tolerances and smooth surfaces. The size is only limited by the capacity of the machine. Most die castings are limited to about 75 lb

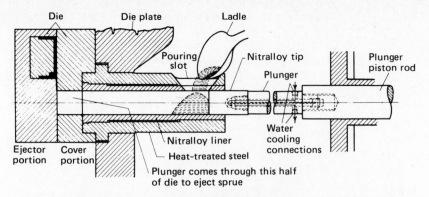

Figure 9.22. Schematic view of the cold-chamber die-casting process. (*Courtesy American Zinc Institute, Inc.*)

Figure 9.23. Examples of some of the variety of products that can be die cast advantageously. The process has the advantage of good detail, close tolerance, good finish, and low labor cost. At the near lower left and lower right are two examples of coffee percolator wells with cast-in-place tubular heating elements. (*Courtesy Vacuum Die Casting Corp.*)

(34 kg) zinc, 65 lb (30 kg) aluminum, and 44 lb (20 kg) of magnesium. Die castings can provide thinner sections than any other casting process. Wall thicknesses as thin as 0.015 in. (0.38 mm) can be achieved with aluminum in small items. However, a more common range for larger sizes will be 0.105 to 0.180 in. (2.67 to 4.57 mm). Examples of some of the variety of parts that can be die cast are shown in Fig. 9-23.

Some difficulty is experienced in getting sound castings in the larger capacities. Gases tend to be entrapped, which results in low strength and annoying leaks. Of course, one way to reduce metal sections without sacrificing strength is to design in ribs and bosses. Another approach to the porosity problem has been to operate the machine under a vacuum. This process is now being developed.

The surface quality is dependent on that of the mold. Parts made from new or repolished dies may have a surface roughness of 24 microinches. The high surface finish available means that, in most cases, coatings such as chrome plating, anodizing, and painting may be applied directly. More recently, decorative finishes of texture as obtained by photoetching have been applied. The technique has been used to simulate woodgrain finishes, as well as textile and leather finishes, and to obtain checkering and crosshatching.

The initial cost of both the die-casting machine (upward of $80,000) and the dies (upward of $1000 to $40,000 or more) limits this process to relatively high production quantities. The die costs are high because of the high accuracy with which they must be made and the high polish required on the surfaces.

Perhaps the sharpest restriction on the die-casting process comes from the limited range of materials that can be cast, namely, zinc-, aluminum-, magnesium-, and copper-base alloys. Recently some progress has been made with ferrous materials. The big problem has been the die life that can be maintained. In 1966 the General Electric Company, in an effort to create a market for the molybdenum produced by its refractory metals department, did some pioneer work in this area. There are now two companies producing ferrous die castings on a limited basis.

Low-Pressure Casting

Low-pressure casting was patented in 1910, but until recently it has not been aggressively promoted. The process may be described as a compromise between gravity-casting processes and high-pressure die casting. Because it borrows from both die casting and permanent-mold low-pressure casting, the process has a semantics problem. In Europe it is known as low-pressure permanent-mold casting and in the United States it is called low-pressure die casting.

In low-pressure casting, the metal is forced up by air pressure from a heated crucible through a feed tube or carrot to fill the die, or mold, from the bottom up (Fig. 9-24). The metal is placed under 5 to 15 lb pressure. It may take half a minute to fill the mold, so any trapped gas has time to escape. The part cools from the top down and directional solidification occurs, which is the same phenomenon that yields good part strength in gravity castings, which cool from the bottom up. After the metal has solidified in the mold, the air pressure is cut off

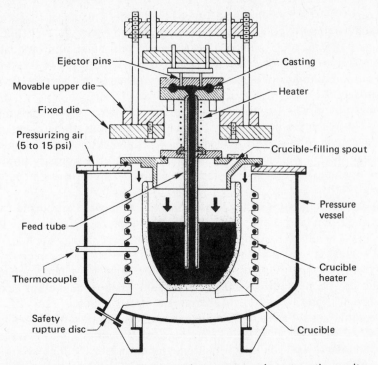

Figure 9.24. In low-pressure casting, pressure is put on the molten metal of the crucible so that it flows up the feed tube to the mold. (*Courtesy* Machine Design.)

and the feed-tube metal returns to the crucible. The upper half of the mold is lifted up with the casting trapped in it. After tilting or moving laterally, ejector pins drop the casting in a catcher mechanism. The whole operation is highly automated and a single operator can handle two machines.

Advantages and Limitations. The closely controlled temperature of the automated low-pressure machines produces better grain structure. Oxides and impurities float on top of the crucible and are not disturbed by the low-pressure air. By maintaining pressure while the casting cools, shrinkage is fed automatically and more accurately than with risers in conventional casting. High-pressure die castings have a porosity problem that is nonexistent in the low-pressure system. Because of the pressure system, the minimum wall thickness can be reduced and the transition from heavy to light sections is more easily made. Production rates are good. As an example, 750-lb (340.19-kg) locomotive wheels can be poured at the rate of one wheel per minute. The majority of low-pressure castings are no larger than 40 lb (18.14 kg); however, some major automobile manufacturers in both the United States and Japan use the low-pressure casting system to produce aluminum engine blocks.

The economic advantages of low-pressure casting are that machines cost about one-third less than high-pressure machines and the dies are about one-third less expensive. Dies are usually cast iron, but sometimes copper, aluminum, or even plaster is used. Die life is longer than for die casting because of the low operating pressure.

Low-pressure dies are simpler in that risers can be eliminated and the runners are less complicated, but ejector pins have to be incorporated.

Tolerances must be wider for low-pressure casting than for die casting because a refractory coating is necessary inside the mold. This coating varies in thickness, depending upon its heat-transfer characteristics. Tooling costs are less than that for die casting, but more than the setup and pattern costs of sand casting. Also casting modifications are harder to make.

Centrifugal Casting

Centrifugal casting consists of having a sand, metal, or ceramic mold that is rotated at high speeds. When the molten metal is poured into the mold it is thrown against the mold wall, where it remains until it cools and solidifies. The process is being increasingly used for such products as cast-iron pipes, cylinder liners, gun barrels, pressure vessels, brake drums, gears, and flywheels. The metals used include almost all castable alloys. Most dental tooth caps are made by a combined lost-wax process and centrifugal casting.

Advantages and Limitations.   Because of the relatively fast cooling time, centrifugal castings have a fine grain size. There is a tendency for the lighter nonmetallic inclusions, slag particles, and dross to segregate toward the inner radius of the casting (Fig. 9-25), where it can be easily removed by machining. Due to the high purity of the outer skin, centrifugally cast pipes have a high resistance to atmospheric corrosion. Figure 9-25(b) shows a schematic sketch of how a pipe would be centrifugally cast in a horizontal mold.

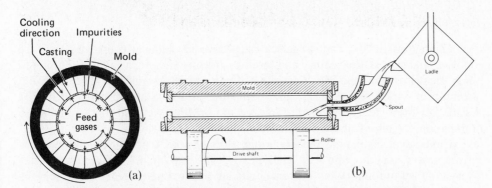

Figure 9.25.   The principle of centrifugal casting is to produce high-grade metal by throwing the heavier metal outward and forcing the impurities to congregate inward (a). Shown at (b) is a schematic of how a horizontal-mold centrifugal casting is made. (*Courtesy Janney Cylinder Co.*)

Axial sections

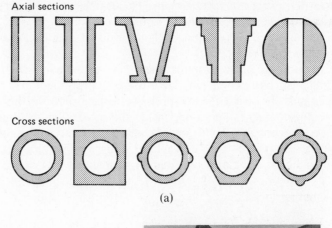

Cross sections

(a)

(b)

Figure 9.26. Some examples of common cross-sectional shapes that can be made by centrifugal casting. Wall thickness is easily varied. Outside diameter may range from several inches to several feet. Also shown is a centrifugal casting used in a liquid-air converter. It is made to very close tolerances. (*Courtesy Sandusky Foundry and Machine Co.*)

Characteristically, centrifugal castings have mechanical properties that are between those of static castings and forgings. As an example, a 3560-T6 aluminum alloy centrifugally cast has a tensile strength of 48 ksi (331.0 MPa), a yield strength of 34 ksi (234.4 MPa), and elongation of 8%. Statically cast, it has 33 ksi (227.5 MPa) tensile strength, 24 ksi (165.5 MPa) yield strength, and 3.5% elongation, a gain of about 30%.

Centrifugal casting has the advantage of being able to produce a wide variety of diameters, lengths, and wall thicknesses. The maximum castable size varies from foundry to foundry but current limits are approximately as follows: lengths of 400 in. (1016 cm), outside diameters of 68 in. (172.72 cm) and total weight about 80,000 lbs. (36,287 kg). Figure 9-26 illustrates common cross-sectional shapes that can be centrifugally cast. The inner surface is round but the outside can vary considerably.

Two dissimilar metal laminates can be made. By adding brazing flux to a solidified, but still spinning, casting, a dissimilar metal can be introduced, as for example a lead-rich surface to cast bearings. This method is also used to conserve more expensive metals and to provide corrosion resistance.

Continuous Casting

Continuous casting is a relatively new process, although it was first patented by Sir Henry Bessemer in 1846. It is used to produce blooms, billets, slabs, and tubing directly from the molten metal. The process can be done vertically or horizontally, as shown in Fig. 9-27. The process is applied to copper and copper alloys, aluminum, steel, and gray and alloy-type cast irons.

The molten metal is poured into a pouring box referred to as a *tundish*. From there it flows at a controlled rate into a water-cooled mold. The mold is rapidly oscillated to facilitate movement. Solidification occurs within 20 to 30 sec. The solidified shape is withdrawn inch by inch by rolls that grip the section as it emerges.

Hollow rods or thick-walled tubing are made by placing a graphite core centrally in the die to a depth below the level at which solidification is complete. Contraction during solidification pulls the metal away from the die and core walls. Dies retain their dimensions for several weeks of operation, depending upon the tolerances required. Convenient lengths are cut with a torch or a circular saw.

Advantages and Limitations. Continuous casting can provide longer continuous lengths than can otherwise be obtained. The process eliminates some of the hot- and cold-rolling operations required in conventional production. The graphite dies used are relatively inexpensive, allowing the process to be used in small or large quantities. Graphite dies used to produce the shapes shown in Fig. 9-28 may range from $50 to $200. Because of the rapid cooling of the metal, the properties are considerably improved over sand casting.

In the case of steel, the high pouring temperature can make severe demands on the materials and design of the mold. Also, the slow solidification rate of steel and the high casting speed make it necessary to have long dies. This increases the chance for bulging of the cast shape due to the deep liquid core. The general restrictions on nonferrous continuous castings are minimum inside diameter for round stock, $\frac{7}{16}$ in. (1.111 cm), and for other shapes, 1 in. (2.54 cm); maximum outside dimensions, $9\frac{1}{4}$ in. (23.5 cm); minimum section thickness, $\frac{5}{32}$ in. (0.396 cm).

Table 9-1 is a review of the principal casting processes and many of their salient points. Of interest here, and not discussed in the text material, are the comparative tooling costs, lead times, secondary operations, and ordering quantities.

The comparative tooling cost will quickly point up which processes may be feasible for a few parts and those where only large quantities will be practical.

Lead time means the amount of time required to produce the part once the design is submitted. You will note the time is given for both sample parts and a production run. Usually, minor changes are required in the sample parts and, when approved, additional patterns are made for the production run.

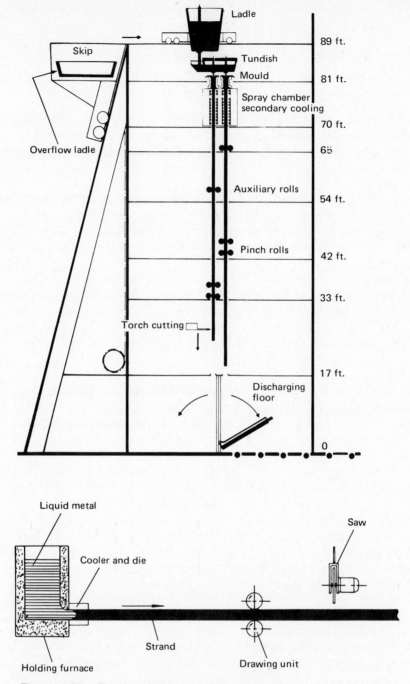

Figure 9.27. The continuous-casting process may be done vertically or horizontally. (*Courtesy* Journal of Metals.)

Figure 9.28.   Some of the many shapes that can be poured by the continuous-casting process. (*Courtesy American Melting and Refining Co.*)

Secondary trimming operations consist of parting line smoothing and gate removal.

Ordering quantities are based on tooling costs and cycle time. Sand castings have low tooling costs. However, the comparatively long cycle time limits the quantities produced for a given time period.

CASTING PROCESS CONSIDERATIONS

There is much more to casting than selecting a process and making the appropriate pattern. During the past decade, research and production experiences have provided scientific principles for better casting techniques. Important considerations are the rate at which a mold cavity is filled, gate placement, riser design, the use of chill blocks, and padding.

Filling the Mold Cavity.   The velocity with which the molten metal fills the mold is determined by the cross-sectional area of the gating system and the mold-pouring rate. Too slow a mold-pouring rate means solidification before filling some parts, allowing surface oxidation. Too high a pouring rate caused by too large a gating system causes sand inclusions by erosion, particularly in green-sand molding, and turbulence. The minimum cross section in the gating system is called a *choke*. In the strict sense, the choke is the section in the gating system where the cross-sectional area times the potential linear velocity is at a minimum. When the gating system is choked at the bottom of the sprue, it is called a

Table 9-1. Comparison of casting processes.

Method	Description	Metals	Size Range	Tolerances	Comparative Tooling Cost, Units
Sand casting	Tempered sand is packed onto wood or metal pattern halves, removed from the pattern, assembled with or without cores, and metal is poured into the resultant cavities.	Most all castable metals	All sizes	$\pm\frac{1}{32}''$ up to 3 in.; $\pm\frac{3}{64}''$ from 3 to 6 in.; $\pm\frac{1}{16}''$ above 6 in. Across parting line add $\pm.055$ for average casting	$2\frac{1}{2}$
Permanent-mold casting	Molten metal is gravity poured into cast iron molds, coated with ceramic mold wash. Cores can be metal, sand, shell, or other materials	Aluminum, some brass, bronze and cast iron	Aluminum: Usually $\frac{1}{2}$ lb to 100 lb Iron and Copper Base: 50 lb	Aluminum: Basic $\pm.015$ to 1''. Add $\pm.002$ for each additional inch. Across parting line add $\pm.020$ for average casting	15
Investment casting	Metal mold makes wax or plastic replica. These are sprayed, then surrounded with investment material, baked out, and metal poured in resultant cavity	Most all castable metals	Fraction of an ounce to 100 lb	$\pm.005''$ in.	4
Plaster-mold casting	Plaster slurry is poured onto pattern halves, allowed to set. The mold is then removed from the pattern, baked, assembled, and the metal is poured into the resultant cavity	Aluminum, brass, bronze, zinc, beryllium copper	Normally up to 500 sq in. area	One side of parting line $\pm.005''$ up to 2 in. Over 2 in. add $\pm.002''$ per inch. Across parting line $\pm.010''$. Allow for parting line shift of .015''	$3\frac{1}{2}$
Die casting	Molten metal is injected, under pressure, into hardened steel dies	Aluminum, zinc, magnesium and limited brass	Not normally over 3 ft square	$\pm.0015''$ per inch. Not less than $\pm.002''$ on any one dimension. Additional .010'' (minimum) on dimensions affected by parting line	24

Method	Surface Finish (AA)	Lead Time Normal	Design Freedom 1 Lowest 5 Highest	Minimum Draft Requirement	Normal Minimum Section Thickness	Secondary Trimming Operations	Ordering Quantities
Sand casting	Nonferrous 150–350 Ferrous 300–700	Samples: 2–6 weeks Production: 2–4 weeks (after approval)	4	1° to 5°	Nonferrous: $\frac{1}{8}$" to $\frac{1}{4}$" Ferrous: $\frac{1}{4}$" to $\frac{3}{8}$"	Grind for gate removal Hand sand parting line	All quantities
Permanent-mold casting	Aluminum: 150–200 Iron: 200–350 Copper base: 125	Samples: 6–14 weeks Production: 2–3 weeks (after approval)	3	Aluminum: 2° min Iron: External—3° Internal—7° Copper Base: External—0° Internal—$\frac{1}{2}$° to 3°	Aluminum: $\frac{9}{64}$ for average areas Iron: $\frac{7}{32}$ for average areas Copper base: 0.030" to .080"	Grind for gate removal	Usually 500 and up
Investment casting	63–125	Samples: 3–8 weeks Production: 3–7 weeks (after approval)	5	None	Aluminum: .030" Beryllium copper: .030" Stainless steel: .060" Carbon steel: .090"	Grind for gate removal	Aluminum: Usually under 2000 Other metals: all quantities
Plaster-mold casting	63–125	Samples: 2–6 weeks Production: 2–4 weeks (after approval)	3–4	External: 0° to $\frac{1}{2}$° Internal: $\frac{1}{2}$° to 2°	.070"	Grind for gate removal Hand sand parting line	Aluminum: Usually under 20.0 lb Copper base: all quantities
Die casting	32–90	Samples: 8–16 weeks Production 2–4 weeks (after approval)	2–3	Aluminum: 1° to 3° Zinc: $\frac{1}{2}$° to 2°	Aluminum: 0.65" for average areas	Die trim for flash and gate removal	Usually 2.0 lb and up

nonpressurized system. This system is somewhat less reliable than a pressurized system in which the choke is at the gate.

The first metal in the pouring basin and down the sprue usually has some turbulence that carries slag into the runner. To avoid slag in the casting, the runner should extend past the last gate to trap the initial slag (Fig. 9-29). By the time the gates become operative, the liquid level should be high enough so that no slag can enter the casting cavity. The runner should be laid out to minimize turbulence, that is, it should be as straight and as smooth as possible. The gate that was shown in Fig. 9-2 is made to enter the cavity at the parting line. Gating arrangements may also be made at the top or bottom of the cavity. The parting-line gate is the easiest for the patternmaker to make; however, the metal drops into the cavity, which may cause some erosion of the sand and some turbulence of the metal. For nonferrous metals, this drop aggravates the dross and entraps air in the metal.

Top-gating is used for simple designs in gray iron but not for nonferrous alloys, since excessive dross would be formed by the agitation.

Bottom-gating provides a smooth flow of metal into the mold. However, it does have the disadvantage of an unfavorable temperature gradient. It cools as it rises, with the result of having cold metal in the riser and hot metal at the gate.

Risers.   Risers are designed and placed so as to ensure filling the cavity during solidification. They also act to relieve gas pressure in the mold and to reduce pressure on the lifting surfaces of the mold. The volume of metal in the riser should be sufficient to retain heat long enough to feed the shrinkage cavity and to equalize the temperature in the mold, avoiding casting strains.

The riser requirements vary with the type of metal being poured. Gray cast iron, for example, needs less feeding than some alloys because a period of graphitization occurs during the final stages of solidification, which causes an expansion that tends to counteract the metal shrinkage. Many nonferrous metals require elaborate feeding systems to obtain sound castings. Two riser designs for the same casting are shown in Fig. 9-30.

Risers are placed near the heavy sections of the casting. The feed metal must be located above the highest point of the casting.

Chill Blocks.   Chill blocks are metal blocks placed in the mold for localized heat dissipation. They may be placed at an intersection or joint where there is a comparatively large volume of metal to cool, thus relieving a hot spot or maintaining a more uniform cooling rate and better microstructure. They may also be placed at the far surface of a mold, away from a riser or sprue. This will help the far end of the mold to freeze rapidly, promoting directional solidification. Chill blocks are also used at points where it is desirable to have localized hardening, as in the case of bearings or wear surfaces.

Padding.   Padding consists of adding to or building up a section to obtain adequate feeding of isolated sections. Figure 9-30 showed two methods of feeding the central and outside boss. Shown at (b) is the plan of using two risers and

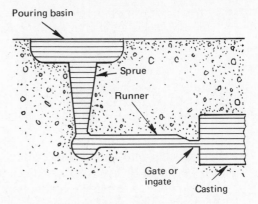

Figure 9.29. The gate acts to reduce the turbulence and erosion as the metal enters the cavity. It also reduces the contact area when the runner is removed.

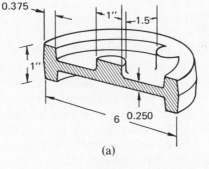

(a)

(b)

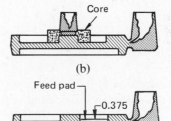

(c)

Figure 9.30. Risers provide a volume of metal to feed the bosses shown at (a). Two individual risers may be used as shown at (b), or a single riser as shown at (c) with the addition of a feed pad. (*Courtesy* Machine Design.)

at (c) one riser with a pad. The second plan provides a yield of 45% of the metal poured, compared to 30% when two risers are used. The feeding distance to the central hub is $^{9/2}t$, where t = the thickness of the feed path. By rule-of-thumb, the total thickness of pad and casting (at the pad location) should not be less than one-fifth of the metal-feeding distance. This rule is not absolute but a good generalization.

Hot Tears. The strength of metal near solidification or freezing point is very low. Stresses imposed at this temperature may lead to incipient flaws that

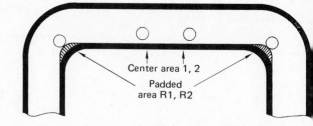

Figure 9.31. Exothermic padding at R1 and R2 to prevent hot tears.

later develop into well-defined cracks or *hot tears* upon cooling. Precautions must be taken to avoid stress concentrations caused by shrinkage stresses at weak points. An example of this type of problem is shown in Fig. 9-31. This high-quality alloy-steel casting consists of a 350-lb (158.75-kg) cradle with a 0.5-in. (12.7-cm) wall. The design is conducive to hot tearing at the natural hot spots in the inside corners. Cracks occur in this radius area since this is the hottest, weakest area and is exposed to the highest stress concentration.

Exothermic Padding.   In this case, an economical means of solving the problem is with a small amount of exothermic padding material as shown. The exothermic mixture consists of mill scale, powdered aluminum, charcoal, and refractory powder. The heat evolves as a result of the exothermic reaction and keeps the metal molten long enough so that excessive strains are not built up during final solidification and shrinkage.

Exothermic materials are also used to cover risers. The heat of the exothermic reaction keeps the riser top open to the atmospheric pressure and improves its efficiency.

CASTING DESIGN PRINCIPLES

Casting design differs from most other fabricating design principles in that it is more intimately associated with the process itself. The effects of solidification, parting-line placement, tolerances, minimum section thickness, and draft will be examined from the design viewpoint.

Solidification and Design

The designer of castings must be aware of how solidification is affected by or affects: part geometry, changes in cross-sectional area, mechanical properties, and heavy isolated areas.

Solidification and Section Geometry.   After casting, and when cooling has started, nucleation first takes place at the mold walls of the thinner sections. Grains form that may be dendritic, columnar, or equiaxed depending upon the cooling rate and alloys. In dendritic structures, some of the arms break off during the normal conditions of growth, thus furnishing nuclei for central, inner portions

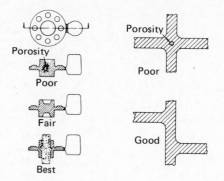

Figure 9.32. Heavy isolated sections cause porosity in the casting. (*Courtesy Central Foundry Division, G.M.*)

of the casting. New dendrites grow from these fragmented parts to form the equiaxed central zone. The cooling of a simple straight section presents no problem if it is not large. However, when two sections cojoin or intersect, a mechanical weakness develops, as shown schematically in Fig. 9-32(a). Rounded sections and fillets tend to minimize these *hot spots* or solidification defects (b).

Solidification and Cross-Sectional Area. Cooling rates are directly related to mass and surface area and may be expressed as a ratio of surface area to volume or mass. The slow cooling of a large mass with a relatively small area will produce a considerably softer material than the fast cooling of a large thin section.

The smallest section that can be cast as gray cast iron depends not only on metal composition but also on foundry practices. The foundryman can adjust the rate at which graphitization takes place by changing the silicon content or by adding innoculants in the ladle.

The mass effect of increased section thickness or decreased cooling rate is much more pronounced in cast iron than in cast steel. In cast steel, the main effect of a larger section is increased grain size. However, in cast iron not only is the grain size affected but also the graphite size and distribution, and they are also dependent on the amount of combined carbon.

The cooling rate from 1200°F (650°C) determines the ratio of combined carbon to graphite, which controls the hardness and strength of the iron.

A method of determining the hardness-to-section size is to use a wedge-shaped bar that has a taper of about 10°. The bar is cast in a sand mold and sectioned near the center of the length. Rockwell hardness tests are made from the tip progressively to the thicker section.

Figure 9-33 shows the relationship between section thickness and hardness. The structure ranges from white cast iron at the tip to a mixture of carbide and pearlite. After a mottled iron, which is a mixture of white iron and gray iron, the hardness decreases sharply. The minimum hardness is reached where large amounts of ferrite and fine flake graphite are formed. With a slightly slower cooling rate, the structure changes to a flake graphite with pearlitic matrix that reaches another maximum hardness on the curve. This structure is the most

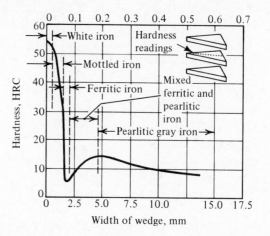

Figure 9.33. The effect of thickness section on hardness and structure. (Metals Handbook, *Vol. 1,* Properties and Selection: Irons and Steels, *Bardes, B., Ed., 9th ed., American Society for Metals, p. 14.*)

desirable for wear resistance and strength. From this point on the graphite flakes become coarser with the pearlitic lamellae more widely spaced, resulting in a slightly softer and weaker structure as shown in Table 9-2.

Table 9-2. Properties of ferritic and pearlitic malleable irons.

Structure	BHN	Tensile ksi	Yield as a % of tensile ksi	Elongation %, 2 in.
Ferritic, maximum ductility	200	80	69	10
Pearlitic, high strength, wear resistant	270	100	80	2
Essentially pearlitic, good strength	250	80	75	3

Most commercial gray cast irons are represented by the downward sloping portion of the curve as shown in Fig. 9-33.

Figure 9-34 shows the average tensile strength of five classes of cast iron versus the section diameters. The curves are similar to those that were shown in the lower curve of Fig. 9-33.

Parting-Line Placement. With high-production molding techniques it is desirable to place the gate and feeder at the parting line. Figure 9-35(a) shows proper and improper parting-line placements. In the upper figure of (a) the parting line has been placed at the flange, causing the heavy boss at the top to become

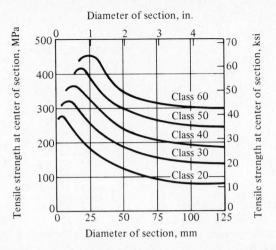

Figure 9.34. Effect of section diameter on tensile strength at center of cast specimen for five classes of gray iron. (Metals Handbook, *Vol. 1,* Properties and Selection: Irons and Steels, *Bardes, B., Ed., 9th ed., American Society for Metals, p. 14.*)

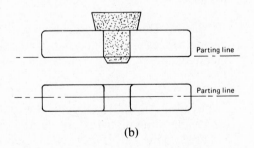

(a)

Figure 9.35. Proper placement of the parting line facilitates gating and in this case helps eliminate porosity (a). Parting relocated to eliminate a core (b). (*Courtesy Central Foundry Division, G.M.*)

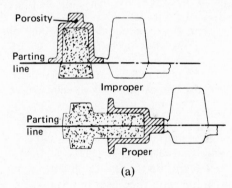

(b)

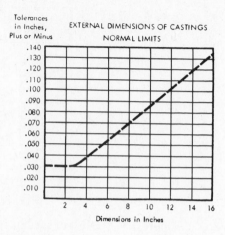

Figure 9.36. General tolerances for high-production malleable castings. (*Courtesy Malleable Research and Development Foundation.*)

isolated from the riser, resulting in porosity. By turning the casting 90° and placing the parting line axially, the riser can feed the heavy boss directly. Parting-line placement can often be used to eliminate a core, as shown at (b). Some general suggestions in regard to the parting line are:

1. Avoid putting the parting line where it may later interfere with machining or other operations.
2. Keep the parting line as even as possible to simplify pattern making and mold making, and to keep fins to a minimum.
3. Use parting lines to eliminate cores when possible.

Tolerances.   The variety of sizes, shapes, and processes makes it difficult to give more than general guidelines as to tolerances. For high-production malleable-iron castings, the general tolerance for green-sand molding is ± 0.030 in. (0.076 cm) up to 2 in. (5.08 cm) in size, with a straight-line increase in tolerance as dimension increases, as shown in Fig. 9-36. Tolerances for other processes are given in Table 9-1.

Section Thickness.   Recommended minimum section thicknesses for various casting processes are given in Table 9-1. Recommended wall thicknesses for aluminum by sand and permanent-mold casting are shown graphically in Fig. 9-37.

Draft.   All walls perpendicular to the parting plane require draft. Normally, the casting drawing does not show draft. The standard foundry practice is to add draft to the part (Fig. 9-38). In general, green-sand molds require 2° of draft. Shell molds are normally made with 1° draft. Deep pockets may require from 3 to 10° draft for green sand and 1 to 2° for shell molding. Draft requirements are also given in Table 9-1.

Machining Allowances.   Drawings showing the casting prior to machining should have the finish allowance included in the dimensions. It is recommended

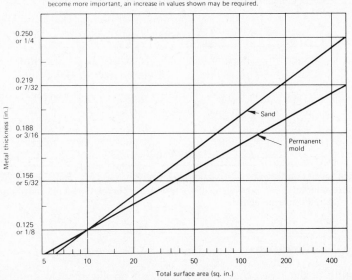

The figures in the graph below show desired minimum metal thickness that should be specified for aluminum castings of minimum complexity. When metal flow is restricted, or other factors become more important, an increase in values shown may be required.

Figure 9.37. The minimum wall thickness recommended for aluminum castings. (*Courtesy Aluminum Association.*)

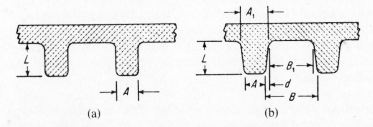

(a)   (b)

Figure 9.38. Normally a casting drawing will be made as shown at (a); standard foundry practice is to add the required draft as shown at (b).

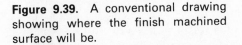

Figure 9.39. A conventional drawing showing where the finish machined surface will be.

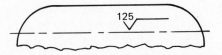

that an undimensioned phantom line be used (Fig. 9-39) to represent surfaces after machining, thus enabling the foundry to recognize the surface for processing reference. Table 9-3 shows the amount of metal normally added for machining aluminum and steel castings. The amount of material added relates to the overall size of the casting rather than the individual dimensions. Thus the finish allowance for any given casting is the same for all surfaces requiring machining.

Table 9-3. Machining allowances for aluminum and steel castings.

Largest Casting Dimension (in.)	Aluminum Allowance (in.)				Steel
	Sand Casting	Permanent Mold	PM with Sand Cores	PM with Shell Cores	Sand Casting
To 6	0.060	0.045	0.060	0.060	$\frac{3}{16}$
6 to 12	0.090	0.060	0.090	0.060	$\frac{3}{16}$
12 to 18	0.120	0.075	0.120	0.090	$\frac{3}{16}$
18 to 24	0.150	0.090	0.180	0.120	$\frac{1}{4}$
Over 24		———————Consult foundry———————			

Values shown are based on maximum buildup or safety factor for average variations in flatness, squareness, concentricity, etc., as well as linear tolerance. There are recommended minimums, and draft, when required, must be added.

Economic Considerations

Good casting designs incorporate savings with no sacrifice of quality. The designer should consider such alternatives as multiple-cavity casting, fabrication, the use of cores, subsequent machining operations, and weight reduction.

Multiple-Cavity Casting. Substantial savings can be realized by incorporating more than one finished part into a single casting. Molding and handling costs can be greatly reduced, since only a fraction of the number of castings need be handled. An example of multiple casting is shown in Fig. 9-40, where five bearing caps are cast together. They are machined as a unit and then cut into separate pieces.

Fabrication. Existing designs should be under constant evaluation. Often a part that has been fabricated by a combination of casting and welding can be made more economically by one process or vice-versa. Figure 9-41 shows a fabricated axle housing. The combination of casting and welding greatly simplified manufacture. Welding the heavy wall tube on each end of the housing was done by automatic submerged-arc welding at a cost of 12.36¢ per weld, exclusive of power and maintenance costs. If both welds are made simultaneously, the cost drops to 7.46¢ per weld.

Figure 9-42 shows a hub used on a rear wheel of a truck. In this case the original design called for forging and welding, as shown at (a). However, a reevaluation of the design proved the steel casting (b) to be lighter, stronger, and more economical.

Cored Holes. Designers are often confronted with the problem of whether to specify cored or drilled holes. The economies are best studied on an individual basis, but it is generally more economical to core larger holes, when appreciable metal can be saved and faster machining results. The universal-joint yoke shown in Fig. 9-43 was originally a solid-steel forging. A rather slow drilling operation

Figure 9.40. Five bearing caps are cast as a single unit, machined, and then cut into separate parts. (*Courtesy Central Foundry Division, G.M.*)

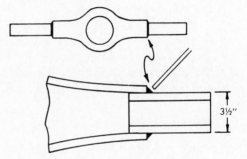

Figure 9.41. A combination of casting and welding, using the advantages of each for economical design.

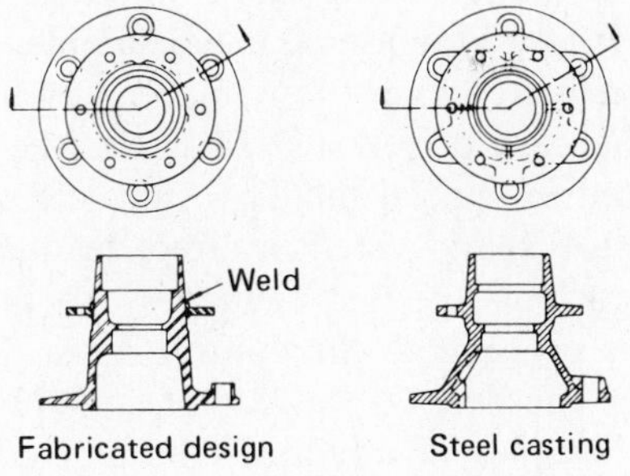

Weld

Fabricated design   Steel casting

Figure 9.42. The casting at the right replaces a two-piece welded forging. (*Courtesy Central Foundry Division, G.M.*)

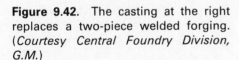

.84" core

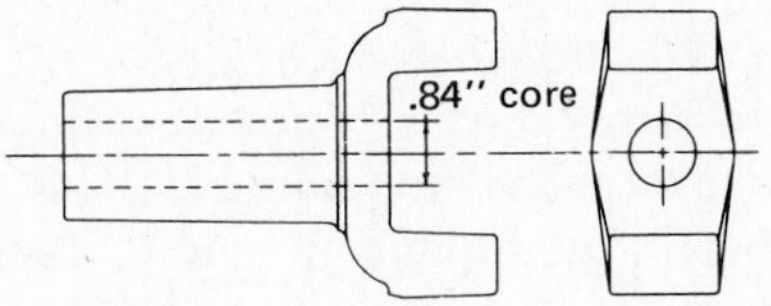

Figure 9.43. Cored holes eliminate or reduce secondary operations of drilling and boring. (*Courtesy Central Foundry Division, G.M.*)

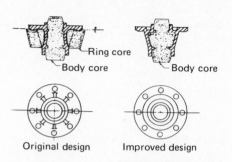

Figure 9.44. The redesign of this truck-wheel hub resulted in being able to make this casting with one core instead of two. (*Courtesy Central Foundry Division, G.M.*)

was required to make the shaft hole. In casting, the hole was cored and was followed by a fast core-drilling operation, since a relatively small amount of metal had to be removed.

The proper use of cores can also result in a less expensive casting. Figure 9-44 shows the original and the improved design of a truck wheel hub. The original casting could not be formed in green sand. Two cores were necessary, a *ring core* and a *body core*. By redesigning to eliminate the ribs and backdraft, the green-sand casting was made with only one core.

Quality Control

Today's modern foundry is essentially a continuous chemical operation with many unique high-temperature problems. Close control of the metal poured is one of the most important functions required to ensure high-quality castings.

In large foundries, metal samples from the melting unit and holding furnaces are delivered every few minutes by a pneumatic tube system to the laboratory. In the case of ferrous materials, the samples are in the form of circular discs poured in special molds to provide a fully chilled structure. In the laboratory they are spark-tested in direct-reading vacuum spectrographs, which determine the various alloying elements except silicon and carbon. Test results are immediately sent to the control room. The control engineer can then rapidly calculate the desired weights of constituents for the next charge.

Samples of molding sand and core sand are also sent via pneumatic tube to the testing laboratory on a periodic basis. The results are returned to process control engineers and are plotted on control charts. Trends are noted and corrections made when necessary.

In large, modern foundries, on-line analog and digital computers are used to control production variables. In addition to control over process variables, computers monitor plant safety, cooling-water temperatures, bearing temperatures, etc.

Although in-process control is important, it does not preclude a constant check of the final product. For example, in the manufacture of gray-iron cylinder blocks, sections are removed after each operating shift. Test specimens are surface ground and checked for hardness, graphite size, and distribution. When the hardness falls outside the specification limits, all blocks made during the suspect period are held up. Further tests show which blocks can be sent on and which scrapped.

Nondestructive testing methods such as ultrasonic testing, sonic testing, eddy current testing, and X rays are also widely used to ensure quality production.

When to Use the Casting Process

Design and manufacturing engineers must often decide when it is advantageous to use the casting process or when it may be better to use some other method of fabrication. Given below is a list of some of the main considerations that may be used in choosing the casting process.

1. Parts that require complex internal cavities, such as asymmetric parts, or those that are quite inaccessible for machining. Also the cavities that are large and may necessitate considerable metal removal.
2. When a large number of parts are to be made out of aluminum or zinc and have rather complex structures.
3. Parts requiring heavy, formed, cross-sectional areas. Heavy sections can be fabricated if the part is relatively simple, but forming poses many problems. Fabricating and machining may be very time-consuming.
4. Castings allow bulk or metal mass to be placed advantageously, as in machine bases.
5. Damping, both sound and mechanical, is often needed in machine tools. Gray cast iron can provide this quality better than any other metal.
6. Modern foundry practices make it feasible, in some cases, to produce one of a kind. Some patterns may be made quickly out of wax or styrofoam. The part may be cast within a few hours of its conception and often with a choice of materials such as steel, cast iron, copper, brass, or aluminum.
7. Several individual parts may be quite easily integrated into one part with a savings of both material and labor.
8. Parts that are extremely difficult to machine such as the refractory material used in turbine blades may be cast to close tolerances.
9. When it is desirable to minimize directional properties of the metal. Castings have better *anisotropic* qualities than forged or wrought materials. Anisotropic refers to the directional qualities of most wrought metals that reduce ductility, impact, and fatigue properties transverse to the direction of rolling.
10. When using precious metals, since there is little or no loss of material.

There are also times when it is not advantageous to use the casting process, as follows:

1. Parts that can be stamped out on a punch press.
2. Parts that can be deep drawn.
3. Parts that can be made by extrusion.
4. Parts that can be made by cold-heading.
5. Parts made from highly reactive metals.

Questions and Problems

9-1. Why are investment castings made only in a relatively limited size range?

9-2. What is the main difference between the lost-wax method of casting and the Shaw process?

9-3. What is the big advantage of making castings by the full-mold process?

9-4. Why are castings made in a permanent mold generally stronger than those made by green-sand molding?

9-5. What is the main limitation of the die-casting process?

9-6. What are some of the advantages of low-pressure casting over conventional casting?

9-7. Why are the properties of a centrifugally cast part generally better than those statically cast?

9-8. (a) What is the purpose of padding in a casting design? (b) What is the purpose of exothermic padding?

9-9. What is the difference in draft requirements between sand casting and shell molding?

9-10. What are some ways a casting designer can cut costs?

9-11. What hardness would you expect in a gray cast iron casting that was about $\frac{5}{8}$ in. (15.87 mm) thick?

9-12. What structure could be expected in an iron casting that had some sections $\frac{1}{4}$ in. (6.10 mm) thick?

9-13. What tensile strength could be expected in 2-in. (50.80-mm) dia. bar of class 40 cast iron?

9-14. Show by sketches how the casting designs (Fig. P9.1) can be improved.

9-15. Shown in the sketch (P9.2) is a sectional view of a sand-cast steel gear housing. Problems were encountered in getting a porosity-free hub. Make a sketch to show where the riser should be placed and how you would ensure adequate metal at the hub.

9-16. Difficulty was encountered in this aluminum-alloy sand casting (Fig. P9.3). Even though individual risers and chills were provided for each of the four internal bosses, they were not adequately filled. Make a sketch or trac-

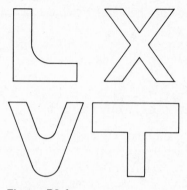

Figure P9.1.

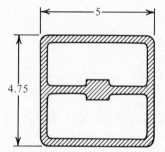

Figure P9.2.

ing to show how this problem might be solved.

9-17. The sand-cast steel lever shown in Fig. P9.4 encountered excessive porosity and hot tears. Explain what the problem is and make a sketch to show how it can be overcome without adding any weight to the casting.

9-18. The malleable-iron gear housing shown in Fig. P9.5 was designed to be fed by six risers. Show how the part may be redesigned with less mass and three risers.

9-19. The malleable-iron gear housing as shown in Fig. P9.6 did not get adequate metal at the hub. Show how it can be corrected without adding any weight to the casting. Express the feeding distance of the boss in terms of t.

9-20. Shown in Fig. P9.7 is a wheel spindle for a truck. Make a sketch to show the following: (a) parting line, (b) core, (c) riser.

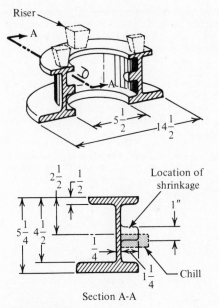

Figure P9.3.

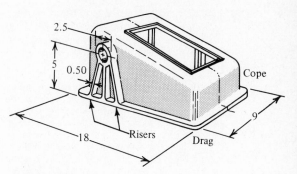

Figure P9.6.

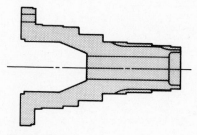

Figure P9.7.

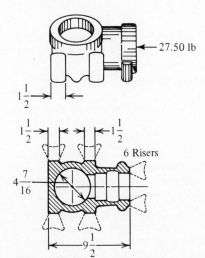

Figure P9.4.

Figure P9.5.

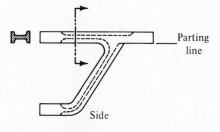

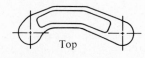

Figure P9.8.

9-21. Shown in Fig. P9.8 is a casting design of a suspension arm. (a) What is the main objection to this design? (b) Make a sketch to show how it can be improved.

9-22. Make a sketch to show schematically how the grain structure would appear in cross section of a 3 in. × 3 in. cast-steel square bar that has had ample time to cool naturally in air.

9-23. Shown in Fig. P9.9 is a small 3 in. × 3 in. aluminum casting. (a) Select a process for casting this part and tell why you chose it. (b) Make a sketch to show the pattern.

9-24. (a) State the difficulty that may be encountered in making the casting as shown in Fig. P9.10. (b) Make a sketch to show how the casting may be redesigned to alleviate the problem. (c) Assume this is an aluminum casting with overall dimensions of 6 in. × 12 in. State the following: 1. Minimum thickness of thin section if made by sand casting and by permanent-mold casting. 2. Machining allowance for the flat surface of the heavy wall section if sand cast. 3. Tolerance expected. 4. If no machining is planned and this is a high-production item, the casting process used. Why?

9-25. A woodworking vise jaw is made out of gray cast iron. The front plate is made with a clamping surface of 4 in. × 6 in., as shown in Fig. P9.11. Make a sketch of how this jaw should be designed for maximum strength and economy.

9-26. Figure P9.12 shows the rib design of a casting. (a) Tell what is wrong with the design. (b) Show two alternative designs that would eliminate the problem.

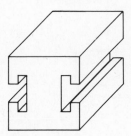

Figure P9.9.

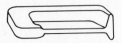

Figure P9.10.

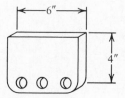

Figure P9.11.

Figure P9.12.

Bibliography

American Society for Metals. *Metals Handbook,* 9th ed. Vol. 1, *Properties and Selection: Irons and Steels.* Metals Park, Ohio, 1978.

——. *Metals Handbook,* 9th ed. Vol. 2, *Properties and Selection: Nonferrous Alloys.* Metals Park, Ohio, 1979.

Beadle, John D., ed. *Castings.* Basingstoke, Hampshire, U.K.: The Macmillan Press Limited, 1971.

Cristman, Douglas R. "Casting Process Selection for Light Metal Parts." *Paper No. 602, Society of Die Casting Engineers,* 1977.

Edwards, F. "The Horizontal Sheet Casting of Aluminum." *Metals and Materials* 6 (February 1972): 70–71.

Gray Iron Castings Handbook. Cleveland: Gray Iron Founders' Society, Inc., 1972.

Guide to Copper Casting Alloys: Their Selection Process and Use. New York: American Smelting and Refining Company, n.d.

Heine, R.; C. Loper, Jr.; and P. Rosenthal. *Principles of Metal Casting.* New York: McGraw-Hill, 1967.

Karsay, Stephen I. "Gating and Risering Gray and Ductile Irons." *Foundry,* Part I 100 (August 1972): 41–43; Part II 100 (September 1972):58–61.

Steel Castings Handbook, 4th ed. Cleveland: Steel Founders' Society of America, 1970.

United Nations. *Engineering Equipment for Foundries.* New York: Pergamon Press, 1979.

CHAPTER TEN

Plastic-Molding Processes

CASTING AND MOLDING PROCESSES

There is a similarity between the molding of plastics and the molding of metals, as discussed in the previous chapter, and even though some of the properties are quite different, many of the same design principles are applicable.

The metal-casting process is based on heating the base material to the liquid state and then pouring or squeezing it into a mold. For most molding operations, plastics are heated to a liquid or a semifluid mass and formed in a mold under pressure. Plastics can also be poured from a liquid state into a mold using either heat or a catalyst for hardening, but this method is not used extensively except for nylon.

The basic structure of plastics was discussed in Chapter 2. You may recall that plastics are essentially giant-sized molecules that pass through a fluid to a "gel" state during the molding process. Those that form strong chemical cross-links at this stage are termed *thermosetting,* since the process cannot be reversed. *Thermoplastics* undergo no permanent chemical change under moderate heating, and therefore the process can be repeated if a change in the product is desired. Common types of both thermoplastic and thermosetting materials and their properties, advantages, and disadvantages are listed in Table 10-1.

Plastics may be obtained for processing in a number of different forms, as shown in Table 10-2. Also shown are the principal processing methods used and discussed in this chapter.

Table 10-1. Common plastics, advantages and disadvantages.

Material	Advantages	Disadvantages
PE polyethylene	Flexible; easily processed; electrically insulating; water barrier; nearly chemically inert	Soft; low temperature service; easily scratched
PVC vinyl	Easily processed; can be plasticized to allow rigid-to-soft flexibility; good weathering characteristics	Some properties such as inertness reduced by plasticizers; will sun-fade; attacked by hydrocarbons
PS polystyrene	Clear; easily processed; easily compounded to improve impact resistance; good stain, abrasion resistance	Not outdoor material; fades; brittle unless compounded; loses some resistance to staining; attacked by hydrocarbons
Polypropylene	Very flexible; very low density; does not stress crack; excellent electrical resistance; good abrasion resistance; can be electroplated; superior weathering characteristics	Cannot be heat sealed; difficult to dye; somewhat soft in unmodified state
Acetals	Low coefficient of friction; easily processed; high abrasion resistance; solvent resistant	Must be copolymer for dimensional stability; some loss in properties during reprocessing of scrap; not suitable for outdoor applications
Polycarbonates	High heat resistance; excellent clarity; self-lubricating	Difficult to process owing to water absorption; difficult to color
Polysulfones	High-temperature applications; high strength; self-extinguishing; easily colored	Attacked by hydrocarbons
TFE Teflon	Chemically inert; low friction coefficient; high-temperature applications	Cannot be extruded or injection molded; soft
Phenolics	Good surface finish; low water absorption; excellent flame resistance	Cannot be colored
Amino resins	Excellent surface characteristics; excellent strength; solvent resistance; chip resistance	Usually filled with alphacellulose; unfilled not easily colored; not easily injection molded
Polyesters	Usually used with glass for reinforcing; easily colored; resin available in variety of forms	Flammable; some attack by ultraviolet light
Urethanes	Easily foamed, good bearing surfaces	Difficult to reinforce
Epoxies	Low heat curing; high strength; excellent adhesive; good chemical resistance; easily filled	Viscous liquid; entrains air bubbles; requires separate catalyst
Diallyl phthalate DAP	High water resistance; good high-temperature properties; excellent strength	Not easily colored; material not easily processed

Table 10-2. Plastic processing methods.

Liquid	Powders, Granules, and Beads	Raw Material Form Powders or Liquids	Sheet	Reinforced Plastics
Casting	Compression molding	Rotomolding	Thermoforming	Hand layup
	Transfer molding		Vacuum	Vacuum bag
	Injection molding		plug assist	Pressure bag
	Blow molding		Pressure	Filament
	Extrusion molding		plug assist	winding
	Expandable-bead molding		Matched mold	Matched die
	Foam molding			

Figure 10.1. Cast nylon outlasted previous phenolic counterparts by two to one and eliminated paper contamination from lubricant throwoff. (*Courtesy Polymer Corp.*)

Casting

Casting of plastics is not used extensively except for epoxies and nylons. Cast epoxies are used for making molds, dies, and bushings, as well as encapsulating electrical coils, transformers, switchgear, etc. Cast nylon, unlike other polymers, is processed directly from a liquid monomer. A catalyst, and other additives, is heated and poured into the mold at atmospheric pressure. The additives in the compound serve to "kick over" the monomer into a polymer form while it is in the mold. The nylon monomer is unique in that it is processed by direct polymerization; other thermoplastic resins cannot be processed in this manner.

Advantages and Limitations. The primary advantage of the casting process is the accurately contoured product, such as the gears shown in Fig. 10-1, that can be made several inches thick, yet free of voids.

Cast-nylon parts are available in two formulations, "Mono Cast" and "Nylatron," which contain molybdenum disulfide for improved lubrication. These materials provide an excellent balance of wear and corrosion resistance, self-lubrication, impact strength, and nongalling characteristics. They are lightweight, about one-seventh that of steel, and have excellent sound- and vibration-damping qualities. Because of the slower cooling cycle in the casting process than in molding, surfaces are harder (more crystalline) and have better wear

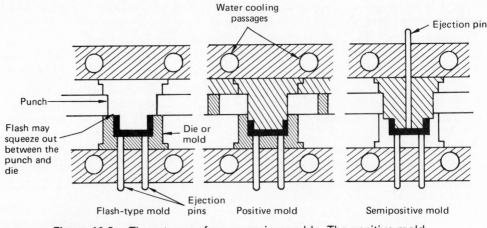

Figure 10.2. Three types of compression molds. The positive mold must have a carefully measured amount of plastic to be sure it fills the cavity but not have an excess since the mold is made to prevent flash from forming.

resistance. Mold cost is quite low compared to that of injection molds. A typical mold for a cast part costs about $500 to $1000. The cost of the material is competitive to that of stainless steels and brasses. Typical applications range from guides and cams for soft handling of bottles to pedestal liners for locomotive service. Other applications include large gears for all types of machinery, elevator buckets for handling foundry sand and other abrasive materials, bearings for steel-mill rollout tables, and roll covers for paper-making machinery.

Nylons cannot be used where the continuous temperature exposure is 275°F (135°C) or more. Nylons absorb water on the outer surface and should not be used for applications requiring continuous immersion in water. Nylons are attacked by strong acids and bases but they are little affected by solvents, detergents, or weak acids. These materials can be machined, but care must be taken not to leave abrupt surface changes since they are notch-sensitive.

Compression Molding

The first method used to form plastics in production quantities was by compression molding. The method has been improved and now provides several mold variations: flash type, positive, and semipositive, as shown in Fig. 10-2. The most common type used in automatic molding presses is the semipositive type. The loading chamber ensures the correct amount of material and enables accurate control of the flash or excess materials that squeeze out at the die-closing line.

The mold must be heated, which is most often done with electric cartridge-type heaters placed as close to the cavities as possible but in a way that will produce uniform heat throughout the mold surface. The die area must also be cooled, which is done by circulating water through passages, as shown in the sketch.

Clamping is done by the action of the press and is usually a straight ram hydraulic arrangement or a toggle arrangement activated by either air or hydraulic cylinders. Pressure varies depending on the material being molded and the configuration of the part. As a rule-of-thumb, parts that are 1 in. or less deep will require a pressure of 3000 psi (20.68 MPa) as calculated on the *projected area* within the outside wall of the loading chamber. Projected area is the area of the part that is parallel to the parting line of the mold. Parts deeper than 1 in. will require 3000 psi plus 700 psi (219 kg/cm^2) for each additional inch or part of an inch.

Parts are removed from the mold with the aid of ejector pins. The shape of the part will determine whether ejection should be done from either the top or bottom of the mold half. In manual operation, the operator may remove the part from the mold. In semiautomatic operation, the operator will remove the part from the ejector pins. In a fully automatic operation, the part is made to fall free after ejection or a combing device may be added to provide a positive release.

Mold temperature varies with conditions but is generally around 320 to 380°F (160–193°C). The hotter the mold can be run, the shorter will be the cure time. However, too hot a mold may cause the plastic to set up before it has had a chance to fill all parts of the mold. As a rough guide, the cure time for phenolics (thermosetting plastics) is 60 sec for $\frac{1}{8}$ in. (.317 cm) of cross section. However, regardless of thickness, no further curing will take place after 3 min. This is because the outside layer will cure but the inside, if too thick, will still be in semigranular form.

Transfer Molding.   Transfer molding is a variation of compression molding. It involves the use of a plunger and heated pot separate from the mold cavity. The material is loaded into the pot and, when hot, is transferred under pressure through runners into the closed mold cavity (Fig. 10-3). A three-part mold is

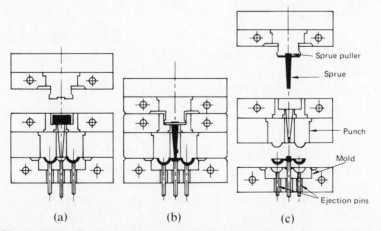

Figure 10.3.   The three stages of transfer molding, preform in place (a), liquification and transfer into mold (b), ejection (c).

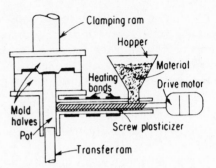

Figure 10.4. A newer method of feeding compression-type molds is with screw arrangement. The preheated slug automatically dropped into position greatly increases production.

used as shown at (c). As the mold is opened, the sprue and a residual disc of material is separated from the part and is removed manually. The molded part is raised from the cavity by ejection pins.

A newer development for regular compression molding and transfer molding is a screw-feeding arrangement, as shown in Fig. 10-4. The screw moves slugs of material, preheated to 290°F (143°C), to where they are dropped into the bottom of the transfer pot and then moved into the mold. This arrangement allows the cure time to be cut 25% over that of using cold powder. Production capacity per mold has increased 400% with this method and heavy-walled sections are cured throughout.

Advantages and Limitations. Compression-molded parts do not have high stresses, since the material is not forced through gates. The process is comparatively simple, which makes initial mold costs low. Transfer molding is particularly good for making complex parts where inserts are needed. In ordinary compression molding, fragile inserts may be damaged or shift before the resin reaches its plastic state. However, transfer molds are more complex and more costly to build.

Injection Molding

The greatest quantity of plastic parts are made by injection molding. The process consists of feeding a plastic compound in powdered or granular form from a hopper through metering and melting stages and then injecting it into a mold. After a brief cooling period, the mold is opened and the solidified part ejected. In most cases, it is ready for immediate use.

Several methods are used to force or inject the melted plastic into the mold. The most commonly used system in the larger machines is the in-line reciprocating screw, as shown in Fig. 10-5. The screw acts as a combination injection and plasticizing unit. As the plastic is fed to the rotating screw, it passes through three zones as shown: feed, compression, and metering. After the feed zone, the screw-flight depth is gradually reduced, forcing the plastic to compress. The work is converted to heat by shearing the plastic, making it a semifluid mass. In the metering zone, additional heat is applied by conduction from the barrel surface. As the chamber in front of the screw becomes filled, it forces the screw

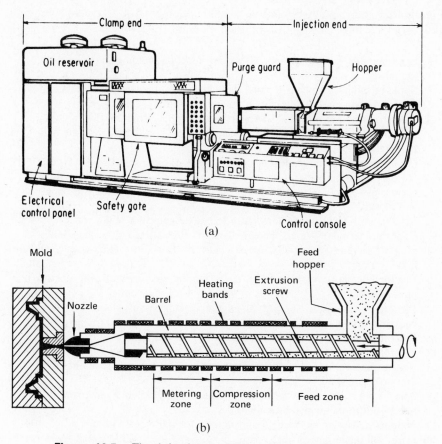

Figure 10.5. The injection-molding machine (a) and the reciprocating-screw injection system (b).

back, tripping a limit switch that activates a hydraulic cylinder that forces the
screw forward and injects the fluid plastic into the closed mold. An antiflowback
valve prevents plastic under pressure from escaping back into the screw flights.

The clamping force that a machine is capable of exerting is part of the size
designation and is measured in tons. A rule-of-thumb can be used to determine
the tonnage required for a particular job. It is based on two tons of clamp force
per square inch of projected area. If the flow pattern is difficult and the parts
are thin, this may have to go to three or four tons.

Many reciprocating-screw machines are capable of handling thermosetting
plastic materials. Previously these materials were handled by compression or
transfer molding. Thermosetting materials cure or polymerize in the mold and
are ejected hot in the range of 375 to 410°F (190–210°C). Thermoplastic parts
must be allowed to cool in the mold in order to remove them without distortion.
Thus thermosetting cycles can be faster. Of course the mold must be heated
rather than chilled, as with thermoplastics.

Figure 10-6 shows an exploded view of the elements of a typical two-plate injection mold. A brief description of each element follows:

1. Locating ring—aligns the nozzle of the injection-molding machine with the mold.
2. Sprue bushing—a conical channel that carries the injected plastic through the top clamp plate to the part or runner system.
3. Clamp plate—secures the stationary side of the mold to the molding machine. Clamps are inserted in the recess between the clamp plate and the cavity-retainer plate.
4. Cavity-retainer plate—holds the cavity inserts (the cavity is the area where the part is formed).
5. Leader pin—provides alignment for the two halves of the mold as it opens and closes.
6. Leader pin bushing—provides close tolerance guide for leader pin.
7. Core-retainer plate—holds the core element, the mating half of the cavity.
8. Support plate—adds rigidity and strength to the plate stackup (not used in all molds).
9. Ejector housing—provides travel space for the ejector plate and ejector pins.
10. Ejector plate—often referred to as the ejector cover plate. Provides backup for pins set into the ejector-retaining plate.
11. Ejector-retainer plate—plate in which the ejector pins are held.
12. Ejector pin—pushes part off the core or out of the cavity.
13. Return pin—returns the ejector plate back to its original position when the mold is closed.
14. Sprue puller pin—pulls sprue out of the bushing when mold opens by means of an undercut (not shown).
15. Support pillar—gives strength and rigidity for the ejector plates.
16. Core—the male part of the cavity.
17. Cavity—place where the plastic is formed. Some molds may have the cavity cut directly in the cavity plate rather than by the use of an insert as shown.
18. Gate—the restricted area of the runner right before the material enters the cavity.
19. Runner—a passageway for the plastic to flow from the sprue to the part.
20. Sprue—a passageway for the plastic from the nozzle to the runner.

Advantages and Limitations.   Injection molding is one of the most economical methods of mass producing a single item. The parts taken from the mold, in most cases, are finished products. There is very little waste of thermoplastic plastics since the runners and sprues can be ground up and reused.

Injection molds must be carefully designed by one who specializes in this type of work. They are expensive if quantities are not large. However, the quantities usually associated with injection molds makes their actual per-piece cost relatively insignificant. As a very rough figure, a one-cavity simple mold may run about $1500. Two- and three-cavity simple molds may run $2000 and $3000, respectively. Amortizing over a run of 100,000 parts would make the mold cost per piece $0.015, $0.02, or $0.03.

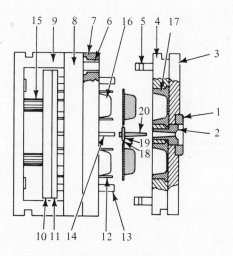

1) Locating ring
2) Sprue bushing
3) Clamp plate
4) Cavity retainer plate
5) Leader pin
6) Leader pin bushing
7) Core retainer plate
8) Support plate
9) Ejector housing
10) Ejector plate
11) Ejector retainer plate
12) Ejector pin
13) Return pin
14) Sprue puller pin
15) Support pillar
16) Core
17) Cavity
18) Gate
19) Runner
20) Sprue

Figure 10.6. Conventional two-plate mold. Reprinted by permission of *Modern Plastics Encyclopedia,* McGraw-Hill, Inc.

Bulk-Injection Molding. Injection-molded plastics are no longer limited to relatively small parts with thin wall sections. A unique bulk-injection molding process has been developed by Eimco Met Plastics that is able to custom mold parts weighing as much as 300 lb, with wall thicknesses of up to $3\frac{1}{2}$ in. (88.90 mm) and outside dimensions to 60×80 in. $(1.52 \times 2.2$ m$)$.

The thicker the wall section of the mold the longer the plastic must be packed into the mold before it is hard enough to be demolded. Few parts conventionally made require more than 2 min. But with the bulk-injection molding process, a part may be packed for 3 hr. The long cycle makes the process suitable for relatively low quantity production, usually fewer than 5000 pieces per year.

Coinjection Molding. Coinjection molding makes it possible to mold articles with a solid skin of one thermoplastic and a core of another thermoplastic. The skin material is usually solid while the core material contains blowing agents.

The basic process may be one, two, or three-channel technology. In one-channel technology, the two melts are injected into the mold, one after the other, by shifting a valve as shown in Fig. 10-7. The skin material cools and adheres to the colder surface; a dense skin is formed under proper parameter settings. The thickness of the skin can be controlled by adjustment of injection speed, stock temperature, mold temperature, and flow compatibility of the two melts.

In the two- and three-channel techniques, both plastic melts may be introduced simultaneously. This allows for better control of wall thickness of the skin, especially in gate areas on both sides of the part.

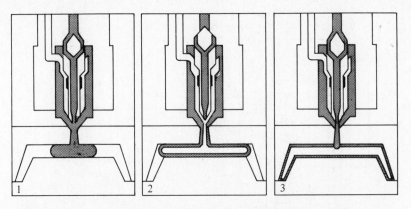

Figure 10.7. Coinjection molding accomplished by a sliding torpedo valve. The skin material is injected first, followed by core material, and then the mold is packed with skin during the cure. Reprinted by permission of *Modern Plastics Magazine,* McGraw-Hill, Inc.

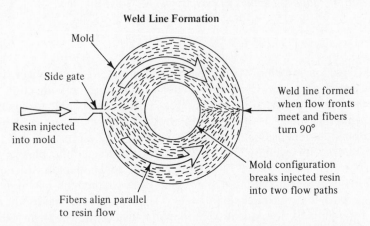

Figure 10.8. Strength is reduced along the weld line in any injection-molded part but particularly when fiber reinforcement is turned at right angles and is able to offer little circumferential reinforcing effect. The change in fiber direction may also cause distortion. (*Courtesy* Machine Design.)

Injection-molded Carbon-fiber Composites.   By mixing carbon or glass fibers in injection-molded plastic parts, they can be made lightweight yet stiffer than steel. However, there are two problems associated with the process, which are weld lines and fiber orientation.

Weld lines are produced in an injection-molded part when two flow fronts meet and fail to adhere properly (Fig. 10-8). Distortion and loss of strength are more pronounced in carbon-reinforced parts because the fibers fail to mesh

Direct Single Gate

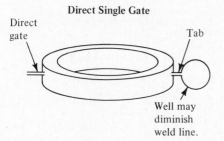

Ring Gate

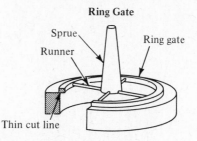

A single, distinct weld line is produced opposite the gate.

Dissipated weld lines are produced between each pair of runners.

Direct Multiple Gates

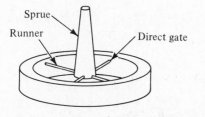

Diaphragm Gate

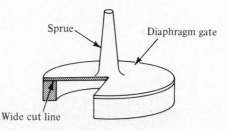

Distinct weld lines that are stronger than that of the single gate are produced between each pair of gates.

Weld lines are totally eliminated.

Figure 10.9. Weld lines may be strengthened, moved, or totally eliminated from parts by using appropriate gating techniques. (*Courtesy* Machine Design.)

properly at the weld line. Strength at the weld line may be as low as 25% of the normal strength of the base material.

Weld lines may be strengthened, moved, or totally eliminated from parts by appropriate gating techniques (Fig. 10-9). Typically, single gating produces the weakest weld lines, while direct multiple gating increases overall part strength by producing several stronger weld lines. Of course, tooling costs are greater for developing more sophisticated gating, and machining off excess plastic in the gate area increases labor costs and produces a larger percentage of scrap. However, as with other thermoplastic parts, the scrap can be cut off along with the sprues and runners and ground into pellets to make more parts.

Since some of the carbon or glass fibers are broken when the scrap is reground, the properties degrade. Usually the first regrind results in only a slight falloff in properties. By the third regrind the original properties will be reduced to about 50% of the original. The allowable percent of reground material is usually specified by the designer.

Rotomolding

In rotational molding, the product is formed inside a closed mold that is rotated about two axes as heat is applied. Liquid or powdered thermoplastic or ther-

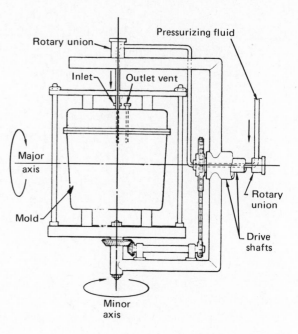

Figure 10.10. Schematic of rotomolding equipment used to form either thermoset or thermoplastic parts.

mosetting plastic is poured into the mold, either manually or automatically. The mold halves are then clamped shut.

The loaded mold is then rolled into an oven where it spins on both axes. Heat causes the powdered materials to become semiliquid or the liquid materials to gel. (Permissible temperature ranges are considerably greater than for injection molding.) As the mold rotates, the material is distributed on mold-cavity walls solely by gravitational force; centrifugal force is not used. The mold and rotating mechanism is shown in the sketch (Fig. 10-10).

When the parts have been properly formed, the molds are cooled by a combination of cold-water spray, forced cold air, and/or cool liquid circulating inside the mold. The mold continues to rotate during the cooling cycle. Unloading is usually a simple manual operation, although forced air or mechanical methods are sometimes used to eject the part.

Advantages and Limitations. Rotomolding can be used to make parts in sizes and shapes that would be difficult by any other process. For example, the process is now being used to handle rectangular shapes 5.5 × 5.5 × 12 ft (1.65 × 1.65 × 3.6 m) and cylindrical parts 15 ft (4.5 m) in diameter with 8 ft (2.4 m) straight walls. Boat hulls can be molded up to 14 ft (4.2 m) long. More common production equipment is for parts that can be contained in a 5-ft (1.5-m) sphere. Part weights may range from 0.2 oz (6 ml) to 250 lb (112.5 kg) and volumes from 1 cu in. to 1000 gallons (16.39 cm^3 to 3800 l) capacity. Figure 10-11 shows a double-walled phonograph case rotomolded of high-density polyethylene. It provides shock and vibration cushioning advantages over the previous vinyl-covered wood version, and at lower cost and less weight.

Figure 10.11. A double-walled phonograph case made of high-density polyethylene by the rotomolding process. It provides the added benefit of shock and vibration cushioning to the basic cost and weight reduction advantages over a prior vinyl-covered wood version. (*Courtesy Arvin Industries and USI Chemicals Co.*)

Wall thicknesses may range from 0.030 to 0.500 in. (0.76 to 12.7 mm) by simply adjusting the amount of charge and cycle time. Heat-insulating plugs can be used to reduce the wall thickness or to eliminate the wall entirely in a given area. Wastebaskets, for example, are made by using an asbestos disc at the end that is to be open to prevent plastic formation.

Rotational molds are relatively inexpensive when compared to molds for injection and blow molding. Some examples of typical mold costs are: a welded-steel prototype mold for making 25-gallon (95-l) refuse cans costs about $375; a single-cavity prototype mold of cast aluminum for an 8-in. (20.32-cm) dia. light globe costs about $600; a 10-cavity production mold for the same light globe in cast aluminum costs about $4000.

As with molds for other processes, the precision required is an important factor in the cost of the mold. A novelty item that has a wide tolerance range can use a much cheaper mold than precision instruments. Rotomolding can often use an existing metal part as the prototype mold. For example, a metal lawn spreader became the prototype mold for a plastic lawn spreader.

In short runs, rotomolding is generally less expensive than most other molding processes. As production runs lengthen, other processes become competitive and must be carefully evaluated. However, long runs do not automatically eliminate rotomolding, as shown in the following example.

A 1-gal (3.8-l) container and lid had a unit cost for 10,000 rotomolded parts of $0.362; by injection mold, $0.942; by blow mold, $0.389. For a production run of 100,000, the costs were: rotational, $0.286; injection, $0.468; blow molding, $0.176. At a run of 1 million cans, rotational costs were $0.203; injection, $0.167; and blow molding, $0.130. For 10 million parts, the cost picture changes because of lower machine and mold factors. Here rotomolding has an advantage, at $0.093, against injection at $0.127 and blow molding at $0.108.

Changing colors during a production run is comparatively easy, whereas in injection and blow molding it is costly and time-consuming. All that is required in rotomolding is changing the color of the charge. Conceivably it can be changed with each cycle.

The basic limitations of rotomolding are that the part must be hollow and typical tolerances are ±5%. The cycle time is slow because of the loading and cumbersome part removal unless the process is highly automated. It is difficult to get varying wall thicknesses within a part, so parts are often designed with reinforcing ribs to stiffen flat areas. The range of moldable materials is not as broad as it is for other molding processes. Basic raw materials run about 5¢ per pound higher than for injection molding, but much of this is saved because of the 100% raw-material usage without regrinding.

Expandable-Bead Molding

The expandable-bead process consists of placing small beads of polystyrene along with a small amount of blowing agent in a tumbling container. The polystyrene beads soften under heat, which allows a blowing agent to expand them. When the beads reach a given size, depending on the density required, they are quickly cooled. This solidifies the polystyrene in its larger, foamed size. The expanded beads are then placed in a mold (usually aluminum, Fig. 10-12) until it is completely filled. The entrance port is then closed and steam is injected, resoftening the beads and fusing them together. After cooling, the finished, expanded part is removed from the mold.

Advantages and Limitations. Comparatively simple molds can be used to make rather large parts such as ice chests, water jugs, water float toys, shipping containers, and display figures. Expanded-bead products have an excellent strength-to-weight ratio and provide good insulation and shock-absorbing qualities.

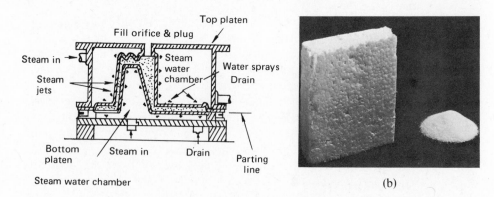

(a)

(b)

Figure 10.12. Expandable-bead molding setup (a). The polystyrene beads and an expanded section (b).

Reaction-Injection Molding (RIM)

Reaction-injection molding involves the impingement mixing of two or more streams of highly reactive liquids (polyol premix and isocyanate) in a small chamber at high pressure, followed by a rapid flow of the mixture into the mold cavity or a forming conveyor as shown schematically in Fig. 10-13. The mixture takes the form of the mold or restraining fixture, reacts, and gels rapidly into a resilient elastomer or rigid plastic depending on the formulation used.

When the gas–resin mixture is shot under pressure into the mold cavity, the gas expands within the plasticized material as it fills the mold, producing an internal cellular structure as well as a tough external skin on the mold face. This structure is usually referred to as integral skin. The rigidity of integral skin and a solid plastic are compared in Fig. 10-14.

Advantages and Limitations.   Structural foams have been used in the past as wood replacements. The detailed woodgrain finish was accurately reproduced on many molded parts such as mirror frames, coffee tables, stereo cabinets, etc. Now, however, structural foams are being used not only to replace wood and metal parts, but also to make distinct structural improvements. As an example, an increase of 25% in wall thickness can result in more than twice the rigidity for a solid plastic part of equal weight. Foam offers 1.2 times the rigidity of an equal weight of aluminum and 2.4 times that of an equal weight of steel.

Parts can be produced in a single shot that weighs up to 100 lb (45 kg) with a cycle time of 2 or 3 min. The processing temperatures are low, raw materials are normally kept at 80–100°F (37.8°C), and mold temperatures normally do not

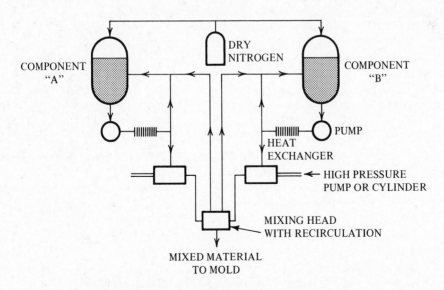

Figure 10.13.   A schematic of the RIM process.

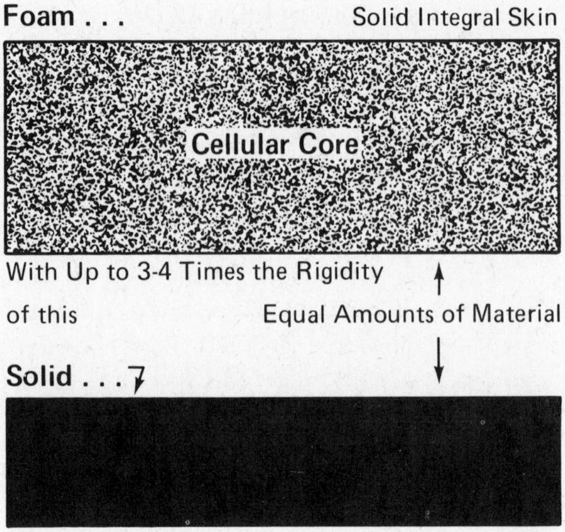

Foam . . .   Solid Integral Skin

With Up to 3-4 Times the Rigidity
of this   Equal Amounts of Material

Solid . . . 7

Figure 10.14. A cross section of a foamed sample shows a rigid cellular core within a solid integral "skin." The equivalent weight of a solid injection-molded part would be thinner with ⅓ to ¼ the rigidity. (*Courtesy General Electric, Plastics Division.*)

exceed 150°F (66°C). Clamping pressures are much less than that for conventional injection molding since the mold pressure is only 15 to 75 psi (1.05 to 5.25 kg/cm²).

Most mass-production molds are made from highly polished nickel-plated steel to impart a high-gloss surface and to release molded parts readily. However, because of the low heat and pressure involved, prototype molds can be made inexpensively from cast aluminum, Kirksite, or epoxy.

Reinforcements may be added to the foam in the form of milled-glass fibers approximately 0.5 mills (0.015 mm) in diameter and typically $\frac{1}{32}$, $\frac{1}{16}$, $\frac{1}{8}$, and $\frac{1}{4}$ in. (0.79, 1.57, 3.17, and 6.35 mm) long. Also used are boron, graphite, and mica flake. With 10% glass, thermal expansion decreases 35%. In general $\frac{1}{16}$–$\frac{1}{8}$-in. (1.57–3.17-mm) milled glass increases the flexural modulus by 25 to 50% at 10% loadings and by 100% at 20% loadings. With flexural modulus up to 300,000 psi (210 kg/sq mm) and advances to allow the use of $\frac{1}{4}$-in. (6.35-mm) glass, it is believed that a flexural modulus of 500,000 psi (350 kg/sq mm) is feasible.

At present RIM, solid and microcellular, is used extensively in automotive fascia, "friendly fenders" and bumpers (Fig. 10-15).

Low-density urethane is used in insulating refrigerator and freezer cabinets where the foam is formed directly in the component.

High-modulus reinforced, solid and microcellular, elastomers are being evaluated for hoods and trunk lids.

Structural foams can be painted, woodgrained, metallized, or printed. Any type of self-tapping screws may be used for assembly. Inserts can be molded in place or installed by ultrasonic insertion. Parts can also be bonded ultrasonically or with adhesives.

Successful RIM molding must allow for venting trapped air. Usually the mold is filled from the bottom with hidden surfaces that allow air to escape, which usually have to be trimmed off later. An ideal mold would be gas-permeable and liquid-tight.

Figure 10.15. The mold and clamping fixture opens up like jaws on the RIM molding machine. (*Courtesy Guide Division, G.M.*)

Mold carriers, or clamping systems that hold the mold halves together, must be able to rotate so that optimum venting can take place. If a mold is held at an "improper" angle, air bubbles may form and cause surface defects. The "proper" angle for optimum processing is different for each tool and has to be determined empirically.

Extruding

Plastic extrusion is similar to metal extrusion in that a hot material (plastic melt) is forced through a die having an opening shaped to produce a desired cross section. Depending on the material used, the barrel is heated anywhere from 250 to 600°F (121–316°C) to transform the thermoplastic from a solid to a melt. At the end of the extruder barrel is a screen pack for filtering and building back pressure (Fig. 10-16). A breaker plate serves to hold the screen pack in place and straighten the helical flow as it comes off the screen.

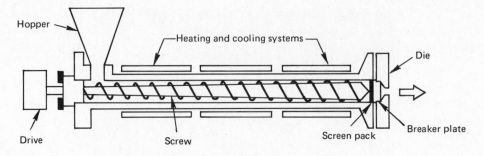

Hopper

Heating and cooling systems

Die

Drive

Screw

Screen pack

Breaker plate

(a)

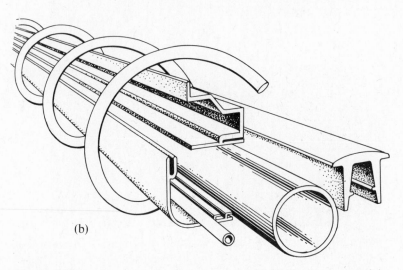

(b)

Figure 10.16. A schematic view of a plastics extrusion press (a) and examples of some plastic extrusions (b).

Figure 10.17. A modification of the extrusion process allows a metal strip to be fed through the die and completely imbedded in the plastic.

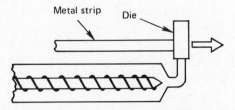

Metal strip   Die

A special modification of the extrusion process is extruding a thin plastic film on a flat substrate such as paper, cloth, or metal. The plastic is immediately and uniformly bonded to the substrate by pressure rolls.

Metal strips, wires, or roll-formed shapes can be incorporated into the extrusion process by means of an offset die, as shown in Fig. 10-17.

Dual extrusion is a process of combining different materials in a single

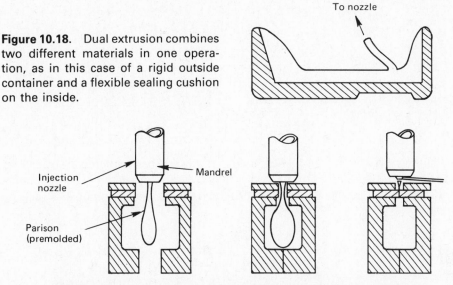

Figure 10.18. Dual extrusion combines two different materials in one operation, as in this case of a rigid outside container and a flexible sealing cushion on the inside.

Figure 10.19. In injection blow molding the parison or tube is formed around the mandrel and while it is still hot it is transferred to the blowing mold where air is injected.

extrusion operation. Thus rigid and flexible vinyl extrusions that offer advantages in both assembly and sealing can be made. The rigid portion is used for shape retention or attachment and the flexible portion for sealing, cushioning, or absorbing impact, as shown in Fig. 10-18.

Blow Molding

Blow molding is a process that is used extensively to make bottles and other lightweight, hollow, plastic parts. Two methods are used: injection blow molding and extrusion blow molding.

Injection blow molding, as shown in Fig. 10-19, is used primarily for small containers up to about 8-oz capacity (224 g). The parison or tube is formed by the injection of plasticized material around a hollow mandrel. While the material is still molten and still on the mandrel, it is transferred into the blowing mold where air is used to inflate it. Usually no finishing operations are required. Accurate threads may be formed at the neck.

Figure 10-20 shows the extrusion-type blow molding. A molten-plastic pipe (parison) is inflated under relatively low pressure inside a split-metal mold. The die closes, pinching the end and closing the top around the mandrel. Air enters through the mandrel and inflates the tube until the plastic contacts the cold wall, where it solidifies. The mold opens, the bottle is ejected, and the tailpiece falls off.

The molds for simple, symmetrical, round shapes can be machined easily. Parts that require extensive ribbing are more easily made by casting. The parting

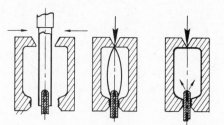

Figure 10.20. The successive steps in extrusion blow molding.

line of the mold should be placed where it will not be aesthetically offensive. The maximum depth of undercut, as in indented handle grips, will be dependent on the flexibility of the material and the amount of wall thinning that can be tolerated.

THERMOFORMING

Thermoforming refers to heating a sheet of plastic material until it becomes soft and pliable and then forming it either by vacuum, air pressure, or between matching mold halves. There are variations of these processes, but the essential elements of each are shown in Fig. 10-21.

In vacuum forming, the plastic sheet is clamped in place and heated. The vacuum beneath the sheet causes atmospheric pressure to push the sheet down into the mold. Areas of the sheet reaching the mold last are usually the thinnest.

Molds may be made out of wood, metal, plastic, plasters, etc. The heated sheet may be clamped over the female die or draped over the male form. As the mold closes, the part is formed. Excellent reproduction of mold details, including lettering and grained surfaces, may be obtained.

In pressure-bubble, plug-assist vacuum forming, the sheet is clamped in place across the female cavity and heated. Air is introduced into the cavity and blows the sheet upward into a bubble. A photocell is often used to control the height and signals the plug to plunge into the plastic sheet. When the plug reaches its lowest position a vacuum draws the sheet against the mold, or in some cases air pressure is used.

Perhaps the most attractive feature of thermoforming is the relatively inexpensive tooling. The process is used in making a wide variety of everyday products such as disposable drinking cups, coffee-cream containers, margarine tubs, meat trays, egg cartons, and picnic plates. Output rates may reach as high as 1500 parts per minute on multiple-cavity dies. Most parts use the plug-assist pressure-forming method. Sheet thicknesses normally used range from 0.025 to 0.500 in. (0.63–12.70 mm).

One of the larger markets for thermoforming is for polystyrene foam sheet. The first of these was for meat and produce trays and later for egg cartons. Currently 70% of the meat and produce tray market utilizes the foamed tray and 60% of the eggs are packaged in thermoformed containers.

Most thermoplastic sheet is thermoformed at temperatures of 225 to 325°F

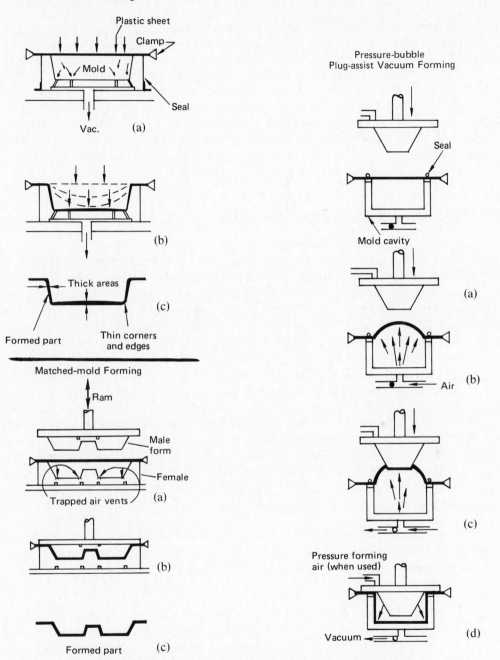

Figure 10.21. Thermoforming methods. Reprinted from the 1974–75 *Modern Plastics Encyclopedia.* Copyright 1974 by McGraw-Hill, Inc.

(115–165°C). The heating is critical since plastics do not have very good thermal conductivity and because temperature affects the ductility or forming characteristics of the materials. Too much heat causes the sheet to flow without drawing and too little heat will cause rupture early in the forming process.

Many different heating systems are used, such as quartz lamps, small ceramic modules, gas-fired ovens, emitter strip panels, and calrod-type resistance heaters.

Recent developments include a microprocessor with computer interfacing. With this control system, a thermoforming machine can be completely programmed for the manufacture of any given part. Once programmed, the part can be easily set up and run whenever needed.

Reinforced-Plastic Molding

Reinforced plastics generally refer to polymers that have been reinforced with glass fibers. Other materials used are asbestos, sisal, synthetic fibers such as nylon and polyvinyl chloride, cotton fibers, paper, and metal filaments. Relatively new are carbon, graphite, and boron fibers, as discussed in Chapter 2 under composite materials. These high-strength composites using graphite fibers are now commercially available with moduli of 50,000,000 psi (344,700,000 MPa) and tensile strengths of about 300,000 psi (2,068,000 MPa). They are as strong as or stronger than the best alloy steels and are lighter than aluminum.

Fibrous-glass reinforcement is available in many different forms: as continuous strand, woven fabric, reinforcing mat, and chopped, short lengths (Fig. 10-22). Glass-fiber fabrics are used where rather high strengths are required. They can be of tight or open weave and have almost equal amounts of glass in each direction, or have most of their fibers running in one direction. Woven cloth is used for heavy reinforcement and mat for light and medium reinforcement. The chopped strands facilitate molding in complex structures.

Molding Reinforced Thermosets

Size, volume, shape, strength requirements, and economics of the product will determine which of the following molding processes will be used: hand layup, matched die molding, rigidized vacuum forming, pultrusion, and filament winding.

Hand Layup. Hand layup is the oldest and simplest method of forming reinforced thermosets. Only one mold is used and it is often referred to as the *open-mold* method. The mat or fabric reinforcement is cut and fitted to the mold and is then saturated with resin, applied by hand with the aid of a roller or brush, or sprayed on with a special gun. The resin is carefully worked into the reinforcement to ensure complete "wet-out" and to eliminate any air pockets. It is not possible to work to close tolerances, particularly in maintaining wall thicknesses.

An extension of the hand layup method is vacuum-bag, pressure-bag, or autoclave molding, as shown in Fig. 10-23.

(a)　　　　　　　　　　　　(b)

(c)　　　　　　　　　　　　(d)

Figure 10.22. Basic forms of Fiberglas used in reinforcing plastics are continuous strand (a), woven fabric (b), reinforcing mat used for medium-strength structures (c), and chopped short lengths of Fiberglas used to facilitate molding complex structures (d). (*Courtesy Owens-Corning Fiberglas Corporation.*)

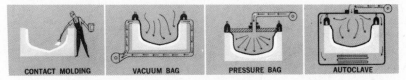

CONTACT MOLDING　　VACUUM BAG　　PRESSURE BAG　　AUTOCLAVE

Figure 10.23. Vacuum bag, pressure bag, and autoclave are extensions of the hand layup or contact molding. (*Courtesy Owens-Corning Fiberglas Corporation.*)

Matched-Die Molding. Matched dies overcome the objections of the hand layup method and provide accuracy, high production, and smooth surfaces both inside and out. The quantity required, however, must be much higher to warrant the die cost. The dies are usually made of steel for small parts and cast iron for larger parts. They may also be made out of reinforced plastics or wood for short runs. Pressures used range from 200 to 300 psi (14–21 kg per cm²) and temperatures from 230 to 260°F (110–127°C).

Some design limitations of this process are that there can be no undercuts, no absolutely square corners, no molded-in openings on vertical walls, and generally there must be a draft of at least 1°.

Three principal categories of resin–glass systems are used today to mold parts by the matched-metal-die and compression methods. These are bulk-molding compound, sheet-molding compound, and preform.

Bulk-molding compound (BMC), which was formerly called premix, is a mixture of short, $\frac{1}{8}$ to $\frac{1}{2}$ in. (3.17–12.70 mm), glass fibers and a polyester resin, along with various mineral fillers, such as calcium carbonate and hydrated alumina, and other additives. The high filler content of BMC materials provides good electrical properties and dimensional stability at low cost.

BMC is available in doughlike bulk form for injection and compression molding and as extruded "logs" that are cut to weight for transfer or compression molding. Figure 10-24 shows the BMC used for compression molding. Typical parts that are compression molded include automobile fender extensions, grills, and electrical components such as distributor caps and circuit-breaker housings.

Sheet-molding compound (SMC) differs from BMC in that longer glass fibers are dispersed in a polyester–filler mixture and then sandwiched between two polyethylene or nylon film sheets (which are removed prior to molding) (Fig. 10-25). Standard SMC contains 25 to 35% (by weight) of chopped, 1 to 2-in

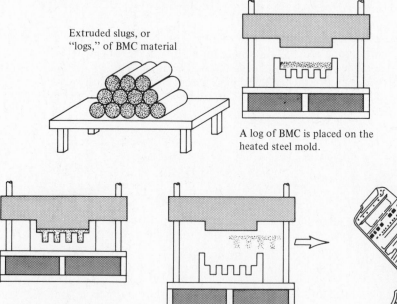

Extruded slugs, or "logs," of BMC material

A log of BMC is placed on the heated steel mold.

The mold is closed, forcing the resin/glass-fiber mixture into the die cavities. Heat and pressure cure the resin.

The mold opens, and the cured part is removed, trimmed, and made ready for assembly.

Figure 10.24. The use of BMC "logs" for compression molding. (*Courtesy* Machine Design.)

(25.40–50.80-mm) randomly oriented glass fibers. Two other forms of SMC are available: One contains 8 to 12-in. (203.20–304.80-mm) chopped glass fibers oriented in one direction; the other contains continuous fibers, also oriented in the same direction.

Because of the longer glass fibers and higher glass loading in SMC, strength of SMC-molded parts is greater than those molded from BMC materials. Typical applications of SMC are automotive parts such as front-end panels, tailgate doors, and exterior body panels; truck fenders and doors; and tractor hoods.

SMC materials are available with glass contents of 60 to 80%. These structural SMC compounds, offering strengths about double those of the standard versions, are being evaluated for load-bearing applications such as automobile wheels, radiator-core supports, and various brackets. SMC is not recommended for deep-drawn parts because of the orientation of the fibers; a considerable difference in strength properties occurs between the flow and normal-to-flow directions.

Preform molding is quite different from BMC and SMC methods, even though the same tooling may be involved. In this method a fiberglass preform of the part is produced by spraying or dropping chopped glass fibers and resin binder in a controlled manner onto a perforated metal screen that has the same

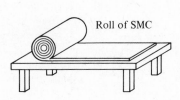

Roll of SMC

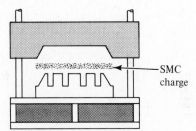

SMC charge

A measured (by weight) charge of SMC is placed on the heated steel mold.

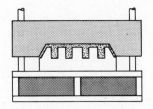

The mold is closed, forcing the resin and glass into the cavities that form ribs and bosses on the part. Heat and pressure cure the resin.

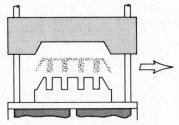

The mold opens and the cured part is removed, trimmed, and made ready for assembly.

Figure 10.25. The steps in using SMC for compression molding. (*Courtesy* Machine Design.)

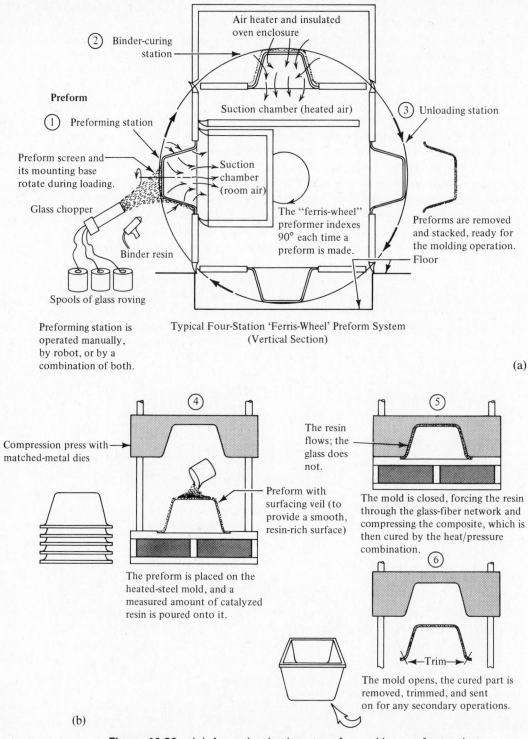

Preform

① Preforming station

② Binder-curing station

Air heater and insulated oven enclosure

Suction chamber (heated air)

③ Unloading station

Preform screen and its mounting base rotate during loading.

Suction chamber (room air)

Glass chopper

Binder resin

Spools of glass roving

The "ferris-wheel" preformer indexes 90° each time a preform is made.

Preforms are removed and stacked, ready for the molding operation.

Floor

Preforming station is operated manually, by robot, or by a combination of both.

Typical Four-Station 'Ferris-Wheel' Preform System (Vertical Section)

(a)

④

Compression press with matched-metal dies

The preform is placed on the heated-steel mold, and a measured amount of catalyzed resin is poured onto it.

Preform with surfacing veil (to provide a smooth, resin-rich surface)

⑤

The resin flows; the glass does not.

The mold is closed, forcing the resin through the glass-fiber network and compressing the composite, which is then cured by the heat/pressure combination.

⑥

←Trim→

The mold opens, the cured part is removed, trimmed, and sent on for any secondary operations.

(b)

418

Figure 10.26. (a) A mechanized system for making preforms that is referred to as a "ferris-wheel" system. (b) The conventional compression molding in closed dies for making preforms. (*Courtesy Machine Design.*)

shape as the male molding die. The glass–resin mixture is held to the screen and compacted by a high-suction fan that draws air from the inside of the screen.

The preform on its screen is then heated in a recirculating-air oven to cure the binder resin. At this stage, the preform is self-supporting and can be handled without damage. It is then positioned on the male die and a measured amount of catalyzed resin containing fillers and pigment is poured onto the preform. The heated die closes and, as with SMC and BMC, the part is cured.

Shown in steps 1 to 3 of Fig. 10-26(a) is a mechanized system of making preforms that is referred to as a "ferris-wheel" system.

Shown in steps 4 to 6 of Fig. 10-26(b) is the conventional compression molding in closed dies for making the finished parts.

The advantage of the preform method is that there is no flow of the fibers and therefore the molded part is evenly reinforced with glass in the original random pattern. The strength is appreciably higher and the properties are uniform in the plane of the wall.

Many preform-making operations are now mechanized, reducing the cost and improving the uniformity and quality of the product. Robot-controlled delivery of chopped glass and resin binder to the preform screens is now being used for some of the larger, relatively simple shapes. These advantages are equalizing the economics between preform operations and those of the "ready-made" SMC and BMC systems.

Rigidized Vacuum Forming.   Rigidized vacuum forming is an intermediate process between basic hand layup and costly matched die molding.

A part is first made by thermoforming and then it is reinforced and rigidized by glass reinforcement. The thermoformed parts are placed on a holding fixture to insure that during the curing of the reinforcing laminate the parts will not change dimensionally. The sprayup process is mostly used for rigidizing because of the higher production capabilities.

Examples of rigidized vacuum-formed parts are bathroom tubs and sinks and shower stalls, automotive body and recreational vehicle panels, motor shrouds, and other items where higher production and uniformity are desired.

Pultrusion.   Pultrusion, as the name implies, is like pulling an extrusion. Constant cross-sectional shapes are produced by pulling resin-impregnated, reinforcing material through a heated-steel shaping die in such a way as to promote adequate polymerization of the resin (Fig. 10-27). The degree of polymerization that occurs will be dependent on resin system reactivity, the die temperature, die length, stock speed, stock dimensions, mass effects, and stock transverse thermal conductivity.

The most common reinforcing material is fiberglass, but graphite and carbon filaments are increasingly being used. The reinforcing need not be in chopped form, as woven and nonwoven fabrics, braids and filaments of metal, nonmetal and ceramics are also used.

Pultrusion is used mainly with thermosetting plastics but it may also be used with thermoplastics. Polyesters represent the bulk of the market; however, epoxies will be used more and more as the basic process is better understood.

From a production standpoint, pultrusion lines have grown more sophisticated. In order to obtain higher running speeds, more effort has been made to reduce the friction of the reinforcement going into the dies. This has been achieved by arranging creel systems to put the reinforcing material more in-line with the plane of the dies, and by using solid fluorocarbon forming blocks to pre-shape the resin-saturated reinforcement before it reaches the dies.

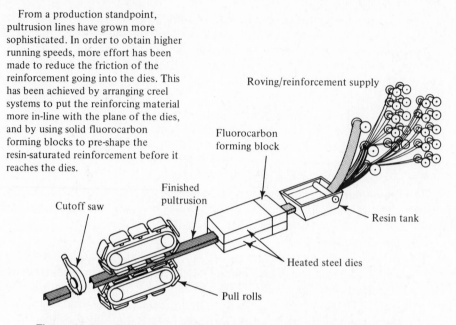

Figure 10.27. Pultrusion consists of pulling continuous-strand woven roving, surfacing mat, woven fabrics, and reinforcing mat through a resin bath to wet out the fibers and then through a heated steel die. The entrance section of the die is water-cooled to prevent premature curing or binding. The pultrusion moves through the die at speeds ranging from 2 to 100 in./min (50–2540 mm), depending upon the cure time, which in turn depends on the section thickness. (*Courtesy* Machine Design.)

The market for pultruded products is wide and growing. The main use has been for electrical, recreational, and construction purposes, and where corrosion resistance is necessary. The transportation industry promises to become the major user of the pultrusion process in the years ahead.

Filament Winding.   Filament winding is a process of wrapping reinforcements consisting of resin-impregnated, tensioned, continuous filaments around a form called a mandrel (Fig. 10-28). After the resin matrix is cured, the mandrel may be removed leaving a hollow, monolithic shell.

The outstanding property of filament-wound structures is their high strength-to-weight ratio, which is better than that of steel or even titanium.

The most common filament used is a continuous glass filament but graphite is also used. The matrix resin is generally epoxy or polyester.

The largest known helically wound product is a railroad freight car. Filament winding is especially suitable for lightweight pressure vessels such as solid rocket cases. Metal-lined compressed gas cylinders, which may be half the weight of an equivalent steel unit, are being made in large quantities for portable life-support systems.

Figure 10.28. Filament winding used in making a lightweight pressure vessel. (*Courtesy* American Machinist.)

Forged-Plastic Parts

The forging of plastic materials is a relatively new process. It was developed to shape materials that are difficult or impossible to mold and is used as a low-cost solution for small production runs.

The forging operation starts with a blank or billet of the required shape and volume for the finished part, as shown in Fig. 10-29. The blank is heated to a preselected temperature and transferred to the forging dies, which are closed to deform the work material and fill the die cavity. The dies are kept in the closed position for a definite period of time, usually 15 to 60 sec. When the dies are opened, the finished forging is removed. Since forging involves deformation of the work material in heated and softened condition, the process is applicable only to thermoplastics.

Advantages and Limitations.   The most important advantage of the plastic-forging process is the capability of producing thick parts with relatively abrupt changes in section. Assuming such parts could be made (to an acceptable quality) by injection molding, the time element would be a factor. Injection-molded parts of this type may require up to 5 min, whereas forging could be done in as little as 30 sec.

In injection molding, where the polymer is forced into the mold cavity in a fluid state and then allowed to solidify, even cooling is extremely difficult to achieve if thin and thick sections are adjacent to each other. As a result, stresses tend to distort the part. The problem is further complicated in crystalline polymers by the large reduction in volume on solidification (mold shrinkage), causing surface depressions and internal porosity. In forging, the solid material is deformed to the shape of the die cavity. The temperature is much lower than in injection molding, and no phase change is involved. Thus forging has few limitations on section thickness, which results in greater design freedom.

Only simple tooling is required for forging. The tooling pressure required is less than 1000 psi (6.89 MPa) as compared to 8000 to 30,000 psi (55.16 to 206.8 MPa) for injection molding. The temperature required is just below the melting point, 250 to 300°F (121–149°C), compared to injection molding, which requires 300 to 600°F (149–316°C). Forging dies do not require cooling passages except to increase production. There is also no need for gates or runners. Forging can be done on a small hydraulic press. For example, the parts shown in Fig. 10-30 can be made on a 5-ton hydraulic press. An injection-molding machine would cost five times as much and require from five to ten times more floor space.

Extruded Blank

Blank is Heated

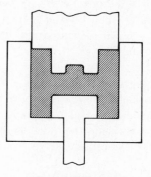

Forging Operation

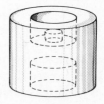

Finished Part

Figure 10.29. The sequence used to forge plastic parts. Forging dies are sometimes equipped with heating and cooling means (not shown).

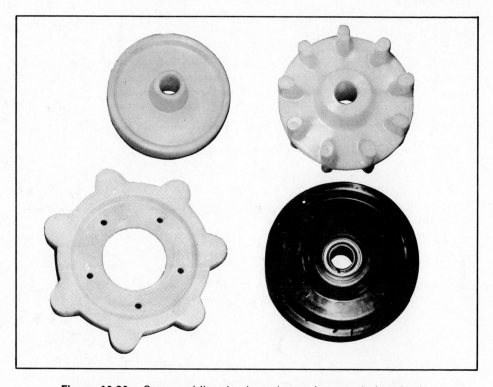

Figure 10.30. Snowmobile wheels and sprockets made by plastic forging. The material is ultrahigh molecular-weight polyethylene, a polymer that cannot be injection molded. The parts have the sheen and finish of molded parts and the properties are equal or superior to machined parts. (*Courtesy Glasrock Products, Inc.*)

Parts that are deformed significantly in the forging process have improved mechanical properties due to molecular orientation. Increases in tensile modulus, impact strength, and abrasion resistance have been observed on some materials.

As yet the number of materials used for forging is limited. Materials used commercially are polypropylene, high-density polyethylene, and ultrahigh-molecular-weight polyethylene. Amorphous polymers such as polyvinylchloride (PVC) and acrylonitrile butadiene styrene (ABS) are less satisfactory because of their rubberlike action. Their high elastic recovery rate makes it difficult to maintain dimensional accuracy.

The forging process is generally suited to small-quantity production but it can be automated for high production rates.

Machining is the only other process besides forging and injection molding that is capable of producing complex, solid parts. Since no special tooling is involved, machining is more cost effective when the quantity is on the order of a few hundred; however, machining produces scrap that may not be reusable. Therefore with an increase in production, volume forging is favored.

Joining Plastics

Present methods of joining plastics include hot-air welding, friction welding, heated metal plate, solvent welding, and the high-energy methods of dielectric, magnetic, and ultrasonic joining. All of these methods are applicable to thermoplastics, but only adhesive bonding and polymerization apply with adequate success to thermosetting materials. Some of these joining processes are discussed briefly.

Hot-Air Welding.   The analogy of hot-air welding to metal welding is close, except the temperatures are much lower and there is no need to apply a flux as a shielding agent. The various types of hot-air welding are tack, hand, and high-speed hand, as shown in Fig. 10-31.

Unlike metal welding, the rod can be made in a triangular shape so that one pass will fill large gaps. If a round rod is used, several passes are used, as shown in Fig. 10-32. Plastic welding can be used to produce a joint of 100% base material tensile strength on certain types of thermoplastics. The strength depends on temperature, the amount of pressure on the rod, preparation of the material, and the skill of the operator. The temperature range should be 400 to 600°F (204–316°C) on the plastic. The filler rod should be the same as the base material.

Solvent Welding.   Solvent welding is a widely used plastics-joining process. The solvent may be applied to the joint by dipping, spraying, and brushing. The most common method is to hold the solvent-soaked pad against the surfaces. The edges dissolve, producing tacky surfaces that are then brought together, moved slightly to obtain mixing, and held in contact, in correct relationship, to dry and harden. Drying time ranges from a few minutes to hours.

Friction Welding.   Spin welding is limited to round parts that can be rotated while in contact to generate heat. The frictional heat melts the interfaces, and when the motion is stopped, the parts fuse together. Spinning speeds of 10 to

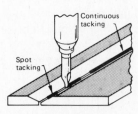

Tack welding is a shallow fusion of the mating surfaces of the base material and produces very little tensile strength. It requires no welding rod. Spot tacking may be sufficient, but continuous tacking may be required for added strength.

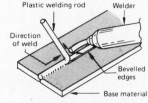

Hand welding can provide tensile strength up to 100% if properly done. It requires a welding rod, and is recommended for welding corners, short runs, or small radii.

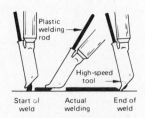

High-speed welding is a variation of hand welding in which the rod is fed automatically. Rod is fed into the preheating tube in the welding tool by the motion of the welder as it is pulled along the joint. Welding speed ranges from 24 to 60 ipm.

Figure 10.31.   Hand welding of plastics using the hot-air method. (*Courtesy* Machine Design.)

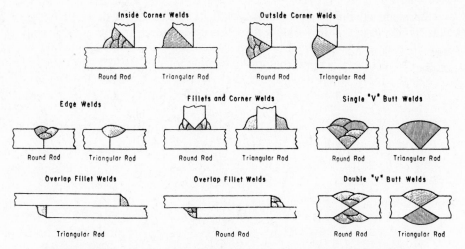

Figure 10.32. Types of welds made with the round and triangular rod in plastics. Triangular rod makes it possible to make most welds in one pass.

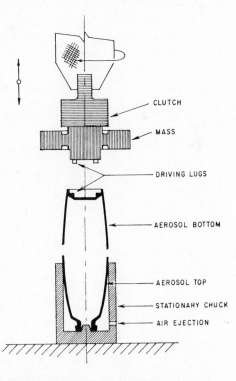

Figure 10.33. A spin weld made with an inertia tool running at a speed of 4000 rpm. The flywheel or mass comes to a stop within 0.5 sec, losing all its kinetic energy in friction and thus causing melt to form in the joint. After the flywheel stops the melt solidifies within a fraction of a second. (*Courtesy E. I. du Pont de Nemours & Co.*)

15 fps at the interface and pressures of 20 to 200 psi (1.4 to 14 kg per cm^2) must be matched to the material to produce just the right amount of heat. Melting and fusion can occur in as little as 0.25 sec. Maintaining pressure for 1 to 2 sec after braking normally completes the cycle. Figure 10-33 is a sketch of the tooling used to spin an aerosol-bottle bottom in place.

Ultrasonic Welding.   Ultrasonic welding is done with high-frequency sound waves (about 20 kHz). It works best on those thermoplastics that have relatively low melting points and moduli of elasticity greater than 200,000 psi (1379 MPa). A sonic tip, or tips, placed at the approximate center of the part transmits sound waves to the joint, where they are converted to frictional heat (Fig. 10-34). The operation takes 3 to 4 sec. Joint strength is close to base-material strength.

Welding With Radio Frequencies.   Dielectric welding with certain FCC-approved frequencies is limited to plastics having a fair degree of polarity, such

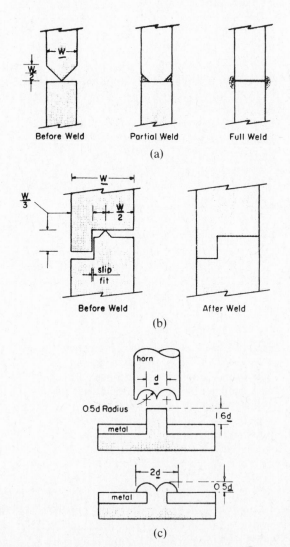

Figure 10.34. Joint designs for ultrasonic welding, (a) and (b). Joint design for ultrasonic staking (c). (*Courtesy Society of Manufacturing Engineers.*)

as saran, polyvinyl chloride, or styrene acrylonitrile. Welding is simple and rapid, but only those surfaces to be joined can be exposed to the rotating electrical field.

Induction welding, although applicable to virtually all thermoplastics, is specialized. Its prime use is for rapid butt welding of pipe. A metal ring inserted in the joint is induction heated and melts the surrounding plastic. A slight pressure completes the joint. The cost is high but the joint quality is good.

Plating Plastics

Plastics may be plated to enhance their appearance and utility. The steps involved are described briefly.

First, the plastic is etched using strong chemicals such as chromic and sulfuric acid mixtures. The etching provides a good surface for adhesion of the plating. After etching a catalyst such as colloidal palladium is applied by dipping. This causes nickel or copper to deposit on the surface. The process is referred to as an "electroless" copper or nickel process. A very thin coating of metal is autocatalytically deposited on the surface, making it electrically conductive. At this point the preplated part can be racked up for plating and treated similarly to metal parts that are to be electroplated. Normally copper, nickel, and chromium are electrodeposited to produce a part that appears to be chromium-plated metal.

Plated plastics have been used extensively in the automotive industry to make such parts as grills and decorative trim as well as knobs and levers for car interiors.

Plastic Design Principles

As stated at the beginning of this section on plastics, there are many similarities in molding metals and plastics. These similarities carry over into design. The more knowledge the designer has of the manufacturing process, the better his designs will be. Often a designer will take a molder into his confidence so that the early stages of design and development will produce the best results. General considerations are material selection, process selection, tolerances, standards, and specifications.

Material Selection. Material selection will be based on the properties sought, such as impact strength, tensile and flexural strengths, maximum and minimum resistance temperatures, weathering qualities, flame and chemical resistance, electrical resistance, wear and scratch resistance, etc.

Process Selection. The part design, material, production requirements, and final price are the main factors to consider in choosing a process. Often more than one method can be used to produce equally acceptable parts but the final decision will be made on the basis of total cost, which includes design, material, production, tooling, etc.

Tolerances and Wall Thickness.   As with metal-casting material, shrinkages must be considered in order to maintain close tolerances. In some plastics there are two shrinkages: a *mold shrinkage* that occurs upon solidification, and shrinkage that occurs in some materials after 24 hrs, *after shrinkage.* For example, the mold shrinkage for a melamine plastic may be 0.007 to 0.009 in./in. and the after shrinkage 0.006 to 0.008 in./in. Thus a total shrinkage of 0.013 to 0.017 in./in. can be figured. Minimum wall thickness recommendations for compression, transfer, and injection molding are shown in Table 10-3.

Table 10-3.   Recommended minimum wall thicknesses for compression, transfer, and injection molding.

Depth (in.)	Wall Thickness
Up to 2	0.060
2 to 4	0.060 to 0.080
4 to 8	0.080 to 0.100

Standards and Specifications.   Final drawings should be made using recognized symbols and standards. Such a system is provided in Military Standard MIL STD-8B and in Engineering and Technical Standards of Plastics and Custom Molders as found in Section 3 of the *Modern Plastics Encyclopedia.*

General Design Considerations.   Design points regarding castings and forgings apply, such as avoiding undercuts, heavy cross sections, adjacent thick and thin sections, sharp corners, and square holes where possible. Gate and parting-line placement are also important. These points and many others are well illustrated in several issues of *Modern Plastics Encyclopedia.*

Plastic Process Selection

The graph (Fig. 10-35) helps to summarize this chapter by showing a rough comparison of process ranges and comparing relative costs versus typical production runs.

In most cases, a very strong determinant in selecting a process is answering the question "how many parts?" As an example, to produce 50 or 100 parts by casting nylon would probably cost only a few hundred dollars for tooling.

The optimum number of cavities to tool for injection molding is also a significant cost variable. While a single cavity mold might be run for 10,000 or 15,000 parts, a run of 100,000 to 500,000 parts may warrant a multicavity mold, perhaps 12 or more, to keep the machine time down.

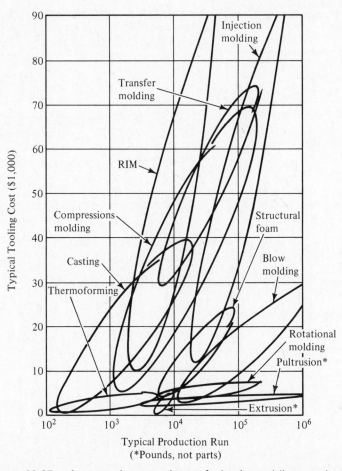

Figure 10.35. A general comparison of plastic molding methods as to tooling cost and capacity. (*Courtesy* Machine Design.)

Questions and Problems

10-1. Why are plastics often referred to as polymers?

10-2. What causes thermoset polymers to have different characteristics from thermoplastic polymers?

10-3. What is a unique feature of the nylon monomer?

10-4. What is the difference between compression molding and transfer molding?

10-5. What happens if a plastic mold is run too hot?

10-6. What mechanism creates the pressure for plastic injection molding?

10-7. How does one go about determining the required clamping force for a particular mold?

10-8. Why may the cost of a mold become relatively insignificant?

10-9. What is the main advantage of rotomolding?

10-10. What is reaction injection molding being used for?

10-11. (a) What is meant by thermoforming? (b) Why are plugs used?

10-12. (a) What is the least expensive way of making a fiberglass product? (b) Why is the use of this method quite limited?

10-13. What advantage does the forging of plastic parts offer?

10-14. What methods are used to join thermosetting plastics?

10-15. What is the function of the return pins in an injection mold?

10-16. What is the purpose of the core in an injection mold?

10-17. Why may coinjection molding produce a better part than standard injection molding?

10-18. What difficulty may be encountered in injection molding of reinforced plastics?

10-19. Name two methods of avoiding weld lines in injection molding.

10-20. Why is RIM extensively used for typewriter frames and cabinets?

10-21. How may the highest flexural modulus be obtained for RIM?

10-22. What is the latest development for thermoforming?

10-23. What is the difference between the structures of a vacuum-formed part and a rigidized vacuum-formed part?

10-24. How does pultrusion differ from extrusion?

10-25. (a) Make sketches to show how each of the molded parts in Fig. P10.1 could be improved and still perform the same function. (b) State the principles by which you changed the designs. Assume all parts are circular.

10-26. (a) Beads or ribs are used to reinforce large areas. Shown in Fig. P10.2 is a rib pattern. Show how it may be improved. (b) State your principle in making the design changes.

10-27. Which of the two parting-line placements shown in Fig. P10.3 is the best, and why?

10-28. (a) Show by sketches how the designs shown in Fig. P10.4 can be improved. (b) What are the principles involved in making your design changes?

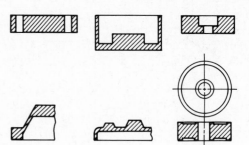
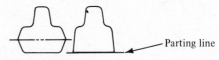

Figure P10.1.

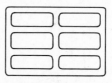

Figure P10.2.

Parting line

Figure P10.3.

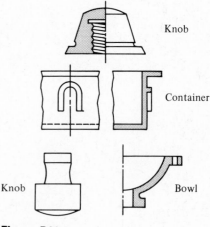

Knob

Container

Knob

Bowl

Figure P10.4.

10-29. Shown in Fig. P10.5 is a full-scale drawing of a knob for a television set. (a) What molding methods could be used for a prototype run of 500? (b) What molding method would most likely be used for 10,000 parts? (c)Why might polystyrene be a good choice of material? (d) What could be done, if anything, to improve the design? (e) Could this knob be made for high production out of a thermosetting plastic? (f) If you wanted this knob to be chromium plated, what material would you choose? (g) What is the minimum recommended wall thickness for this part if injection molded?

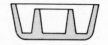

Figure P10.5.

Case Study

Plastics Plant A was well equipped to do a wide variety of molding, including compression molding, RIM, transfer molding, injection molding, and blow molding.

A contract became available for molding plastic containers of about two-gallon capacity (7.6 l). The outside dimensions of the container were to be 7 in. dia. (177.80 mm) at the bottom and 9 in. dia. (228.60 mm) at the top. The height was 12 in. (304.80 mm), (Fig. CS10.1). The containers would be used for storing liquids so that a snap-on cover with a ¼-in. (6.10-mm) lip would be required. The plant managers were assured that if they got the contract they could expect orders totaling 50,000 containers per year. If injection molded, the wall thickness would have to be $\frac{3}{32}$ in. (2.36 mm) thick.

Various alternatives were investigated. The cost of tooling would run between $75,000 and $80,000.

Select a molding method that you feel

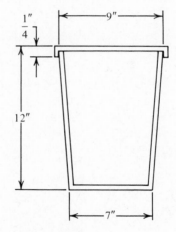

Figure CS10.1.

would be the most appropriate based on the amount of material used annually and the tooling cost.

Bibliography

Becker, Walter E., ed. *Reaction Injection Molding.* New York: Van Nostrand Reinhold Co., 1979.

Dreger, Donald R., ed. "Massive Plastic Parts." *Machine Design* 52 (January 24, 1980): 58–64.

Dreger, Donald R., ed. "Selecting a Process for Plastic Parts." *Machine Design* 52 (February 21, 1980): 76–80.

DuBois, J. H., and J. W. Frederick. *Plastics,* 5th ed. New York: Van Nostrand Reinhold Co., 1974.

Editor. "Coinjection: New Molding Technology." *Modern Plastics* 56 (July 1976): 40–42.

———. "What's Exciting About Stretch Blow Molding?" *Modern Plastics* 53 (February 1976).

Krause, John K., ed. "Injection Molded Carbon-Fiber Composites." *Machine Design* 51 (August 9, 1979): 79–85.

Modern Plastics Encyclopedia. New York: McGraw-Hill, 1978–1979.

Seymour, William B. *Modern Plastics Technology.* Reston, Va.: Reston Publishing Co., 1975.

Vaccari, John A., ed. "Winding Plastics for New Jobs." *American Machinist* 124 (May 1981): 125–130.

CHAPTER ELEVEN

Powdered Metallurgy

Powdered metallurgy (PM), like forging, casting, and plastic molding, requires a mold or a die to control the shape of the finished product. In powdered metallurgy, finely divided powders are pressed into a steel or carbide die of the desired shape. In the first stage, pressing is done at room temperature. The powdered particles interlock and have sufficient strength to make what is known as a "green compact." This compact is removed from the die and heated or *sintered* at a high temperature. This takes place in a neutral or reducing atmosphere at some temperature near, but below, the melting point of the metal. An exchange of atoms between the individual particles welds them together. The result is a more or less porous piece of metal (high densities can be achieved) of the approximate size and shape of the die cavity. A schematic of the basic process is shown in Fig. 11-1, including the secondary operation of coining and infiltration.

A considerable knowledge of techniques is necessary in such matters as the selection of the right combinations of metal powders, die design, sintering conditions, and any secondary operations.

METAL-POWDER PRODUCTION

Metal powders are produced in several ways, such as atomization, reduction, and electrolysis.

433

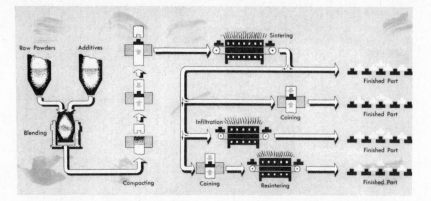

Figure 11.1.   Schematic diagram of the powder-metallurgy process. (*Courtesy* Machine Design.)

Atomization.   In this process, the molten metal is forced through a nozzle, where it is atomized with compressed air and, upon solidification, a wide range of particle sizes and shapes are formed. The fineness of the powder depends on the pressure of gas and the fluidity and rate of flow of the metal. The process was first applied to aluminum, zinc, lead, and tin, but now includes steels. Pure iron powder is produced by first atomizing molten cast iron. The process produces fine particles but with oxidized surfaces. The carbon and oxygen contents of the powder are controlled however, and by heating to 1740°F (950°C) decarburization occurs by mutual reaction.

Reduction.   The reducing agents used for common metals are carbon monoxide and hydrogen. The resulting sponge metal is then crushed and ground to produce a powder.

Electrolysis.   The electrolysis method of producing metal powders is similar to electroplating. In this process, the metal plates are placed in a tank of electrolyte. The plates act as anodes, while other metal plates are placed in the electrolyte to act as cathodes. High-amperage current produces a powdery deposit on the cathodes. After a buildup, the cathode plates are removed from the tank, scraped off, and the deposit pulverized to produce powder of the desired grain size. An annealing process follows pulverization to remove work-hardening effects of scraping.

SUPERALLOY-POWDER PRODUCTION

Demand for high-performance components operating at elevated temperatures has resulted in the development of nickel-type superalloys. The production of these alloy powders is done by three major methods: inert-gas atomization, rotating electrode, and soluble-gas atomization, as shown schematically in Fig.

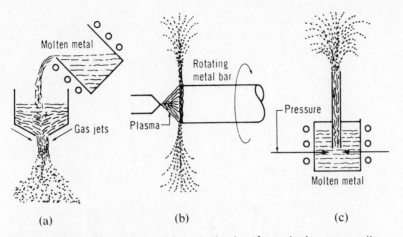

Figure 11.2. The three major methods of producing superalloy powder are (a) inert-gas atomization, (b) rotating electrode, and (c) soluble-gas atomization. (*Courtesy Climax Molybdenum.*)

11-2. The inert-gas process is usually done with argon-gas jets. The rotating-electrode process is done in a closed chamber filled with an inert gas. The molten droplets that are thrown off from the rotating electrode are collected to make up the powder. In the soluble-gas process, molten metal is pressurized in a vacuum and the liquid erupts into particles.

THE POWDER-METALLURGY PROCESS

The concept of making parts from powdered metals is simple and straightforward; however, the techniques employed can be very sophisticated, requiring a high level of technical competence and a substantial investment in capital equipment. The process consists of three basic steps: blending, compacting, and sintering.

Blending.   Blending refers to mixing the metal powder to obtain the desired properties. Lubricants are added to the powder to reduce friction between the grains as they are being compacted as well as to reduce die wear. The blending may be done wet or dry. Wet mixing has the advantage of reducing dust and the danger of explosion, which is present in some finely divided powders.

Compacting.   Metal powders present problems of internal friction. There-fore, when placed in a die, the pressures are not distributed uniformly throughout the compact. Both hardness and density decrease as the distance from the punch increases. To rectify this condition, punches are used at both ends (Fig. 11-3) and a lubricant is mixed with the metal powder.

The use of lubricants improves the density, minimizes the load required, and increases die life. However, lubricants can create problems in feeding the

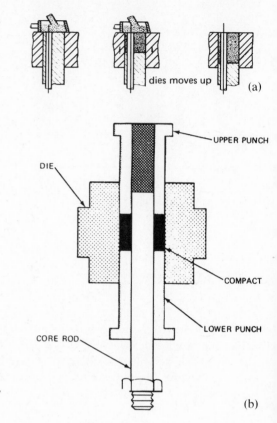

Figure 11.3. Filling the PM die (a),
compacting a bushing (b).

powder into the die and in lubricant reaction. Lubricants must be driven off by
slow heating before sintering.

For high-volume production, tungsten carbide is used as the die material.
Although the cost is higher, it will outwear the normally used tool steels by a
ratio of about 10:1. Some carbide dies can be used to produce a million parts
before the tolerances are exceeded. High pressures, sometimes in excess of 50
tons per sq in. (689.5 MPa), are used to cold bond the powder particles. Some
powders such as brass, bronze, and aluminum compact at relatively low pressures
ranging from 12 to 25 tsi (10.8 to 22.5 metric tons per sq in.). A 95% theoretical
density can be achieved with aluminum powders at 25 tsi.

Isostatic Pressing.   To overcome the difficulty of variations in density due
to compacting in just two directions, a method, *isostatic pressing,* has been
developed to produce equal pressure on all sides of the powder. The process
can be achieved by two methods: wet bag and dry bag (Fig. 11-4).

In the wet-bag process, a plastic or rubber mold is filled with powder and
is tightly sealed after being evacuated with a hypodermic-needle vacuum connection. The bag is then placed in a vessel filled with oil and pressure is increased

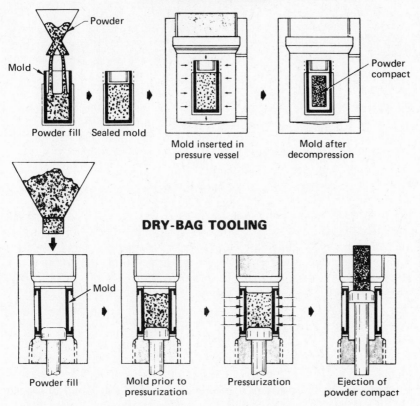

WET-BAG TOOLING

Powder

Mold

Powder fill   Sealed mold

Mold inserted in
pressure vessel

Powder
compact

Mold after
decompression

DRY-BAG TOOLING

Mold

Powder fill

Mold prior to
pressurization

Pressurization

Ejection of
powder compact

Figure 11.4. Isostatic powder pressing may be done wet or dry as shown. The wet-bag process has a separate container or mold that is loaded outside of the press, whereas in the dry-bag process the mold is an integral part of the press. The wet-bag process is better suited to forming intricate parts, but the dry-bag process is simpler and better suited to high production. (*Courtesy* Machine Design.)

to the desired value, usually 26 to 44 tsi (23.4 to 39.6 metric tons per square inch).

The dry-bag process employs a rigid rubber die that can maintain its shape. The bag is filled with powder and is closed on top by a punch of the press. The pressure is then applied to the rubber die.

Owing to an absence of die-wall friction, and the application of uniform pressure, isostatic compacts show almost equal density and strength in all directions, regardless of size. Some variation in density exists from the surface to the center of the compact, however. The maximum pressure in die pressing is limited for a given press by the surface area of the compact. For example, 10,000 psi (68.95 MPa) pressure (typical for iron powder) requires a 500-ton (450 metric tons) press for a shape with 10 sq in. (25.4 sq cm) of surface area. Such

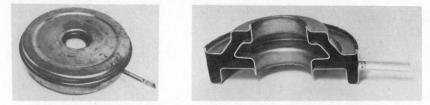

Figure 11.5. The powdered metals are poured into the metal container through the pouring spout that is now crimped. The right photo shows the cross-sectional view of a turbine wheel that has been hot isostatically pressed. (*Courtesy Wyman Gordon Co.*)

equipment is rarely found in die-pressing facilities because of the large production runs required to amortize the equipment and tooling costs. In isostatic compaction, pressure is not affected by part size. Consequently, large powder-metal parts have been made, such as blocks 24 × 32 × 26 in. (60 × 80 × 63 cm), a 1700-lb (770-kg) tungsten crucible, and an oxide insulator 24 in. (60 cm) in diameter and 70 in. (175 cm) long.

The chief limitation of isostatic pressing is that compacts with close dimensional tolerances cannot be made.

Near Net Shape.   Near net shape is a more recent term used to describe isothermal forging–Gatorizing* and hot isostatic pressing (HIP).

In this process the powder is placed in a disposable metal container and sealed (Fig. 11-5). The sealed powder is heated to about 2050°F (1120°C) and forged at a very closely controlled strain rate in dies that are also heated to the same temperature. A vacuum chamber is required around the die and the press platen to keep them from oxidizing. Conventionally pressed and sintered parts that have 90 to 94% density can be fully densified by the HIP process.

Figure 11-6 illustrates the difference between isothermal forging–Gatorizing and conventional forging.

Gatorizing has been used to produce large quantities of turbine components for jet engines out of high temperature alloys such as René 95 and Ti–6A1–4V. Nondestructive inspection has shown an extremely low rejection rate. The powdered metal isostatically hot-pressed forgings are used in the new generation of jet engines that burn hotter and produce more thrust.

Sintering.   After compacting, the *green compact,* as it is often referred to, is considered fragile but can be handled. To achieve strength and hardness, it must be sintered in a vacuum or in an atmospherically controlled furnace. Time and temperature in the furnace are closely regulated to achieve the desired properties.

* Gatorizing—tradename of Pratt & Whitney Aircraft Company.

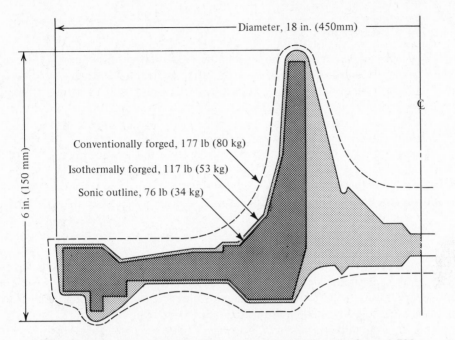

Diameter, 18 in. (450mm)

6 in. (150 mm)

Conventionally forged, 177 lb (80 kg)

Isothermally forged, 117 lb (53 kg)

Sonic outline, 76 lb (34 kg)

Figure 11.6. A comparison between a conventionally forged PM part and an isothermally forged part. The part is a turbine disc for Pratt & Whitney's F-100 engine. (*Courtesy Metal Progress.*)

Spark Sintering.   A newer approach is that of spark sintering. In this process the powders are hot-pressed and a combined ac and dc current is passed directly through the powder charge and the mold that contains it (if it is a conductor). At the same time, the powder is compacted with a programmed variation in force that minimizes the time required to bring the powder charge to a sintered density. The voltage varies, starting with 10 v and decreasing to 5 v during the 5-min cycle required to produce the part. A total current flow (ac and dc) of 1500 amp per sq in. is also typical. The use of both alternating and direct current speeds up the bonding process. The average part is compacted at a starting pressure of 400 to 500 psi (28 to 5.5 MPa), which rises to 2000 psi (13.79 MPa) at the end of the cycle.

There are several advantages in the use of spark sintering. Dies are relatively inexpensive. For example, a steel die for a given part was estimated at $1000. The same die in graphite as used in spark machining cost $50. Large parts can be made, such as a 135-lb (60-kg) iron pressing and a 32-lb (14.5-kg) beryllium pressing. The cycle time is short, in the case of the large pressings mentioned, the time was 30 to 45 min per part.

Spark sintering does not compete with traditional high-volume production of small iron, copper, aluminum, or cemented-carbide parts. It is particularly good for larger parts that must have a higher density than that produced by conventional powder-metallurgy methods.

Secondary Operations

Many powder-metal parts may be used in the "as-sintered" condition. However, when the desired surface finish, tolerance, or metal structures cannot be obtained at this stage, additional finishing operations are used, such as sizing, coining, sinter-forging, machining, impregnation, infiltration, plating, and heat treatment.

Sizing.   When a part must meet close tolerances, sintering is followed by a sizing operation. The sintered part is placed in a die and repressed.

Coining.   Coining is similar to sizing except the part is repressed in a die to reduce the void space and impart greater density. After coining, the part is usually resintered for stress relief. Often sizing and coining operations are combined in the same die.

Machining.   The principal object of powdered metallurgy is to produce a product that is dimensionally accurate. However, certain features such as threads, reentrant angles, grooves, and side holes are usually not practical. These features are generally machined on the sintered blanks.

All the conventional metal-cutting operations are performed easily on sintered metal parts. Very sharp tools and fine feeds are necessary for PM filters to maintain the open pore structure and for self-lubricating bearings.

Impregnation.   When self-lubricating properties are desired, as in "lifetime bearings," the sintered parts are impregnated with oil, grease, or other lubricants. The parts are placed in tanks of specified lubricants and heated to approximately 200°F (93°C) for about 10 to 20 min. The lubricant is retained in the part due to capillary action until pressure or heat draws it to the surface.

Infiltration.   An infiltrated part is made by first pressing and sintering the metal powder to about 77% of theoretical density. For an iron-powder part that is to be infiltrated with copper, it is sent through the furnace a second time with a copper blank placed on it. The copper melts and soaks into the porous structure, producing close to 100% density. The process also provides increased strength, hardness, and corrosion resistance.

Plating.   Prior to plating, PM parts are peened, tumbled, or given other treatments that will make the surface smooth and dense. Plating may get into the granular structure of the part and cause a galvanic action to be set up; therefore it is better to impregnate the part first. Plastic resins are usually used for impregnation as they have a low coefficient of expansion, a relatively low cost, and good filling properties, and they do not react galvanically with the metal. After impregnation, regular plating procedures are used.

Heat Treatment.   Just as with wrought or cast metals, PM parts are heat-treated to improve grain structure, strength, and hardness. Conventional heat-treating steps can be used but care must be taken in several steps of the process. Porosity decreases heat conductivity, and therefore longer heating and shorter cooling periods are required. A controlled atmosphere or vacuum furnace must be used.

Properties of Powdered-Metal Parts

In the past, the deficiency of engineering data for sintering metals fostered some doubts about the predictability of their mechanical properties and other design data.

The mechanical properties of certain powdered metals are shown in Table 11-1. In general, sintered alloys have impact strengths and ductility comparable to those of ferrous castings but somewhat less than that of wrought metals.

Figure 11-7 shows a comparison between high-strength sintered and heat-treated aluminum and iron PM parts. Aluminum premixes with 90 to 95% den-

Table 11-1. Mechanical properties of PM parts. (*Courtesy* Machine Design.)

Material	Density (g/cm^3)	Tensile Strength (psi)	Elongation (%)
90-10 bronze	6.4–7.2	14,000–20,000	1–3
80-20 brass	7.2–8.0	20,000–37,000	10–21
Nickel silver	7.2–7.9	25,000–42,000	15–14
316 stainless steel	6.2–6.6	38,500–60,000	2–10
601AB aluminum	2.42–2.55	20,000–34,500*	6–2
201AB aluminum	2.50–2.64	29,200–48,100*	3–2

* Heat treated

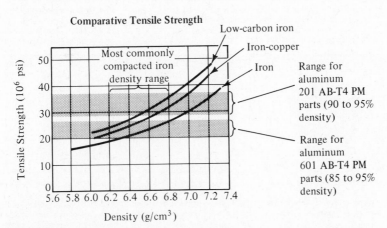

Figure 11.7. The tensile strength and densities of commonly compacted iron powders compared with aluminum. (*Courtesy* Machine Design.)

sity are shown to be comparable to those of iron premixes with densities of 6.8 g/cm³.

In recent years, tool steels have been made by the PM process. Conventionally produced high-alloy tool steels have a tendency toward carbide segregation because of the slow cooling of the ingots. The segregation persists even after hot-working of the ingot and billet, causing undesirable effects in tool performance.

PM tools provide very fine carbide size and complete homogeneity in bar stock and in the final tools. Tools include spade drills, knife blades for gear cutters, reamer blades, and cutting-tool inserts.

Hot-Forming.   Hot-forming or inter-forging is a comparatively new PM process, in which the part is placed in a furnace and heated to the upper critical temperature and then placed in a forging die where a forging blow forms the part with fine detail (Fig. 11-8). Forging may be accomplished in one blow if the preform has adequate porosity and shape.

Some advantages of hot-forming of PM parts may be listed briefly as follows:

1. The number of normal forging steps are reduced; ideally only two dies are needed: one for compacting and one for forging.

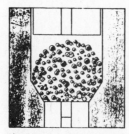

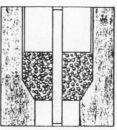

Metal powder—blended to any alloy depending on strength and density requirements—is loaded into compacting press.

Powder is compacted. The resulting preform is in the "green" state.

Preforms are sintered and coated with a lubricant. Up to this point, process parallels conventional PM compacting.

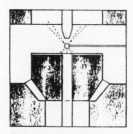

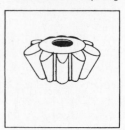

Preforms are reheated to hot-forming temperatures in a forming press. Lubricant is applied to the forming dies.

Preform is struck in the hot-forming press. Metal flow in the dies gives optimum strength and density throughout the finished part.

Finished part after hot forming can be as strong as parts produced by any other process. As-formed parts are accurate and uniform.

Figure 11.8.   The basic steps used in hot-forming powdered-metal parts. (*Courtesy* Machine Design.)

Figure 11.9. Gear cutting and broaching are eliminated by hot-forming this intricate gear. Material displacement during hot-forming is planned to distribute high density and strength where it is needed most. (*Courtesy PM Equipment and Powder Metal Products, Inc.*)

2. Forging pressures are less, thus smaller presses can be used.
3. The forging temperature is lower than for conventional forging.
4. Less skill is required at the forging press or hammer.
5. Secondary operations are eliminated which are necessary in conventional forging, such as removal of flash and machining. The dies can be made with little or no draft. Figure 11-9 illustrates finished gears made by this process.
6. Tooling costs are lower due to lower temperature and pressure requirements.

Advantages.   Perhaps the most unique advantage of the PM process is the precise control that can be exercised over the powders. This permits variation in physical and mechanical properties while assuring consistent performance characteristics.

Practically any desired alloy or mixture of metals, including those not available in wrought form such as tungsten carbide, can be produced. If need be, a single part can be made hard and dense in one area and soft and porous in another.

Parts can be produced in a variety of shapes that include irregularly shaped holes, eccentrics, splines, counterbores, gear teeth, etc., that in most cases require no machining (Fig. 11-10). As an example, a bore in a gear may be molded to a diameter between 0.5870 and 0.5875 in. and no machining is necessary.

Figure 11.10. A variety of shapes may be produced by the powder-metallurgy process to tolerances that require no machining. (*Courtesy Wickes Engineering Materials.*)

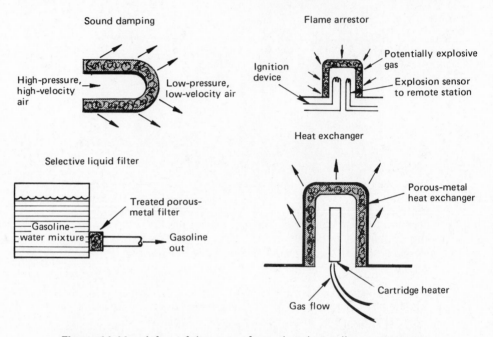

Figure 11.11. A few of the uses of powdered-metallurgy structures of the porous type.

The self-lubrication through a network of small infiltrated pores can simplify design and ensure trouble-free maintenance.

The controlled porosity of the PM process is also important in the creation and application of filters. These filters may be used to separate or selectively diffuse the flow of gas or liquids, dampen sound, act as a flame arrestor, serve as a heat exchanger, etc., as shown schematically in Fig. 11-11.

The excellent damping characteristics of PM parts make them important components in dictating machines, business machines, air-conditioning blowers, etc.

The more recent development of hot-forming has brought the PM process as a reasonable alternative to several other manufacturing processes, including shell-mold casting, lost-wax casting, and die casting (especially the newer ferrous-die casting that eliminates machining, gear cutting, and forging).

If forging requires considerable secondary machining, then PM hot-forming may offer a cost advantage. Hot forming can produce a wide range of mechanical

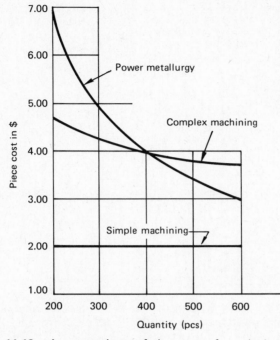

Figure 11.12. A comparison of the cost of producing parts by machining or by powdered metals.

properties all the way up to ultimate values. Tensile strength, impact resistance, and elongation can be made to match equivalent properties of forgings.

Forgings may involve four to five die setups, whereas the same details can be formed with one strike in PM hot-forming.

Limitations.   Initial tooling costs are relatively high. Production volumes of less than 10,000 identical parts are normally not practical. However, there are some exceptions. There are times when even 50 pieces may prove economical, depending upon the design. The design may require characteristics easily produced by the PM process but costly by any other process.

A graphical comparison of the break-even point between PM and complex machining is shown in Fig. 11-12. Complex machining involves numerous critical dimensions, skewed surfaces, compound curves, tangent radii, intricate surface features, etc.

Weak, thin sections should be avoided, as should feather edges and deep, narrow slots.

Corrosion protection requires special attention and precautions.

Spray Forging

A newer process combining the best features of traditional forging and powder metallurgy is spray forging.

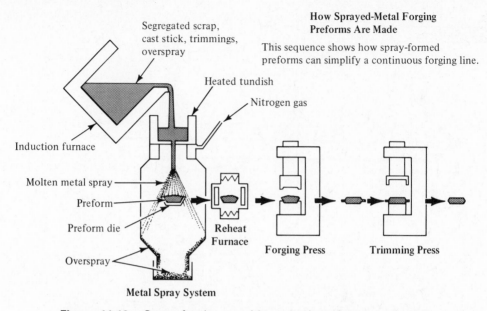

Figure 11.13. Spray forging combines the best features of traditional hot-forging and powder metallurgy. (*Courtesy* Machine Design.)

The Spray or Osprey Process. The spray-forging process was developed by Osprey Metals Ltd., of the United Kingdom. In this process, the metal, flash, overspray, scrap parts, and cast sticks are melted in an induction furnace and are then transferred to a heated tundish. The molten metal, streaming from the tundish, is atomized in high-velocity nitrogen jets that direct it into shaped molds or dies. The particles weld together on impact, forming a dense preform shape that is ready for forging.

The rate of metal spray is about 25 lb/min (11.32 kg) regardless of the preform shape or weight. The amount of overspray (Fig. 11-13) is a function of the symmetry of the mold receiving the metal. For symmetrical parts, the overspray is about 30%. The overspray is collected, either as a dense agglomerate or as a powder, and is returned to the furnace for recycling. It is not oxidized since it was atomized in nitrogen.

Advantages and Limitations. Any metal that can be atomized can be spray formed; however, the metals most often used are carbon steel, alloys, stainless, and tool steels.

The sprayed preforms are 95 to 99% dense, and the porosity, which is uniform throughout the part, does not cause internal oxidation.

The preforms can be forged, with one stroke of a press, immediately after forming or later, with or without additional heating.

Reduction in metal-processing costs are claimed to be as high as 50% of the costs of conventionally forged parts. The initial plant built in 1976 in England had a range of 2 to 3 lb at about 400 pieces/hr. However, the facility can handle preforms of up to 15 lb (6.8 kg).

Questions

11-1. Why is a plastics injection molding machine not feasible to use when making PM preforms?

11-2. What are the principal advantages of making high-speed tool steel by the PM process rather than by the conventional process of casting as an ingot and subjecting it to a series of hot deformation operations to eliminate the as-cast microstructure?

11-3. Would small (micron size in diameter) slivers of metal be practical for powder metallurgy? Tell why or why not.

11-4. What is the main difference between hot-forming and hot spray forging?

11-5. A 0.500-in. (12.70-mm) dia. fastener is subjected to a tensile load of 50,000 psi (344.7 MPa). Could this part be made out of a PM aluminum alloy? Why or why not?

11-6. What advantage would PM bearings have over a standard roller bearing?

11-7. The fastener mentioned in Question 11-5 measured 2.010 in. (51.05 mm) long in the 0.5-in. (12.70-mm) dia. section after loading and 2 in. (50.80 mm) before loading. Would this elongation be permissible if the aluminum PM were used. Tell why or why not?

Case Study

Company X had produced small-investment cast-connecting rods as shown in Fig. CS11.1. They decided to investigate the possibilities of making the same part by powdered metallurgy, which could increase production considerably and eliminate machining operations.

The properties desired would be a 30-ksi tensile strength and a 3% elongation. Assume the part shown has closely fitted pins and is subjected to a tensile load from these pins. Would a PM structure be satisfactory?

Typical PM Part

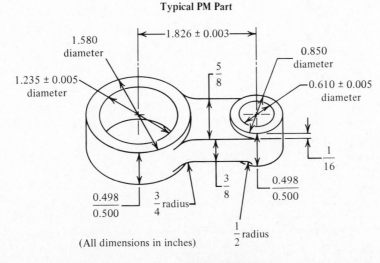

(All dimensions in inches)

Figure CS11.1.

Bibliography

Aluminum Powder Metallurgy. Pittsburgh: Aluminum Company of America, 1970.

Creating with Metal Powders. Riverton, N.J.: Hoeganaes Corporation, 1971.

Dreger, Donald R., ed. "Strong, Low-Cost Forgings from Sprayed-Metal Preforms." *Machine Design* 48 (August 12, 1976):86–87.

Hirchhorn, J. S. *Introduction to Powder Metallurgy.* New York: American Powder Metallurgy Institute, 1969.

Mocarski, S., and D. W. Hall. "High Temperature Sintering of Ferrous Powder Parts Substantially Improves Properties." *Metal Progress* 116 (December 1979): 50–56.

Powder Metallurgy Design Guidebook, 11th ed. New York: American Powder Metallurgy Institute, 1980.

CHAPTER TWELVE

Arc-Welding Processes

Today's manufacturing engineer must be able to assess the value of a wide variety of materials-joining processes. Tradition may dictate a process such as welding or brazing, but modern technology may show that adhesives or mechanical-joining methods would be more efficient or that a combination of adhesives and mechanical fasteners would serve more effectively.

In the next several chapters, materials-joining processes will be discussed. As shown in the diagram (Fig. 12-1) welding is only one of three main methods of materials joining.

Even though welding dates to the age of the Pharoahs, modern welding technology is a relatively young science that began at the turn of the century.

Forge welding, the oldest process, was in all likelihood discovered by goldsmiths, who found they could join pieces of gold by hammering them together, or forge welding. The process carried over into iron and helped build the Colossus at Rhodes in 280 B.C. The framework covered with bronze sheets was made of iron and, according to the mathematician Philo, showed signs of having been joined in suitable places by "hammering of Cyclopean force."

For centuries the only method available for joining metals metallurgically (fusion welding) was forge welding, in which heated metals were pounded or

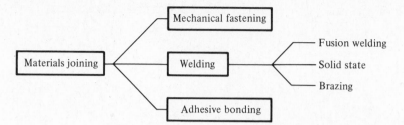

Figure 12.1.   A broad classification of materials-joining processes.

rammed together. Smithing, die welding, and roll welding are mostly processes of the past except for a resurgence of ornamental iron work.

Welding as we think of it today began with the almost concurrent development of three new processes just before 1900: arc welding, resistance, and oxyacetylene welding. Each process represented a new source of thermal energy. As different joining needs arose, new variations of basic processes, as well as totally new concepts, were developed, and today there are more than 50 separately definable welding processes in use. Figure 12-2 is a master chart of these welding processes.

Each of the many welding processes has a significant advantage for a particular application. The next several chapters will provide background information that will help today's manufacturing engineer recognize the individual advantages of the various materials-joining processes and, together with practical experience, be able to make a better selection.

Welding is defined by the American Welding Society (AWS) as a "localized coalescence of metals or nonmetals produced by either heating of the materials to a suitable temperature, with or without the application of pressure, or by the application of pressure alone, and with or without the use of filler metal."

This is a broad statement to cover all processes; however, in this chapter only *fusion* welding will be discussed. In fusion welding coalescence is achieved by melting together either filler metal and base metal or base metal alone.

Specifically, the arc-welding processes discussed in this chapter are as follows:

Shielded metal-arc welding (SMAW), also referred to as stick electrode welding.

Flux-cored arc welding (FCAW).

Gas metal-arc welding (GMAW), formerly known as MIG (metal inert-gas) welding.

Gas tungsten-arc welding (GTAW), formerly known as TIG (tungsten inert-gas) welding.

Plasma-arc welding (PAW).

Submerged-arc welding (SAW).

Electroslag welding (ESW).

Stud-arc welding (SW).

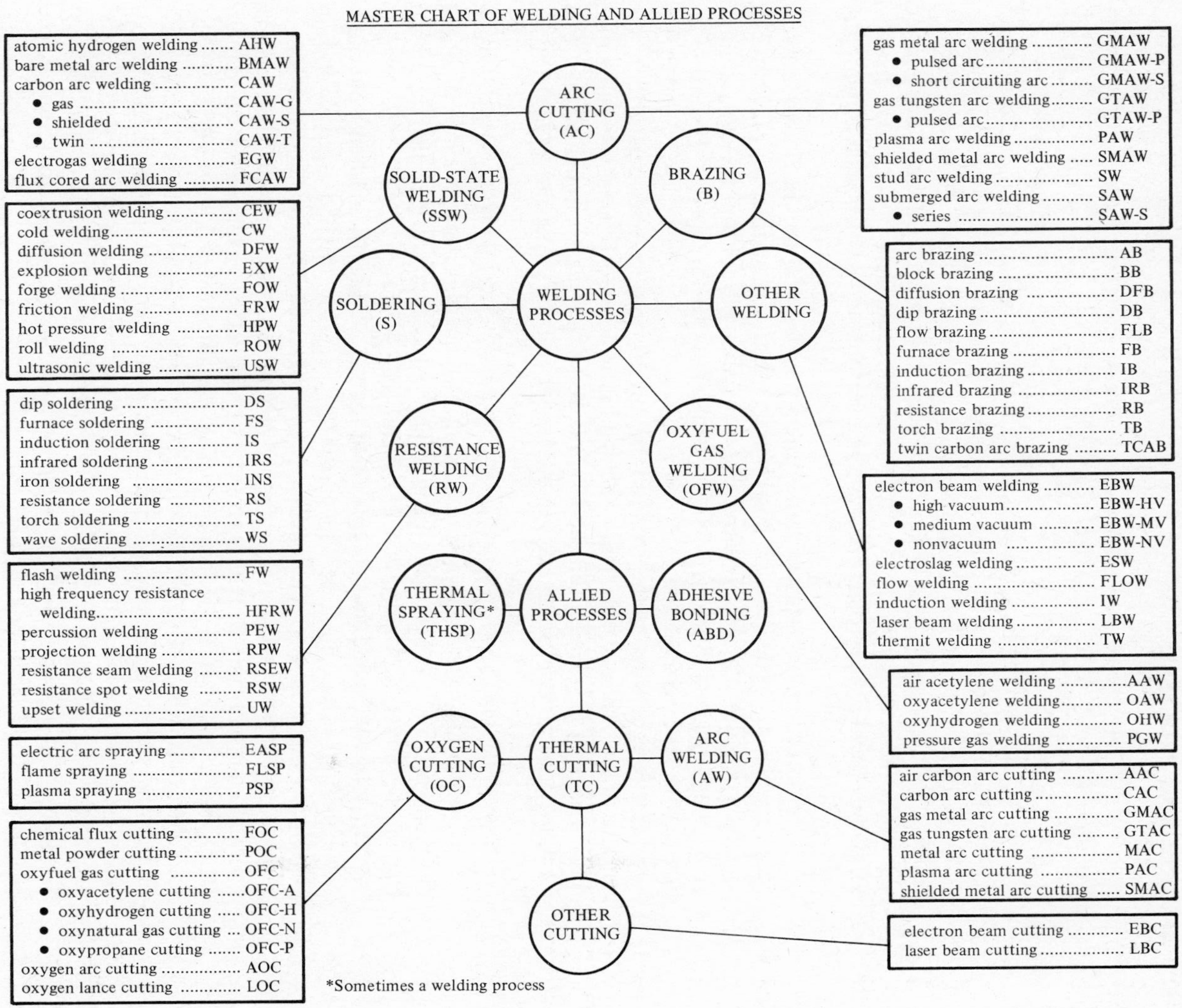

gas metal arc welding .... GMAW
• pulsed arc .... GMAW-P
• short circuiting arc .... GMAW-S
gas tungsten arc welding ..... GTAW
• pulsed arc .... GTAW-P
plasma arc welding .... PAW
shielded metal arc welding .... SMAW
stud arc welding .... SW
submerged arc welding .... SAW
• series .... SAW-S

arc brazing .... AB
block brazing .... BB
diffusion brazing .... DFB
dip brazing .... DB
flow brazing .... FLB
furnace brazing .... FB
induction brazing .... IB
infrared brazing .... IRB
resistance brazing .... RB
torch brazing .... TB
twin carbon arc brazing .... TCAB

electron beam welding .... EBW
• high vacuum .... EBW-HV
• medium vacuum .... EBW-MV
• nonvacuum .... EBW-NV
electroslag welding .... ESW
flow welding .... FLOW
induction welding .... IW
laser beam welding .... LBW
thermit welding .... TW

air acetylene welding .... AAW
oxyacetylene welding .... OAW
oxyhydrogen welding .... OHW
pressure gas welding .... PGW

air carbon arc cutting .... AAC
carbon arc cutting .... CAC
gas metal arc cutting .... GMAC
gas tungsten arc cutting .... GTAC
metal arc cutting .... MAC
plasma arc cutting .... PAC
shielded metal arc cutting .... SMAC

electron beam cutting .... EBC
laser beam cutting .... LBC

ARC
CUTTING
(AC)

BRAZING
(B)

OTHER
WELDING

WELDING
PROCESSES

SOLID-STATE
WELDING
(SSW)

OXYFUEL
GAS
WELDING
(OFW)

ADHESIVE
BONDING
(ABD)

ARC
WELDING
(AW)

SOLDERING
(S)

RESISTANCE
WELDING
(RW)

ALLIED
PROCESSES

THERMAL
CUTTING
(TC)

THERMAL
SPRAYING*
(THSP)

OXYGEN
CUTTING
(OC)

OTHER
CUTTING

atomic hydrogen welding ...... AHW
bare metal arc welding .... BMAW
carbon arc welding .... CAW
• gas .... CAW-G
• shielded .... CAW-S
• twin .... CAW-T
electrogas welding .... EGW
flux cored arc welding .... FCAW

coextrusion welding .... CEW
cold welding .... CW
diffusion welding .... DFW
explosion welding .... EXW
forge welding .... FOW
friction welding .... FRW
hot pressure welding .... HPW
roll welding .... ROW
ultrasonic welding .... USW

dip soldering .... DS
furnace soldering .... FS
induction soldering .... IS
infrared soldering .... IRS
iron soldering .... INS
resistance soldering .... RS
torch soldering .... TS
wave soldering .... WS

flash welding .... FW
high frequency resistance
   welding .... HFRW
percussion welding .... PEW
projection welding .... RPW
resistance seam welding .... RSEW
resistance spot welding .... RSW
upset welding .... UW

electric arc spraying .... EASP
flame spraying .... FLSP
plasma spraying .... PSP

chemical flux cutting .... FOC
metal powder cutting .... POC
oxyfuel gas cutting .... OFC
• oxyacetylene cutting .... OFC-A
• oxyhydrogen cutting .... OFC-H
• oxynatural gas cutting ..... OFC-N
• oxypropane cutting .... OFC-P
oxygen arc cutting .... AOC
oxygen lance cutting .... LOC

*Sometimes a welding process

Figure 12.2. A master chart of welding processes. © 1979 by the American Welding Society.

ARC WELDING

Basically, arc welding is a process in which an electric arc is produced between a metal electrode or wire carrying high-amperage current and the workpiece (Fig. 12-3). Under the intense heat of the arc, ranging from 5000 to 10,000°F (2760–5537°C), a small part of the base metal is brought to the melting temperature. At the same time, the end of the metal electrode is melted and droplets of molten metal pass through the arc to the base metal—which is called *globular transfer*.

The preceding brief description of the arc is very basic. The flow of current between the electrode and the workpiece is better described as a column of ionized gas called plasma. Negative electrons are emitted from the cathode, and along with the negative ions of the plasma, flow to the anode. Positive ions flow in the reverse direction.

Heat is generated at the cathode, mostly by the positive ions striking the surface of the cathode. Heat at the anode is generated primarily by electrons that have been accelerated by the arc voltage as they pass through the plasma. The electrons give off their energy as heat when they strike the anode.

Arc Shielding.   An intense arc and filler metal is not enough to make a satisfactory weld. Metals at high temperatures react chemically with oxygen and nitrogen, the main constituents of air. The oxides and nitrides that form can change the physical properties of the weld.

A protective shield must be provided for the arc and the molten-weld pool. Arc shielding is accomplished in various ways. One method is to provide a coating on the electrode that vaporizes by the heat of the arc and provides a gaseous shield, as was shown in Fig. 12-3. Other methods are discussed later.

Shielding not only provides protection from the atmosphere for the molten metal, but it also affects the stability and metal-transfer characteristics of the arc. The extruded coating on the electrode supplies ingredients that react with deleterious substances, such as oxides and salts, and ties up these substances chemically into a slag. The slag is lighter than the weld pool and therefore it rises to the top and crusts over the newly solidified metal. Even after solidification the slag continues to protect the weld metal until it cools to the point where reaction with the air is negligible.

Polarity.   In conventional dc welding where the workpiece is the positive pole and the electrode is negative, the hookup is referred to as being straight polarity (dcsp) (Fig. 12-4). When the arrangement is reversed, it is referred to as reverse polarity (dcrp).

Polarity can be used to control the location of the liberated heat. Usually it is preferable to have more of the heat at the workpiece because that is the larger area. Thus if large deposits are to be made on heavy workpieces, straight polarity would be the most effective hookup. On the other hand, in overhead welding where the weld pool should be kept relatively small and a fast freeze helps hold the metal in place, reverse polarity would be best. Where it is nec-

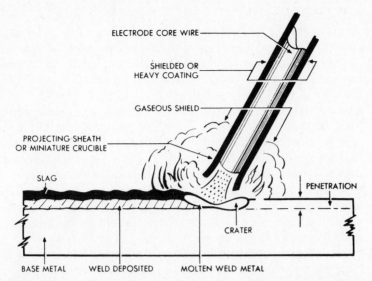

Figure 12.3. The welding action of a cellulosic-coated "stick electrode."

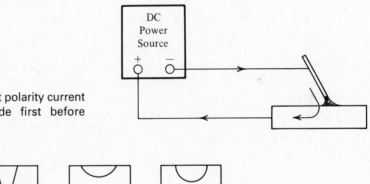

Figure 12.4. DC straight polarity current flows through electrode first before work.

Figure 12.5. The effects of polarity in welding.

essary to keep the workpiece as cool as possible, as in arc welding cast iron, reverse polarity is the best choice.

In general the use of dcsp will result in a higher deposition rate (more metal deposited per unit time) but less penetration. The converse is also true. The effects of polarity on bead width and penetration are shown in Fig. 12-5.

Current Density.   Current density is the amps per unit of electrode cross-sectional area. The deposition rate varies directly with current density, that is, the higher the current density, the higher the deposition rate.

The use of a consumable electrode generally results in higher thermal efficiencies and a narrower heat-affected zone than those obtained with nonconsumable electrodes because more heat is transferred directly into the weld pool. Typically, thermal efficiencies for metal-arc welding range from 75 to 80%, and for nonconsumable electrodes (usually tungsten), 50 to 60%.

Arc Blow.   Arc blow refers to the metal stream being deflected from its intended path. Rather than follow the shortest path from the electrode to the work, the metal is deflected forward or backward from the direction of travel, or less frequently to the side. Arc blow results in excessive spatter, incomplete fusion, reduced welding speed, and generally lowered weld quality.

Arc blow is caused by either magnetic or thermal conditions in the arc area. Magnetic blow comes from an unbalanced magnetic field surrounding the arc. Magnetic fields are set up in a continuous series of circles perpendicular to the current path. When the fields are unbalanced, the arc bends away from the greater concentration, resulting in arc blow. This condition exists particularly with direct current because the induced fields are constant in direction. It is only minor for ac because the magnetic fields collapse as the current reverses.

Thermal arc blow results from the physics of the electric arc, which requires a hot spot on both the electrode and the workpiece to maintain a continuous flow of current in the arc stream. As the electrode moves relative to the work, the arc will tend to lag behind because it is reluctant to move to a cold spot. With manual weld, thermal arc blow is not likely to become a problem, but with higher speeds of automatic welding thermal and magnetic arc blow are likely to become more frequent.

Methods of controlling arc blow:

1. Changing to ac current.
2. Reducing welding current and keeping arc length to a minimum.
3. Placing the ground connection as far as possible from the weld.
4. Wrapping the ground cable around the workpiece so that the current will flow to establish a second magnetic field that neutralizes the magnetic field of the arc blow.

Arc-Welding Energy Sources

Energy for arc welding is obtained by generators, transformers, or rectifiers as shown in Fig. 12-6. The main criteria is that there be a constant current supply. In arc welding, there is a great deal of fluctuation in current requirements. When the arc is struck, the electrode is essentially in short circuit, which would immediately require a sudden surge of current unless the machine is designed to prevent this. In addition, as the globules of metal travel across the arc stream, they tend to cause short circuiting. A constant current machine is designed to minimize these sudden surges.

Traditionally, power sources for metal-arc welding have a *drooping* volt-ampere characteristic as shown in Fig. 12-7. In drooping the terminal voltages

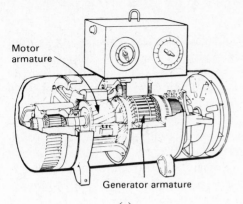

Motor armature

Generator armature

(a)

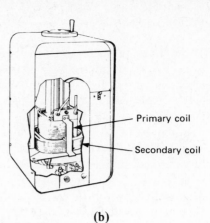

Primary coil

Secondary coil

(b)

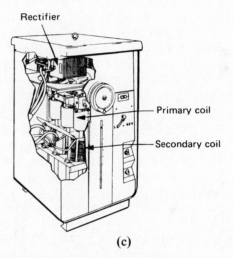

Rectifier

Primary coil

Secondary coil

(c)

Figure 12.6. The energy for arc welding is obtained by generators, transformers, or rectifiers.

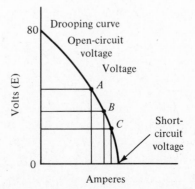

Figure 12.7. The amp-volt relationship (power curve) associated with stick electrode (SMAW).

455

of the machine decrease as the welding current increases. This type of current is necessary when welding manually with covered electrodes. Since the arc length is varied, it causes the voltage to change. The open-circuit voltage (no arc) is with the machine running at its maximum and represents usually 60 to 80 v. When the arc is operating under normal conditions, the voltage will be between 20 to 40 v. Point B indicates the optimum machine setting and arc length. Dropping vertically from this point intersects the amperage being used. Moving horizontally from B indicates the voltage being used. If the arc length is decreased, with no change in power setting, the amp-volt relationship is indicated by position C, increased amperage and decreased voltage. Position A shows the results of increased arc length.

This curve shows maximum amperage is available at 0 voltage and vice versa. If the electrode is short-circuited to the work, the amperage will go to maximum, which will heat up the entire electrode rapidly.

Not all processes use the "drooping arc voltage" curve shown here. Other current curves will be discussed later.

Alternating-current welders are theoretically both direct-current, straight-polarity (dcsp) and direct-current, reverse-polarity (dcrp). In some cases where moisture scale or oxides are present on the material welded, the flow of current in the reverse-polarity direction is reduced. To prevent this from becoming a problem, it is common practice by some manufacturers of welding equipment to provide for high-frequency, low-power current, as shown in Fig. 12-8. The high-frequency current jumps the gap between the electrode and the workpiece and pierces the oxide film, making a path for the welding current. Superimposing high-voltage, high-frequency current on the ac welding current has several advantages: (1) the arc can be started without actually touching the electrode to the workpiece; (2) a longer arc is possible, which is useful in some applications discussed later; (3) it is possible to use a wider range of welding currents for a specific diameter electrode; and (4) the arc is more stable.

Welding Machine Sizes.   Sizes of welding machines are designated according to their output rating, which may range from 150 to 1000 amp. The output rating has usually been based on a 60% duty cycle. This means that the power supply can deliver its rated load output for 6 min out of every 10 min. In manual welding particularly, a power source is not required to deliver the current continuously as in other electrical equipment. Fully automatic power-supply units are usually rated at 100% duty cycle.

It is possible to draw more than the rated output current from a power supply at reduced duty cycles. For example, a power supply rated to produce 100 amp on a 60% duty cycle might be operated at 150 amp on a 27% duty cycle. It is also possible to extend the duty cycle by drawing less than the rated current. For example, a power supply rated to produce 100 amp on a 60% duty cycle could be used at 77 amp on a 100% duty cycle.

Percentage duty cycle is defined as the ratio of the square of the rated current to the load current multiplied by rated duty cycle.

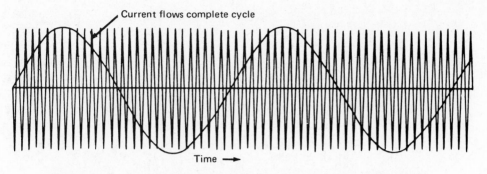

Current flows complete cycle

Time ⟶

Figure 12.8. A superimposed high-frequency current along with capacitors stabilizes the welding current and produces a balanced wave form. (*Courtesy Airco Welding Products.*)

$$\% \text{ duty cycle } = \frac{(C_r)^2}{(l_c)^2} \times \text{rated duty cycle}$$

where C_r = rated current
         l_c = current load

Example: A machine rated at 100 amp at 60%-duty cycle is accelerated to 150 amp (150% load). The output cycle would be reduced to 27%.

$$\% \text{ duty cycle } = \frac{(100)^2}{(150)^2} \times 60 = 27\%$$

Conversely, if the 100-amp 60%-duty cycle were to be operated at 100%-duty cycle, the output current should be reduced to 77 amp.

$$100 = \frac{(100)^2}{(l_c)^2} = 60\,(l_c)^2 = \frac{100^2}{100} \times 60 = 6000 \quad l_c = 77 \text{ amp}$$

The size and type of energy source is dependent on the average range of work to be done. For light and medium work, and general repair and maintenance, a 150- to 200-amp machine is usually suitable. For average production work, and plant maintenance and repair, a 250- to 300-amp machine is used. For large, heavy-duty, structural-type welding, machines of 400 to 1000 amp are used. Transformers and transformer rectifiers have grown in popularity and have replaced many of the motor generator sets for general use.

A new generation of solid-state power supplies are now available, generally in the form of transformers that step down ac line voltage to welding voltage and a controllable semiconductor bridge to convert ac to dc. The bridge includes diodes and SCRs (silicon-controlled rectifiers). A solid-state diode is a two-terminal device for current rectification. It conducts current in one direction

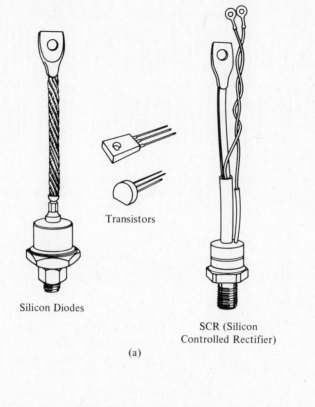

Transistors

Silicon Diodes

SCR (Silicon
Controlled Rectifier)

(a)

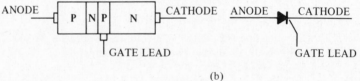

(b)

Figure 12.9. Typical semiconductors (a) and a silicon-controlled rectifier and schematic symbol. (*Courtesy Power Publications, Appleton, Wis.*)

only when it is forward biased, with the anode electrically positive to the cathode (Fig. 12-9).

A silicon-controlled rectifier is a three-terminal device. SCRs are also called thyristors, a contraction of the word *thyratron*, a type of vacuum tube whose characteristics are similar to SCRs, and transistors.

Electrode Types

Electrodes used in manual welding are usually covered with a flux coating that vaporizes in the heat of the arc to form a protective gas (CO_2) as was shown in Fig. 12-3. This gas excludes nitrogen and oxygen from the molten metal, thus preventing the formation of undesirable oxides and promoting a smooth flow of molten metal. This form of shielding, now referred to as *shielded metal-arc welding* (SMAW) or *stick-electrode welding,* may consist of cellulosics or minerals or a combination of both.

Cellulosic Coatings. Cellulosic coatings derive their name from the cellulosic materials from which they are made, such as word pulp, sawdust, cotton, and various other compositions obtained in the manufacture of rayon. The expanding gas furnished by the burning cellulosic material acts to give a forceful digging action to the weld. Properly made, the electrode coating will burn just a bit more slowly than the core wire, forming a cup or crucible at the end of the rod that makes it easier to direct the arc. Cellulosic-coated electrodes can be used in any position, but are particularly useful in vertical, horizontal, and overhead positions.

Mineral Coatings. Mineral coatings are manufactured from natural silicates such as asbestos and clay. By adding oxides of certain refractory metals such as titanium, the harsh digging action of the arc is modified to produce one that is softer and less penetrating. This type of electrode is used to advantage when the fit-up is poor and on sheet metal where shallow penetration is desired. A large amount of slag is produced, which serves to protect and control the quality of the deposited metal as it cools. The heavy slag retards the cooling rate of the deposited metal, allowing gas to escape and slag particles to rise to the top. Cooling stresses are reduced and a more homogeneous microstructure results. Because of the large amount of slag produced, the mineral-coated electrode is used most advantageously on *downhand* welding. Downhand welding is used in making flat welds and those inclined up to 45°.

Iron-Powder Coatings. The use of iron powder has been a later addition to the electrode-coating field. The use of iron powder brought several desirable effects: after the arc is struck, the metal melts away, leaving a well-defined crucible at the end of the rod that effectively concentrates the heat and gives an automatically consistent arc length. In fact, the electrode may be just slowly dragged over the work since the coating will maintain the proper arc length. The slag that forms over the weld is often self-removing and the appearance of the bead is improved. It also furnishes more inches of weld per electrode because the iron in the coating goes into the weld. However, because of the heavy coating, there are about half as many rods per pound.

Low-Hydrogen Electrodes. In low-hydrogen electrodes the coating contains a high proportion of calcium carbonate or calcium fluoride. Cellulose, clays, asbestos, and other minerals that contain moisture are not used. This is to insure the lowest possible hydrogen content in the arc atmosphere. The low-hydrogen electrodes provide excellent ductility and have about a medium depth of penetration. Low-hydrogen electrodes operate best on dcrp, but some types can be used with ac. The electrodes were developed for welding higher-strength, high-carbon and alloy steels in which the ordinary electrodes are subject to underbead cracking. The underbead cracks occur in the parent metal near the line of fusion between the weld and the parent metal. It is an excellent maintenance electrode since it can usually produce sound welds in steels whose analysis is unknown. Low-hydrogen electrodes are specified as EXX15, 16, 18, and 28. Welding electrode specifications are discussed in the next paragraph.

Welding-Electrode Specifications and Selection

The American Welding Society (AWS) has twelve classes of covered electrodes for mild and low-alloy steel. Each series of numbers is prefixed by "E," which designates an electrode. The next two or three digits designate mechanical properties. The third (or fourth) digit designates applicable welding positions. The last digit is for electrode stability. Some common electrode designations are shown in Table 12-1. Note that EXXX2 and EXXX4 have an iron-powder coating.

Selection of electrodes is usually based on chemical composition, mechanical properties, and operating characteristics, in that order. Operating characteristics of electrodes have to do mainly with how rapidly the weld metal *fills* (deposition rate), how fast it *freezes* (for use in various positions), and how fast the electrode *follows* (welding speed). The type and position of the joint determines whether the electrode should primarily have fill, freeze, or follow characteristics.

Fill electrodes are used primarily for easy-to-weld joints in the flat position. Examples of fill-type electrodes are E6027, E7024, and E7028. The covering of these electrodes contains as much as 50% iron powder, which increases both deposition rate and current requirements.

Fast-freeze electrodes have relatively low deposition rates and are used for vertical and overhead or steeply inclined joints. Because their covering contains less than 10% iron powder, fast-freeze electrodes require less current than fill electrodes. Examples are E6010 and E6011.

Electrodes such as E7014 and E7018 can be classified as fill-freeze types since they have some of the characteristics of both classes.

Follow electrodes can be used at high travel speeds with a minimum of skips

Table 12-1. Some common mild and low-carbon steel electrode designations.

Class Number	Current	Arc	Penetration	Iron Powder % (by weight)
EXX10	dcrp	digging	deep	0.10
EXXX1	ac, dcrp	digging	deep	0
EXXX2	ac, dcsp	medium	medium	0–10
EXXX3	ac, dcsp	soft	light	0–10
EXXX4	ac, dcsp dcrp	soft	light	25–40
EXXX5	dcrp	medium	medium	0
EXXX6	ac, dcrp	medium	medium	0
EXXX8	ac, dcrp	medium	medium	25–40

and misses. Their principal use is on all types of joints in 10- to 18-gage (0.1345 to 0.0478-in. thick) sheet metal. See Appendix H for metric conversion factors. Typical follow electrodes are E6012 and E6013.

Examples of joint types along with the choice of electrode are shown in Fig. 12-10. Code designations for electrodes of three common materials are shown in Table 12-2.

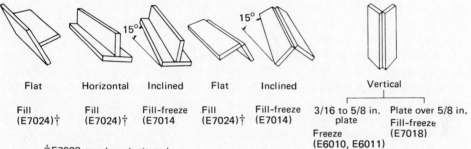

Fillet welds over 10 to 12 in. in length on 3/16 in. or thicker plate

| Flat | Horizontal | Inclined | Flat | Inclined | Vertical | |
| Fill (E7024)† | Fill (E7024)† | Fill-freeze (E7014) | Fill (E7024)† | Fill-freeze (E7014) | 3/16 to 5/8 in. plate Freeze (E6010, E6011) | Plate over 5/8 in. Fill-freeze (E7018) |

† E7028 may be substituted

Figure 12.10. Examples of joints and the corresponding selection of electrodes.

Table 12-2. Classification and marking of electrodes.

	E designates electrode	These digits designate minimum strength in ksi.		These digits designate weld metal composition.
Low-alloy	↓	↓		↓
E7018	E7018	E7018	E7018	
E9018B-3	E̲9018B-3	E9̲018B-3	E90̲1̲8B-3	E9018B-3̲
Stainless				
E308-15	E̲308-15		E308-15̲	E308-15̲
E316-16	E̲316-16		E316-1̲6	E3̲16-16
Nickel Alloy				
ENiCrFe-3	ENiCrFe-3			ENiCrFe-3̲
E9N10	E̲9N10			E9N10̲

These digits designate useability of the electrode—welding position and current:

Low-hydrogen Types

15 = all position DC
16 = all position DC or AC
18 = all position iron powder DC or AC

Flux-Cored Continuous Welding Wire

As arc welding grew in popularity, particularly as a production means of fabrication, it was only natural that methods would be studied to make the process automatic or at least semiautomatic. One disadvantage of the stick electrode is the need to stop and change it as it gets short. Continuous wire seemed to be the answer, but there were different approaches as to how the arc shielding could be accomplished. A continuous covered electrode was tried and was marketed for a short time, but was not well accepted. A more feasible approach was to put the flux on the inside of a hollow wire, as shown in Fig. 12-11. This process became commercially available in 1954. At that time a CO_2 gas was added for auxiliary shielding. There are two processes used today, a "nongas shield" and a "gas shield." The nongas shield is designed to develop sufficient gas shielding from the flux itself. The welds have somewhat lower ductility and notch toughness, but it has proven satisfactory for most mild-steel applications.

The most versatile energy source for semiautomatic, flux-cored wire welding is a 500-amp, dc, constant voltage machine as shown in Fig. 12-12. Many types of wire feeders are available, but the constant-speed type as shown is the most satisfactory and easiest to use. Generally $\frac{3}{32}$- or $\frac{1}{8}$-in. dia. (2.36 or 3.17-mm) wire is used and the operating conditions are 375 to 500 amp with approximately 32 v. Most of the semiautomatic flux-cored welding is done with air- or gas-cooled guns and cable, even though water-cooled guns and cable are available.

Fully automatic units have been developed that will feed one, two, or three flux-cored wires simultaneously. These multihead units are frequently used to give larger fillet welds as required in heavy bridge and girder work.

Advantages and Limitations. The main advantages of the flux-cored wire are the speed at which the weld metal can be deposited and the depth of penetration obtained. It is especially fast in downhand welding of fillets and grooves where heavy deposits are required. A deposition rate of 35 lb/hr (15.75 kg/hr) can be maintained on automatic machines with a 0.120-in. dia. (3.05-mm) electrode. This is about three times faster than that obtained with the fastest stick electrode.

The flux-cored wire offers better control over penetration, as was shown in Fig. 12-11. Not only is it deeper than with the manual electrode when desired, but also it has better control when poor fit-up is encountered. In this case, the operator allows the electrode to stick out of the gun farther. The longer "stick-out" automatically reduces the energy input into the arc, and thus into the heat that enters the base metal, and produces a higher preheat to the electrode. The immediate effect is to increase the deposition rate. Often a weld with a poor fit-up will require several passes with intermediate cleaning. With the added deposition rate, it can be made in one pass. The flux-cored wire is particularly useful where there is a large volume of work to be welded in the downhand or flat position.

The flux-cored wire has been limited to flat and horizontal welding. Without additional gas shielding, the welds are not quite as clean as with the coated electrode.

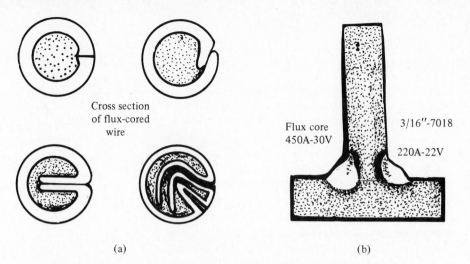

Cross section
of flux-cored
wire

Flux core
450A-30V

3/16"-7018

220A-22V

(a)

(b)

Figure 12.11. Some cross-sectional configurations of flux-cored wire (a). A fillet weld made with a ⅛ in. flux-cored wire at 450 amp compared to that of ³⁄₁₆ in. manual weld with a stick electrode at 200 amp (b). (*Courtesy Arcos Corporation.*)

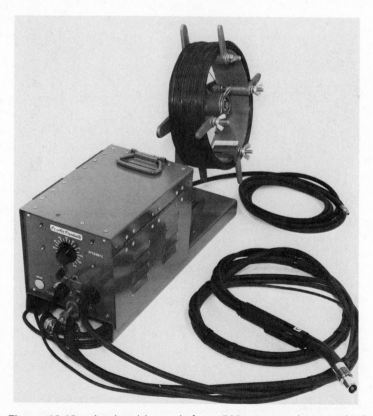

Figure 12.12. A wire drive unit for a 500-amp semiautomatic dc-cv power source. (*Courtesy Air Products and Chemicals, Inc.*)

Solid Continuous-Welding Wires

Welding wires are made for use in many kinds of automatic and semiautomatic machines. Mild-steel wires are lightly coated with copper for good electrical contact, for corrosion resistance, and for consistent operating conditions. The wires are packaged in spools of 4 in. dia. for hand-held guns and may range up to spools weighing 750 lb for larger machines.

Mild-steel wires for gas metal-arc welding are covered by AWS specification A5.18-65 and ASTM A559-65. The classification designation consists of the letter "E" for electrode, followed by two digits, 60 or 70, representing the minimum tensile strength as welded in ksi. Next there is one of three letters—S, T, or U—signifying solid-, composite-, and emissive-coated filler metals, respectively. A typical classification is E70S-1. This is read as: an electrode of 70 ksi minimum TS, of solid construction, to be used with argon plus 1 to 5% oxygen. The numbers following S and the dash indicate the shielding gas used as 1—AO; 2 and 3—AO and CO_2; 4, 5, and 6—CO_2.

Flux-Cored Electrode Classification.   Flux-cored electrodes are classified in AWS A5.29.80 specification as follows.

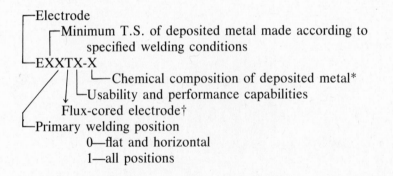

```
┌─Electrode
│   ┌─Minimum T.S. of deposited metal made according to
│   │     specified welding conditions
└─EXXTX-X
  ╱ │ │   └─Chemical composition of deposited metal*
 ╱  │ └─Usability and performance capabilities
╱   Flux-cored electrode†
└─Primary welding position
        0—flat and horizontal
        1—all positions
```

An example of a flux-cored electrode designation would be E81T1–B1, which represents 8–80,000 psi T.S.; 1—all positions; T—flux-cored electrode; 1—used with CO_2 or CO_2 and argon†, B1—Cr–Mo.

The methods of metal transfer through the arc are discussed later.

Stainless steel wires are available in diameters ranging from 0.035 to 0.300

---

* The chemical composition is divided into five groups designated by the following letters: carbon–molybdenum—A; chromium–molybdenum—B; nickel–Ni, manganese–molybdenum—D; all other alloy steels—K, G, and W. The letters are followed by numbers to designate various percentages of the alloy composition.

† T1—Co_2 or CO_2 and Argon shielding, larger diameters, $\frac{5}{64}$ (2.0 mm) are used for flat and horizontal position, small diameters, $\frac{1}{16}$ in. (1.6 mm) and smaller are used in all positions. T1 electrodes are characterized by spray transfer.

T4—Self-shielding, dcrp, globular transfer.

T5—CO_2 shielding or CO_2 and Argon mix, flat and horizontal position, globular transfer.

T8—Self-shielding, dcsp, all positions.

in. (0.90 to 7.62 mm) for gas-shielded metal-arc welding and in several sizes from 0.045 to 0.15625-in. dia. (1.14 to 3.96 mm) for submerged-arc welding. Stainless steel wires are used for joining chromium-grade stainless steels, stainless steel to mild steel, and for stainless steel overlay work.

Aluminum wires are available in many different aluminum alloys. The wire size ranges from 0.020 to 0.125 in. (0.51 to 3.17 mm) in diameter. These high-quality wires are prepared with extremely clean surfaces by a shaving process that removes the surface oxides. The wire coils are then individually foil-covered to prevent exposure in shipment and storage.

Wire Feeding

Arc length is critical. If the arc length is too long or the voltage too high the metal tends to melt off the end of the electrode in large globules that wander from side to side. This results in a wide spattered, irregular weld deposit or bead with poor fusion between the base metal and the filler. If the arc is too short or the voltage too low, there is not enough heat to melt the base metal properly, resulting in a high, uneven bead with poor fusion. In manual welding, the arc length, which should be about the diameter of the electrode, is controlled entirely by the skill of the operator.

Automatic and semiautomatic arc-welding systems use wire feeders to feed the wire continuously into the arc. Basically there are two wire feeding systems: constant speed and voltage sensing.

Constant-speed feeders operate on the basis of fixed burnoff rate of the electrode for a given current level and can therefore only be used with a constant-voltage-type welding-power supply. In operation the wire feed is preset and the power source supplies the amount of current needed to melt the wire at the proper rate to maintain the arc. These machines are termed *constant potential* (cp).

Voltage-sensing feeders sense the voltage across the arc and regulate the wire feed rate to compensate for changes in that voltage. These machines are termed *constant current* or *constant voltage* (cv). A comparison between volt-ampere curves of the constant current and the constant potential machines is shown in Fig. 12-13.

With a constant potential machine, as the wire feed increases a higher current will be demanded in order to maintain the constant voltage. Wire speed and welding current are therefore directly proportional.

Stated another way, when the wire is fed into the arc at a specific rate, a proportionate amount of current is automatically drawn. The cp welder provides the current required by the load imposed. When the wire is fed faster, the current increases; if it is fed slower, the current decreases.

On most gas metal-arc welding (GMAW) machines, there is a provision for *slope* control. As was shown in Fig. 12-13, the cp curve has a slightly downward slope. By altering the normally flat shape of the slope, it is possible to control the pinch force on the consumable wire, which is particularly important in the short-circuiting transfer method of welding as discussed later.

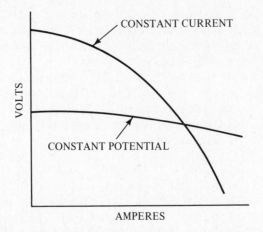

Figure 12.13. A comparison of the volt-ampere curves for constant current and constant potential machines.

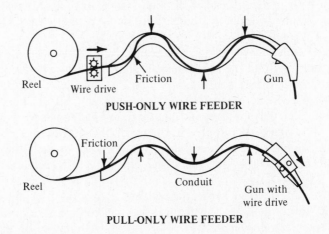

Figure 12.14. A comparison of two wire feed systems. (*Courtesy* Welding Design and Fabrication.)

Push or Pull Wire Feeders. Wire feeders may be either of the push type or pull type. In a push type, the electrode wire is pulled from the reel by feed rolls and then pushed through a flexible conduit to the contact tube and into the arc. Push-type wire feeders are usually mounted on the power supply. Both types of feeders are shown schematically in Fig. 12-14.

Soft, small diameter wire can cause foulups, often referred to as "bird-nesting." For this reason, pull-type systems with the feed rollers directly in the portable welding gun, as shown in Fig. 12-15, are now available. Their main disadvantage is that the added feed mechanism makes the guns large, heavy, and cumbersome.

Figure 12.15.   Portable hand gun for GMAW. (*Courtesy Air Products and Chemicals, Inc.*)

Arc-Welding Processes

The arc-welding processes most widely used are shielded metal-arc welding (SMAW), flux-cored welding (FCW), gas metal-arc welding (GMAW), gas tungsten-arc welding (GTAW), plasma-arc welding (PAW), submerged-arc welding (SAW), electroslag welding (ESW), and stud arc welding (SW).

SHIELDED METAL-ARC WELDING (SMAW)

SMAW, commonly called stick welding, is one of the oldest of the modern arc-welding processes. It is also the most widely used. Basically, it is a manual process in which the heat is generated by an arc that is established between the flux-covered consumable electrode and the work as was shown in Fig. 12-3. Typically, the electrode stick is 9 to 18 in. (228.60–457.20 mm) long.

Advantages.   SMAW is widely used because the equipment is relatively simple, very portable, and less expensive than other arc-welding equipment. All that is needed is an adequate power supply, a relatively simple electrode holder, and a set of cables as shown in Fig. 12-16. Welding can be done in virtually any position. The process is well adapted to making repairs even in out-of-position locations (other than flat).

Limitations.   The rate at which metal can be deposited is not as high as in certain other arc-welding processes. Welding must be interrupted each time an electrode is consumed, causing downtime and a wasted stub. Slag must be chipped and wire brushed from the weld surface. Electrode selection and care are more critical for welding hardenable steels than by some other processes. Softer metals such as zinc, lead, and tin, which have low melting and boiling temperatures, do not lend themselves to SMAW.

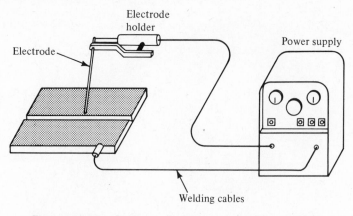

Figure 12.16.   Basic equipment required for SMAW.

FLUX-CORED ARC WELDING (FCAW)

FCAW is a relatively new development dating from the late 1950s. It was developed to overcome the deficiency of SMAW in which the weld had to be interrupted to replenish the electrode. Since a covered wire could not be coiled without breaking the covering, the solution was to put the self-shielding on the inside of the wire, as was shown in Fig. 12-11. The continuous wire led to the development of welding "guns" and fully automatic welding. Two FCAW processes developed—one with self-shielding, the other with auxilary-gas shielding. In the self-shielding process, all the protection for the weld deposit comes from the core compounds that generate a gaseous shield as the weld metal is being deposited.

Auxiliary-gas shielding can be provided by special equipment to provide extra protection during welding. The gas most often used is carbon dioxide and it is often referred to as the CO_2 process.

CO_2 helps provide a broad, deep penetration pattern with less spatter and less chance of surface porosity. The weld is also generally more ductile than if only self-shielding is used.

The basic equipment required for FCAW is a power supply, a wire feeder, an electrode holder, and, when appropriate, a means of supplying auxiliary shielding gas. If auxiliary shielding gas is used, a different type of electrode holder is needed, as shown in Fig. 12-17. The electrical contact tube on this welding gun extends nearly to the end of the nozzle so that the electrical stickout is nearly the same as the visible-electrode stickout (Fig. 12-18). As was mentioned previously, the amount of current that can be used for a given size electrode is limited; otherwise the wire will get too hot. With only a short, usually about $\frac{3}{4}$ to 1 in. (19.05 to 25.40 mm) stickout, this is not a problem.

For self-shielding the stickout is much greater, usually about $2\frac{1}{2}$ in. (63.50 mm) or more, since no provision has to be made for gas shielding. Therefore the gas-cup space at the end of the electrode holder can be occupied by an

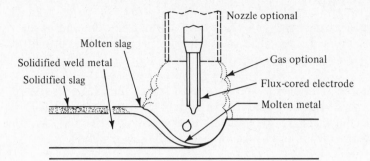

Nozzle optional

Molten slag

Solidified weld metal

Solidified slag

Gas optional

Flux-cored electrode

Molten metal

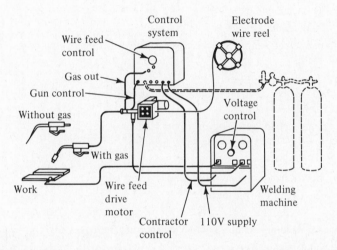

Control system

Electrode wire reel

Wire feed control

Gas out

Gun control

Without gas

With gas

Work

Wire feed drive motor

Contractor control

110V supply

Voltage control

Welding machine

Figure 12.17. A schematic of the flux-cored arc welding (FCAW) process and equipment.

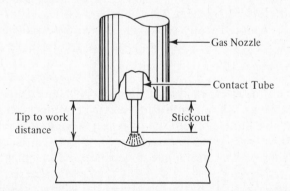

Gas Nozzle

Contact Tube

Tip to work distance

Stickout

Figure 12.18. The stickout is the length of the wire protruding beyond the nozzle.

insulated wire guide, which prevents the wire from touching the end or nozzle portion of the electrode holder. The use of longer electrical stickout allows a higher deposition rate since the wire will get more preheat.

FCAW Operating Variables.   The most used feeder for FCAW is the constant-speed type used with constant-voltage current supply. The principal variables that can be controlled are arc voltage, travel speed, and electrode stickout. Listed below are the results of changes in each of these variables:

Voltage variations:
Excessive—heavy splatter and porous welds.
Increasing—flattens and widens weld bead.
Decreasing—may cause convex bead.
Extremely low—causes electrode to "stub"; electrode goes through weld pool and sticks to base metal.
Current variations:
Increased—increased voltage can be used without porosity, and usually a good bead results.
Excessive—convex weld bead, resulting in a poor appearance.
Low—large droplet transfer, with a nonuniform bead.
Travel Speed:
Too high—convex bead, shallow penetration and uneven edges.
Too slow—weld high, rough uneven bead.
Electrode stickout:
Increasing—decreases welding current; decreasing stickout increases current.
Excessive—spatter and irregular arc.
Short—greater weld penetration than long stickout. Stickout that is too short causes spatter buildup on nozzle and contact tube.

A welding gun made for gas shielding may also be used for self-shielding when auxiliary gas is not required. Water-cooled guns have the disadvantage of requiring a continuous supply of water, which causes the gun to become quite complex.

Advantages.   Because the electrical contact can be made close to the work, high deposition rates can be achieved with small-diameter welding wires. As shown in Fig. 12-19, a 0.045-in. (1.45-mm) dia. wire can be fed at about 1000 in. (25.4 m) per minute.

FCAW can be used for a wide range of metal thicknesses, beginning as thin as $\frac{1}{16}$ in. (1.57 mm).

The self-shielding process is relatively simple and the welding gun is relatively light and easy to use. Because of the high deposition rate possible, the process is used extensively in building up worn surfaces either in carbon steel or stainless steel. The deposition rate is about twice that of SMAW for a comparable setup.

FCAW can be used in any position in the smaller diameter wires.

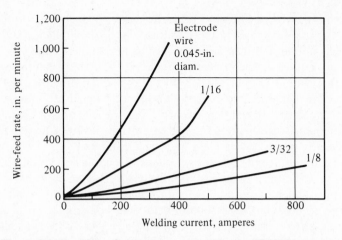

Figure 12.19. Wire feed rates used with the FCAW process.

Disadvantages.   The flux-cored wire emits quite a bit of welding fume during operations. In most cases the operator needs a gun that is equipped to draw off exhaust fumes and improve visibility, especially indoors.

GAS METAL-ARC WELDING (GMAW)

A bare, continuous wire is used and arc shielding is provided entirely by an externally supplied gas mixture. The shielding gases may be helium, argon, carbon dioxide, oxygen, or their mixtures. The equipment used is the same as that used for FCAW with auxiliary gas shielding.

GMAW Electrodes.   The electrode wire for GMAW is of high purity, closely controlled both chemically and in physical characteristics. Most of the wires are protected with a flash coating of copper. The selection of the wire depends on the weld position, service requirements, and condition of the material being welded.

Composite wires are available for welding alloy steels. These electrodes are similar in construction to the flux-cored electrode except that the inner core of the tubular electrode wire is filled with appropriate alloying material in the form of metal powder.

Shielding Gases.   Originally only inert gases, argon and helium, were used for shielding, hence the term MIG for metal inert-gas welding. Now carbon dioxide is used extensively, and oxygen and carbon dioxide are often mixed with inert gases. Generally argon, helium, or mixtures of these gases are used when welding nonferrous metals, CO_2 for steels, and CO_2 with argon and sometimes helium for steel and stainless steel, also argon and small amounts of oxygen for steels and stainless steels.

The chief factor influencing shielding effectiveness is the gas density. Argon is approximately one and one-third times as heavy as air and ten times as heavy as helium. Argon, after leaving the torch nozzle, forms a blanket over the weld area, whereas helium tends to rise in a turbulent fashion around the nozzle. To get the same equivalent shielding effectiveness, the flow of helium will have to be two to three times that of argon. On the other hand, helium has somewhat different electrical characteristics such that it makes for faster automatic welding and a hotter arc for wide and deep welds. Thus the higher available heat favors the use of helium over argon at the higher current densities as used on thicker plates, especially those of relatively high heat conductivity.

Both argon and helium provide excellent arc stability with dc power. However, with ac, which is almost exclusively used on aluminum and magnesium, argon yields excellent arc stability and good cleaning action while helium is slightly more erratic and the resultant weld-bead appearance is usually dirtier. Thus, with ac power argon finds wide acceptance, but helium is definitely preferred for overhead welding.

Carbon dioxide is the newest of the shielding gases. It produces a bowl-shaped weld penetration. A comparison of the type of bead obtained with argon, helium, and carbon dioxide is shown in Fig. 12-20. In welding, the carbon dioxide breaks down into carbon monoxide and oxygen. High deposition rates result when the carbon monoxide and oxygen recombine to add heat during welding. The carbon dioxide breakdown requires balancing deoxidizers in the filler metal to prevent porosity. To keep spatter within bounds and to get the best welds with carbon dioxide a short uniform arc is needed.

CO_2 has become widely used in welding of steel by the short-circuiting arc (discussed in next paragraph) as a mode of metal transfer. A true spray arc is not obtained with a shielding gas composed entirely of CO_2. In many cases, CO_2 provides good welding speed and good penetration and has proven to be less expensive than argon-shielded methods.

Metal-Transfer Modes.   Metal is transferred across the arc mainly by three modes: short circuiting, globular transfer, and spray transfer.

In regular GMAW, molten metal is transferred across the arc in the form of drops. The size of the drops, the number per unit time, and the direction they travel, other than in flat welding, is determined by the magnitude of the current and the gas used.

For currents ranging from the lowest value at which a stable arc can be maintained, called *transition current,* the drops continue to grow in size until they are several times the wire diameter size before they are detached at the rate of several drops per second. As they grow, they wobble around and disturb the plasma arc so that it moves around the work. Below the transition current the drops may grow large enough to cause a short circuit with attendant blasts of liquid metal.

At the transition current, the size of the detached drops decreases abruptly to equal or smaller than the electrode diameter, and the rate of detachment increases to several hundred per second. Above the transition current, the rate of detachment increases as the current increases but at a very slow rate compared

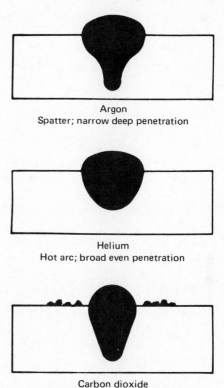

Argon
Spatter; narrow deep penetration

Helium
Hot arc; broad even penetration

Carbon dioxide
Stable arc; finger penetration

Figure 12.20.   A comparison of the effect of various shielding gases on the weld bead. (*Courtesy* Welding Design and Fabrication.)

with the current. The plasma-cone size and penetration into the metal also increase.

The filler metal transfer below the transition current is called *globular* transfer; above the transition current, it is called *spray* transfer. The term spray was adopted because the metal emanating from the tip of the electrode looks like a spray of water from a nozzle.

In spray transfer, the electromagnetic force is strong enough to propel the metal in line with the electrode tip regardless of the position in which the electrode is pointed. Weld spatter is minimized and weld-deposit efficiencies as high as 98% can be achieved.

Owing to the high current level and resulting high heat input, spray transfer does not lend itself to welding thin metals. In addition, overhead welding becomes difficult because of the large weld puddle formed.

To help overcome these difficulties, *pulsed-spray transfer* was developed. The pulsed-spray method provides a dual-level welding current (Fig. 12-21) generated by a special power source. A high level of current, or "pulse-peak" current, is provided, which is above the nonpulsing transition current, after which the current falls below the average transition arc current. The metal is transferred

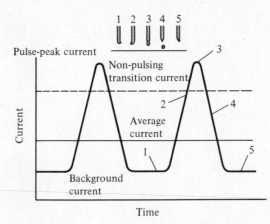

Figure 12.21. Pulsed-spray transfer sequence.

during the pulse when most of the energy for melting the work and the electrode is supplied.

The pulsed-arc mode is capable of welding thinner sections than are practical with conventional spray transfer because the heat input is less. Also out-of-position welding is feasible because the spray transfer can fight gravity and the pulsed sequence keeps the weld pool from becoming too large.

GMAW Power Source and Wire Feeders. Most GMAW is done with dcrp because it produces deeper penetration than straight polarity. Also straight polarity tends to produce an unstable arc with considerable spatter. It is sometimes used when penetration must be at a minimum. Constant voltage power supplies are generally preferred.

Although various types of wire feeders are available for GMAW, the most popular is the constant-feed type because it matches the constant voltage power supply. Since small-diameter wire is often used, a pull-type feed is best.

Advantages and Limitations. The main advantage of GMAW welding is that it can produce high-quality welds at high speeds and there is no flux to remove. It is a very versatile process that is used on both light- and heavy-gauge structural plates. It is used with comparative ease on carbon steels, low-alloy steels, stainless steels, aluminum, magnesium, titanium, copper, nickel, and zirconium. With the use of the pulsed spray-arc a relatively cool weld pool is produced, reducing burn-through on thin materials and also making it possible to weld in all positions.

With spray transfer, heavy wire electrodes will melt readily, making deep penetration possible. Since the individual droplets are small, they can be easily directed as required. Since the metal transfer is produced by an axial force that is stronger than gravity, the process can be effectively used on out-of-position welding.

GAS TUNGSTEN-ARC WELDING (GTAW)

GTAW is commonly referred to as TIG since it was originally tungsten inert-gas welding. As with GMAW, shielding is provided by an external gas supply. Filler metal may or may not be used depending upon the requirements of the job.

Gas tungsten-arc welding is quite similar to GMAW except that the electrode is like a torch that produces the necessary heat for the weld (Fig. 12-22). If the work is too heavy for fusion of the abutting edges, and if groove joints or reinforcement are required, filler metal is added manually or by a wire feed system.

Tungsten is used as the nonconsumable electrode because it has the highest melting temperature of all metals. It is also a strong emitter of electrons, which help ionize the arc path and thus generate a stable arc. Commercially pure tungsten may be used as the electrode but tungsten electrodes alloyed with thoria or zirconia are used to stabilize the arc and provide easier starting.

The tungsten electrode end may be pointed, or completely hemispherical, as a bulb. The pointed end is ideal for welding in restricted locations. The hemispherical end profile electrode will handle the greatest current density because less of the electrode is in contact with the arc. Thoriated and zirconiated tungsten electrodes hold their shaped ends better than pure tungsten over a large temperature range.

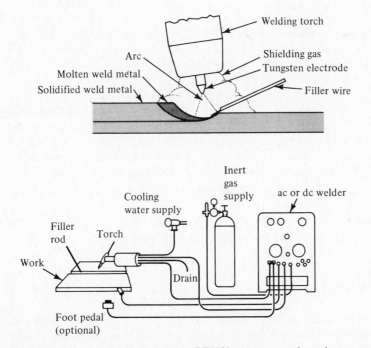

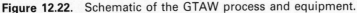

Figure 12.22. Schematic of the GTAW process and equipment.

When welding mirrorlike surface finishes, reflected radiation reduces the current-carrying capacity of the arc. If the current range is too low for a given electrode, the arc will wander over the end of the electrode. Excessive current causes the tip of the electrode to ball up and to vibrate at high frequencies. When this happens, tungsten starts to transfer across the arc. Such contamination of the weld puddle may require removal before welding can be resumed.

Shielding and Power Supplies. Argon is the most used shielding gas for GTAW, although helium may be used or a mixture of argon and helium. Argon is preferred because it is more readily available; and because it is heavier than helium, it provides better shielding at low flow rates. Argon also produces a "softer" arc, which is smooth and stable. Helium is used when fast hot-welding is desired and it is ideal for overhead welding since it is lighter than air.

Direct-current straight polarity (dcsp) is most often used for carbon and stainless steels since 70% of the heat in all dc arcs is generated in the anode. This allows the electrode to carry maximum current levels.

Pulsed-current power supplies are also used for GTAW. These provide a rapid current rise and decay and a high- or low-pulse repetition rate. The advantages of the pulsed-current GTAW are listed as follows:

1. Increased depth-to-width ratio of weld beads.
2. Reduction of drop-through. With high-current and short-duration pulses the root or bottom pass of a weld will melt the metal and it will solidify before the puddle becomes large enough to sag (drop through).
3. Minimal heat-affected zone.
4. Stirring the weld puddle. The high pulse of the current develops arc and electromagnetic forces much greater than those with constant-current welding. These forces agitate the weld puddle, thereby reducing porosity and incomplete fusion at the bottom or the joint.

Advantages and Limitations. GTAW can be used to make top-quality welds in almost all metals and alloys including high-temperature alloys, many hard-facing alloys, titanium, zirconium, gold, and silver. Almost no weld cleanup is required since no flux is used. There is very little if any weld spatter since the weld metal is not carried across the arc. Welding can be done in all positions. The process is particularly well suited for welding thin materials where a high-quality finish is desired.

The main disadvantage of GTAW is that it is relatively slow and for that reason is not well suited to welding heavier metals.

PLASMA-ARC WELDING (PAW)

Plasma-arc welding is very similar to GTAW except that the arc is forced to pass through a restriction before reaching the workpiece. This is accomplished by surrounding the electrode with a nozzle that has a small orifice and forcing the inert gas through the orifice. The result is a jet of intensely hot and fast-moving plasma caused by the concentrated ionization of the arc. Plasma-arc torches regularly develop temperatures as high as 30,000°F (16,490°C).

In practice, two separate streams of gas are supplied to the welding torch. One stream surrounds the electrode within the orifice body and passes through the orifice to form the plasma jet. This gas must be inert and is usually argon. The other stream of gas passes between the orifice body and the outer shell of the torch to act as a shielding gas for the arc and weld puddle (Fig. 12-23). An inert gas such as argon can also be used for the shielding, but nonoxidizing gas mixtures such as argon with 5% hydrogen have often been used to advantage. As with GTAW, filler metal may or may not be used depending upon the requirements of the joint.

There are two main types of plasma torches: transferred arc and nontransferred arc. The nontransferred arc has both electrodes in the torch. The transferred arc, in which the workpiece is positively charged, is far superior since it transfers more energy and is less susceptible to magnetic deflection. It is of course limited to conductive materials. The two types of plasma torches are shown, and compared to the GTAW torch (Fig. 12-24).

PAW Power Supplies. PAW is done almost exclusively with dcsp from a constant-current power supply. A provision is made for starting the arc because the electrode tip is within the orifice body and therefore the arc cannot be started by touching it to the workpiece.

For low-current welding, a pilot-arc system is typically used to initiate the arc and maintain it during brief downtimes. The pilot arc is fed from a separate power supply that is connected to the electrode and the orifice. To start the pilot arc, the electrode is advanced within the torch until it makes contact with the orifice and is then retracted to a specified distance.

For high-current welding, the arc is usually started by a high-frequency current superimposed on the main welding current. The effect is to produce a

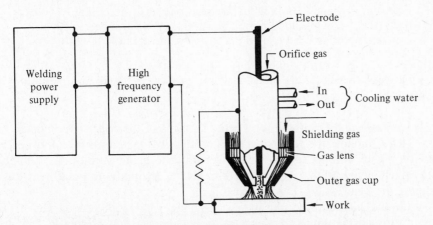

Figure 12.23. Sectional view of the plasma-welding torch.

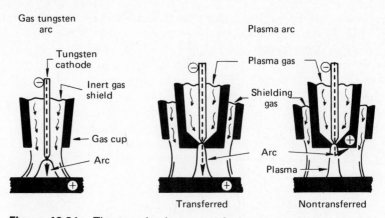

Figure 12.24. The two basic types of plasma torches compared with the GTA torch. Note that the arc shape in the GTA torch is conical and expands rapidly. (*Courtesy Union Carbide Corp., Linde Division.*)

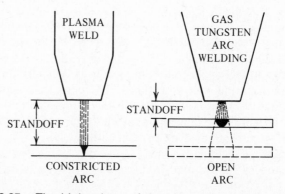

Figure 12.25. The higher heat of the plasma arc results in a narrower weld, less heat-affected zone, deeper penetration, and lower current demand. The plasma arc also provides the operator with greater flexibility since minor variations in the standoff distance have little effect on bead width or heat concentration at the work.

spark between the electrode and the workpiece that initiates and maintains the plasma arc.

Advantages and Limitations. The advantages of the constricted plasma arc over GTAW include the following: greater energy concentration, improved arc stability, higher welding speeds, and lower width-to-depth ratio of the weld bead for a given penetration. Also more arc length flexibility is possible without changing the bead width (Fig. 12-25).

With the high-current plasma process, the weld puddle area has a hole pierced through the entire thickness of the weldment [Fig. 12-26(a)]. This void

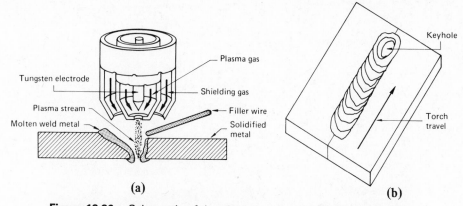

Figure 12.26. Schematic of the plasma-arc torch utilizing the keyhole technique (a). The keyhole effect seen throughout the plasma-welding operation indicates complete penetration (b). This type of weld can be made on thin gages of titanium, stainless steel, or aluminum at high speeds.

is known as the "keyhole" effect. The keyhole is formed at the leading edge of the weld puddle, where the forces of the plasma jet displace the molten metal to permit the arc to pass completely through the workpiece. As the torch progresses, the molten metal, supported by surface tension, flows in behind the keyhole to form the bead. The keyhole ensures complete penetration and weld uniformity. Since the weld puddle is supported by surface tension, close-fitting backing bars are not required but underbead protection should be provided by a channel filled with shielding gas.

Disadvantages of PAW compared to GMAW are higher equipment cost, short life of the orifice body, need for greater welder knowledge, and a high rate of inert gas consumption. PAW equipment usually costs two to five times as much as GMAW equipment. Although more knowledge is required by the welder, much of PAW is automated so the actual operation can be done with relatively little skill.

SUBMERGED-ARC WELDING (SAW)

Submerged-arc welding derives its name from the fact that the arc is hidden under a heavy coating of granular mineral material or flux (Fig. 12-27). A bare wire is used and the fusible flux blanket provides protection for the molten metal from the atmosphere. The dry granular flux is fed continuously from a hopper through a pipe placed slightly ahead of the arc zone. The arc is struck in the submerged area and the weld is completed without the usual sparks, spatter, and smoke. The process may be manual, semiautomatic, or fully automatic, although its main application today is with fully automatic systems, primarily for plate and structural work. In semiautomatic welding, the welding gun itself

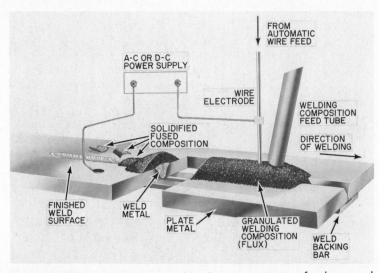

Figure 12.27. A cutaway showing the components of submerged-arc welding. The arc is kept under a layer of granular flux. (*Courtesy Plant Engineering.*)

may be equipped with a flux-feeding device, a small hopper on top of the gun, with gravity flow through the nozzle concentric with the electrode or a concentric nozzle connected to an air-pressurized flux tank. Automatic operations are usually equipped with a vacuum system to pick up the unfused flux to be used again.

Current Supply. The current used for SAW can be either ac or dc. By using dcrp deeper penetration can be obtained; however, with dcsp a higher deposition rate with a broader bead can be achieved. A flat-characteristic power source has the advantage of using a fairly constant voltage. Above 1000 amp, ac current is preferred to reduce arc blow. If one machine does not supply enough power, duplicate machines may be connected in parallel.

As with the flux-cored process, multiple electrodes may be used. Common arrangements are with two electrodes in tandem or in transverse (side by side or parallel) arrangement.

Advantages and Limitations. Submerged-arc welding has several distinct advantages. Very high currents can be used, ranging from 200 to 2000 amp. It may be noted that in conventional welding, where the arc is exposed, currents above 300 amp must be used with great care due to the intensity of high infrared and ultraviolet light rays. The ability to use high current in submerged-arc welding brings with it high deposition rates and good penetration. The process is thermally efficient since much of the heat is kept under a blanket of slag. There is a high dilution rate of base metal with weld deposit, being on the order of twice as much plate as weld metal. Each size of wire can be used over a wide range of

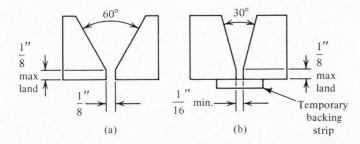

Figure 12.28. A joint as prepared for manual arc welding in 0.75-in. thick plate (a), and for submerged-arc welding (b). The filler wire required per lineal foot is 1.1 and 0.7 lb, respectively. A temporary backup strip is used with the submerged-arc process due to the high current density. If the joints are to be highly stressed, back gouging and running a bead on the back side are recommended.

current settings but, for any given current, the smallest wire size will give the best penetration. Because of the deep penetration possible, the use of smaller Vs is practical, as shown by the example in Fig. 12-28. The weld beads are extremely smooth and most of the fused flux pops off by itself as the metal cools and contracts.

Because of the high deposition rate and smooth bead, SAW is a favorite process used in rebuilding, hard surfacing, crushing rolls, and in manufacturing a wide variety of equipment.

One of the disadvantages inherent in the process is that since the weld cannot be seen, it is more difficult to guide it. Therefore the welding head must be accurately preset with respect to the joint or the operator must be able to adjust the head by observing an indicator, usually a pointer or light beam, focused on the joint ahead of the flux. In the case of grooved joints, rollers can be made to guide the welding head by riding along the edges of the plate.

Another disadvantage of the submerged-arc process is that it is largely limited to flat-position welding. Overhead welding is quite impractical due to the high fluidity of the weld pool and the granular flux.

The SAW process is usually not suitable for metal less than about $\frac{3}{16}$ in. (7.92 mm) thick because burn-through is likely.

ELECTROSLAG WELDING (ESW)

Electroslag welding is the specialized adoption of submerged-arc welding for joining thick materials in a vertical position. Strictly speaking, it is not an arc-welding process because it depends on the electrical resistivity of a molten flux to produce the heat necessary to melt both the filler and base metals (Fig. 12-29).

In this process, a granular flux is placed in the gap between the plates being welded, and, as the current is turned on, welding takes place in a U-shaped

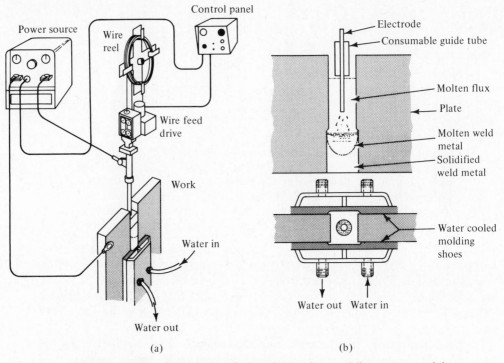

Figure 12.29. Schematic of the electroslag welding process (a); cross-sectional and plain view (b).

starting block, tack-welded to the bottom of the joint. As the flux melts, a slag blanket from 1 to 1.5 in. (2.54 to 3.81 cm) thick is formed. At this point the arc goes out, and current is conducted directly from the electrode wire through the slag. The high resistance of the slag causes most of the heating for the remainder of the weld.

This process is used to weld metals from 1.5 to 15 in. (3.81 to 38.1 cm) thick. The maximum reported thickness is 24 ft (7.3 m). The molten metal and slag are retained in the joint by means of copper shoes that automatically move upward, as the weld progresses, by means of a temperature-sensitive mechanism. Two probes control the flow of granular flux from a hopper over the weld zone.

A variation of electroslag welding is *electrogas welding*. The concept is much the same. The main difference is that an inert gas is used to shield the molten metal. The process is used on thinner materials than those associated with electroslag welding.

Advantages and Limitations.   The electroslag welding process is quite automatic, and once started it will keep going until the job is completed. Heating is uniform, which keeps warpage to a minimum. Runoff tabs can be used to ensure a complete weld. Joint preparation is not required except to remove any heavy mill scale that may be present. The scale interferes with heat transfer to the copper shoes and may prevent a good fit-up, allowing some of the weld metal to run out. Each wire used deposits about 50 lb (22.5 kg) of weld metal per hour. The process can be used for welding hot-rolled carbon steels; low-alloy, high-strength steel; and quenched-and-tempered, low-alloy steels.

The notch toughness of the coarse-grained, heat-affected zone of heat-treated steels may not be as good as that of the base metal. However, if the part is heat-treated afterward, the as-cast structure may be recrystallized and refined. Most electroslag systems are restricted to plates thicker than 0.5 in. (12.70 mm) and cost savings are seldom realized when they are less than 0.75 in. (19.05 mm) thick.

STUD-ARC WELDING (SW)

Stud welding is basically an arc process. An arc is generated between the stud or similar part and the base metal. The operation sequence begins with the stud being placed in the stud gun (Fig. 12-30). The gun is then positioned over the spot where the stud is to be placed. When the trigger, or switch, is depressed, current flows through the stud, which at the same time is lifted back slightly from the work, creating an arc (Fig. 12-31). After a short arcing period, the stud is plunged into the molten pool created on the base plate and the gun is withdrawn.

Stud welding may be done with the conventional dc power supply or with a capacitor-discharge power supply. The latter process uses a low-voltage electrostatic storage system. In capacitor-discharge stud welding, a small tip or projection on the end of the stud is suddenly disintegrated to produce a flux or deoxidizer for the weld pool.

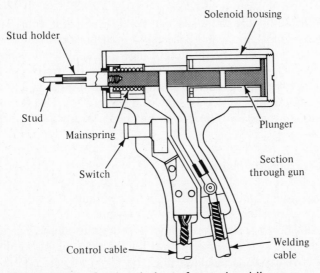

Figure 12.30. Sectional view of a stud welding gun.

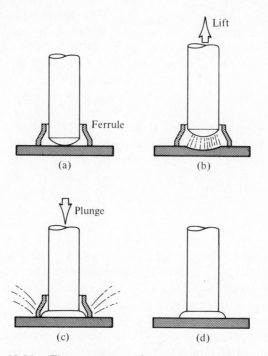

Ferrule

Lift

Plunge

(a)

(b)

(c)

(d)

Figure 12.31.   The sequence of steps used in stud welding.

Low-carbon and austenitic stainless steel are welded by either the arc or the capacitor-discharge method. Some non-heat-treatable aluminum alloys can be welded by the arc method but most aluminum alloys and copper, brass, galvanized sheet, and metal thinner than 0.036 in. (0.91 mm) are welded only by the capacitor-discharge method.

The weld base for capacitor-discharge stud welding is nearly always round. Square or rectangular pins, collard studs, or split pins and studs over $\frac{3}{8}$ in. (9.52 mm) are stud-arc welded. The diameter range for capacitor-discharge stud welding is $\frac{1}{16}$ to $\frac{3}{8}$ in. (1.57 to 9.52 mm). The stud may be used with a ceramic-arc shielding ferrule, which excludes air from the weld pool area and acts as a dam to retain the molten metal as was shown in Fig. 12-31. The ferrule is not used when capacitance discharge current is available because of the rapid welding action.

Advantages and Limitations.   Stud welding provides a quick and easy way for securing fasteners to plate or round stock. It is typically used for securing air, water, hydraulic, and electrical lines to buildings, vehicles, and large appliances and securing feet and handles to appliances.

Studs must be of a size and shape that permits chucking, and the cross-sectional area of the weld base must be within the range of the welding-equipment power supply. Areas to be welded must be clean and free from paint, scale, rust, grease, oil, dirt, zinc, or cadmium plating. Aluminum surfaces may also need oxide removal.

ARC-WELDING SUMMARY

The number of options in arc welding are many and varied. As an aid in bringing these variables together, the summary chart (Fig. 12-32) shows the various welding processes, metals commonly welded by the process, position that can be welded, and power source.

Adaptive Control for GMAW and GTAW

Automatic welding has been available for a long time but it is based on preset controls. If variables change outside of the anticipated limits, bad welds result. In recent years work has been done on *adaptive controls*. That is, sensing devices are incorporated into the welding system to produce signals that will correct the operating parameters for changes in direction, metal thickness, gap-width misalignment, joint design, etc.

The most important sensors being developed are:

Search for the beginning point of the weld "path."
Following the path.
Positioning the torch.
Controlling and regulating the weld parameters.

Since, in practice, there are many variables involved in GMAW welding, not all of them can be taken care of by one sensor.

Sensors may be divided into two main types, those concerned with weld process and those concerned with weld geometry.

Weld Process.   The weld process may be observed from without or within. If from without, observing the weld pool and arc is possible. If within (direct sensing), the current used may be a good indicator of the process. Internal measurements are much easier to accomplish than external ones.

Weld Geometry.   Sensing-joint geometry may be of the contact or noncontact type. Contactless sensors have the advantage of being able to be used near the torch and do not pose the problems of errors in measurement due to wear, soil, and being misaligned by obstructions. However, it is difficult to determine dislocation of edges and varying gap widths.

Basically, the following physical measuring principles are available: mechanical, electrical, inductive-capacitive, optical (including infrared measurements), acoustical, and pneumatic.

Mechanical sensors are suited for following the joint track and steering the electrode to the center of the joint. They are usually simple, inexpensive, and easily changed to various profiles. They usually consist of a pin or feeler wire that runs in the joint or along the edge of a workpiece. The sensor may be rigidly attached to the torch or indirectly attached through electrical signals that control the torch by means of motors.

Welding Conditions	SMAW	Flux Cored		GMAW	GTAW	PAW	SAW	ESW	SW
		With Gas	Without Gas						
Metals Commonly Welded	Most	Nonferrous and stainless steel	Steel	Nonferrous and stainless	Most	Most	Steel	Steel	Most
Power Source and Type	AC or DC CC	AC or DC CV	DC CV or CC	DC CV	AC or DC CC with HF	DC CC	AC or DC CV or CC	DC CV preferred	DC CC
Wire Feeder and Type	Not required	Constant speed	Constant speed or volt sensitive	Constant speed	For automatic only	For automatic only	Match power source	Match power source	Not used
Welding Position	All	All	Flat and horizontal	All	All	All	Flat and horizontal fillet	Vertical	Flat and horizontal
Normal Use	Manual	Semi and automatic	Semi and automatic	Semi and automatic	Manual and automatic	Manual and automatic	Automatic	Automatic	Semi
Base Metal Thickness Normally Welded	All	Medium to thick	Medium to thick	All	Thin	Very thin to medium	Medium to thick	½ in. (12.70 mm) to unlimited	Thin to medium
Major Advantage	Portable and flexible	High deposition	High deposition	Welds most metals	Welds most metals	Flexible, easier, faster	High deposition	High deposition	Unique
Major Limitation	High skill required	Can weld steels only	Reduced impact properties	Fairly expensive; uses gas	Fairly slow	Fairly expensive and complex	Weld not visible	Vertical position	Stud welding only

Figure 12-32. Summary chart of arc-welding processes.

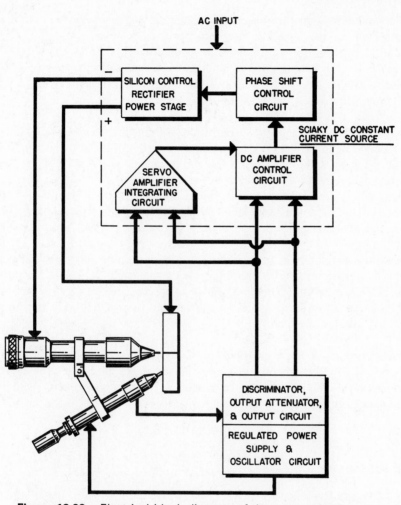

Figure 12.33. Electrical block diagram of the arc-penetration control system. (*Courtesy* Welding Engineer.)

Electrical sensors are based on the length of the arc. In specific cases, information can be gained as to the center of the joint and the width of the joint.

Inductive sensors in the form of magnetic fluxes (both high and low frequency) are induced on both sides of a joint. The two signals are compared and the differences indicate the deviation from the joint center.

Optical sensors can either use the light of the arc or measure the reflection of incident light. Diodes, diode fields, or television cameras can be used to search for the beginning point of the weld, the track, and positioning the torch. The disadvantage of optical sensors is their sensitivity to brightness and their high cost. Despite the difficult environment, there are a large number of systems that have been tested.

Pneumatic systems are based on measuring variations in dynamic pressure on both sides of the joint but have not been widely used. Acoustical sensors utilize differences in transit times of ultra-sound signals. In practice, there have been difficulties in coupling (or attaching) the sound sources.

Two systems of adaptive control for weld penetration have been worked out: (1) arc penetration, and (2) melt-through.

Arc Penetration. The arc penetration system has a sensor control on the same side as the arc (Fig. 12-33). In GTAW the tungsten electrode will advance into the work or withdraw from it due to variances in fit-up or changes in the heat sink. Movement of the electrode with respect to the work is detected by a sensor, which produces a signal automatically increasing or decreasing welding current to maintain a constant electrode to work position. The result is constant penetration of the arc. Dual probes are also used to average out the mismatch in butt joints.

Another development in the use of arc penetration control is opposed-arc welding in which two GTAW welding torches are used, one on each side of the butt joint. Experimental work has shown energy requirements may be reduced to 33% of that required to make the joint in the normal method. Furthermore, distortion is greatly reduced.

Melt-Through Weld Penetration Control System. This system, although similar to the arc penetration control system, is based on a more direct measurement of penetration by having a photocell sensor on the back side of the weld. It has proven successful in making aluminum welded joints from 0.050 in. to 0.750 in. thick.

Questions and Problems

12-1. Suppose an E6024 electrode was used for overhead or vertical welding. What difficulties might be experienced?

12-2. State the relationship between:

(a) arc length and voltage,

(b) increasing the wire speed and voltage (on an automatic machine),

(c) the effect of (a) and (b) on current on a drooping volt-ampere machine.

12-3. A 250-amp 60% duty cycle machine is being used at 300 amp. How much must the duty cycle be reduced to operate at 300 amp?

12-4. What is the significance of polarity? How is it used to advantage?

12-5. When is it particularly advantageous to have a superimposed high-frequency current on a regular ac welding current?

12-6. If you encountered a strong arc blow in attempting to weld the inside corner of a box, what would you do to try and overcome this condition?

12-7. How can a welder do a reasonably good job of welding plates that have a poor fit-up if he is using a flux-cored process?

12-8. Is there any advantage in using CO_2 together with the flux-cored electrode while welding mild steel?

12-9. Select the electrodes for the joints shown in Fig. P12.1 and give a reason for your choice of each.

12-10. (a) What is the advantage of using helium over argon in some welding applications?

(b) Why isn't helium used very much for

Plate 3/16in. (7.92 mm) or Thicker

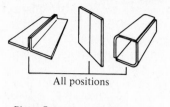

All positions

Plate, flat
position
3/8 in.
and
thicker

Structural
joints, all
positions
3/16 to 5/8
in. plate

Figure P12.1.

a shielding gas for welding aluminum or magnesium?

(c) What is the special requirement for CO_2 GMAW?

12-11. What would be the difference in the amount of filler wire required for 10 ft. of weld made in $\frac{3}{4}$ in. thick plate with stick electrode and submerged arc? Assume conditions as given in Fig. 12-28.

12-12. (a) Make a sketch of the current characteristics of a drooping-arc-voltage machine with a long arc, normal arc, and short arc. Shade the area that shows the variation in current.

(b) Why may a machine of this type be thought of as "constant current"?

12-13. When may the terms TIG and MIG be used?

12-14. Contrast the use of spray transfer with pulsed-spray transfer as to general use.

12-15. What is the meaning of plasma arc?

12-16. What would be the main advantages of us-

ing high-current plasma welding for making a butt weld in $\frac{1}{2}$ in. thick aluminum plates?

12-17. Can a constant-speed wire feed be used with a constant-voltage power source? Why or why not?

12-18. When would a voltage-sensing type of wire feeder be used?

12-19. When can a hand-held wire feeder be used to good advantage?

12-20. If a weld made by FCAW has heavy spatter and porous appearance, what is the probable cause?

12-21. Why is the amount of stickout used in FCAW and GMAW very important?

12-22. What is the difference between FCAW and GMAW?

12-23. What type of current and shielding gas are used for welding aluminum with the GMAW process?

12-24. What is the main factor that determines the type of metal transfer through the arc?

12-25. What are the advantages and disadvantages of spray transfer?

12-26. Why is the pulsed-spray transfer better for out-of-position welding?

12-27. Is a pure tungsten electrode preferred for GTAW? Why or why not?

12-28. What are the main advantages of PAW?

12-29. Is a flux required for stud welding with a capacitor-discharge power source?

12-30. What welding process would be best for joining 2-in. (50.0-mm) plates in the flat position? Why?

Bibliography

AWS Welding Handbook, 7th ed. Miami, 1976.

Brosilow, Rosalie. "Gases for Shielded Metal Arc Welding." *Welding Design and Fabrication* (October 1978).

Brown, Stuart C. "Power Supplies in the Age of Electronics." *Welding Design and Fabrication* (December 1979).

Cary, Howard B. *Modern Welding Technology.* Englewood Cliffs, N.J.: Prentice-Hall, 1979.

Foith, J. P. "Sensors for Fully Mechanical Arc Welding Problems, Requirements, Classification." *Schweissen und Schneiden,* Thirty-First Year (1979), Vol. 7.

Giachino, J. W.; W. Weeks; and G. S. Johnson. *Welding Technology,* 2nd ed. Chicago: American Technical Society, 1973.

Kennedy, Gower A. *Welding Technology.* New York: Howard W. Sams & Co., 1974.

Lindberg, R. A., and N. R. Braton. *Welding and Other Joining Processes.* Boston: Allyn and Bacon, 1976.

Metals and How to Weld Them. Cleveland: The James F. Lincoln Arc Welding Foundation, 1962.

Pierre, Edward R. *Welding Processes and Power Sources.* Appleton, Wisc.: Power Publications Co., 1973.

Schaffer, George, ed. "Welding." *Special Report 698 American Machinist* 121 (September 1977): 83–106.

Uttrachi, Gerald D. "Selecting Fluxes for Submerged Arc Welding." *Welding Design and Fabrication* 51 (February 1978): 79–85.

CHAPTER 13

Gas Welding and Related Processes

Gas welding usually refers to oxyacetylene welding or, as the name implies, the burning of acetylene, the fuel gas, with the addition of oxygen to create a high-temperature flame. The flame temperature normally obtained with $2\frac{1}{2}$ parts oxygen to 1 part of acetylene is about 5850°F (3232°C). The gases are fed through the torch on a 1:1 ratio with the remaining $1\frac{1}{2}$ parts of oxygen coming from the surrounding air. The chemical reaction may be shown as $3Fe + 2O_2 \rightarrow Fe_3O_4 +$ heat.

Other fuel gases used in welding, brazing, soldering, or metal cutting are propane, natural gas, propylene, and MAPP, a stabilized methylacetylene propadiene gas manufactured by the Dow Chemical Company.

Table 13-1 shows a comparison of fuel gases used in welding and cutting. Data indicate that MAPP flame cutting is about 15% faster.

OXYACETYLENE WELDING

The oxyacetylene torch is widely used in making repairs, cutting, brazing, and soldering. The equipment is relatively low in cost, readily available, and portable. Gas welding is slower than arc welding, which gives the operator more control over the weld pool.

Table 13-1. Oxygen consumed per cubic foot of fuel burned. (*Courtesy* Modern Machine Shop)

Fuel	Flame Temperature	Total Heat (Btu/cu ft)	Cu Ft of Oxygen Per Cu Ft of Fuel
Acetylene	5589	1470	1.4
MAPP gas	5301	2406	2.5
Propylene	5193	2371	3.5
Natural gas	4600	1000	1.9
Propane	4570	2498	4.3

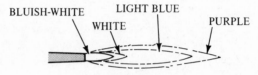

REDUCING FLAME
5700°F (3149°C)

NEUTRAL FLAME
5850°F (3232°C)

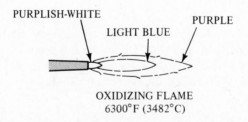

OXIDIZING FLAME
6300°F (3482°C)

Figure 13.1. The three types of oxyacetylene flames.

Oxyacetylene Welding Flame. Three distinct flame variations are produced with an oxyacetylene gas mixture, reducing or carburizing, neutral, and oxidizing.

The acetylene flame is long and bushy. As the oxygen is turned on, the long flame reduces in size and a feathery inner cone appears, producing a reducing flame. When the oxygen proportion is further adjusted, the feathery, white cone disappears and a rounded inner cone and outer envelope indicate a neutral flame.

The length of the luminous inner cone is usually between $\frac{1}{16}$ and $\frac{5}{8}$ in. (1.57–15.87 mm) long.

If more oxygen is turned on, the flame as a whole grows smaller and the inner cone is reduced in size to produce an oxidizing flame. All three flames are shown in Fig. 13-1.

Each of the three flames may be used as follows:

Carburizing flame may be used to advantage in welding high-carbon steels, for hard-facing operations, and for welding such nonferrous alloys as nickel and monel. It is also used in silver-brazing operations. In these operations, only the intermediate and outer flames are used so that a low-temperature soaking heat will be imparted to the parts being joined by silver solder. If a carburizing flame is used in welding steel, the excess carbon will enter the metal, causing porosity in solidified metal.

Neutral flame is used for most welding operations.

Oxidizing flame is used for fusion welding of brass and bronze; however, the flame is only slightly oxidizing. An oxidizing flame used on steel will cause the metal to foam and spark.

Equipment

Torches.   The oxyacetylene welding torch is used to mix the gases in the right proportions (Fig. 13-2). The torch tips are interchangeable and provide a variety of sizes to handle a range of metal thicknesses.

Tanks.   Figure 13-3 is an illustration of a portable oxyacetylene welding outfit. The oxygen cylinder is of seamless drawn steel with a capacity of 220 cu ft at a pressure of 2200 psi (16.8 MPa) at a temperature of 70°F (21°C).

Acetylene can be compressed into a steel cylinder under a pressure of 250 psi (17 kg per cm²). The acetylene is dissolved in acetone, which in turn is absorbed by the porous material in the cylinder, such as balsa wood, charcoal, finely shredded asbestos, infusorial earth, corn pith, and portland cement. This is necessary because acetylene, when stored in a free state under pressure greater than 32 psi (2.2 kg per cm²), can be made to dissociate or break down by heat or shock and possibly explode.

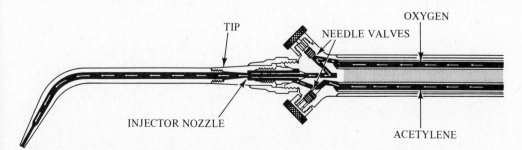

Figure 13.2. Sectional view of the oxyacetylene welding torch.

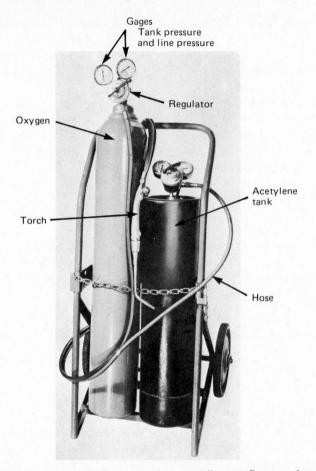

Figure 13.3.   A portable oxyacetylene welding outfit, complete with carrying cart. (*Courtesy Union Carbide Corp., Linde Division.*)

The acetylene should not be drawn off at a continuous rate faster than 50 cu ft (9834 cm^3) per hour from a standard cylinder. When volumes greater than this are required the cylinders should be manifolded together. Acetylene cylinders should never be used in the horizontal position since some of the acetone would be drawn off.

Regulators.   The regulators at the top of each tank are for reducing the high tank pressure [2200 psi for oxygen and 220 psi (15.17 MPa) for acetylene] to working pressure and to provide a smooth flow of gas.

Gas Welding Procedure

There are several methods of oxyacetylene welding that produce good results. Both forehand and backhand welding are used. In forehand welding the welding

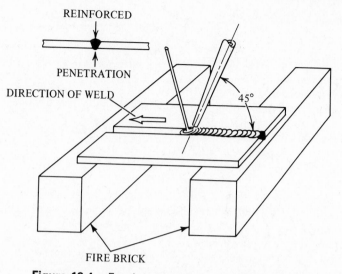

Figure 13.4. Forehand welding of a butt joint.

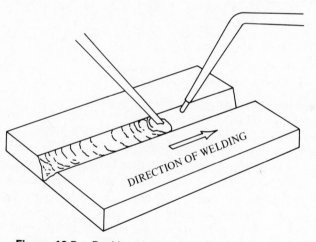

Figure 13.5. Backhand welding of a butt joint.

rod precedes the tip in the direction the weld is being made (Fig. 13-4). This position, with the torch at about 45° to the plate, permits uniform preheating of the plate edges immediately ahead of the molten puddle. In backhand welding, the torch is pointed back at the molten puddle with the torch held at about 60° to the plate (Fig. 13-5). In either case, the welding rod is kept within the outer flame envelope to protect it from the atmosphere and manipulated to provide the desired melt-off rate. If large deposits of metal are required, the end of the rod is moved so that it will describe full circles in the molten pool. The torch is moved to and fro across the weld while slowly advancing in the direction of the weld.

Advantages of gas welding:

1. The gas flame is generally more easily controlled and is not as piercing as shielded-arc welding. Therefore it is used extensively for sheet-metal fabrication and repairs.
2. The oxyacetylene or oxy-MAPP flame is versatile. It may be used for brazing, preheating, postheating, heating for forming, heat treatment, flame cutting, etc.
3. The gas welding outfit is very portable (see Fig. 13-3).

Limitations of gas welding:

1. The process is much slower than arc welding.
2. Harmful thermal effects are aggravated by prolonged heat. This often results in increased grain growth, more distortion, and, in some cases, a loss of corrosion resistance.
3. Most oxyacetylene tanks are leased, and when kept beyond the specified time, demurrage charges are incurred.
4. There are safety problems in handling gases.
5. Flux applications and the shielding provided by the secondary envelope of the flame are not nearly as positive as that supplied by the gases used for GTA and GMA welding.

BRAZING

Brazing is a process of joining metals with a nonferrous filler metal that has a melting point below that of the metals being joined. By AWS definition the melting point of the filler metal will be above 800°F (427°C). Below this temperature are the solders. The filler metal must *wet* the surfaces to be joined, that is, there must be a molecular attraction between the molten filler material and the materials being joined. The brazing alloy, when heated to the proper temperature, flows into the small joint clearances by capillary action. A limited amount of alloying occurs between the filler metal and the base metal at elevated temperatures (Fig. 13-6). As a result, the strength of the joint when properly made may exceed that of the base material. The strength is attributed to three sources: atomic forces between the metals at the interface; alloying, which comes from diffusion of the metals at the interface; and intergranular penetration.

The heat for brazing may be provided in many different ways, the most common of which are by torch, induction, furnace, and hot dipping.

Torch Brazing. Torch brazing is one of the oldest and most widely used methods of heating for brazing, being versatile and adaptable to most jobs. It is especially good for repair work in the field and for small-lot jobs in the shop. As it is a manual operation, torch brazing may have high labor costs and the skill and judgment of the torch operator will determine the quality of the work.

Torch brazing may be done with the use of natural gas and oxygen, air and acetylene, butane, propane, MAPP, and the regular oxyacetylene. If the latter

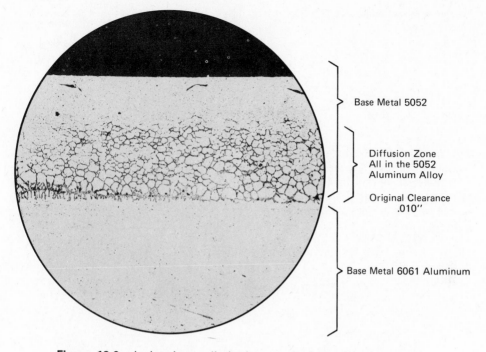

Base Metal 5052

Diffusion Zone
All in the 5052
Aluminum Alloy

Original Clearance
.010"

Base Metal 6061 Aluminum

Figure 13.6. In brazing, a limited amount of alloying occurs between the filler metal and the base metal at elevated temperatures. (*Courtesy Wall Colmonoy.*)

is used, a neutral flame is recommended, which is a balanced condition of acetylene and oxygen. A slightly reducing flame (excess acetylene) may be used to avoid oxidation of the metal. The filler metal is touched to the joint when the flux becomes a clear liquid. It is the parent metal, not the flame, that transfers the heat to the brazing alloy.

Induction Brazing. The heat for induction brazing is furnished by an ac coil placed in close proximity to the joint (Fig. 13-7). The high-frequency current is usually provided by a solid-state oscillator that produces a frequency of 200,000 to 5,000,000 Hz. These alternating high-frequency currents induce opposing currents in the work, which, by electrical resistance, develop the heat. This method has the advantage of providing: (1) good heat distribution, (2) accurate heat control, (3) uniformity of results, and (4) speed. It is especially good for certain types of repetitive work that require close control.

Furnace Brazing. Parts that are to be furnace brazed have the flux and brazing material preplaced in the joint. If the furnace has a neutral or shielding atmosphere the flux may not be necessary. The furnace may be batch or conveyor type. Automatic controls regulate both time and temperature and, where appli-

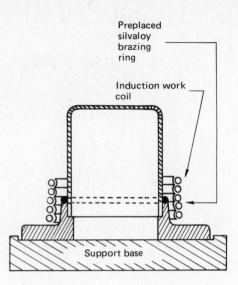

Figure 13.7. An example of external induction brazing. The preform brazing filler material is preplaced in close contact to the work to assure melting at the proper temperature. (*Courtesy Engelhard Industries.*)

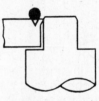

Staked and pasted

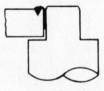

Heat to 2000°F
under hydrogen atmosphere

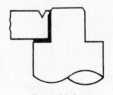

Brazed joint

Figure 13.8. The three phases of furnace brazing. The hydrogen acts as a reducing agent removing any light oxide film from the surfaces of the components. This technique eliminates the need for fluxing agents. In addition, the hydrogen protects the surface from oxidation during the heating and cooling cycle. Thus clean, bright assemblies with a high joint strength are produced. (*Courtesy Society of Manufacturing Engineers.*)

cable, atmosphere. The three steps used in copper-hydrogen furnace brazing are shown in Fig. 13-8.

Dip Brazing.   Dip brazing gets its name from the fact that parts are jigged (or in some cases the parts are self-jigging, as shown in Fig. 13-9) and are placed in a chemical or molten-metal bath maintained at the correct brazing temperature. In a chemical bath the parts are first thoroughly cleaned, then assembled with a filler-metal preform placed on or in the joint. The bath, usually molten salts, is maintained at a higher temperature than the filler metal being used. After dipping, the parts are removed and immediately cleaned to remove the flux. The number and size of the parts to be brazed are limited only by the capacity of the bath and the handling facilities.

Brazing-Joint Design

Two factors are especially important in brazing-joint design—area and clearance. An example of the area principle can be shown by designing a butt or lap joint (Fig. 13-10). Lap joints are used in preference to the butt joint. An overlap of three or four times the thinnest member will usually give the maximum efficiency. Overlaps greater than this lead to insufficient penetration, inclusions, etc. This joint is also recommended for the leak tightness and good electrical conductivity. Some principles of joint design are illustrated in Fig. 13-11.

Brazed-Joint Clearance.   The joint clearance determines the thickness of the alloy film that will be formed between the parts. This has an important influence on the strength of the joint. Maximum strength is obtained when there is just enough room for the alloy to flow through the joint when heated and the flux and gases are allowed to escape. The ideal clearance for production work ranges between 0.001 and 0.005 in. (0.030 to 0.127 mm). Joints below 0.001 in. and above 0.008 in. (0.025 and 0.203 mm) should be avoided if at all possible. The effect of joint clearance on tensile strength is shown in Fig. 13-12.

In determining the clearance, the thermal expansion of the parts must be

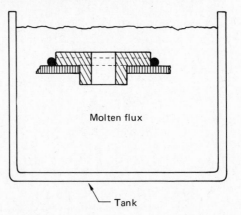

Molten flux

Tank

Figure 13.9.   Parts to be dip brazed are first placed in jigs or, as shown here, the parts may be self-jigging.

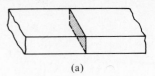

(a)

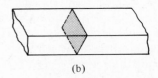

(b)

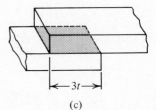

Figure 13.10. The area principle in braze joints. The butt joint (a) is very poor. A scarf butt joint (b) is much better. A lap joint (c) should provide an overlap of 3*t*.

(c)

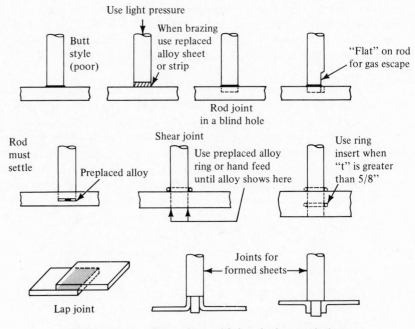

Figure 13.11. Some brazed-joint design variations.

(Based on brazing butt joints of stainless steel to stainless steel, using Easy-Flo* filler metal.)

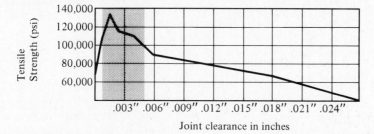

Joint clearance in inches

Figure 13.12. The effect of joint clearance on tensile strength (based on brazing butt joints of stainless steel to stainless steel, using Easy-Flo filler metal). (*Courtesy Handy and Harman.*)

considered and sufficient clearance allowed so that at brazing temperatures there will still be sufficient room for the entrance of the alloy. In Appendix A is a table showing the thermal expansion of the various metals that can be used to determine brazing-joint clearance.

> **Example:** A 3-in. dia. stainless steel plug is to be silver brazed into a recessed hole in a steel plate. Determine the clearance required. Assume a temperature increase of 1250°F.

> **Solution:** Coefficient of expansion for stainless steel and steel, respectively: 9.3 in. × 10^{-6}/in./°F − 6.7 in. × 10^{-6}/in./°F = 3(0.0000093 − 0.0000067)1250 = 0.00975. Therefore the total brazing clearance before heating is 0.0035 in. + 0.00975 in. = 0.01325 in.

Brazed-Joint Strength. The following simple formula is useful in determining the lap length (depth of shear) on tubular or flat joints.

$$L_l = \frac{(S_t)t(F_s)}{S_s(D)}$$

where: L_l = lap length

             S_t = tensile strength of weakest member

             t = thickness of weakest member

             F_s = factor of safety

             S_s = shear strength of brazing alloy

             D = diameter of shear area. *Note: D* is omitted on flat joints.

> **Example:** A $\frac{1}{4}$ in. × 2 in. strap is brazed to a $\frac{1}{2}$ in. thick steel plate. The brazing alloy used has an S_s of 36,000 psi. The desired safety factor is 2. The tensile strength of the steel is 50,000 psi.

$$L_1 = \frac{50,000 \times 0.250 \times 2}{36,000}$$
$$= 1.694$$

Therefore the lap length should be about $3t$.

Brazing Fluxes.   The primary purpose of brazing fluxes is to dissolve and absorb oxides that heating tends to form. Therefore the parts to be brazed must first be cleaned of all existing oxides, oil, grease, and other foreign matter either by degreasing, grinding, pickling, or a combination of these cleaning processes. When the metal is heated to the brazing temperature, the flux becomes a clear liquid that wets the surface and aids in the flow of the filler metal. Some of the most common ingredients of fluxes are sodium, potassium, and lithium. They are used in making up chemical compounds such as borates, fluorides, chlorides, boric acids, alkalies, and wetting agents.

Generally, fluxes are available in three forms: powder, paste, or liquid. Of the three, paste is most commonly used, although powdered flux is frequently mixed with water or alcohol to give a pastelike consistency.

Brazing Filler Metals.   Brazing filler metals are divided into seven classifications. In order of popularity, these are: silver, copper, copper zinc, copper phosphorus, aluminum silicon, copper gold, and magnesium. These alloys are produced in many forms such as wire, rods, coated sheets, and powder. Frequently brazing alloys are preformed into rings or special shapes to simplify placement of the correct amount at the joint.

The service temperature of brazing varies widely. Most are for room temperature but some have been developed for stainless steel and have a service temperature of up to 2200°F (1204°C).

Advantages and Limitations.   Brazing is well suited to mass-production techniques for joining both ferrous and nonferrous metals. Some of the principal advantages are:

1. Dissimilar metals can be joined easily.
2. Assemblies can be joined in virtually a stress-free condition.
3. Complex assemblies can be joined in several steps by using filler metals with progressively lower melting temperatures.
4. Materials of different thicknesses can be joined easily.
5. Brazed joints require little or no finishing other than flux removal.

Limitations:

1. Joint design is somewhat limited if strength is a factor.
2. Joining is generally limited to sheet-metal thicknesses and relatively small assemblies.
3. Cost of joint preparation can be high.

Braze Welding

Braze welding is similar to brazing in that the base metal is not melted but joined by an alloy of lower melting point. The main difference is that in braze welding the alloy is not drawn into the joint by capillary action. A braze-welded joint is prepared very much like a joint is prepared for welding except that an effort should be made to avoid sharp corners because they are easily overheated and may also be points of stress concentration.

Braze welding is used extensively for repair work as well as some fabrication on such metals as cast iron, malleable iron, wrought iron, and steel. It is used, but to a lesser extent, on copper, nickel, and high-melting-point brasses.

SOLDERING

Soldering is a way of joining two or more pieces of metal by means of a fusible alloy or metal, called solder, applied in the molten state. Soldering can be distinguished from brazing in that it is done at a lower temperature (below 800°F, 427°C) and there is less alloying of the filler metal with the base metal.

Solder Alloys. The four metals used as the principal alloying elements of all nonproprietary solder formulations are tin, lead, cadmium, and zinc. These alloying elements fall into three thermal groups. Those that melt below 596°F (313°C) are called low-melting-point solders. Those that melt between 596 and 700°F (313–371°C) are called intermediate solders. Those that melt between 700 and 800°F (371–427°C) are termed high-temperature solders. The strongest and most corrosion-resistant are the high-temperature solders. The strength of low-temperature solders in shear is a bit in excess of 5000 psi (35 MPa).

The low-melting-point solders (378–596°F, 190–313°C) are the most widely used and are of the tin-lead or tin-antimony type. The tin-lead alloys provide good strength at low temperatures. A combination of 62% lead and 38% tin produces the lowest melting point and in practice is referred to as 60–40 solder. Because it is an eutectic alloy, it behaves like a pure metal and has only one melting temperature, 362°F (183°C), at which it is completely liquid. Increasing the tin content produces better wetting and flow characteristics.

The tin-lead-antimony solders are generally used for the same types of applications as the tin-lead alloys except that they are not recommended for use on aluminum, zinc, or galvanized steel.

The tin-zinc group of alloys is used mainly for soldering aluminum, primarily where a lower soldering temperature than that of a zinc-aluminum solder is required. Zinc-aluminum solders are designed specifically for soldering aluminum.

Indium solder, 50% indium and 50% tin, is particularly suitable for products subjected to cryogenic temperatures and for glass-to-metal bonds.

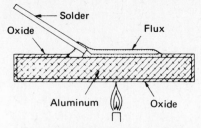

Solder

Oxide

Flux

Aluminum

Oxide

Removing aluminum oxide by fluxing.
Chemical action of the flux removes
the oxide.

(a)

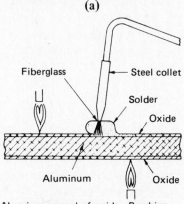

Fiberglass

Steel collet

Solder

Oxide

Aluminum

Oxide

Abrasive removal of oxide. Brushing
action abrades oxide from the aluminum
surface, allowing solder to tin the surface.

(b)

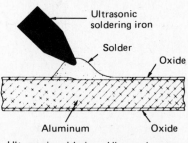

Ultrasonic
soldering iron

Solder

Oxide

Aluminum

Oxide

Figure 13.13. Three methods of removing the oxide film from aluminum
during the soldering process. (*Courtesy
Reynolds Metals Company.*)

Ultrasonic soldering. Ultrasonic waves
cause cavitation, breaking up the oxide
and floating it to the top of the solder
puddle. A frequency of 20 kc is often
used to cause active cavitation.

(c)

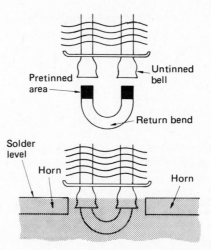

Figure 13.14. Ultrasonic energy is used to remove the oxide as the aluminum tubing is immersed in the liquid solder. (*Courtesy* Product Engineering.)

Solder Design Considerations. Soldered joints should not be placed under great stress. In sheet metal, the joint should take advantage of the mechanical properties of the base metal by using interlocking seams, edge reinforcing, or rivets. Butt joints should be avoided. Lap joints loaded in pure shear may be used if the proper clearance is allowed. Clearances of 0.003–0.010 in. (0.08–2.54 mm) are permissible but the best strength and ease of soldering is obtained with about 0.005 in. (0.13 mm) clearance.

Soldering Methods. There are many ways of transmitting heat to the joint area. These include the well-known soldering "irons" or "coppers," torches, resistance heating, hot plates, ovens, induction heating, dip soldering, and wave soldering.

Aluminum Soldering. Aluminum is often thought of as being very difficult to solder. One reason is that an oxide film forms quickly over the surface after cleaning. Now, however, there are several methods of removing the oxide as the soldering is being done. These methods are flux soldering, friction soldering, and ultrasonic soldering, as shown in Fig. 13-13. Ultrasonic soldering is now being used to solder aluminum tubing in heat-exchanger coils without flux. The return bends, as shown in Fig. 13-14, are pretinned (coated with solder) and placed in the untinned bell. The assembled joint is then immersed in the ultrasonically activated solder. The sonic energy removes the oxides. The process lends itself to automation.

Of the low-temperature solders used for aluminum, a zinz-lead eutectic, which melts at 400°F (204°C), has an advantage over brazing because less heat is required, which results in a minimum loss of strength and hardness and the least distortion. This is particularly important in complex structures. If greater strength is required, the intermediate- or high-temperature solders may be used. The high-temperature solders contain 90 to 100% zinc and are the least expensive and the strongest of all aluminum solders.

Corrosion in aluminum-soldered joints can be particularly troublesome. It may be either chemical or galvanic in nature. In any case, moisture must be present. By coating the joint with a waterproof coating, the problem can be eliminated.

FLAME CUTTING

Flame cutting is a material-separation process, but since it is a necessary step in fabrication and is accomplished with welding equipment, it is included in this chapter.

The burning of red-hot iron or steel in an atmosphere of pure oxygen was a common laboratory experiment in the early 1800s. Theoretically, 4.6 cu ft (904 cm^3) of oxygen is required to oxidize 1 lb (0.45 kg) of iron. This principle has been built into a cutting torch (Fig. 13-15). The torch somewhat resembles the welding torch but it has a provision for several neutral flames and a jet of high-pressure oxygen in the center. The preheating flames are brought close to the metal as in welding, and when the metal becomes cherry red in color, about 1500°F (816°C), the oxygen jet is released by a quick acting valve.

The quality of the cut is governed by tip size, oxygen pressure, cutting speed, and preheating flames. Cuts that are improperly made will produce ragged and irregular edges, with slag adhering at the bottom of the plates. One indication of the proper speed is the drag lines, as shown in Fig. 13-16. Drag is the horizontal lag that exists between the entrance and exit points of the cutting oxygen stream. It is generally specified as a ratio or as a percentage as follows:

$$\text{Drag} = \frac{d \text{ in.}}{t \text{ in.}}$$

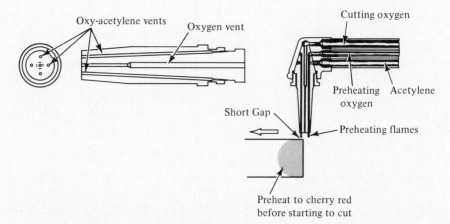

Figure 13.15. The cutting torch has an oxygen jet surrounded by heating flames.

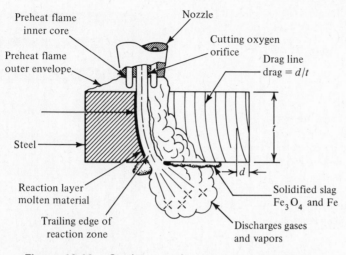

Figure 13.16. Cutting reaction zone and drag lines.

$$\text{Drag } \% = \frac{d}{t} \times 100$$

where: d = amount of horizontal lag
t = the thickness of the metal being cut

The amount of drag that is permissible is dependent upon the quality of the part required. For precision machine cutting, nearly vertical drag lines are used. Generally a drag of 10% is the uppermost limit, but for straight cutting it may go as high as 20%. Defects that arise in metal cutting are shown and explained in Fig. 13-17.

Although acetylene is widely used as a fuel gas, other gases that are used are MAPP, propylene, natural gas, and propane. Tests have shown that propane and MAPP have a considerably longer preheat time than acetylene.

Flame Machining.   Although manual cutting is used extensively for salvage and repair work, machine flame cutting provides greater speed, accuracy, and economy. With a cutting torch mounted on a variable-speed electric motor, referred to as a radiograph, straight cuts can be made with the addition of a track and circular cuts can be made with a compass attachment.

Other methods of guiding the cutting torch are templates, photoelectric tracers, and numerical control. Hand-cutting and machine-cutting tolerances are compared in Table 13-2.

Computer Numerical Control (CNC) Flame Machining.   In recent years the optical tracer used in flame cutting is gradually being replaced by CNC flame cutters (Fig. 13-18). The first significant use of NC flame cutting was in the

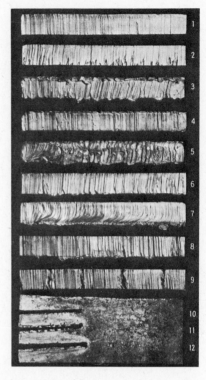

1. *Correct cut*
2. *Preheat flames too small*
3. *Preheat flames too long*
4. *Slow cut; low oxygen pressure*
5. *Oxygen pressure too high; nozzle too small*
6. *Cut too small*
7. *Cut too fast*
8. *Torch travel unsteady*
9. *Cut lost; poor restarts*
10. *Correct*
11. *Too much preheat*
12. *Nozzle too high*

Figure 13.17. Faults to avoid in torch cutting. (*Courtesy Union Carbide Corp., Linde Division.*)

Table 13-2. Oxyfuel flame-cutting tolerances and speeds.

Method	Thickness (in.)	Tolerance (in.)	Cutting Speed
Hand	$\frac{1}{8}$ to 18	$\pm \frac{1}{32}$	10 to 14 ipm on 2 in. thick
			18 to 26 ipm on $\frac{1}{8}$ in. thick
Machine	$\frac{1}{8}$ to 18	± 0.005	12 to 15 ipm on 2 in. thick

shipyards, where cutting very large and often complicated shapes with great accuracy and high speeds was needed.

CNC torch control is now used to obtain accuracies in the range of ± 0.005 in. (0.13 mm), or for one-of-a-kind or short-run parts, gear teeth, and other complicated shapes if they can be defined by a series of lines, radii, and arcs. Only CNC control can follow a precision contour at the speed of 60 to 250 in./min (152 to 635 cm) required by plasma-arc machining systems.

Expert operators can occasionally achieve as high as 60% "torch-on" time with photocell machines. However, the average is closer to 40% versus 80 to 85% for CNC flame cutters. More sophisticated computer control systems, often

Figure 13.18.   A CNC flame-machining setup. (*Courtesy MG Cutting Systems.*)

in combination with DNC (direct numerical control), permit the development of parts "nesting programs," which maximize the utilization of material and minimize scrap as shown in Fig. 13-19. A hard-copy plotter makes a scale drawing of the actual cuts that will be made by the flame cutter. Layout lines for subsequent bending or hole drilling may also be placed on the plate by a pneumatic punch head. It is comparatively easy to shift workpieces around on the plotter. The manufacturing engineer can specify the minimum webbing allowed between components on the plate and the computer programs can then determine the optimum figure.

The key to the success of a flame-cutting program described here is the programmer who is responsible, with the aid of the computer, for nesting, layout, and data handling.

Advantages.   NC and CNC provide the following advantages for flame machining:

1. Simple, quick plate layout that can be checked on the x–y plotter.
2. Minimum scrap allowance due to optimum nesting of multiple cuts.
3. Elimination of plate or template alignment for each row speeds multiple part production.
4. Standard shapes that can be cut without waiting for templates.
5. Elimination of template preparation, storage, and maintenance.

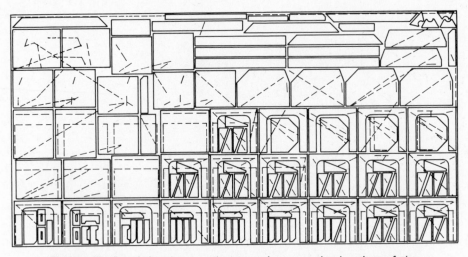

Figure 13.19. A hard-copy plotter makes a scale drawing of the actual parts that will be made by the flame cutter. Layout lines for subsequent bending or hole drilling may also be placed on the plate by a pneumatic punch head. The punch makes a series of indentations to indicate the bend line or hole to be drilled. (*Courtesy Modern Machine Shop.*)

6. Self-correcting path control, precise restarting after cut interruption, and elimination of human error in torch setting.
7. Indirect advantages of scheduling more work (60 to 85%) per machine.

Limitations.   To use CNC profitably, procedures must be exact. Each step must be thought through in a precise, disciplined manner.

Although various programming languages and a minicomputer system may be used, much of the software (program to suit in-house machines) must be developed for maximum performance.

PLASMA-ARC CUTTING (PAC)

The main difference between the plasma torch used in welding and that used in cutting is that the cutting torch does not require shielding gas. Plasma-arc cutting differs from oxyacetylene cutting in that plasma is not dependent on a chemical reaction with the metal as in the rapid oxidation of steel. The plasma-arc torch consists of a constricted arc that operates between the tungsten electrode and the workpiece (transferred arc). The arc furnishes the highly concentrated heat to melt the metal and a high-velocity gas stream removes the metal from the kerf. The cutting gas also protects the kerf walls from oxidation. Since the cutting process is independent of chemical reaction, it can be used to sever either ferrous or nonferrous metals.

As shown schematically in Fig. 13-20, a pilot-arc circuit is used to facilitate

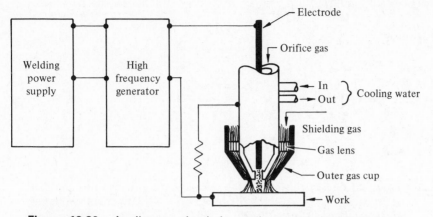

Figure 13.20. A pilot-arc circuit is used to start the plasma-arc torch.

starting the arc. The pilot-arc circuit connects the shielding cup or nozzle to the ground side of the power supply through a current-limiting resistor. In making a mechanized cut, the torch is first positioned about ⅛ in. (3.17 mm) above the starting edge of the work surface. When the start button is depressed, high-frequency current flows to establish the pilot arc between the electrode and the workpiece. The pilot arc sets up an ionized path for the cutting arc. As soon as the arc is established the high-frequency current is shut off and the carriage moves forward for the cut. Water cooling is necessary to prolong the life of the nozzle.

More recently water-injection plasma cutting has been developed. The water-injection method utilizes a radial flow of water directed at the plasma from an annular slot in the nozzle to constrict the arc and cool the workpiece. The constriction and cooling effects provide a narrower kerf, cleaner cut face, longer-lasting torch parts, and a narrow heat-affected zone. In this torch (Fig. 13-21) the arc can be directed to the cut side of the metal, producing a square cut as shown. The other side, or scrap side, has more of a chamfer.

The orifice gas, fed to the torch at the rate of 50 to 350 cu ft/hr, is determined by the particular application. Aluminum and magnesium are cut with nitrogen, nitrogen-hydrogen, and argon-hydrogen. Stainless steel thicker than 2 in. (50.8 mm) is best cut with argon and hydrogen.

Oxygen is an excellent cutting gas although its corrosive effect on the electrode makes it expensive. The best way of using oxygen is to introduce it downstream in the flow orifice, as shown in Fig. 13-22.

Advantages   The plasma-arc torch has several advantages over the conventional oxyacetylene cutting torch:

1. It can cut faster and leave a narrower kerf (width of cut). Mild steel, 0.75 in. (19.05 mm) thick, can be cut at 70 ipm (1778 mm/min), compared to oxyacetylene cutting at 20 ipm (508 mm/min) (Fig. 13-23).

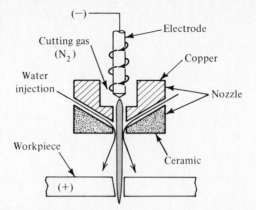

Figure 13.21. Schematic of the water-injection plasma torch used in metal cutting. Note the straight cut on the right side of the flame and the beveled cut on the left side. (*Courtesy* Welding Design and Fabrication.)

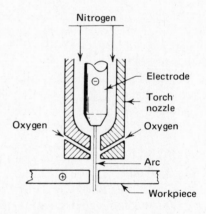

Figure 13.22. Oxygen introduced down-stream in the flow orifice eliminates oxidation and erosion of the electrode. Plasma gas mixtures of 80 nitrogen and 20 oxygen cut mild steel 25% faster than does the conventional plasma torch.

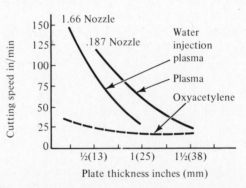

Figure 13.23. The water-injection plasma torch with a smaller diameter nozzle shows a rapid reduction in speed with the plate thickness. (*Courtesy* Welding Design and Fabrication.)

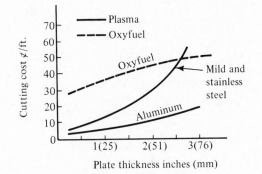

Figure 13.24. The plasma arc shows a cost advantage of 4:1 over oxyfuel on ¼-in. (6.4-mm) mild steel plate, and a cost advantage of 2:1 over oxyfuel on 1½-in. (38-mm) mild steel plate. (*Courtesy Welding Design and Fabrication.*)

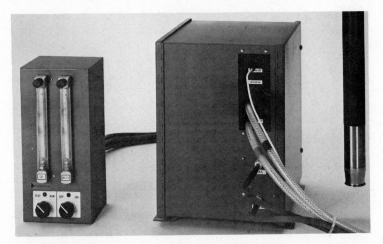

Figure 13.25. A plasma torch with a water muffler. (*Courtesy Hypertherm, Inc.*)

2. Operating costs for PAC are about ¼ that of the oxyfuel on ¼-in. (6.4-mm) mild steel and it has a cost advantage of 2:1 over oxyfuel on 1½-in. (38-mm) mild steel plate (Fig. 13-24).

3. Cutting is quite automated. The power control unit performs all on-off and sequencing functions. Height of the torch above the work is automatically adjusted by a capacitance circuit. Once the right height is reached, the circuit "locks in" to a voltage circuit that maintains the proper height.

4. The heat-affected zone is narrow.

5. High-quality cuts, especially with water injection, are obtained.

6. Any conductive metal can be cut.

Disadvantages:

1. High initial investment.
2. The thickness of the metal that can be cut is limited by current capacity of the torch. A heavy-duty mechanized torch rated at 1000 amp will cut 7-in. (180-mm) thick stainless steel or 8-in. (200-mm) thick aluminum.
3. Formerly the noise and smoke associated with the process were considered a severe disadvantage. Now, however, with cutting tables that have water up to the bottom of the cut surface, smoke is no longer a problem. Also a water muffler (Fig. 13-25) is used to reduce the noise level from 110 db to 90–95 db.

OXYGEN-ASSISTED GAS-LASER CUTTING

A comparatively recent addition to the metal-cutting field is the oxygen-assisted gas-laser cutting torch as shown in Fig. 13-26. The laser beam is directed downward by the focusing lens. High-pressure oxygen enters the chamber and goes out the jet at high speed. The 525 watts of power is focused onto the surface, which, with the oxygen, is enough to produce a cutting action on the steel. The advantages of the laser-cutting process are that it typically produces accurate narrow widths of cut (kerf), smooth edges, and a significantly reduced heat zone, thus minimizing distortion.

Cutting speeds vary with the material and the thickness and complexity of

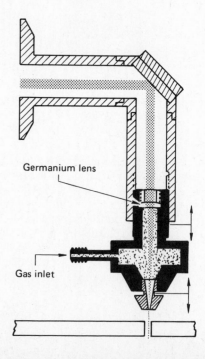

Germanium lens

Gas inlet

Figure 13.26. An oxygen-assisted laser beam used for cutting steel. (*Courtesy Society of Manufacturing Engineers.*)

the contour. Straight cuts in stainless steels 0.063 in. (1.6 mm) thick can be made at 60 in./min (1.5 m/min); 0.025-in. (0.6-mm) titanium at 300 in./min (7.6 m/min).

When the oxygen-assist cutting torch is NC controlled, large savings can result. As an example, stainless steel sheets 0.035 in. (0.89 mm) thick were to be profile cut. The dies needed to make the cut would have cost $2200. It took less than an hour to program the desired shape, resulting in a total cost of $250. The parts were cut on the oxygen-assist laser torch at 100 in./min (2.5 m/min), and the cost per part was $1.25.

HARD FACING

Hard surfacing or hard facing is a process of applying an overlay of metal on a part to provide a better, more wear-resistance surface.

Hard-facing alloys come in four classes: tungsten, cobalt, nickel, and iron base.

Tungsten.   Tungsten has excellent resistance to abrasive wear. Tungsten-base alloys are usually in the form of steel-tube rods containing tungsten powder. Particle size, shape, and weight can be varied as can the matrix material used to hold the particles in place.

Cobalt-base Alloys.   Cobalt-base alloys are characterized by excellent resistance to adhesive wear, heat, and corrosion. Alloys in this system are essentially Co–30 Cr alloys with varying amounts of tungsten and carbon.

Nickel-base Alloys.   Nickel-base alloys are noted for their excellent resistance to severe environments. Depending upon their composition, they are capable of balanced adhesive and abrasive wear resistance. The most common nickel-base alloys for hard facing contain chromium and boron: borides and carbides in a strong nickel matrix develop an extremely hard deposit (Rc 67) with excellent resistance to low-stress abrasion.

Iron-base Alloys.   Iron-base alloys generally are characterized by either excellent resistance to abrasive wear or excellent resistance to impact, depending upon their composition. They also cost less than other hard-facing alloys in other systems. Three hardening mechanisms are used (carbide formation, work hardening, and martensitic transformation) to achieve hardnesses ranging from Rb 80 to Rc 60.

Hard-facing Processes

Hard facing may be applied by most welding processes; oxyfuel, GTAW, GMAW, SAW, and FCAW.

Oxyfuel.   The oxyfuel method of applying hard surfacing has several advantages; penetration is limited; deposits are low in thermal stresses as well as decarburization; and they are smooth and free of slag. The deposition is slow, however, and the welder needs much skill.

GTAW.   The arc provides high-intensity heat, and therefore the base-metal dilution and thermal gradients are relatively great. If the base metal is allowed to dilute the hard-surfacing deposit, the wear resistance will be reduced. GTAW, however, provides a good method of overlaying small surfaces with a smooth deposit.

SMAW.   Even though the arc provides for high heat intensity, preheating is required because the thermal gradients are steep. Dilution is relatively high and deposits tend to be rough.

SAW.   The SAW process provides for high deposition rates, easy slag removal, and smooth welds of excellent quality. However, SAW is generally limited to flat or round parts that can be rotated in the flat position. Because of the high heat and deep penetration, dilution is a problem. Three or more layers are usually required for maximum hardness.

FCAW.   The flux-cored wire process (Fig. 13-27) is at present used primarily with iron-base alloys for buildup and hard facing. However, this process is gradually being used with tungsten- and cobalt-base alloys.

Surfacing with Powders and Wire

Hard facing with metal powders and wire has grown rapidly, particularly for coating original manufacture (OEM). Methods include flame spraying, plasma spraying, and plasma-arc welding.

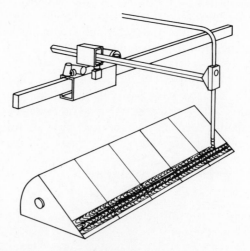

Figure 13.27.   Depositing hard surfacing on shovel teeth using the flux-cored electrode. (*Courtesy Stoody Company.*)

Flame Spraying. Flame spraying can be manual or automatic. Surfacing with oxyacetylene or oxyhydrogen gases applies coatings quickly and uniformly. The powder is injected into the torch, as shown in Fig. 13-28. Flame spraying is most effective on small cylindrical parts. Deposits usually have densities of 80 to 85%. The bond between the hard-facing alloys and the base metal is mechanical. Fusing (postheating) improves deposit density and bond.

Plasma-spray Process. In the plasma-spray process, the powder is fed into the arc and flows out as high-velocity droplets (Fig. 13-29). Though this metal produces a high mechanical bond, particles are molten, developing dense deposits that do not require fusing. Because of its high arc temperatures, the process can apply refractory coatings.

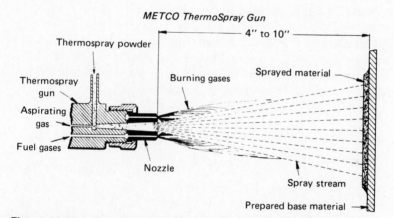

Figure 13.28. Hard-facing powders are injected into the gas stream and are carried out through the flame. (*Courtesy Metco Inc.*)

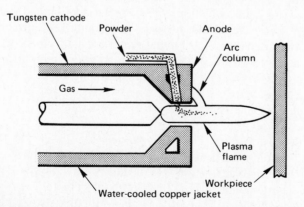

Figure 13.29. Hard facing using powder and the plasma-arc torch. (*Courtesy* Machine Design.)

A single pass lays deposits from 0.010 to 0.250 in. (0.25–6.3 mm) thick. The method is well suited to high-speed production requiring thin overlays, but heavy layers can also be made. The plasma-arc method is significantly lower in total costs for hard facing than other deposition methods.

Types of Bonds

Hard surfacing has two types of bonds, fusion and mechanical. The fusion bond is obtained in the various welding processes. Mechanical bonds are obtained in the powder- or wire-spray processes. For strong mechanical bonds the substrate should be as rough as possible. The surface may be roughed up by first using a sharp, pointed or threading tool at a relatively shallow cut and then going over this surface with a knurling tool. The knurling tool will press a diamond pattern on the threaded part. Mechanical bond strengths range from below 1000 psi (6.89 MPa) for a flame-sprayed coating to about 10,000 psi (68.95 MPa) with a plasma torch and a molybdenum undercoat.

Applications.   In the past, hard surfacing was thought of as a repair process. Now, however, the process is used extensively to achieve a better surface, either for wear or corrosion, to items in the process of being manufactured, for example, valves and valve seats, bearing surfaces, and large crushing rolls.

STRIP OVERLAY WELDING

Strip overlay welding, as shown in Fig. 13-30, is an SAW process that employs a strip of metal instead of the usual wire. The strip is 1.2 to 4.7 in. (30 to 120 mm) wide. Thus a wide path of material can be laid down in one pass. The process, which is two to five times faster than conventional hard surfacing, was introduced into this country in 1977 although it has been popular in Europe since the early 1960s.

Strip overlay welding has the advantage of eliminating high-intensity arc forces at any single point, which minimizes metal dilution and penetration. The transition zone from base to filler metal is only about 0.002 in. (0.05 mm) thick. Minimal dilution reduces the filler metal requirements from 10 to 20% over the conventional multiple-wire SAW process. In addition, the surface is relatively smooth in the as-welded condition. The smoother surface reduces machining and it also has less defects. Rework is reduced to about a fourth of what is normally expected in the submerged-arc wire process.

The basic limitation of the strip overlay process is that it can be used only on relatively thick substrates. For example, to use a 2.4-in. (60-mm) wide strip, the base metal should be at least 1.6 in. (40 mm) thick and have enough surface area to warrant the high deposition rate of the strip overlay welding method and to handle the large amount of heat that will be generated.

A variety of alloys are currently available for the strip process, including 99.5% Fe, 300-series stainless steels and some 400 series, several nickel-base alloys, and a number of proprietary Sandvik alloys.

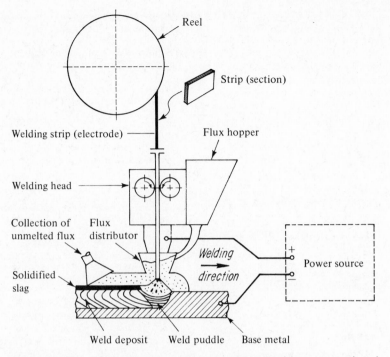

Figure 13.30. Strip overlay welding closely resembles conventional submerged-arc welding except for the electrode—it's a thin strip rather than a wire. (*Courtesy* Metal Progress.)

Questions and Problems

13-1. (a) Compare the cost of cutting 10 ft. of 1-in. thick plate between that of plasma and oxyacetylene cutting under the conditions given in Fig. 13-24.

(b) What is another main advantage of plasma cutting?

13-2. A 2-in. thick steel plate was machine cut with an oxyacetylene torch (radiograph). In examining the cut surface you determine the drag to be about 0.050 in. Would this be classified as a high-quality cut? Explain.

13-3. What is wrong with a flame cut that is irregular and with the top edge showing some evidence of melting?

13-4. Two pieces of aluminum tubing are to be joined as a brazed lap joint. The larger tube, 2 in. in outside dia., has a wall thickness of $\frac{1}{8}$ in. The nearest standard size $\frac{1}{8}$-in. thick tube wall that will fit inside is 1.750 in. in dia.

(a) What do you recommend be done to make a proper fit for brazing?

(b) What size should the smaller tube be?

(c) Both tubes are made out of 2024-T3 aluminum with a tensile strength of 64,000 psi, and a shear strength of 15,000 psi. How much overlap should be used? The shear strength of the brazing alloy is approximately 20,000 psi. The desired factor of safety is 3.

13-5. A stainless-steel pipe 2 in. in dia. is to be joined with a copper pipe. The inside diameter of the stainless steel pipe is 1.750 in. The copper pipe will have to be machined slightly on the outside diameter to obtain the desired size for a brazed joint. The coefficient of expansion for stainless steel is 8.5×10^{-6}/in./deg. F, and for copper, it is 9.8×10^{-6}. What size should the outside diameter of the copper pipe be? A nickel-alloy filler material will be used with a brazing temperature of 2100°F.

13-6. (a) What is meant by a eutectic solder?

(b) What is the disadvantage of using a strictly eutectic solder?

13-7. What are two advantages of using MAPP and propane over acetylene as a fuel gas?

13-8. What are some special applications of the carburizing flame?

13-9. Why can acetylene be compressed in a cylinder at 250 psi when it is self-explosive at 32 psi?

13-10. What is the main disadvantage of gas welding?

13-11. What would probably be the fastest method of brazing small assemblies on a production basis?

13-12. What is the main difference between brazing and braze welding?

13-13. Why is aluminum usually considered difficult to solder?

13-14. How can corrosion be eliminated in an aluminum-soldered joint?

13-15. What type of parts are particularly easy to program for NC flame machining?

13-16. What is the main advantage of NC over photocell tracers in flame machining?

13-17. What is the main difference in principle between an oxyfuel torch and a plasma-arc torch for cutting metal?

13-18. Why is a pilot arc needed for PAC?

13-19. What is the advantage of water injection for PAC?

13-20. Compare the cutting cost of 5 10-in. dia. 1.5-in. thick steel parts as cut by oxyfuel and a plasma torch.

13-21. What would be the difference in cutting time between plasma cutting and oxyfuel for cutting four 4 × 8 ft plates to make eight 2 × 8 ft plates $\frac{3}{4}$ in. thick?

13-22. How much time could be saved in Problem 13-20 if a water-injection plasma jet were used instead of the conventional plasma torch?

13-23. A long-run metal-forming die is made from medium carbon steel. To get good wear qualities the die was hard surfaced. What material and method of application should be used?

Case Study (Covers Chapters 13 and 15)

A fabricating company was asked to supply four stainless steel rings for large nuclear waste storage vessels. The rings were to be 157 in. (3.9 m) in dia., 12 in. (304.80 mm) wide, and $\frac{7}{16}$ in. (11.10 mm) thick. The problem is to figure out how these rings could be most economically fabricated. The largest stainless steel plates available were 110 in. (274 cm) wide by 298 in. (740 cm) long. The material is 304L stainless.

(a) How could these rings or segments

be economically arranged on the plate?

(b) How would they be best cut out?

(c) What type of joint preparation would be needed? Make a sketch showing the joint geometry and the possible passes required.

(d) What welding processes could be used?

(e) How can carbide precipitation be avoided in the welded joint?

Bibliography

Budinski, Kenneth G. "Surfacing I, What a Designer Needs to Know; Surfacing II, Weld Overlays That Resist Corrosion." *Welding Design and Fabrication* 49 (July 1976): 49–59.

Editor. "Computerized Cutting Needs No Programmer." *Welding Design and Fabrication* 52 (June 1979): 80–83.

Gettelman, Ken, ed. "Computer Graphics Organize the Nest."*Modern Machine Shop* (June 1978).

Hogan, John A. "Plasma Arc Cutting with Water Injection." *Welding Design and Fabrication* 51 (March 1978): 98–101.

Kapus, Bernard G. "Computer Control Takes the Variations Out of Hardfacing." *Welding Design and Fabrication* (June 1980).

Kirchner, R. W. "Choose the Right Surfacing Method." *Welding Design and Fabrication* 51 (August 1978): 53–57.

Lindberg, R. A., and N. R. Braton. *Welding and Other Joining Processes.* Boston: Allyn and Bacon, 1976.

Lutes, W. L., and H. F. Reid. "Hardfacing Helps New Equipment Last Longer." *Welding Design and Fabrication* 51 (August 1978): 51–52.

Metcalfe, J. C., and M. B. C. Quigley. "Keyhole Stability in Plasma Arc Welding." *Welding Research Supplement 401-S-404-s* (November 1975).

CHAPTER FOURTEEN

Resistance and Specialized Welding Processes

Resistance welding is based on the well-known principle that as a metal impedes the flow of electric current, heat is generated. The amount of heat generated is related to the magnitude of the electric current, the resistance of the current-conducting path, and the time the current is allowed to flow. Of course the metal makes a much better path for the current than the arc in arc welding, and therefore the current must be higher. The interface between the two surfaces of the workpiece offers the greatest resistance to current flow in comparison to the balance of the circuit and is therefore the area of greatest heat. The heat generation is directly proportional to the square of the current times the resistance and is expressed as:

$$H = I^2RT$$

where: H = heat generated in watt hours
T = time in hours
I = current in amperes
R = resistance in ohms

The heat may be changed from watt hours to Btus by multiplying the current by 3.412, since there are 3412 Btus in 1 kilowatt hour.

The principal types of resistance welding are:

Spot welding	Butt welding
Seam welding	Upset welding
Projection welding	Flash welding
High-frequency welding	

SPOT WELDING

Basic spot welding consists of clamping two or more pieces of metal between two copper electrodes, applying pressure, and then passing sufficient current through the metal to make the weld.

Pressure, weld time, and hold time are recognized as the fundamental variations in spot welding. For most spot welding, these variables must be kept within very close limits.

Pressure. Spot welds are made by cleaning the two pieces of metal to be lapped and placing them between the copper electrodes of the spot welder [Fig. 14-1(a)]. The squeeze time is used to bring the two workpieces together in intimate contact just prior to current flow [Fig. 14-1(b)]. It may be considered the most important variable in resistance welding. The resistance at the joint interface is inversely proportional to the pressure ($R = 1/P$). The effect of high

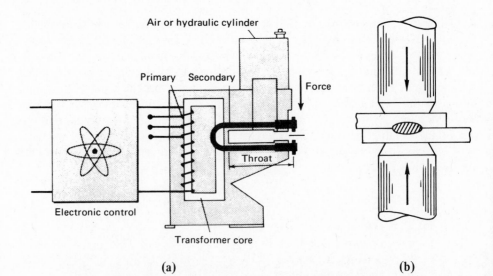

(a)   (b)

Figure 14.1. The fundamental electrical components usually employed in resistance-welding machines (a). The two pieces of metal are placed between the two copper electrodes for spot welding (b).

and low pressures on resistance are shown graphically in Fig. 14-2. Variations at low pressure have a great effect on resistance and vice versa. On welding equipment requiring high heat and pressure, the equipment is usually massive and air-operated.

Weld Time.   The weld time is the period when the current flows. It is usually timed in fractions of a second or Hz. For example, a spot weld can be made between two $\frac{1}{16}$-in. (1.6-mm) thick pieces of mild steel in 11 cycles [Fig. 14-3(a)].

Low-carbon steels offer little or no metallurgical problems when spot welded. Once the weld nugget is formed, the metal passes from austenite back to pearlite. If medium- or high-carbon steels were to be treated in the same manner, the rapid cooling of the spot would result in some embrittlement. This does not mean that medium- or high-carbon steels cannot be resistance welded, but other processes such as quench and postheat are needed to permit the proper heat treatment of the spot immediately after it is made.

Hold Time.   Hold time is basically a cooling period. It is the interval from the end of the current flow until the electrodes part. Water-cooled electrodes serve to transfer the heat away from the weld rapidly.

Other variables that may be sequenced along with the steps already described are low current for preheating; either high or low current as required for spot heat treatment; and a forging action to help refine the grain of the weld. This is done by increasing the electrode pressure during the cool time. Some of these added functions are particularly useful in joining hardenable alloy steels that may crack if welded with one surge of current and allowed to cool.

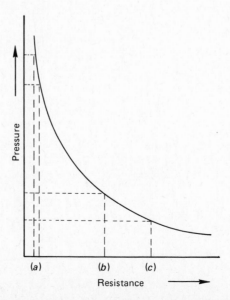

Figure 14.2. High welding pressures (a) create less resistance variation than medium (b) and low pressures (c). Since heat is proportional to resistance, there is less heat variation.

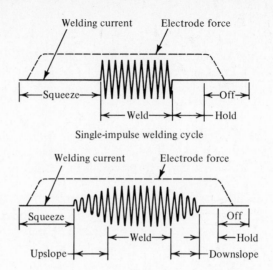

Single-impulse welding cycle

Single-impulse welding cycle with upslope and downslope heat control

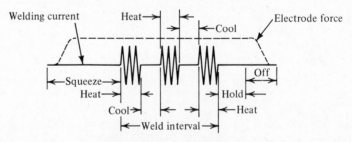

Single-impulse welding cycle with forging force and postheat

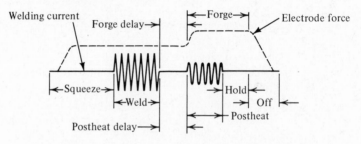

Multiple-impulse welding cycle

Figure 14.3. Basic-resistance spot-welding cycles show the relationship of time, current, and electrode force for each segment. (*Courtesy* American Machinist.)

Slope Control.   Slope control refers to the rise and fall of the welding current, as shown in Fig. 14-3(b). The current may be made to build up gradually (up-slope) or to decrease gradually (down-slope or current decay).

Up-slope gives the electrodes a few impulses of time to sink into the metal and seat before they have to carry the full welding current. If this is not done in welding aluminum, overheating occurs at the electrode tip, resulting in unwanted metal on the tip. The down-slope or decay helps to reduce internal cracking of the weld nugget. Retarding the cooling action ensures a proper temperature during the forging action. An increase in electrode pressure as the weld solidifies produces the forging effect [Fig. 14-3(c)]. Multiple-impulse welding cycles [Fig. 14-3(d)] provide short (a few cycles) heating periods alternating with short cooling periods for better control of the weld nugget, which reduces spitting. Spitting refers to the molten metal being propelled out from the weld nugget.

Spot-Welding Power Supplies

There are two basic types of resistance-welding power supplies: *stored* and *direct energy*. The stored-energy machines draw current at a relatively constant rate and store it so it can be discharged instantaneously when needed to make a weld. The direct-energy machines discharge the power to make the weld just as they receive it.

The stored-energy machine has the advantage of low-current input, which makes for a more balanced load on the plant's three-phase electrical system. In addition, variations in line voltage have relatively little effect on the efficiency of the machine. The most widely accepted method of energy storage for welders is by means of a bank of capacitors.

Direct-energy machines have the advantage of both lower initial cost and lower maintenance cost. The wave shape of the current is also more controllable, making it very versatile for a greater variety of metals.

Spot-Weld Design Considerations.   For spot welds to obtain maximum strength, the following design factors should be considered: the amount of overlap, the spot spacing, and accessibility.

Overlap.   The amount of metal overlap is determined by the weld size, which in turn is determined by the metal thickness. As an example, an acceptable weld size for a thickness range of 0.032 to 0.188 in. (0.812–4.775 mm) is roughly estimated at 0.10 in. (2.540 mm) plus two times the thickness of the thinnest member. The overlap should be equal to two times the weld size plus $\frac{1}{8}$ in. (3.2 mm), ($\frac{1}{8}$ in. is the tolerance for positioning the weld). If fixturing is used so that the spot will be positioned in the center of the overlap distance, the $\frac{1}{8}$ in. can be disregarded.

Welds placed too close to the edge will often squirt molten metal from the joint. Edge expulsion or "spitting" results in weak welds.

Spot Spacing.   The spots should not be so close that current is shunted to the previous weld, thus reducing the size of the weld being made. A general rule is to allow $16t$ between the welds, where t is the thickness of the material. If distortion becomes a larger factor than strength, then $48t$ is used. If it becomes

Table 14-1. Spot-welding data for mild steel.

Sheet Thickness	Tip Diameter*	Electrode Force	Weld Cycles	Amperes	Diameter of Fused Zone	Minimum Spot-Weld Spacing, in.
0.020	$\frac{3}{8}$	300	6	6,500	0.13	$\frac{3}{8}$
0.035	$\frac{3}{8}$	500	8	9,500	0.17	$\frac{1}{2}$
0.047	$\frac{1}{2}$	650	10	10,500	0.19	$\frac{3}{4}$
0.059	$\frac{1}{2}$	800	14	12,000	0.25	1
0.074	$\frac{5}{8}$	1,100	17	14,000	0.28	$1\frac{1}{4}$
0.089	$\frac{5}{8}$	1,300	20	15,000	0.30	$1\frac{1}{2}$
0.104	$\frac{5}{8}$	1,600	23	17,500	0.31	$1\frac{5}{8}$
0.119	$\frac{7}{8}$	1,800	26	19,000	0.32	$1\frac{3}{4}$

* Outside diameter of the tip dressed down to a 20-deg. bevel. The resulting diameter of the cone (d) will be $2t + 0.1$ in.

necessary to place spot welds where there is a likelihood of current shunting, the current must be increased to compensate. Table 14-1 lists typical spot-welding data for mild steel.

Spot Strength.   Spot strength varies directly with area. It is generally safe to assume that the strength will be equal to the weld area times the tensile strength of the metal in the annealed state.

Testing of Spot Welds.   The appearance of a spot weld can be very deceiving. A weld may appear good even though, in essence, there is no weld.

Three common destructive tests for spot welds are the peel test, tensile test, and cross-section test, as shown in Fig. 14-4.

The peel test is probably the simplest test to execute that gives effective results. All that is required is a sample strip and a means of rolling the top edge back. If the weld is good, the weld nugget will pull a hole in either piece. This is true in materials up to 0.094 in. (2.387 mm) thick. Greater thicknesses may only pull out a slug of metal, leaving a crater in the other piece.

Portable tensile-testing machines are often used for spot welds. These machines are of 10,000 psi capacity and will pull single spot welds in mild steel up to 0.094 in. (2.387 mm) thick.

The cross-section test consists of cutting through the middle of the weld, polishing it, etching it in a suitable acid, and inspecting it under a microscope. The penetration of the weld nugget into the base sheet should be from 40 to 70% of the sheet thickness. In addition to penetration, the metallurgical structure of the weld can be examined for grain size, microstructure, and any harmful effects such as carbide precipitation that may occur in stainless steel.

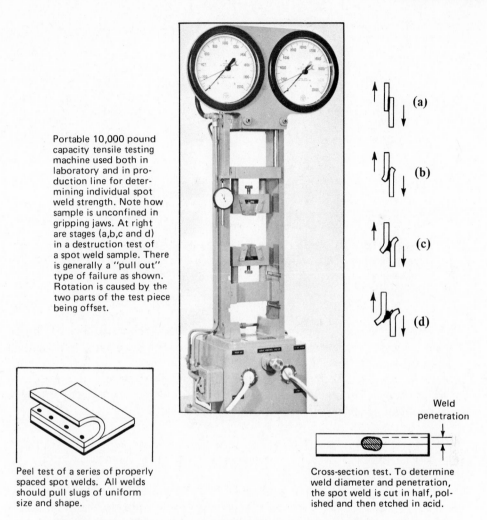

Portable 10,000 pound capacity tensile testing machine used both in laboratory and in production line for determining individual spot weld strength. Note how sample is unconfined in gripping jaws. At right are stages (a,b,c and d) in a destruction test of a spot weld sample. There is generally a "pull out" type of failure as shown. Rotation is caused by the two parts of the test piece being offset.

Peel test of a series of properly spaced spot welds. All welds should pull slugs of uniform size and shape.

Cross-section test. To determine weld diameter and penetration, the spot weld is cut in half, polished and then etched in acid.

Weld penetration

Figure 14.4.   Three common destructive tests for spot welds. (*Courtesy Acco, Wilson Instrument Division.*)

SEAM WELDING

Seam welding is a continuous type of spot welding. Instead of using regular rod-type pointed electrodes, the work is passed between copper wheels or rollers that act as electrodes. The appearance of the completed weld is that of a series of overlapping spot welds that resemble stitches. Seam welding can be used to produce highly efficient water- and gas-tight joints. A variation of seam welding used to produce a series of intermittent spots is called *roll welding* (Fig. 14-5).

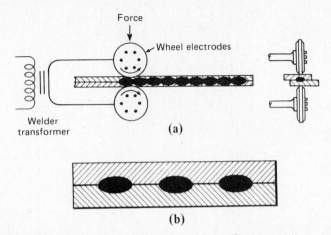

Figure 14.5. Seam and roll spot welding. Seam welding can be done as shown, or one of the wheels can be replaced with a flat backing electrode that supports the work for the entire length of the seam.

Seam-Weld Testing.   Seam welds can be tested as mentioned for spot welds but "pillow tests" can also be made. Two rectangular or square sheets are joined together by welding a seam all around the edges. A pipe connection is welded to the center of one sheet. Hydraulic pressure is applied through the pipe until the sheets expand and burst. The bursting pressure is recorded as is the length of the weld line. Failure should occur in the parent metal rather than in the weld.

PROJECTION WELDING

Projection welding is another variation of spot welding. Small projections are raised on one side of the sheet or plate with a punch and die. The projections act to localize the heat of the welding circuit. During the welding process, the projections collapse, owing to heat and pressure, and the parts to be joined are brought in close contact (Fig. 14-6).

One of the most common applications of projection welding is for attaching small fasteners, nuts, special bolts, studs, and similar parts to larger components. A wide variety of these parts are made with preformed projections.

HIGH-FREQUENCY RESISTANCE WELDING

High-frequency resistance welding of tubes and strip material is an adaptation of an old idea, that of causing current to flow across a metallic joint, heating the edges to the melting point, and then joining them together. What is new about the process is that a high frequency (450 kHz) is used instead of 60 Hz.

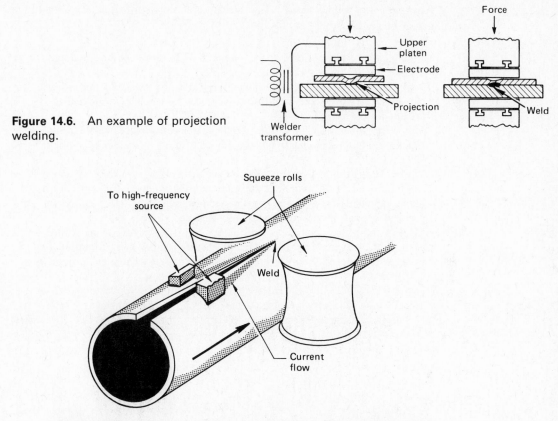

Figure 14.6. An example of projection welding.

Figure 14.7. A working section of a high-frequency welding unit. High-frequency current heats the metal edges to the welding temperature and at the same moment the squeeze rolls pinch the moving edges of the tubing together. (*Courtesy* Metal Progress.)

The current does not follow a direct low-resistance path between the two contacts; rather it follows a long, low inductance path as shown in Fig. 14-7. This circuitous path of the current allows adjustment of the heating on either side of the weld (an asset when joining dissimilar metals). Also the shallow depth of the inductance path produces high-intensity heating in shallow zones, resulting in a minimum heat-affected zone. The localized high-intensity heating allows the whole edge of the material to reach welding temperature at the same rate as the strip progresses through the welding machine. Pressure applied at the root of the "V" brings the two edges together in a forge weld.

The oxide films on the metal surface act as a high-frequency conductor by virtue of their peculiar properties at high frequency.

Because the metals are pinched together (in tube welding) at the point of the "V," molten material, if present, is squeezed out (upset) and the weld is essentially forged.

Advantages and Limitations.   The advantages of high-frequency resistance welding can be listed briefly as follows:

1. Joining can be done at high speed—up to 1000 fpm in light-gage materials up to 0.012 in. thick (0.56 mm).
2. Very thin sheets, ranging from 0.004 to 0.020 in. (0.10 to 0.51 mm) thick, can be lap welded. The contact pressure on the seam is so light that it creates no problem.
3. Metals up to 0.25 in. (6.35 mm) thick have been welded satisfactorily. Thicker metals sometimes require a device to preheat the edges to about 1000°F (537°C).
4. The power required is a fraction of that needed for standard resistance welding, the reason being that the effective resistance of the metal is so much higher at 450 kHz that, by the I^2R law, the amperage required for a given amount of heat input is sharply reduced.
5. Most of the process variables that affect normal resistance welding, such as contact pressure, contact resistance, surface oxides, etc., do not apply because the joint faces are heated without actually touching each other.
6. The heat-affected zone is very small.
7. Dissimilar metals can be welded together. An unusual application is that of band-saw blades, where high-speed steel teeth are welded to a high-carbon-steel backing strip at 50 to 100 fpm.

The main disadvantage is that the process is limited to high-production operations.

RESISTANCE BUTT WELDS

Upset.   The material to be welded, usually bar stock, is clamped so that the ends are in contact with each other. High-density current, usually from 2000 to 5000 amp per sq in., is used to make the weld. The high resistance of the joint causes fusion to take place at the interface. Just enough pressure is applied to keep the joint from arcing. As the metal becomes plastic, the force is enough to make a large, symmetrical upset that eliminates oxidized metal from the joint [Fig. 14-8(a)]. The metal is not melted and no spatter results. For most applications, the upset must be machined off before use.

Flash.   In flash butt welding, the ends of the stock are clamped with a slight separation before welding. As the current is turned on, it jumps the gap and creates a great deal of heat. Some of the metal burns away. The two pieces are moved toward each other in an accelerated motion, and as they reach the proper temperature, they are forced together under high pressure and the current is cut off [Fig. 14-8(b)].

Pipe-Line Flash Butt Welding.   The process of flash butt welding has recently been expanded for pipe-line welding. Traditionally, the process had been limited by the current-carrying capacity of portable equipment. As an example, work-pieces with a cross-sectional area up to $15\frac{1}{2}$ in.^{2} (10,000 mm^2) require 500 to

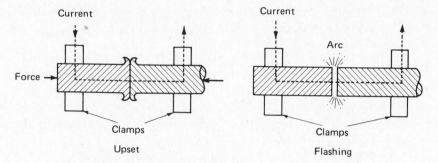

Figure 14.8. Resistance butt welding: upset and flashing.

WORK VIBRATES AXIALLY DURING FLASHING

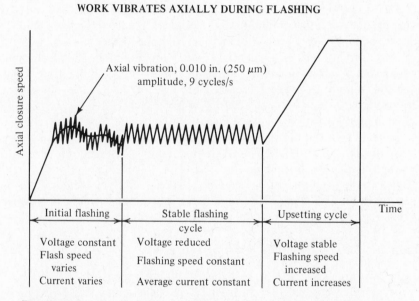

Figure 14.9. The flash butt welding current-time sequence showing the effect of axial vibration of the pipe. (*Courtesy* Welding Design and Fabrication.)

800 kW and weigh 20 to 30 tons (18–27 Mg). Researchers at Paton Welding Institute, Kiev, U.S.S.R., have developed a system of vibrating the workpieces axially as they slowly come together to achieve optimum electrical resistance.

Another problem that was overcome was reduction of the impedance in the welding circuit. In a conventional butt welder with a power rating of 200 to 800 kW, impedance ranges between 170×10^{-6} and 300×10^{-6} ohm. A new circuit design reduces impedance to $\frac{1}{2}$ to $\frac{1}{15}$ of that normally expected. As a result, power requirements are only 20 to 25% of that of a conventional flash butt welder. The diagram in Fig. 14-9 illustrates the current–time–vibration relationship.

This system is used to flash butt weld 56-in. (140-cm) dia. pipe sections.

Normally a pipe-line crew consists of 48 or more welding-related people. The flash-butt-welding crew consists of 12 to 14 men. Flash butt welding is consistently done at eight welds an hour.

Advantages and Limitations of Resistance Welding.   Spot welding can be used for fast, economical joining of steels up to $\frac{1}{8}$ in. (3.17 mm) thick and occasionally up to $\frac{1}{4}$ in. (6.35 mm) thick. The process is well suited to high-speed mass-production setups. The use of automatic current control, timing, and electrode force makes spot welding a reliable process.

Seam welders can be used to make gas- or liquid-tight joints. Seam welding also requires less overlap than spot welds and can in fact be made with less overlap than the diameter of the spot weld.

Flash welding can be used to produce assemblies that otherwise would require forgings or castings. Flash welding is used in producing such widely varying products as aircraft landing gear, bandsaw blades, pipe lines, and welding tap and reamer bodies to low-carbon steel and alloy steel shanks.

Projection welding is particularly useful in mounting metal stampings in material from 0.022 to 0.135 in. (0.53–3.43 mm) thick. Projection welds also reduce the amount of current required because of the greater current density where needed.

The major limitation of resistance welding is the heavy current-level requirements, which impose a high kVA demand on the plant. Even though the duration of the current flow is short, the peak power demand may constitute an undesirable load.

Spot-welding equipment is relatively expensive. Spot welders can be purchased for less than $1000 but to produce satisfactory welds in a variety of materials, larger machines with more sophisticated controls are usually needed.

Homopolar-Pulse Resistance Welding (HPRW)

The HPRW process is a new resistance-welding process in which the faying surfaces (surfaces to be joined) are held in solid contact and are heated by resistance to electric current. In this respect the process is very similar to upset butt welding. However, there are several process characteristics that are different.

The homopolar generator can be modeled electrically as a large variable capacitor. As the motor generator gets up to speed, a switch is opened that allows the generator to coast. When the speed has been reduced to the appropriate point, the pneumatically actuated brushes are lowered onto the rotor and a fraction of a second later the discharge switch closes to form a very high-current, relatively low-voltage dc pulse for the weld.

The important point is that at the time the pulse is delivered there is no connection between the homopolar generator and the electric line. This is a departure from conventional resistance welding, which is usually subject to high-power demand charges because of the sudden loads imposed on the utility power lines.

Very little if any time is lost in waiting for the system to become ready to deliver the next weld pulse and the electrical load on the utility is essentially constant. Usually, before the workpiece is removed and the next piece inserted, the power required is available. For large, low-production welds, the machine charging (or rotor accelerating) time could be relatively long, requiring only a small power input but still producing the megawatt-level output pulses needed for the weld.

Although homopolar generators have existed since the days of Michael Faraday, they have not been used extensively because they are made for high-current low voltage, which is not suited for transmission lines. However, for resistance welding the electrical characteristics are ideal, particularly now because manufacturers are trying to avoid peak loads.

SPECIALIZED WELDING PROCESSES

Technological advances have produced many new welding processes since World War II. Some of them have been developed as a response to better methods of dealing with reactive materials. Some of the processes are new and others are modifications of older processes. These may be classified into three areas: fusion, resistance, and solid state. Significant newer specialized processes in the fusion area are electron-beam welding and laser welding.

Electron-Beam Welding (EBW)

Electron-beam welding is perhaps the most spectacular of the more recent welding developments to become commercially feasible. Electrons, under normal circumstances, are contained within the molecular structure of the parent material by inherent electrical forces. If the metal (usually titanium) is subjected to high temperatures, free electrons will "boil off" the surface of the materials. In many electron beam welders, a wire filament is heated to thermionic-emission temperatures by an electrical current to produce electrons (Fig. 14-10). Another system uses an indirectly heated tungsten rod that is encircled by an auxiliary filament. In any case, free electrons are simultaneously accelerated and collimated into a beam by exposure to an electrical field. As the stream of electrons leaves the anode, it diverges due to the repulsion of individual electrons. At this point, they are exposed to a magnetic field in a focusing coil that refocuses the beam without affecting the velocity of the electrons. The electrons travel up to about 60% of the speed of light but do not give off any heat until they strike the target. The intensity of the beam can easily be controlled by the focusing coil. Sharp focusing is desirable for most welding, but the out-of-focus broad circle can be extremely useful in scouring contaminants from the face of the work prior to welding.

In generating free electrons, the emitters must be operated within a high-vacuum environment to minimize oxidation. A vacuum is also needed in the high-voltage gap, where the electrons are accelerated to prevent a gas discharge or arc. A vacuum or partial vacuum is desirable along the entire path of beam transmission to reduce the frequency of collisions between electrons and air

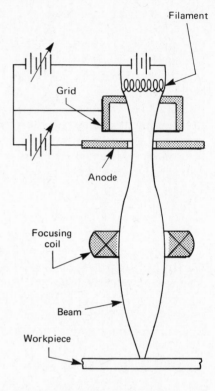

Figure 14.10. A schematic view of the electron-beam welder. A wire filament is heated to thermionic emission temperatures by an electric current to produce electrons. The free electrons are collimated by exposure to an electrical field. As the stream of electrons leaves the anode, they are exposed to a magnetic field in a focusing coil for high concentration of energy.

molecules. The collisions cause the electrons to scatter, which increases the diameter of the beam and reduces its effectiveness. For high production, the necessity of a hard vacuum, 0.0001 Torr (as compared to atmospheric pressure of 760 Torr), presents problems. The nonproductive time required to pump down the vacuum can take anywhere from 3 to 30 min, depending on chamber size, pump size, and vacuum level required.

Production can be increased considerably by performing the welding operation at atmospheric pressure [Fig. 14-11(a)]. In this type of arrangement, the beam-producing elements are surrounded by a hard vacuum produced by a diffusion pump. A neutral gas is made to flow past the exit orifice to prevent metal vapor from being drawn up into the beam and to protect the fusion zone. The weld quality is not greatly reduced, but the distance through which the beam can penetrate is reduced somewhat. Helium used at the beam outlet increases both the speed and penetration of the weld [Fig. 14-11(b)].

The out-of-vacuum system frees the welding process from restrictions on production cycling caused by the time required to pump down the welding chamber. Beam penetration is restricted to about 1 in. (2.54 cm) (in larger systems), with a correspondingly wider cross section of weld. The workpiece must be positioned approximately 0.5 to 0.75 in. (12.70 to 19.05 mm) from the beam-exit orifice.

When high-energy electrons impinge on a solid material and are suddenly

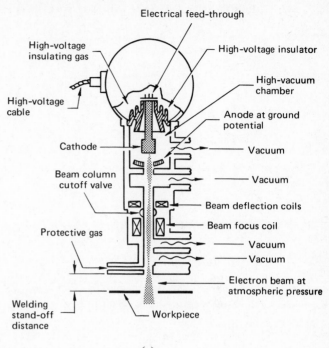

(a)

Figure 14.11. The electron-beam welder can be operated at atmospheric pressure (a). In this case, the electron-beam-producing elements are maintained at a hard vacuum by a separate diffusion pump and a semivacuum is maintained along the beam transmission path to limit the spread of the beam caused by collisions between electrons and air molecules. The very end of the beam-exit nozzle is slightly pressurized by an effluent gas to prevent metal vapors from getting into the transfer column. With air, the practical distance between the exit nozzle and the workpiece is limited to about 1.2 in. Using helium, this distance can be increased to 1.8 in. The speed also can be increased as shown at (b). (*Courtesy* Machine Design.)

(b)

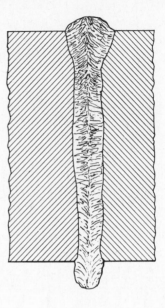

Figure 14.12. High depth to width welds are made with the electron beam.

decelerated, X rays are produced. When welding is done within the evacuated chamber, the X rays are of no consequence; but in the out-of-vacuum system the personnel must be protected. Because X rays travel in a straight line, a labyrinthine arrangement of lead shielding is usually effective.

Advantages and Limitations. Electron-beam welding confines heat to a narrow area, making it possible to have a depth-to-width penetration of about 25:1 (Fig. 14-12). Postwelding operations are negligible or unnecessary since there is no contamination of the weld. Welding speeds can be high, up to about 200 in. per min (508 cm/min) depending upon depth of penetration required and other factors. Electron-beam systems are available with power ratings ranging from 6 to 60 kW to handle a wide variety of work.

Numerical control has been combined with the electron beam to obtain the best speed and beam path. A lower-power scanning beam can be used to obtain detailed path description (Fig. 14-13). Once the path has been defined, the beam power is brought back to the welding level and made to traverse it at a constant speed, maintaining the point of impingement on the joint to close tolerances. Information concerning path variances can also be fed to the computer, which will cause predetermined corrective actions to be taken.

Edge and butt joints can be made in metals as thin as 0.001 in. (0.03 mm). In addition, small, thin parts can be welded to heavy sections. Dissimilar metals are joined as effectively as similar metals.

Electron-beam welding is not practical where a wide gap filling is necessary. For butt welding parts thicker than 0.100 in. (0.254 mm), the air gap should be not more than 0.005 in. (0.13 mm).

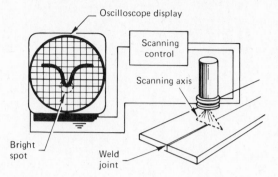

Figure 14.13. The electron beam is made to traverse the weld area with a scanning beam to ascertain the exact location of the joint. This information is then programmed to direct the beam for the actual welding. (*Courtesy* Automation.)

Electron-beam equipment is relatively expensive although small systems are now commercially available at affordable prices and contract work is also readily available.

Laser Welding

The term *laser* stands for "light amplification by simulated emission of radiation." The idea of the laser goes back to Niels Bohr's theory that electrons are given off or absorbed as the energy level of the material changes. Electrons exist only in definite shells. The energy the electron possesses is determined by its distance from the nucleus. Figure 14-14(a) shows a schematic of an electron in shell E_2, which has more energy than electrons in shells E_1 and E_0. Energy is given off as the electron falls back to a lower shell. Thus the electron can absorb or give off energy as it changes levels. A phenomenon that makes the laser possible is that in certain materials an electron can be in a semistable level [Fig. 14-14(b)], so that when a photon of radiation of exactly the right frequency triggers the electron, the energy can cause the electron to fall to a lower level and in the process give off a photon of radiation that is exactly in phase with the photon that caused the triggering action. As shown schematically by the arrows, the triggered photon has twice as much radiant energy.

For laser action to occur, the majority of the electrons must be at the upper energy level. This condition is referred to as a population inversion and is produced through external excitation of the lasing medium, commonly referred to as a "pumping" action that may be achieved optically or electronically.

Essentially, a laser consists of a reservoir of active atoms that can be excited to an upper energy level, a pumping source to excite the available active atoms, and a resonant cavity to ensure that the radiation will make many passes through the active laser medium, thereby obtaining the maximum propagation of stimulated emissions.

Lasers can be classified as solid-state, liquid, or gas types. Only solid-state and gas lasers (Fig. 14-15) are presently used in welding and metalworking.

A typical solid-state lasing medium consists of a single crystal rod of yttrium–aluminum–garnet (called yag), which contains ions of neodymium. Neodymium–doped yag (Nd : yag) has replaced most other materials, including the

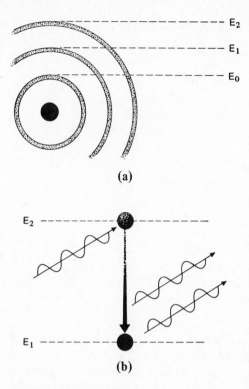

(a)

Figure 14.14. When electrons are pumped or stimulated they go to a higher energy level (a). Energy is emitted when electrons move to a lower level and population inversion takes place as shown at (b). The stimulated emission produces an in-phase or coherent light beam.

(b)

original ruby rod, for high-energy, crystal-laser applications because it allows high pulse rates, has relatively good efficiency, and can be operated with simple cooling systems.

The solid-state laser can be pumped by a flashlamp mounted in a reflecting cavity that also contains the laser rod. The setup as shown in Fig. 14-15(a) utilizes a linear flashlamp parallel to the laser rod inside an elliptical gold-plated enclosure. Flashlamps are usually xenon- or krypton-filled.

At one end of the laser rod is a totally reflective mirror and at the other end is a partially transmitting mirror. The area between the mirrors forms a resonating cavity, causing a preferential buildup of energy until it bursts out of the transmitting end.

Gas lasers consist of an optically transparent tube filled with either a single gas or a gas mixture as the lasing medium. The principal gas lasers are He–Ne, Ar, and CO_2 lasers. The pumping source is some form of electrical discharge applied by electrodes or a special electron-beam arrangement. A typical CO_2 laser actually uses a mixture of three gases: carbon dioxide, helium, and nitrogen for high-power operation. In flowing-gas systems, this mixture is constantly pumped through the resonator to sustain the lasing action.

The cross-flow gas laser eliminates the gas tube, which can become quite cumbersome. The working gas is pumped at high speed at the cross flow to the direction of the laser beam. The system operates somewhat like a closed-cycle

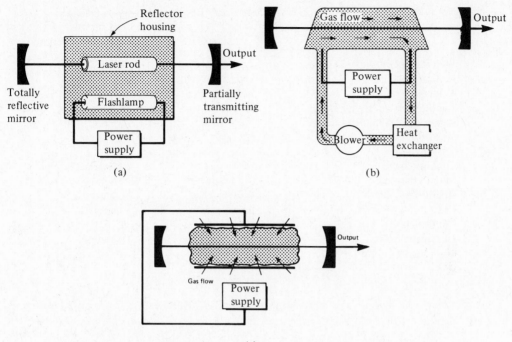

Figure 14.15. A schematic comparison of three laser systems: solid state (a), axial-flow gas (b), and cross-flow gas (c). (*Courtesy American Machinist.*)

wind tunnel, with heated gas passing through a heat exchanger before it is returned to the cavity area. The discharge is stabilized by preionization of gas and with a broad-area electron beam. This cross-flow–electron-beam preionization combination allows power levels of over 15 kW with a laser-cavity length of approximately 4 ft.

For fabrication purposes, the total energy available represents the laser capacity for work. Energy is expressed in joules (J), which are equivalent to watts × seconds. A laser capable of delivering 24 J at a rate of one pulse per second is considered to be a 25-W laser. But if the 25 J are emitted in a single pulse of only 1×10^{-3} sec, then the laser would achieve a peak power of 25,000 W.

Lasers are operated in either a continuous wave (CW) or a pulse mode. Solid-state lasers, such as ND : yag, are usually operated in the pulsed mode because of flashlamp limitations, but continuous operation is also available. Gas lasers are usually operated in a CW mode and are capable of developing higher continuous average power than the solid-state lasers.

Historically, some of the first applications of lasers to welding involved the joining of very small wires. Now, however, solid-state lasers producing 50 J and more are common. Seam welds can be made by overlapping spots at the rate

Table 14-2. Laser vs. electron-beam welding. (*Courtesy* Machine Design.)

Characteristic	Laser	Electron Beam
Welding speed	Low	High
Material range	Extremely wide	Wide
Thickness range	Narrow	Wide
Joint-design versatility	Excellent	Excellent
Range of dissimilar metals	Very wide	Wide
Need for vacuum chamber	No	Yes
Gas coverage	No	No
Tooling costs	Low	Moderately high
Initial cost	Moderate	High
Operating costs	Low	Moderate
Ease of automation	Excellent	Excellent
Controllability	Very good	Good
Heat generation	Low	Moderate
Miniature welding	Limited	Excellent

of 12 in. (30.48 cm) per min. High-power CW solid-state lasers are being used to produce full weld penetration up to 0.125 in. (3.175 mm) at speeds of 10 in. (25.4 cm) per min. Shallow welds are made at faster rates.

Welding at high power levels in thick materials is performed similarly to keyholing with the previously described plasma-arc torch. A hole is opened up so that the laser can penetrate deep into the workpiece. Although there is some initial surface vaporization, the relatively thick section of the base material forms a puddle and the flow of molten material recloses the hole as the beam rapidly advances along the weld path.

Laser welding is most often compared to electron-beam welding since both use a stream of energy to fuse the metals. The actual processes are quite different, as shown in Table 14-2.

Advantages and Limitations. Laser welding has the advantage of controllability, which makes it ideal for tiny precision parts. Intense energy can be brought to bear on a tiny area. For this reason high-melting-point materials and refractories that are virtually unweldable by any other process can be fused together. In addition, it is a noncontacting technique, which makes it possible to weld two pieces of material together at the bottom of a small hole or any other inaccessible spot. Most lasers operate in the visible-light range and thus only affect what they can "see." This makes it possible to weld components inside a clear tube envelope, as in vacuum tubes.

The heat generated in laser welding is very concentrated, thus the physical and chemical properties of the materials are not significantly affected. A 10-in. (25.4-cm) seam weld feels cool to the touch immediately after welding.

The laser power source is only 30% as expensive as that of EB (electron beam). As an example, a complete EB system may cost a minimum of $100,000, whereas a good 60-joule laser capable of welding materials up to 0.040 in. thick (1.02 mm) can be obtained for $20,000.

One of the drawbacks to the use of lasers in welding has been comparatively short life span. The ruby and Nd : yag units are good for about 10,000 shots at maximum rating before the flash tube must be replaced. If operated well below the maximum, a flash tube may last for 100,000 shots.

The laser welding is basically a slow process since only a certain pulse rate is possible. At present, only thin materials are being welded. The depth of penetration is normally restricted to about $\frac{1}{8}$ in. (3.17 mm).

High-quality welds up to about 0.080 in. (2.03 mm) deep can be produced in stainless steels at a rate of 2 in. (50.80 mm) per second using the Nd : YAG laser.

Although the laser is an excellent means of welding dissimilar materials, there are some limitations. For example, it is not practical to weld titanium to stainless steel as the two metals are incompatible.

SOLID-STATE WELDING PROCESSES

Theoretically, it is possible to place two perfectly clean metallic surfaces into such intimate contact that the cohesive forces between the atoms would be sufficient to hold them together. This condition is approached in "wringing in" two precision gage blocks as described in Chapter 21. As good as this method is, it does not approach a perfect bond. The astronauts experienced some welding of parts in outer space due to the nonoxidizing atmosphere.

Despite difficulties, some progress has been made toward solid-state bonding. Welding processes that may be classified under this heading are ultrasonic, explosive, diffusion, and friction processes.

Ultrasonic Welding

Man's ability to hear sound extends from frequencies of 20 to about 14,000 cycles per second. It is quite easy to produce high-frequency electrical pulsations by several commercially available types of power supplies. A device is then needed to convert the electrical pulsations into mechanical vibrations. This device is called a transducer. There are two common types of ultrasonic transducers. One depends upon a piezoelectric effect in a quartz or ceramic crystal of particular geometry; a second type depends upon a magnetostrictive effect in a metal stack as used in welding. A coupler and horn are attached to the transducer to provide the physical displacements at the welding tip. The tip displacement is usually between 50 and 600 microinches at the excitation frequency of 60 kHz. The horn, which is tuned to vibrate at exactly 20,000 Hz, transmits the vibratory energy into the workpiece (Fig. 14-16). The vibratory action first erodes the oxides and other contaminants on the interface surfaces. Once they are clean, the surfaces come into intimate contact and a solid-state bond takes place.

The tips are usually contoured to conform to the shape of the parts being assembled. Horns are usually constructed from a special titanium alloy that has an exceptionally high strength-to-weight ratio, or they may also be made from laminated nickel.

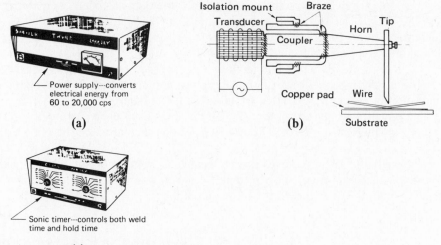

Power supply---converts electrical energy from 60 to 20,000 cps

(a)

(b)

(c)

Sonic timer---controls both weld time and hold time

Figure 14.16. A power source (a) converts conventional 60-Hz electrical energy to 20,000-Hz electrical energy. A transducer assembly then changes the electrical energy to mechanical vibrations at the same frequency (b). The high-frequency mechanical vibrations are then transmitted into the workpiece through a resonant section called a horn. All functions are controlled by a timing mechanism (c) since successful results depend on the right combination of time, pressure, and energy output.

The sonic timer controls the duration of the ultrasonic exposure and the duration of the pressure.

When used for automatic assembly, the sonic converter and horn are usually mounted on a pneumatically operated assembly stand that operates at regulated pressure to bring the horn in contact with the work surface.

Automatic parts handling to increase production includes a means of fixturing the part to prevent lateral or other movement during the ultrasonic exposure cycle. A variety of rotary indexing tables are used that include nesting devices for the use of single or multiple horns. Examples of recommended joint designs and a sketch of an integrated closed-loop system are shown in Fig. 14-17. Parts are automatically oriented and fed from the vibratory feeder to the conveyor where they are welded. Several ultrasonic-welding stations permit welding of different components at different stages of assembly.

Advantages and Limitations. Ultrasonic power is especially useful for the assembly of plastic parts. The process eliminates the need for solvents, heat, or adhesives. In addition it may be used for metal inserting and staking of plastic parts, as shown in Fig. 14-18. A further assembly application is in the reactivation of adhesives. Thus components may have the contacting surfaces coated with the desired adhesive but permanent assembly may be delayed until a convenient time.

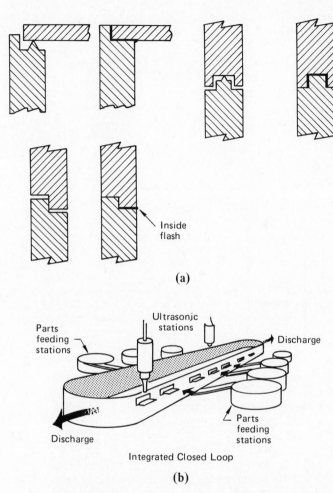

Inside
flash

(a)

Parts
feeding
stations

Ultrasonic
stations

Discharge

Discharge

Parts
feeding
stations

Integrated Closed Loop

(b)

Figure 14.17. Some recommended joint configurations for ultrasonic welding (a). An integrated, closed-loop system used to automate the assembly of parts by ultrasonic welding (b). (*Courtesy* Automation.)

The heat produced by the vibratory energy is not enough to affect the metallurgical properties of the materials joined. Bonds have from 65 to over 100% of parent-metal strength. The reduction of the joint thickness is usually less than 5%. The process is ideally suited for welding thin sheet and foil to itself or to any thickness. Several thin pieces can be joined simultaneously. Dissimilar metals may be joined, but aluminum and its alloys are the most weldable by this process. The ultrasonic process is especially good because the vibratory action breaks up and disperses thin layers of oxides that form, thus providing sound metal-to-metal welds. Normally, degreasing is all that is necessary. The time required to make a spot weld in aluminum sheet stock is 1 sec or less, in foil it is practically instantaneous.

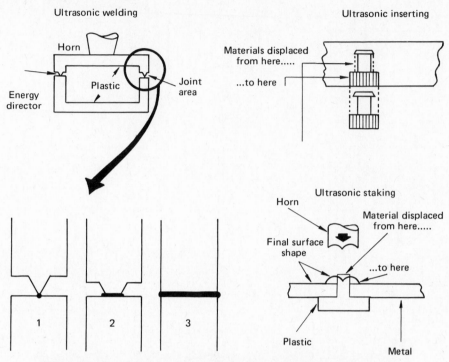

Figure 14.18. Three basic methods used in the assembly of plastic parts with ultrasonic energy. Note that in inserting, the plastic melts momentarily, permitting the part to be forced into place. The plastic then flows around the insert and encapsulates it. (*Courtesy* Automation.)

The maximum thickness of aluminum presently weldable is approximately 0.080 in. (2.03 mm) for high-strength aluminum alloys, e.g., 7075-T6, and about 0.120 in. (3.05 mm) for softer aluminum alloys, e.g., 1100-O.

Equipment life is relatively short due to fatigue encountered by the horn. Some materials being welded tend to weld to the coupling anvil.

Explosive Welding

Explosive bonding is the technique of using explosive charges to form a metallurgical bond between two pieces of metal at pressures of up to several million psi under conditions that produce a minimum of melt at the interface.

The basic technique involves placing two pieces of metal parallel to each other and separated by a small space. One of the metals is usually placed on a firm platform; the other will be impacted against the fixed piece by igniting a layered explosive charge (Fig. 14-19). As may be expected, this description is oversimplified. Usually the surfaces have oxides, nitrides, or absorbed gases. In explosive welding, the high pressure developed at the angled collision surface causes a portion of the surfaces, called a jet, to become fluid and be expelled.

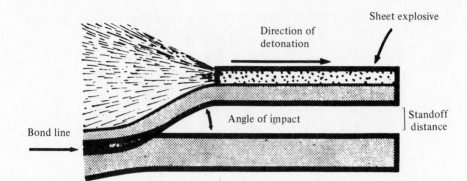

Figure 14.19.   The basic mechanism of explosive bonding is simple. An explosive charge is detonated, driving two or more plates against each other at high velocity. The resulting collision produces a strong, permanent bond.

The severe deformation of the colliding surfaces and the resulting jet break up any surface film, forcing the surfaces into intimate contact. The jetting phenomenon, required for bonding, causes the collision point to become plastic and flow into the space between the two plates. By this mechanism, the necessary conditions for the formation of a direct metal-to-metal bond occur. Melted zones are formed, but at discrete intervals along the bond rather than as a continuous layer. This surface jetting contributes greatly to the strength of the weld, providing a mechanical lock in addition to the metallurgical bond (Fig. 14-20).

Advantages and Limitations.   The major advantage of explosive bonding is that it is essentially a "cold-welding" process. The heat generated is considered insignificant and is rapidly dissipated. Explosive welding is especially useful in joining dissimilar materials and metals that are metallurgically incompatible. As an example, normal welding processes may cause two metals to melt but a brittle alloy forms at the fusion zone rendering it useless. Explosive welding eliminates this problem. Also, some materials are considered unweldable because the strain-hardening properties will be destroyed by the elevated temperatures. Except for small hardness and elongation changes, explosive bonding leaves the materials in the as-received condition.

The process is relatively inexpensive since very little equipment is required. Also it is very "portable."

One problem associated with explosive bonding is that it requires so many disciplines (chemical, metallurgical, mechanical, etc.). The process cannot be carried out in an ordinary factory due to the high noise level, so a remote location must be used unless adequate soundproofing can be found.

Theoretically there is no size limitation on the process; plates as large as 10 × 30 ft have been clad. However, there are difficulties when the parts or sections become very small. It is difficult to keep the effects of the explosion confined to the area desired. Despite this difficulty, most of the current interest is directed toward smaller-scale operations.

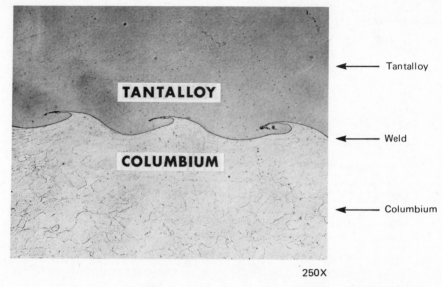

250X

Etchant: Lactic, HNO$_3$, HF

Figure 14.20. Photomicrograph of tantalloy explosively welded to columbium C103 alloy. Note the "rippled" effect at the interface, providing a mechanical lock in addition to the metallurgical bond. (*Courtesy Aeroject Ordinance and Manufacturing Co.*)

Flat plates are relatively easy to weld. However, as the configuration complexity increases, difficulties mount. In some cases positioning of the parts may require expensive jigs and fixtures. Also the charges must be shaped to fit the configuration.

Diffusion Bonding

Diffusion bonding refers to the metallurgical joining of metal surfaces by the application of heat and pressure to cause co-mingling of atoms at the joint interface. The process presupposes that the surfaces to be joined are clean and free of contaminants.

Diffusion bonding differs from fusion welding in that fusion welding is dependent on melting of the metals at the joint interface whereas, in most instances, diffusion bonding is accomplished entirely in the solid state, at about one-half to two-thirds the melting point of the metal.

There are three major conditions that require careful control for successful diffusion bonding. These are pressure, temperature, and time (at temperature and pressure).

The temperature accelerates the co-mingling of atoms at the joint interface and provides metal softening, which aids in surface deformation and more intimate contact. The temperature also aids in breaking up surface oxides.

The time is controlled to be at a minimum. The only time allowed is that sufficient to assure that surfaces are in intimate contact and some atomic movement has occurred across the interface.

Other variables that influence the quality of the bond are atmosphere and surface finish. A good vacuum ($>10^{-3}$ Torr) or oxygen-free inert-gas environments are best suited for diffusion bonding. Moisture-free argon or helium are commonly employed at less than $-70°F$, dew point.

It is essential to provide smooth conforming surfaces at the bond face. Soft materials conform readily and usually do not require other than normal, commercially obtained finishes. Hard materials such as refractory metals, tool steels, and super alloys require special attention. A surface finish no rougher than 16 microinches is recommended unless softer intermediate bonding films that can deform are used.

Surface cleanliness is mandatory. However, this does not imply that unduly restrictive handling is necessary. It merely means that good practice in surface preparation is essential. Vapor degreasing followed by pickling, rinsing, and thorough drying has been used effectively in preparing certain steels, copper, titanium, and other materials for diffusion bonding.

Diffusion Bonding Methods.   Common techniques used in diffusion bonding are hot-press bonding, isostatic hot-gas pressure bonding, vacuum-furnace bonding, explosive bonding, and friction bonding.

Hot-Press Bonding.   Hot-press bonding is a convenient method of diffusion bonding large, flat sheets or where the part configuration requires application of pressure in one axial direction. Usually a hydraulic press equipped with heated platens is used. The parts to be welded are sealed in a flexible metal envelope before inserting between the press platens. Bonding is accomplished in about 20 min at the required temperature and pressure.

The first aircraft to use diffusion-bonded titanium for primary load-carrying structures was the B-1. A total of 66 diffusion-bonded parts in critical fracture areas were used. A joint made by diffusion bonding is shown in Fig. 14-21(a). The 18,000-ton diffusion-bonding press is shown at (b).

Isostatic Hot-Gas Pressure Bonding.   Isostatic pressure refers to the uniform application of pressure to all surfaces of a body at once. This may be accomplished with gases or fluids. For example, the bonding may be accomplished by placing the part in an autoclave and subjecting it to 15,000-psi pressure of inert gas at 1000°F (538°C). Another method may employ 60,000-psi pressure at 3632°F (2000°C) with the pressurizing media being hydraulic fluid in combination with silica sand.

Vacuum-Furnace Bonding.   Radiant-heated vacuum furnaces are used for some types of diffusion bonding. A vacuum of less than 10^{-4} Torr is adequate for bonding such metals as copper–aluminum, copper–copper, iron–copper, and titanium–titanium A pressure may be applied by a hydraulic ram, dead-weight loading, or clamping.

Explosive Bonding.   This type of welding was discussed previously in this chapter.

(a)   (b)

DIFFUSION BONDING PROCESS

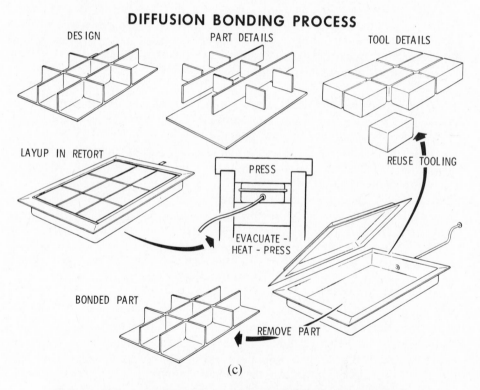

DESIGN   PART DETAILS   TOOL DETAILS

LAYUP IN RETORT   REUSE TOOLING

PRESS

EVACUATE -
HEAT - PRESS

BONDED PART

REMOVE PART

(c)

Figure 14.21. Shown at (a) is a macroetched section of a diffusion-bonded titanium T-joint. The 18,000-ton diffusion-bonding press is shown at (b). At diffusion-bonding temperature, titanium acts like a viscous superplastic liquid. By proper adjustment of time, temperature, and pressure, the metal is made to flow into a myriad of shapes within the dies. A schematic of the entire process is shown at (c). (*Courtesy Rockwell International.*)

Friction Bonding

Friction bonding is often termed friction welding or inertia welding. The principle is that of rubbing surfaces to both clean them and provide heat. In practice, the process is limited in that one of the parts to be joined must be cylindrical or easily mounted in the chuck or holding fixture of the friction-welding machine.

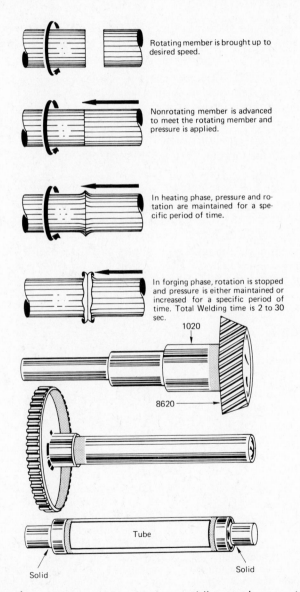

Rotating member is brought up to desired speed.

Nonrotating member is advanced to meet the rotating member and pressure is applied.

In heating phase, pressure and rotation are maintained for a specific period of time.

In forging phase, rotation is stopped and pressure is either maintained or increased for a specific period of time. Total Welding time is 2 to 30 sec.

1020

8620

Tube

Solid

Solid

Figure 14.22. Basic steps in friction welding and some typical applications.

The equipment, which has been developed by several different companies, usually includes the following features:

1. A stationary holding fixture to hold one of the components for welding.
2. A rotating holding fixture that is coupled directly to a flywheel and driving source.
3. A constant-load source (hydraulic cylinder) to bring parts together for joining.

The basic steps are shown in Fig. 14-22 along with typical applications. The joint is generally accomplished without the occurrence of melting. The process has been found effective, repeatable, and efficient.

Advantages and Limitations.   Diffusion bonding results in complete metallurgical-joint continuity without brittle phases or voids, even over comparatively large areas. The problems of dissimilar-metal fusion welds are eliminated. Diffusion bonding permits the utilization of materials based on their condition before bonding. The isostatic process is not limited by complex exterior or interior surfaces.

In friction welding, the energy requirements are only a fraction of those needed for other welding processes. The welding process is relatively clean; there is little spatter, and no arc, fumes, or scale. The heat-affected zone is very narrow and has a grain size that is often smaller than the base metal.

Friction welding is limited to parts specifically designed for its unique requirements. They must be essentially round and must be able to withstand the high torque developed during welding.

Questions and Problems

14-1. Make a sketch of how the spot welds should be laid out when joining two pieces of 16-gauge metal. Show overlap distance, spot size, and spot spacing. No fixturing is used and no distortion is anticipated.

14-2. What would the spot-welding requirements be for welding two sheets of 12-gauge (0.1046-in.) mild steel? State amperage, spot spacing, and spot size.

14-3. (a) How are the electrons obtained for electron-beam welding?

(b) What causes the electrons to have so much force?

14-4. (a) If two pieces of 2-in. thick aluminum plate are welded together by the EB process, what would the approximate width of the weld be?

(b) What should the air gap be between these plates before welding?

14-5. (a) In general terms compare the initial and operating cost of an EB welder with that of a laser welder.

(b) How do they compare for the thickness of the material they can be used to weld?

14-6. How can an EB welder be made to weld an irregular path automatically?

14-7. At the present time what are the main applications of ultrasonic welding?

14-8. Does any melting take place in the metal during explosive welding?

14-9. Why is explosive bonding especially good for joining dissimilar metals?

14-10. What are the conditions that make for good diffusion bonding?

14-11. How may dissimilar metals be successfully joined together by using the friction process?

14-12. Why doesn't the oxide film on the surfaces to be joined in high-frequency resistance welding of tubing interfere with the weld?

14-13. Why is so little current required for high-frequency resistance welding?

14-14. Why may homopolar-resistance butt welding be a process whose time has come?

14-15. Why is a vacuum necessary for EB welding?

14-16. Why is ultrasonic welding classed as a solid-state welding process?

14-17. Why is explosive welding especially good for joining dissimilar metals?

14-18. An aluminum bracket is welded to an aluminum stand. The bracket is fabricated from 0.089-in. thick 2024-T3 aluminum, Su = 55 ksi. Three spot welds are used. Approximately how much weight would the bracket support in a shear load?

Case Study

Automotive manufacturers, in an effort to reduce weight, shifted their attention from low-carbon sheet stock to HSLA steels. They soon found, however, that these materials did not respond to the spot welding techniques in the same way that most of the low-carbon steels had responded. Most of the welds were cracking. As a manufacturing engineer, how would you solve this problem?

Bibliography

Editor. "Explosion Welding." *Mechanical Engineering* 100 (May 1978): 28–35.

Grant, G. B., W. M. Geatherston, et al., "Homopolar Pulse Resistance Welding—A New Welding Process." *Welding Journal* 58 (May 1979): 24–36.

Highland, Donald E., ed. "Lasers in Action." *Production Engineering* (February 1979).

Jefferson, Ted. B. "Flash Butt Welding a 56-Inch Pipeline." *Welding Design and Fabrication* (February 1979).

Matyazic, John. "How Good Is Friction Welding?" *Welding Design and Fabrication* 49 (December 1976): 76–77.

Schaffer, George, ed. "Welding." *American Machinist Special Report 698* 121 (September 1977): 83–106.

Wilcox, R. G., and Walton R. Yerger. "Factors Affecting the Quality of EB Welding." *Metal Progress* 103 (April 1978): 28–35.

Williams, Robert N., ed. "A Likely Pair: EB Welding and Job Shops." *Welding Design and Fabrication* (March 1980).

CHAPTER FIFTEEN

Welding Design, Metallurgy, and Inspection

Welding offers designers a wide variety of processes and techniques that can be used to advantage in joining materials for fabrication. The main considerations are being able to obtain the desired physical and mechanical properties, having the equipment available, and the final appearance of the part.

Both the American Welding Society (AWS) and the American Institute of Steel Construction (AISC) issue codes or standards to be followed in design, fabrication, and erection of structural steel for buildings and bridges. The AWS code number for welding steel structures is listed as AWS D1.1-XX, in which buildings, bridges, and tubular structures are covered. The XX is given here to represent the year the code was reviewed. The AISC does not use an overall code number but is designated by the year in which it was issued.

WELDED-JOINT DESIGN

From the standpoint of geometry, metal plates may be joined together in five main types of joints, as shown in Fig. 15-1. The two types of welds most often used are butt and fillet.

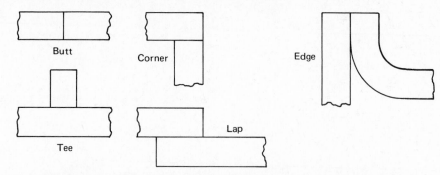

Figure 15.1. The basic types of welded joints. There are many variations of each of these. Variations of the butt joints are shown in Fig. 15-4.

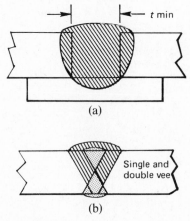

Figure 15.2. The minimum preparation may lead to excessive weld deposit as in the case of the square butt joint in heavy plate (a). Two times as much weld deposit is required for a single-V butt joint as compared to a double-V butt joint (b).

Butt-Weld Joint Design

Butt welds, also known as groove welds, are made between abutting plates in the same plane and are generally described by the way the edges are prepared. In general, the least preparation required, the more economical. However, when plates are thick, the spacing required to obtain complete penetration for manual welding may cause the amount of weld metal to be excessive (Fig. 15-2). The type of edge preparation is largely dependent on how complete penetration can be achieved with the minimum amount of weld metal. A single-V butt joint, together with common terminology, is shown in Fig. 15-3.

Standard Butt-Weld-Edge Preparation and Spacing. The square-edge butt joint requires no preparation other than providing a straight edge and a means of retaining the proper spacing for the weld. This is usually done with tack welds at frequent intervals. Often a square-edge butt joint will be made with a backing

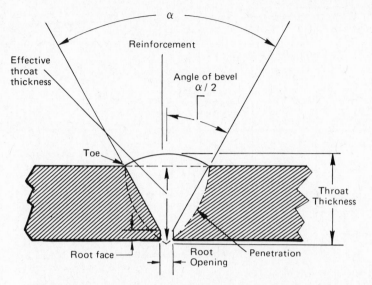

Figure 15.3. A single-V butt weld with common terms.

strip. This strip may remain as part of the joint or it may be machined off. The backing strip is commonly used when all welding must be done from one side or when the root opening is excessive. Standard butt-joint types and spacing for various thicknesses of metal are shown in Fig. 15-4.

A *backing bar* is similar to a backing strip; however, it is intended to be only temporary. It consists of a heavy copper bar with a groove that provides room for full penetration of the weld metal. It is also used in channel form with a flow of inert gas to protect the back side of the weld from the atmosphere. More recently a tape has been developed that acts as a backing strip for closed butt joints. It consists of a flexible, granular refractory layer mounted with a wider pressure-sensitive adhesive foil protected by a paper liner (Fig. 15-5).

Butt Weld Width-to-Depth Ratio.   Butt welds that are designed to maintain a balance between width and depth are generally less susceptible to weld cracks, as shown in Fig. 15-6. The critical width-to-depth ratio is almost entirely limited to the first pass. The second and subsequent passes are almost never subject to the same magnitude of stress. Thus, because of the importance of the width-to-depth factor, the joint preparation may be inherently crack-sensitive.

Butt Weld Strength.   Full-penetration butt welds are generally regarded as having the same load-carrying capacity as the base metal. Hence there is no need to calculate the strength of the weld. This assumes that the weld deposit corresponds to that of the base plate and that either a backing bar was used or a *sealing run*. A sealing run refers to backgouging the back side of the weld by grinding or with an arc-air torch and then depositing a weld bead.

WELD TYPE	Sides Welded	Thickness (in.)	Gap (in.)	Root Face (in.)	Min Included Angle (deg)
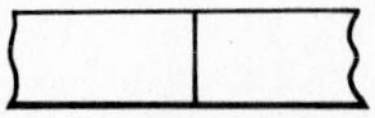 **Closed Square Butt**	One Both	Up to 1/16 Up to 1/8 Up to 5/8 With DP electrodes			
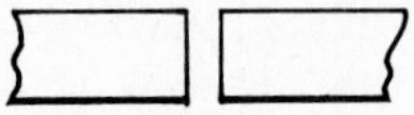 **Open Square Butt**	Both	Up to 3/16	1/16		
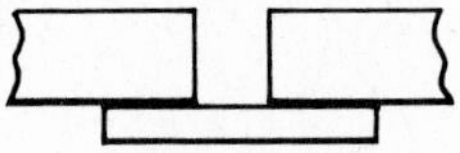 **Square Butt with Backing Bar**	One	Up to 3/16 Up to 1/2 Up to 1/2 With DP electrodes	3/16 5/16 1/4		
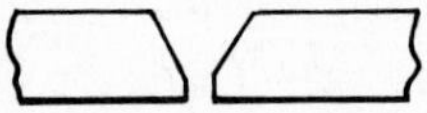 **Single "V" Butt Weld**	One or Both	Over 3/16 and up to 1 If made with backing strip	0 to 1/8 1/8 to 3/16	0 to 1/8	60
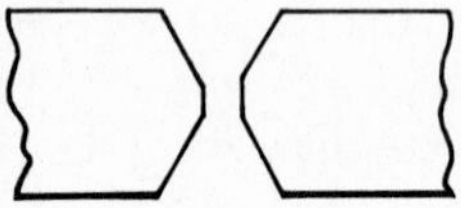 **Double "V" Butt Weld**	Both	Over 1/2	0 to 1/8	0 to 1/8	60
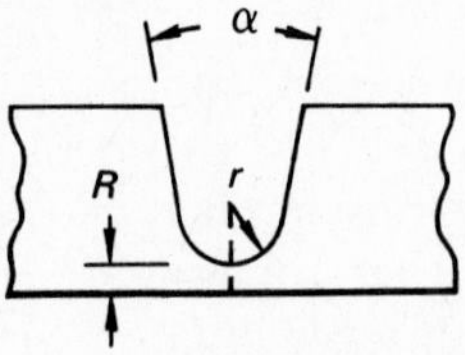 **Single "U" Butt Weld**	One or Both	Over 3/4	0 to 1/16	1/8 to 3/16	10 to 30

Figure 15.4. Standard types of welded butt joints.

WELD TYPE	Sides Welded	Thickness (in.)	Gap (in.)	Root Face (in.)	Min Included Angle (deg)
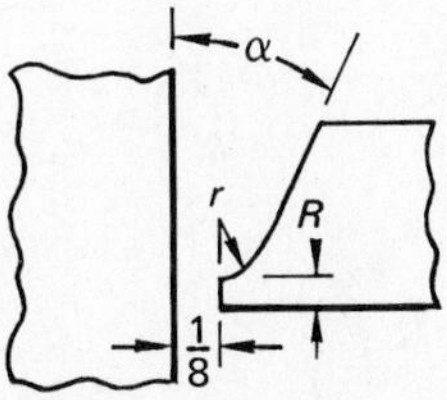 Single "J" Butt Weld	Both	Over 3/4	1/8	1/8 to 3/16	20 to 30
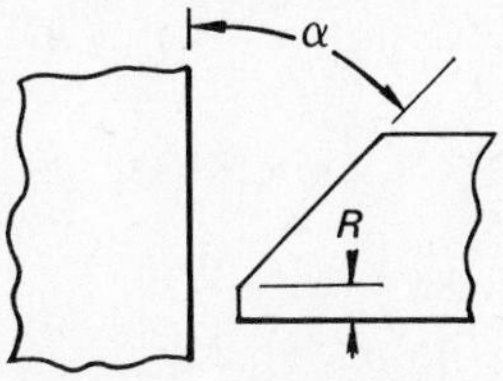 Single Bevel Butt Weld	Both	Up to 1 Unlimited with backing strip	0 to 1/8	0 to 1/8	45 to 50
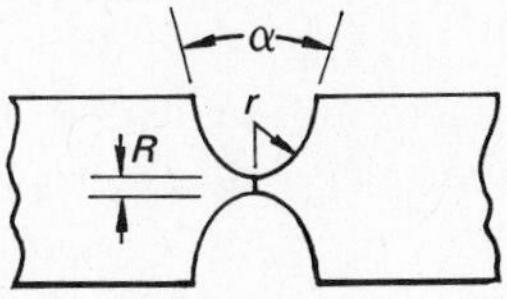 Double "U" Butt Joint	Both	Over 1-1/2		1/8 to 3/16	10 to 30
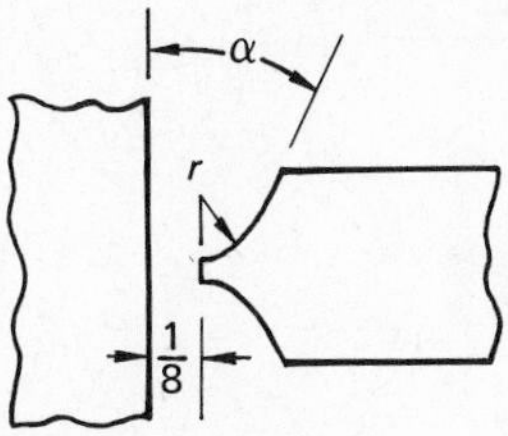 Double "J" Butt Weld	Both	Over 1-1/2	1/8	1/8 to 3/16	20 to 30
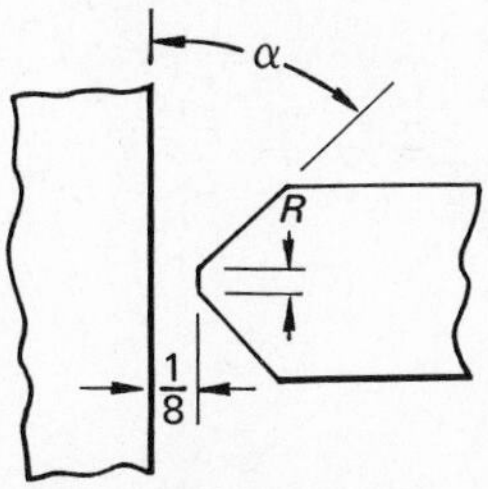 Double Bevel Butt Weld	Both	Over 1/2	0 to 1/8	0 to 1/8	45 to 50

Figure 15.4 (continued)

(a)   (b)

Figure 15.5. A tape-type backing strip utilizes refractory granules on a carrier (a) and in use on an aqueduct (b). (*Courtesy 3M.*)

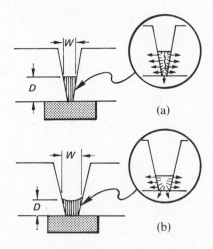

Figure 15.6. The W/D ratio of weld preparation is related to crack sensitivity. The wider root opening (b) provides a better W/D ratio than (a) and is not crack-sensitive. (*Courtesy The James F. Lincoln Arc Welding Foundation.*)

Fillet-Weld Joint Design

Fillet welds are used to fill in a corner. They are the most common welds in structural work. Fillet welds with a detailed view of a T-joint are shown in Fig. 15-7. The leg of a fillet weld is usually equal to the height of the base material (h). The nominal weld outline shown as dashed lines in the detail view makes a 45° triangle when each leg is equal to h. The effect throat (t_e) of the fillet is

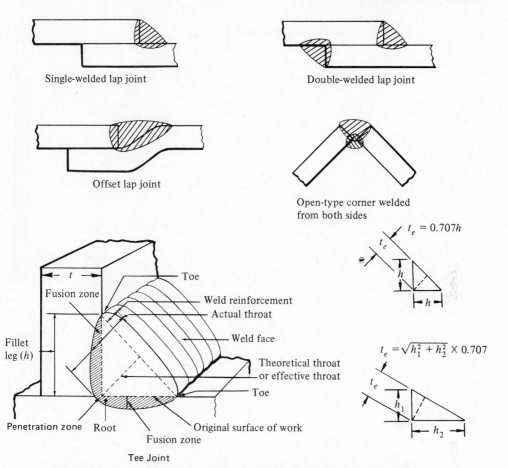

Figure 15.7. Lap joints, open corner joints, and tee joints all use fillet-type welds. The fillet weld may have equal or unequal legs but the effective throat (t_e) is taken perpendicular to the weld face from the root of the joint, as shown by the dashed line.

the distance from the root of the joint to a point perpendicular to the hypotenuse. For a 45° fillet weld, the throat is equal to 0.707 (sine of 45°) times the leg dimension.

Fillet welds are not always made with leg sizes equal to h. The *maximum* size permitted by AISC specification 1.17.6, for the design, fabrication, and erection of structural steel for buildings, states:

1. Along edges of material less than $\frac{1}{4}$ in. thick, the maximum size may be equal to the thickness of the material.
2. Along edges of material $\frac{1}{4}$ in. or more in thickness, the maximum vertical leg may be $\frac{1}{16}$ in. less than the thickness of the material unless the weld is especially designated on drawings to be built out to obtain full throat thickness.

Table 15-1. Minimum* fillet weld size (AWS Structural Welding Code D1.1–72).

Material Thickness of Thicker Part Joined in Inches (mm)		Fillet Weld (h) in Inches (mm)	
To $\frac{1}{4}$ incl.	(6.35 mm)	$\frac{1}{8}$	(3.175 mm)
Over $\frac{1}{4}$	to $\frac{1}{2}$ (6.35–12.70 mm)	$\frac{3}{16}$	(4.7625 mm)
Over $\frac{1}{2}$	to $\frac{3}{4}$ (12.70–19.05 mm)	$\frac{1}{4}$	(6.35 mm)
Over $\frac{3}{4}$	to $1\frac{1}{2}$ (19.05–38.10 mm)	$\frac{5}{16}$	(7.9375 mm)
Over $1\frac{1}{2}$	to $2\frac{1}{4}$ (38.10–57.15 mm)	$\frac{3}{8}$	(9.525 mm)
Over $2\frac{1}{4}$	to 6 (57.15–152.40 mm)	$\frac{1}{2}$	(12.70 mm)
Over 6	>(152.40 mm)	$\frac{5}{8}$	(15.875 mm)

* Except that the weld size need not exceed the thickness of the thinner joined part.

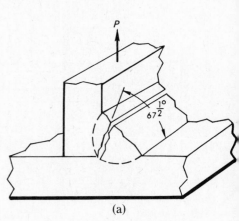

(a)

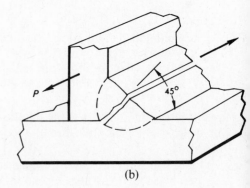

(b)

Figure 15.8. The maximum shear stress occurs at an angle of 67.5° for transverse loading (a) and a 45° for parallel loading (b).

The *minimum-size* fillet weld is governed by the amount of heat, and therefore the size of the weld, required to ensure fusion. The minimum-size fillet welds permitted by the AWS Structural Welding Code are shown in Table 15-1.

Strength of Fillet Welds

The strength of a fillet weld varies with the direction of the applied load, depending upon whether it is parallel or transverse to the weld. In either case, the weld will fail in the shear plane that has the maximum shear stress. For loading parallel to the weld, the plane of rupture is at 45°, as shown in Fig. 15-8(b). Transverse loading in the plane of maximum shear occurs at 67.5° to the horizontal, as shown at (a). Because of this and the fact that the stress distribution for this type of loading is more uniform, the weld will carry about one-third more load in transverse shear than in parallel shear. Two 45° fillet welds with leg dimensions equal to $\frac{3}{4}t$ will develop the full strength of the plate for either type of loading, assuming that the weld metal is equal to that of the base plate and average penetration is obtained.

ALLOWABLE STRESS OF WELDED JOINTS

Allowable stress refers to the allowable unit force per lineal inch of weld. The AWS and AISC code specifies the allowable unit force per inch of weld as being 30% of the minimum specified electrode tensile strength as shown in Table 15-2. The allowable unit force per lineal inch is used to calculate the size fillet weld required for a given application according to the electrode used.

> **Example:** Determine the allowable unit stress (σ) per inch for a 0.5-in fillet weld made with an E70 electrode.

$$\sigma = 0.707t(0.30)(EXX)$$
$$= 0.707 \times 0.5 \times 0.30 \times 70$$
$$= 7.42 \text{ kips/lineal inch (see Table 15-2)}$$

The AISC code (1.14.7) gives limited credit for penetration beyond the root of a fillet weld made by the submerged-arc process. Thus for welds $\frac{3}{8}$ in. or smaller, the t_e is now equal to the leg size of the weld. For submerged-arc fillet welds, larger than $\frac{3}{8}$ in., the t_e is obtained by adding 0.11 to $0.707h$.

The strength of transverse fillet welds loaded in tension (transverse shear) can be found by:

$$\tau = \frac{P}{0.707hl}$$

where: τ = shear stress, psi
P = load, pounds
h = leg of weld, inches
l = total length of weld, inches

Table 15-2. AISC and AWS allowable loads for various sizes of fillet welds.

	Strength Level of Weld Metal (EXX), ksi (in MPa)					
	60 (413.7)	70 (482.6)	80 (551.6)	90 (620.5)	100 (689.5)	110 (758.4)
	Allowable Shear Stress on Throat, ksi, of Fillet Weld or Partial Penetration Groove Weld					
$T =$	18.0 (124.1)	21.0 (144.8)	24.0 (165.5)	27.0 (186.2)	30.0 (206.8)	33.0 (227.5)
	Allowable Force on Fillet Weld, Kips/Linear Inch					
$f =$	12.73h (87.71)	14.85h (102.32)	16.97h (116.92)	19.09h (131.53)	21.21h (146.14)	23.33h (160.74)
Leg Size (h) in	Allowable Force for Various Sizes of Fillet Welds, Kips/Linear Inch (in MPa/linear cm)					
(cm)						
1″	12.73	14.85	16.97	19.09	21.21	23.33
(25.4)	(87.71)	(102.32)	(116.92)	(131.53)	(146.14)	(160.74)
$\frac{7}{8}$″	11.14	12.99	14.85	16.70	18.57	20.41
(22.22)	(76.75)	(89.50)	(102.32)	(115.06)	(127.95)	(140.62)
$\frac{3}{4}$″	9.55	11.14	12.73	14.32	15.92	17.50
(19.05)	(65.79)	(76.75)	(87.71)	(98.66)	(109.69)	(120.58)
$\frac{5}{8}$″	7.96	9.28	10.61	11.93	13.27	14.58
(15.87)	(54.84)	(63.94)	(73.10)	(82.20)	(91.43)	(100.46)
$\frac{1}{2}$″	6.37	7.42	8.48	9.54	10.61	11.67
(12.70)	(43.89)	(51.12)	(58.43)	(65.73)	(73.10)	(80.41)
$\frac{7}{16}$″	5.57	6.50	7.42	8.35	9.28	10.21
(11.11)	(38.38)	(44.79)	(51.12)	(57.53)	(63.94)	(70.35)
$\frac{3}{8}$″	4.77	5.57	6.36	7.16	7.95	8.75
(9.52)	(32.87)	(38.38)	(43.82)	(49.33)	(54.78)	(60.29)
$\frac{5}{16}$″	3.98	4.64	5.30	5.97	6.63	7.29
(7.93)	(27.42)	(31.97)	(36.52)	(41.13)	(45.68)	(50.23)
$\frac{1}{4}$″	3.18	3.71	4.24	4.77	5.30	5.83
(6.35)	(21.91)	(25.56)	(29.21)	(32.87)	(36.52)	(40.17)
$\frac{3}{16}$″	2.39	2.78	3.18	3.58	3.98	4.38
(4.76)	(16.47)	(19.15)	(21.91)	(24.67)	(27.42)	(30.18)
$\frac{1}{8}$″	1.59	1.86	2.12	2.39	2.65	2.92
(3.17)	(10.96)	(12.82)	(14.61)	(16.47)	(18.26)	(20.12)
$\frac{1}{16}$″	0.795	0.93	1.06	1.19	1.33	1.46
(1.58)	(5.48)	(6.41)	(7.30)	(8.20)	(9.16)	(10.06)

WELDING SYMBOLS

This brief presentation of welding symbols is for the purpose of showing how easily they can be used to convey the exact desired information. A more complete presentation of standard welding symbols is given in Appendix 1 and in the second edition of the AWS book, *Weld Inspection*.

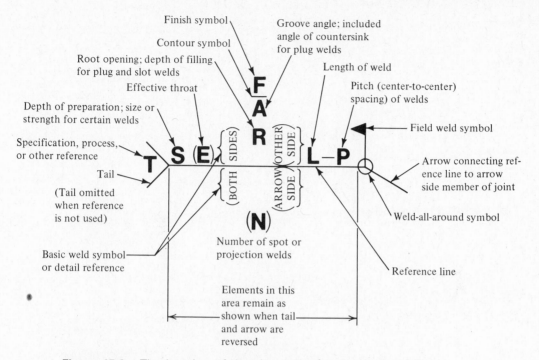

Finish symbol
Contour symbol
Root opening; depth of filling for plug and slot welds
Effective throat
Depth of preparation; size or strength for certain welds
Specification, process, or other reference
Tail
(Tail omitted when reference is not used)
Basic weld symbol or detail reference

Groove angle; included angle of countersink for plug welds
Length of weld
Pitch (center-to-center) spacing of welds
Field weld symbol
Arrow connecting reference line to arrow side member of joint
Weld-all-around symbol
Reference line

F
A
R
T S (E) (BOTH SIDES) (ARROW SIDE)(OTHER SIDE) L–P
(N)

Number of spot or projection welds

Elements in this area remain as shown when tail and arrow are reversed

Figure 15.9. The location of the elements of a standard welding symbol. (*Courtesy American Welding Society.*)

Figure 15-9 shows a weld reference line with the standard placement of information concerning the weld. In actual practice a reference would never have this much information since there is usually not that much to say about a particular weld. Note, for example, that the tail may be omitted when the specification of the process used is omitted. A further breakdown of the symbols is shown in Fig. 15-10. Here, as one example, is a chain intermittent fillet welding symbol showing the desired leg size along with the length and increment spacing of the weld. Thus symbols allow the designer to give rather complete information to the welder without resorting to lengthy notes. Typical examples of the use of welding symbols in structural work are shown in Fig. 15-11.

WELDED CONNECTIONS

The term *welded connections* refers to the design and method of joining standard structural members such as angles, channels, beams, and columns, some of which are shown in Fig. 15-12.

A simple direct method proposed by Blodgett for finding the properties of welded connections is to treat the welds as lines having neither leg nor throat dimensions. The forces, which can be taken from Table 15-2, are considered on a unit-length basis, thus eliminating the knotty problems of combined stresses.

Back or Backing Weld Symbol

Any applicable single
groove weld symbol

Surfacing Weld Symbol Indicating Built-up Surface

Size (height of deposit).
Omission indicates no
specific height desired

$\frac{1}{8}$

Orientation. Location
and all dimensions
other than size are
shown on the drawing

Double Fillet Welding Symbol

Size (length of leg)
Specification. Process
or other reference

IG

$\frac{1}{4}$ 12

Length. Omission
indicates that weld
extends between
abrupt changes in
direction or as
dimensioned

Chain Intermittent Fillet Welding Symbol

Size (length of leg)

$\frac{5}{16}$ 2-6

$\frac{5}{16}$ 2-6

Length of increments

Pitch (distance between
centers) of increments

Staggered Intermittent Fillet Welding Symbol

Size (length of leg)

$\frac{1}{2}$ 3-8

$\frac{1}{2}$ 3-8

Pitch (distance between
centers) of increments

Length of increments

Single-Vee-Groove Welding Symbol

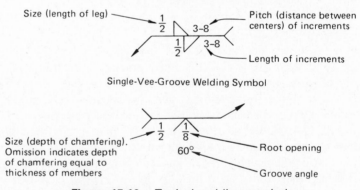

Size (depth of chamfering).
Omission indicates depth
of chamfering equal to
thickness of members

$\frac{1}{2}$ $\frac{1}{8}$

60°

Root opening

Groove angle

Figure 15.10. Typical welding symbols.

This is not to neglect stress distributions within a weld, which can be very complex, but the actual fillet welds tested and on which the unit forces are based have these same conditions.

When a weld is treated as a line, the property of the welded connection is inserted into the standard design formula for that particular type of load and the force on the weld is found in terms of pounds per lineal inch. A sample problem will serve to illustrate the use of this method.

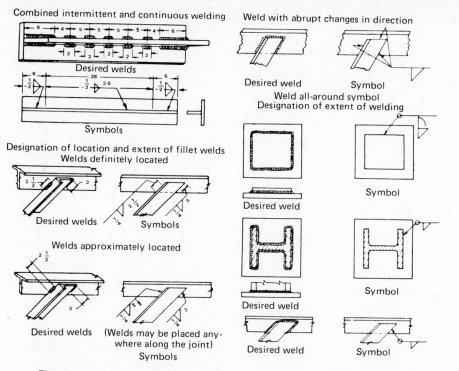

Figure 15.11. An example of the use of welding symbols in structural work.

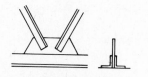

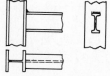

Roof Truss
with Gusset Plate

Column and Beam
Connection (Rigid)

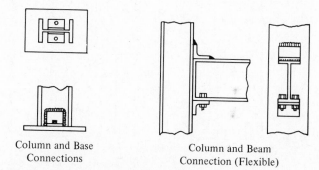

Column and Base
Connections

Column and Beam
Connection (Flexible)

Figure 15.12. Welded-type structural-steel connections.

Example: A clip that has been welded to a supporting structure is shown in Fig. 15-13. Will this clip be able to sustain a load of 2 tons that at times also has some impact force?

Solution: Choose a matching stress equation from Table 15-3. What is the maximum load the clip will hold? The maximum allowable stress for the weld is 3710 psi with an E70 electrode (Table 15-2).

$$\sigma = \frac{0.707P}{hl}$$

Solving for σ,

$$\sigma = \frac{\sigma hl}{0.707} = \frac{3710 \times \frac{1}{4} \times 8}{0.707} = \frac{7420}{0.707} = 10,495 \text{ lb}$$

Applying a safety factor for occasional impact (Table 15-4), use 1.5.

$$P = 10,495 \times 1.5 = 15,742 \text{ lb}$$

Therefore the welded clip will be able to withstand a load of 8 tons plus occasional impact.

DISTORTION

Distortion in weldments is the result of nonuniform heating and cooling. Expansion and contraction of the weld metal and adjacent base metal may be compared to heating a metal bar between vise jaws (Fig. 15-14). The restricted

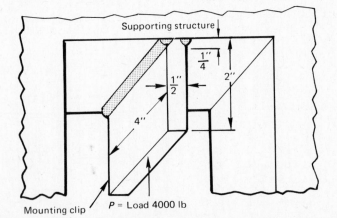

Figure 15.13. A mounting clip is welded to a supporting structure.

Table 15-3. Stress formulas for weld joints.

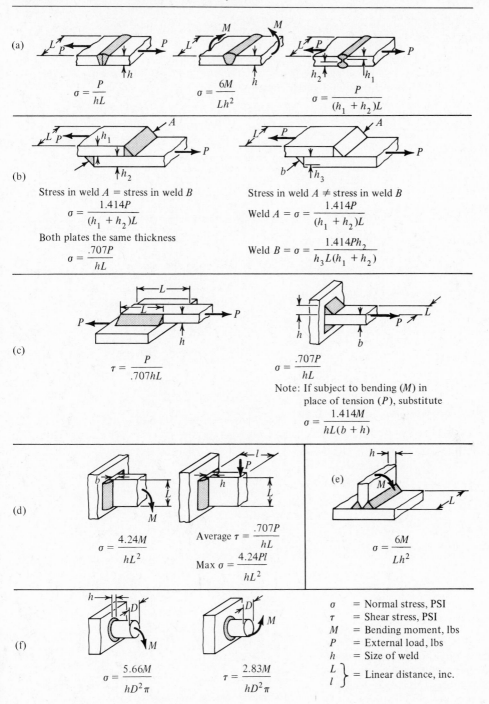

(a)

$$\sigma = \frac{P}{hL}$$

$$\sigma = \frac{6M}{Lh^2}$$

$$\sigma = \frac{P}{(h_1 + h_2)L}$$

(b)

Stress in weld A = stress in weld B

$$\sigma = \frac{1.414P}{(h_1 + h_2)L}$$

Both plates the same thickness

$$\sigma = \frac{.707P}{hL}$$

Stress in weld $A \neq$ stress in weld B

$$\text{Weld } A = \sigma = \frac{1.414P}{(h_1 + h_2)L}$$

$$\text{Weld } B = \sigma = \frac{1.414Ph_2}{h_3 L(h_1 + h_2)}$$

(c)

$$\tau = \frac{P}{.707hL}$$

$$\sigma = \frac{.707P}{hL}$$

Note: If subject to bending (M) in place of tension (P), substitute

$$\sigma = \frac{1.414M}{hL(b + h)}$$

(d)

$$\sigma = \frac{4.24M}{hL^2}$$

$$\text{Average } \tau = \frac{.707P}{hL}$$

$$\text{Max } \sigma = \frac{4.24Pl}{hL^2}$$

(e)

$$\sigma = \frac{6M}{Lh^2}$$

(f)

$$\sigma = \frac{5.66M}{hD^2 \pi}$$

$$\tau = \frac{2.83M}{hD^2 \pi}$$

σ	= Normal stress, PSI
τ	= Shear stress, PSI
M	= Bending moment, lbs
P	= External load, lbs
h	= Size of weld
$\left.\begin{array}{c}L\\l\end{array}\right\}$	= Linear distance, inc.

569

Table 15-4. Weld safety factors.

Type of Weld	Factor
Reinforced butt	1.2
Toe of transverse fillet, occasional impact	1.5
End of parallel fillet	2.7
T-butt with sharp corners	2.0

Unrestrained bar before heating and
after cooling to room temperature

Unrestrained bar when
uniformly heated to a
specific temperature

(a)

Restrained bar at
room temperature

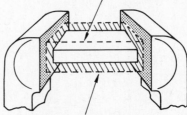

Bar when uniformly
heated while restrained

(b)

Figure 15.14. A schematic presentation of how the heating and cooling cycle, as used in the welding of metals, causes distortion. (*Courtesy The James F. Lincoln Arc Welding Foundation.*)

Restrained bar after heating and
cooling to room temperature is
shorter, thicker and wider

(c)

Figure 15.15. Tensile stresses are set up between the weld metal and the cooler base metal as solidification and contraction take place.

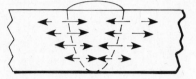

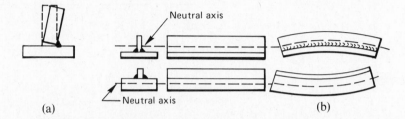

(a)

Neutral axis

Neutral axis

(b)

Figure 15.16. Distortion types: angular (a), longitudinal (b).

bar is not able to expand uniformly, therefore it is upset in the area of greatest heat. As it cools it will contract or shrink in all directions, with the result that the bar is now shorter and thicker than it was before heating.

A weld made on restricted members cannot shrink when cooled, and therefore residual stresses are set up in the weld and in the heat-affected zone (HAZ) adjacent to it (Fig. 15-15).

In an unrestricted weld, the members are free to expand and contract. Since the members move, the degree of stress in the weld and HAZ will not be as great as that of a restricted weld. Some of the locked-in stresses are relieved as the base metal shifts or distorts.

Distortion is evidenced in many different forms, but basically there are two main types: angular and longitudinal (Fig. 15-16). Transverse shrinkage is the main cause of angular distortion. It is also directly related to the volume of weld deposit, as shown in Fig. 15-17.

Distortion Control. Shrinkage of hot metal cannot be prevented, but a knowledge of how it operates is useful in minimizing distortion. The following principles are used in minimizing distortion:

1. *Minimize shrinkage forces.* Use only the amount of weld metal required. Overwelding not only adds to the shrinkage forces but it is also uneconomical. Proper edge preparation and fit-up help keep weld deposit to the required minimum. Intermittent welds can also be used where strength requirements are not critical. However, in this regard, the AISC Code 1.1717 limits the minimum effective length of fillet welds to four times their nominal size. If this condition cannot be met, the size of the weld is considered to be one-quarter of its effective length.

   Where transverse shrinkage forces cause distortion, as in the butt weld, the number of passes should be kept to a minimum. The shrinkage of each pass tends to be cumulative. The use of larger electrodes allows

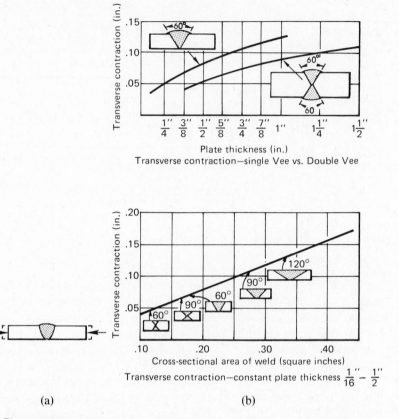

Transverse contraction—single Vee vs. Double Vee

Transverse contraction—constant plate thickness $\frac{1}{16}'' - \frac{1}{2}''$

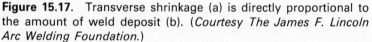

(a)   (b)

Figure 15.17. Transverse shrinkage (a) is directly proportional to the amount of weld deposit (b). (*Courtesy The James F. Lincoln Arc Welding Foundation.*)

more weld deposit per pass with a smaller total volume of heat input into the base metal.

2. *Help shrinkage forces work in the desired direction.* Parts may be positioned or preset out of position before welding so that shrinkage forces will bring them into alignment, as illustrated by the T-joint and butt weld in Fig. 15-18.

3. *Balance shrinkage forces with other forces.* A common example of balancing shrinkage forces is the use of fixtures and clamps. The components are locked in the desired position and held until the weld is finished. This will of course cause shrinkage forces to build up until the yield point of the metal is reached. For typical welds on low-carbon steel plate this would be about 45,000 psi. You may visualize considerable distortion taking place as soon as the clamps are removed. This does not happen due to the small amount of unit strain (ε) compared to the amount of movement that would occur if no restraint existed during the welding process. The strain for steel may be calculated as follows:

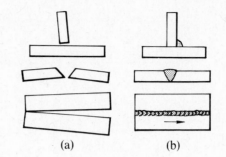

Figure 15.18. Prepositioning (with tack welds) (a) allows weld shrinkage forces to bring parts into alignment (b).

(a)   (b)

$$\varepsilon = \frac{\sigma}{E}$$
$$= \frac{45,000}{30 \times 10^6}$$
$$= 0.0015 \text{ in./in.}$$

where: E = modulus of elasticity

Another common example of balancing shrinkage forces is a butt joint welded alternately on each side.

4. *Remove shrinkage forces after welding.* Peening is often used as a means of stress-relieving after welding. In theory the metal is made to stretch slightly under the force of each blow, thus relieving the stress. Care must be exercised in peening to avoid concealing a crack or work-hardening the metal. First passes are never peened.

In more specialized cases, such as alloy steels, preheating and post-heating are used to minimize the residual stress.

5. *Place welds near the center of gravity.* It is not always possible to place welds on or near the center of gravity; however, even the amount of penetration as shown in Fig. 15-19 can change the amount of distortion. Shown at (a) is a T-weld made by manual welding and at (b) by the submerged arc.

Metal Properties and Distortion.   Distortion is related to the coefficient of expansion, thermal conductivity, modulus of elasticity, and yield strength of the material welded. A high coefficient of expansion tends to increase the shrinkage of the weld metal and the adjacent metal, thus increasing distortion.

A metal with a low thermal conductivity such as stainless steel (Appendix A) retains the heat in the weld area. This results in a steep temperature gradient and greater distortion.

As a weld deposit cools and contracts, the adjacent metal "stretches" to satisfy the volume demand of the weld joint. Thus the higher the yield strength of the base metal, the greater the distortion. If, however, the modulus of elasticity is high, the material is more likely to resist distortion.

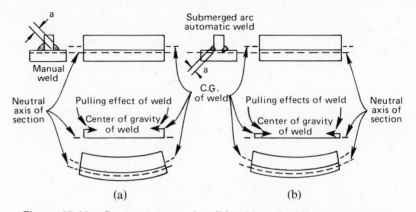

Figure 15.19. Deeper penetration (b) achieved with automatic submerged arc results in welds that are nearer the center of gravity with less distortion than the manual deposit (a). (*Courtesy The James F. Lincoln Arc Welding Foundation.*)

WELDING POSITIONERS AND FIXTURES

Welding positioners and fixtures are used extensively in production welding to reduce costs and improve quality. By just tilting a fillet weld 10° in the downhand position, an increase in speed of over 60% can be realized, as shown in Fig. 15-20.

Weld positioners are also important in improving the operating factor. A high operating factor will be dependent on convenient positioning of the weld, as shown in Fig. 15-21, as well as having all needed supplies convenient and operating controls readily accessible.

Power rolls and idlers are a convenient means of moving the work when making circumferential welds. Figure 15-22 shows a ram-type manipulator that provides a mount for the welding head that allows it to be raised, lowered, extended, retracted, or swung in an arc of 360°. The manipulator can be used with either rolls or positioners as mentioned or with headstocks and tailstocks, as shown in Fig. 15-23.

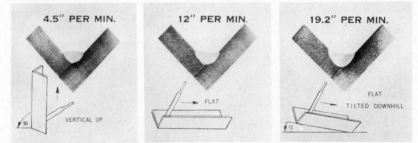

Figure 15.20. Effect of position on speed of welding fillet welds in ⅜-in. plate or thinner. (*Courtesy The James F. Lincoln Welding Arc Foundation.*)

WELDING METALLURGY   575

Figure 15.21. This 12.5-ton welding positioner is able to put all welds on the large rotor housing in a downhand position. (*Courtesy Harnischfeger Corp.*)

Figure 15.22. A ram-type manipulator being used with power rolls.

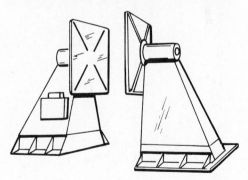

Figure 15.23. A headstock and tailstock as used for positioning work for welding.

WELDING METALLURGY

Grain-Size and Hardness Control. From the molten metal of the weld to the edge of the heat-affected zone (HAZ), there will be a wide variation of temperature. Some of the metal has been heated far above the upper critical temperature, some just at critical, some not up to critical, and all the way down to the unaffected base metal. Therefore, as can be expected, the grain size of

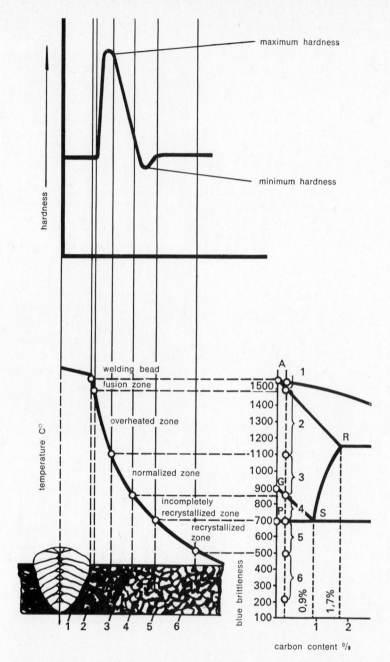

Figure 15.24. The influence of welding heat on the distribution of hardness in the heat-affected zone of quenched and tempered structural steels (schematic). (*Courtesy ESAB.*)

the weld will be rather large, becoming gradually smaller until the recrystallization temperature is reached. Here the grain size will be at a minimum and then will advance gradually larger again until it blends with the unaffected parent metal (Fig. 15-24).

The butt weld shown in Fig. 15-24 was made of high tensile stength steel. These steels are made by adding boron, aluminum, and vanadium for fine grain structure. The carbon content is limited to produce good weldability. These steels, in growing demand today, are delivered in their quenched and tempered condition.

The fact that these steels are hardenable to some extent can create problems. The heat of the weld in the transformation zone will be similar to annealing. Therefore the initial strength of the steel cannot be fully utilized. As shown schematically (Fig. 15-24), an increase in hardness appears in the zones that are heated above the GS line (zones 2 and 3) due to the quenching effect or rapid heat transfer to the cold parent metal. The minimum hardness is found in areas that were heated just above the lower critical temperature (zone 4).

Where and to what extent hardness peaks occur largely depends on the preheating of the parent metal and the heat input at the time the weld is made. By careful control of these factors, hardness in the softer part of the heat-affected zone can be kept above the guaranteed strength of the steel.

Quality welds require the proper proportion of heat input, preheat temperature, and welding speed. By applying certain heat flow and hardenability concepts to individual welding situations, these parameters can be approximated theoretically rather than by trial and error, resulting in considerable savings in time, effort, and expensive samples.

WELDABILITY

Weldability has in the past been regarded as the ease with which a material can be welded. Weldability now includes how well the weldment performs in service. Thus there are two main factors; service weldability and fabrication weldability.

Service Weldability. Service weldability refers to the ability of the process–material combination to turn out a weld that will stand up to thermal and mechanical stresses, corrosion, and other service requirements.

Fabrication Weldability. Fabrication weldability is the capacity of the combination of materials and processes to produce proper fusion, limited porosity, absence of cracks, and proper weld geometry. Fabrication weldability factors include: melting point of the base metal, thermal conductivity, thermal expansion and contraction, electrical resistance, and surface condition.

Melting Point. Metals that have a low melting point, such as aluminum and lead, are more difficult to control and keep from burning through.

Thermal Conductivity and Expansion.   Metals that have a high thermal conductivity, such as aluminum and copper, are more difficult to bring to the fusion temperature; in addition, the rapid expansion and contraction of the HAZ causes stresses and distortion.

Electrical Resistance.   Metals that have a low electrical resistance will require more energy to weld.

Surface Conditions.   The surfaces to be welded should be free of dirt, grease, paint, scale, and corrosion. Although some welding processes are said to be self-cleaning, there is always the possibility of getting contaminants into the weld. See Chapter 2 for weldability of various metals.

CONTROLLING WELD DEFECTS

One of the most common problems encountered in otherwise properly made welds are cracks. Weld cracks are classified as hot cracks and cold cracks.

Hot Cracks.   Hot cracks occur in the weld and fusion zone as the metal solidifies. The main causes are:

a. Tramp elements such as phosphorous and sulphur in the base plate.
b. Joint restraint, which causes high stresses in the weld.
c. Bead shape. As the metal solidifies, a slightly convex bead will provide material for shrinkage but a very convex bead will result in high tensile stresses that may result in longitudinal cracks.
d. A deep, narrow weld, as was shown in Fig. 15-6, under conditions of high restraint may cause internal cracks.
e. The higher the carbon and alloy content of the base metal, the greater the hardness that may result in cracking.
f. Moisture in the joint or electrode coating may cause hydrogen pickup and embrittlement of the weld deposit.
g. Rapid cooling increases the effect of items e and f.

Cold Cracks.   Cold cracks are usually in the HAZ and are caused by:

a. High carbon or alloy content as affected by cooling.
b. Hydrogen embrittlement through migration of hydrogen liberated from the weld.
c. Rate of cooling, which controls items a and b.

CONTROLLING WELD CRACKS

Cold cracks can be minimized or prevented by correcting the causes listed previously, such as having proper bead shape, reducing joint restraint, avoiding hydrogen pickup, and by preheating, postheating, knowing the carbon equivalent of the steel, and controlling the cooling rate and heat input.

Preheating. A wide variety of cooling conditions in an arc weld made without preheat are shown in Fig. 15-24. When a medium- or high-carbon alloy steel is welded, martensite is caused to form in the HAZ, resulting in a weldment having cold cracks. The problem can be avoided by slowing down the cooling rate and the most common method is by preheating.

Of the six zones shown in Fig. 15-24, of main concern is the overheated zone where maximum hardness occurs. This is because the coarse-grained austenite region is subjected to the highest cooling rate and has the highest hardenability. With preheat the HAZ becomes wider, resulting in a slower cooling rate.

Postheating. The martensite or bainite that formed during the cooling of the weld may be softened by tempering. This can be done by heating the HAZ in the range of 900 to 1200°F (427–649°C). This will restore ductility to the zone, but it will not offset any cracks that may have occurred in the HAZ because of the shrinkage stresses in the excessively brittle areas. However, postheat stress-relieving performs a double function: reduction of the residual stress, and tempering any martensite that may have formed in the HAZ.

Carbon Equivalent. A carbon equivalent formula is often used to rate the weldability of steel since carbon has more effect on the weld than any other one element. The carbon equivalent (CE) of a metal can be found by the use of one of several similar formulas as follows:

$$CE = \%C + Mn/4 + \%Ni/20 + \%/Cr/10 + \%Mo/5 + \%Cu/40 - \%V/10$$

A steel with a carbon equivalent of less than 0.50 is generally considered weldable, except for 1045, 1070, and 980X with yield strengths of approximately 78 ksi (55 MPa) or higher. Table 15-5 lists the carbon equivalent of a number of common steels.

If the carbon equivalent is less than 0.40, it is relatively insensitive to hydrogen cracking. Over 0.50 cracking is virtually a certainty unless low-hydrogen welding procedures are used.

Table 15-5. Carbon equivalents of selected steels from AWS D14.3-77.

| Steel | Composition % maximum | | | | | |
	C	Mn	Ni	Cr	Mo	CE
1020	0.23	0.60	—	—	—	0.33
1045	0.50	0.90	—	—	—	0.65
1070	0.75	0.90	—	—	—	0.90
4130	0.33	0.60	—	1.10	0.25	0.565
8260	0.23	0.90	0.70	0.60	0.25	0.585
SAE 950X	0.23	1.35	—	—	—	0.455
SAE 980X	0.26	1.65	—	—	—	0.535

Cooling Rate. The faster the HAZ cools, the harder the microstructure and the more susceptible it is to cracking. The main factor that controls the cooling rate of the HAZ other than preheating is the metal thickness. The thicker the metal, the faster the weld cools. As the cooling rate increases, hardness rises in the HAZ. With large heat inputs, metals less than 0.984 in. thick (25.00 mm) generally weld without problems, but thicker plates may need preheating. Table 15-6 lists the preheat and interpass temperatures for different steels of varying plate thicknesses. A rule-of-thumb is that preheating is unnecessary if the carbon equivalent is less than 0.50 and the weld size to thickness ratio is greater than 1:2 and the steel plate is 1 in. (25.00 mm) thick at most.

The cooling rate of HSLA steels is particularly important. The welding engineer has control over the travel speed, current, preheat, and some of the factors that affect voltage. The cooling rate will vary directly with travel speed. Of course the travel speed must be slow enough to achieve good fusion. The cooling rate is also inversely proportional to the current so that doubling the current reduces the cooling rate to half the original value. Similarly, increasing the preheat, for example, from 100 to 300°F (38 to 149°C), can decrease the cooling rate by 40%.

Assuming proper preparation of the surfaces to be welded, 2-in. (50.80-mm) thick plates of HSLA steel have a recommended preheat of 250°F (120°C) and an energy input of 40 to 60 kilojoules per inch. This combination produces welds that normally have a yield strength of 50,000 psi (344 MPa).

Heat Input. The heat input to the weld also affects the cooling rate. The heat input can be calculated by the following formula:

$$H = 60 \ VI/S \text{ or } H = I^2Rt$$

where:
H = heat input in joules per millimeter
V = voltage

Table 15-6. Typical preheat and interpass temperature. (*Courtesy Lincoln Electric Co.*)

	Plate Thickness (mm)				
Steel	0.250(6.3)	0.500(12.7)	1.00(25.4)	2.00(50.8)	4.00(102)
	°F (°C)	°F (°C)	°F (°C)	°F (°C)	°F (°C)
1020	—	—	—	—	150 (60)
1030	—	—	—	100 (40)	200 (95)
1040	—	—	300 (150)	300 (150)	300 (150)
1050	200 (95)	300 (150)	350 (175)	400 (200)	400 (200)
1060	300 (150)	400 (200)	400 (200)	450 (230)	550 (290)
1080	400 (200)	500 (260)	550 (290)	550 (290)	600 (315)

S = travel speed in millimeters per minute
I = current in amperes
R = resistance in ohms
t = time of current flow in seconds

Welding processes vary in heat input.

FCAW and SAW methods leave flux covers on the welds which, when combined with higher heat inputs, generally benefit welds on carbon steels and most high-strength low-alloy (HSLA) steels by providing protection during solidification and slower cooling.

WELD INSPECTION

Weld inspection methods can be broadly divided into two groups; nondestructive testing (NDT) and destructive testing.

Nondestructive Testing (NDT)

Nondestructive testing includes visual examination, liquid penetrant, magnetic particle, eddy current, ultrasonic, radiography, acoustic emission, thermal, and optical methods. Space permits only a brief discussion of these most often used NDT methods. For a more complete presentation see the references listed at the end of this chapter.

Visual Inspection.   An experienced welder or inspector can detect most of the weld defects by careful examination. The following defects can be observed: undercut, overlap, surface checks, cracks, slag inclusions, penetration, and the extent of reinforcement. Some of these defects are shown in Fig. 15-25.

Liquid Penetrant.   The liquid-penetrant method involves flooding the surface with a light oil-like penetrant solution that is drawn into the surface discontinuities by capillary action. After the excess liquid has been removed from the surface, a thin coating of absorbant material is applied to draw the traces of penetrant from the defects to the surface for observation. Brightly colored dyes of fluorescent materials are added to the penetrant solutions to make the traces more visible.

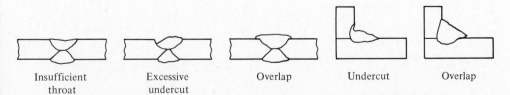

| Insufficient throat | Excessive undercut | Overlap | Undercut | Overlap |

Figure 15.25.   Some welding defects that can be checked visually.

Simple systems consist of spray can kits, including cleaner, penetrant, and developer in pressurized cans. Mass-production systems have been developed using fluorescent-dye penetrant, video-scanning techniques, and computer control and analysis.

Magnetic-Particle Inspection. Magnetic-particle inspection is based on the principle that ferromagnetic materials, when magnetized, will have distorted magnetic fields in which there are material flaws and that these anomalies can be clearly shown with the application of magnetic particles.

The magnetic field can be set up by passing an electric current through all or a portion of the part. The current may be passed through the part or through a conductor in close proximity to the part (Fig. 15-26).

To be effective, the direction of the induced field should be almost perpendicular to the expected flaw.

Either ac or dc can be used to generate the magnetic field. Magnetization is better with ac for surface discontinuities while dc is used to locate subsurface discontinuities or nonmetallic inclusions.

Magnetic-Particle Application. The magnetic particles can be applied either when the current is applied, which is the continuous technique, or after the current has been shut off, which is the residual technique. The sensitivity of the residual technique is lower but there is less chance of false indications produced by current leaks.

After the particles have been sprayed or sprinkled on the surface, the excess is gently removed by blowing or sweeping—leaving only the magnetic pattern.

The magnetic particles are available in several colors or treated with fluorescent material for observation under ultraviolet light.

Eddy-Current Testing. When electrically conductive material is subjected to an alternating magnetic field, small circulating electric currents are generated in the material (Fig. 15-27). These so-called eddy currents are affected by variations in conductivity, magnetic permeability, mass, and homogeneity of the host material. Conditions that affect these characteristics can be sensed by measuring the eddy-current response of the part.

The eddy currents induced into the part interact with the magnetic field of the exciting coil, thereby influencing the impedance, which is the total opposition to the flow of current from the combined effect of resistance, inductance, and capacitance of the coil.

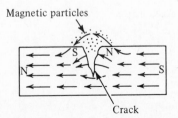

Figure 15.26. Magnetic-particle inspection.

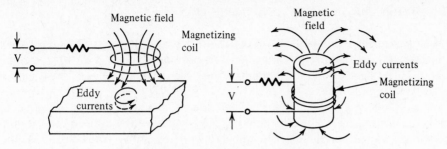

Figure 15.27. Eddy-current testing.

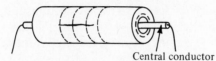

Pole piece   Pole piece

**Circular magnetization with current
passing directly through part**

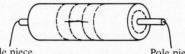

Central conductor

**Circular magnetization with current
passing through central conductor**

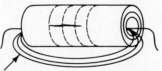

Cable wrappings

**Circular magnetization with
conductor threaded through part**

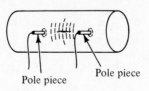

Pole piece   Pole piece

**Circular magnetization with
prod-type contacts**

Figure 15.28. Magnetization of the workpiece may vary depending
upon the application, but for optimum indications the direction of
the magnetic fields should be nearly right angles to the fault. (*Courtesy* American Machinist.)

By measuring the impedance of the exciting coil, or in some cases a separate indicating coil, eddy-current testing can detect cracks, voids, inclusions, seams, and laps. The best results are obtained when the current flow is at right angles to the flaw (Fig. 15-28).

According to a NASA report, eddy-current testing is not as sensitive to small, open flaws as is liquid penetrant. However, it does not require cleanup operations and is generally faster. Compared with magnetic-particle testing, eddy-current tests are not as sensitive to small flaws but they work equally as well on ferromagnetic and nonferromagnetic material.

Ultrasonic Inspection.   Ultrasonic inspection consists of sending a high-frequency vibration (beyond 20 kHz) through a component and observing what happens when the beam hits a discontinuity or a change in density. The altered ultrasonic signal can be used to detect flaws within the material, to measure thickness from one side, and to characterize metallurgical structure.

The sending transducer (usually a piezoelectric crystal) transforms a voltage burst into ultrasonic vibrations. The transducer is coupled to the workpiece by a liquid medium, such as water. A receiving transducer converts the received ultrasonic wave into a corresponding electrical signal [Fig. 15-29(a)]. With appropriate instrumentation, the same transducer alternately serves both functions in a so-called pitch-catch mode [Fig. 15-29(b)]. The signals are sent through the part and the time intervals that elapse between the initial pulse and the arrival of the various echoes are displayed on an oscilloscope screen. A flaw is recognized by the relative position (on the time scale) and amplitude of the echo. Where contact directly above the part is impractical, an angle beam is used as shown in Fig. 15-30.

Immersion inspection provides excellent coupling. Both the part and transducer are immersed, thus permitting reasonably fast scanning by using a traveling

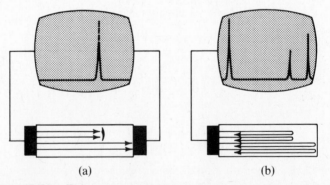

(a)   (b)

Figure 15.29. Through transmission uses separate sending and receiving transducers (a). The pulsed beam is sent through the part and the amplitude is attenuated by intervening flaws. The pulse-echo method (b) often uses the same transducer to send and receive the signal. The first signal is the beginning of the part, and the last the far side. The presence of a flaw may be anywhere in between.

bridge arrangement over the work. NC and computers can be used to control and coordinate the motions for a quick display of defects.

Radiography. Radiography is essentially a shadow pattern created when certain types of radiation penetrate an object and are differentially absorbed depending on variations of thickness, density, or chemical composition of the material. A schematic of a typical radiographic testing system is shown in Fig. 15-31. The shadowgraph is commonly registered on a photographic film to provide a permanent record. Other methods of registering the image include fluoroscopy, xerography, and closed-circuit television scanning.

Three types of penetrating radiation are presently used for industrial radiography: X rays, gamma rays, and neutron beams.

X Rays. X rays are a form of electromagnetic radiation similar to that of light, heat, or radio waves. A distinguishing feature of X rays is their extremely short wavelength—only about 1/10,000 that of light—and it is this characteristic that enables X rays to penetrate materials that absorb or reflect ordinary light. The X rays are generated when electrons, traveling at high speeds, collide with matter. They are generally produced in an evacuated X-ray tube, which contains a heated filament and a target. A high voltage across the filament (cathode) causes it to emit electrons, which are then driven to the target (anode) where the sudden deceleration results in X rays. The higher the tube voltage, the greater the energy and penetrating power of the X rays.

Gamma Rays. Gamma rays are emitted by disintegrating nuclei of radioactive substances. In industrial radiography, artificially produced radioactive isotopes, such as cobalt 60, are used almost exclusively.

Neutron Rays. Neutrons are derived from nuclear reactors, nuclear accelerators, or radioactive isotopes. For most applications, it is necessary to moderate the neutron energy and to collimate the beam.

In general, the X-ray and gamma-ray absorption of the material tested depends on the thickness, density, and, most importantly, the atomic nature of the material. For example, lead is about 1.5 times as dense as steel, but at 220 kV, 0.1 in. (2.54 mm) of lead absorbs as much as 1.2 in. (30.48 mm) of steel.

Although the absorption of neutrons by matter is comparable with that of high-energy X rays or gamma rays, the relative absorption by various elements is quite different. There is no obvious relationship between neutron absorption

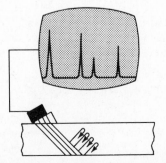

Figure 15.30. The ultrasonic beam can be used in an angular position by inserting a plastic wedge under the transducer. The beam bounces through the part until it uncovers a flaw, which causes an echo.

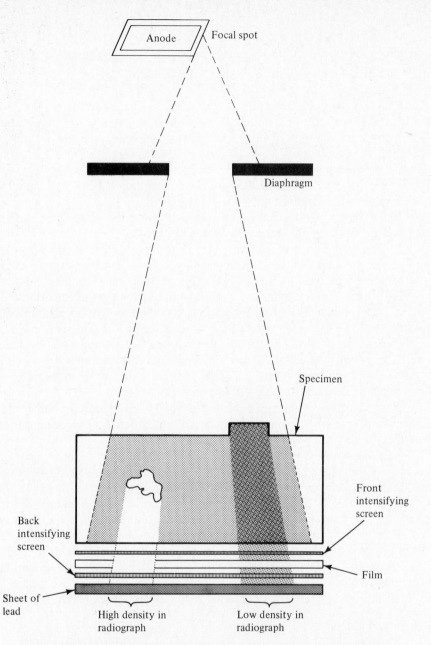

Figure 15.31. A schematic of a typical radiographic setup, which may include the use of a diaphragm and screens to reduce scatter effect and enhance the final image. (*Courtesy* American Machinist.)

and the atomic number of a material. Neutron radiography has been useful in obtaining images of organic material within a metallic housing, and also in the detection of hydrogen embrittlement in engineering materials.

Acoustic-Emission Monitoring.   Engineering materials undergoing stress or plastic deformation emit sound. The acoustic emission is in the form of short bursts or trains of fast impulses in the ultrasonic range. These acoustic emissions can be related to the physical integrity of the material or structure in which they are generated, and the monitoring of these events permits detection and location of flaws as well as prediction of impending failure. The pulse rate and amplitude of acoustic emission bursts are usually very high compared to most natural or man-made noises, and therefore it is possible to isolate the significant signals by careful measurement of emission rates and amplitudes.

Thermal Testing.   Temperature measurement can be used to detect defective components or devices that are themselves active heat sources; or components can be artificially heated, and the dynamic temperature distribution can provide significant flaw data. In any event, NDT testing involves detecting atypical temperatures rather than the absolute temperature of the part.

Contact methods of measuring, in addition to the use of direct-reading thermometers or thermocouples, can be affected by coating the surface to be tested with a material that reacts to temperature changes by altering its color or appearance. Coating materials include special paints with temperature-sensitive pigments, thermal phosphors or organic compounds that emit visible light when excited by ultraviolet radiation, treated papers for bonding to test surfaces, and liquid crystals.

For noncontact thermal testing, infrared detection systems are most frequently used. Infrared scanning systems produce a thermal picture of a part by representing various temperatures as different shades of gray or shades of significant colors.

Table 15-7 summarizes the NDT techniques described as well as indicating the applications, advantages, and disadvantages.

DESTRUCTIVE TESTING

Destructive tests have been developed to test the skill of the welder as well as to check the quality of welded joints for the various types of metal. Common tests are tensile, Izod, hardness, bend, and nick break. Tensile and Izod tests were described in Chapter 2. The destructive tests used on the Alaskan pipeline are shown in Fig. 15-32.

Hardness Evaluation Tests.   The best method of evaluating a weld and HAZ without making a microstructure is by hardness testing the area. Typical Bhn hardness ranges of acceptable welds in some commonly used steel are 111 to 150 for low-carbon steels and 212 to 321 for high-carbon and alloy steels. In

Table 15-7. Nondestructive testing methods. (*Courtesy Metal Progress.*)

Method	Measures or Defects	Applications	Advantages	Limitations
Acoustic emission	Crack initiation and growth rate Internal cracking in welds during cooling Boiling or cavitation Friction or wear	Pressure vessels Stressed structures Turbine or gear boxes Fracture mechanics research Weldments	Remote and continuous surveillance Permanent record Dynamic (rather than static) detection of cracks Portable	Transducers must be placed on part surface Highly ductile materials yield low amplitude emissions Part must be stressed or operating Test system noise needs to be filtered out
Eddy current	Surface and subsurface cracks and seams Alloy content Heat treatment variations Wall thickness, coating thickness Crack depth Metal sorting	Tubing Wire Ball bearings "Spot checks" on all types of surfaces Proximity gauge Metal detector	No special operator skills required High speed, low cost Automation possible for symmetrical parts Permanent record capability for symmetrical parts No couplant or probe contact required	Conductive materials Shallow depth of penetration (thin walls only) Masked or false indications caused by sensitivity to variations, such as part geometry Reference standards required Permeability
Magnetic particles	Surface and slightly subsurface defects; cracks, seams, porosity, inclusions Permeability variations Extremely sensitive for locating small tight cracks	Ferromagnetic materials; bar, forgings, weldments, extrusions, etc.	Advantage over penetrant in that it indicates subsurface defects, particularly inclusions Relatively fast and low cost May be portable	Alignment of magnetic field is critical Demagnetization of parts required after tests Parts must be cleaned before and after inspection Masking by surface coatings

Method	Detects	Applications	Advantages	Limitations
Penetrants	Defects open to surface of parts; cracks, porosity, seams, laps, etc. Through-wall leaks	All parts with nonabsorbing surfaces (forgings, weldments, castings, etc.). Note: Bleed-out from porous surfaces can mask indications of defects	Low cost Portable Indications may be further examined visually Results easily interpreted	Surface films, such as coatings, scale, and smeared metal, may prevent detection of defects Parts must be cleaned before and after inspection Defects must be open to surface
Radiography (thermal neutron)	Hydrogen contamination of titanium or zirconium alloys Defective or improperly loaded pyrotechnic devices Improper assembly of metal, nonmetal parts	Pyrotechnic devices Metallic, nonmetallic assemblies Biological specimens	High neutron absorption by hydrogen, boron, lithium, cadmium, uranium, plutonium Low neutron absorption by most metals Complement to X-ray or gamma-ray radiography	Very costly equipment Nuclear reactor or accelerator required Trained physicists required Radiation hazard Nonportable Indium or gadolinium screens required
Radiography (gamma rays)	Internal defects and variations, porosity, inclusions, cracks, lack of fusion, geometry variations, corrosion	Usually where X-ray machines are not suitable because source cannot be placed in part with small openings and/or power source not available	Low initial cost Permanent records; film Small sources can be placed in parts with small openings Portable Low contrast	One energy level per source Source decay Radiation hazard Trained operators needed Lower image resolution Cost related to energy range

Table 15-7 (continued)

Method	Measures or Defects	Applications	Advantages	Limitations
Radiography (X rays—film and image tubes)	Internal defects and variations; porosity, inclusions, cracks, lack of fusion, geometry variations, corrosion Density variations	Castings Electrical assemblies Weldments Small, thin, complex wrought products Nonmetallics Solid propellant rocket motors	Permanent records; film Adjustable energy levels High sensitivity to density changes No couplant required Geometry variations do not affect direction of X-ray beam	High initial costs Orientation of linear defects in part may not be favorable Radiation hazard Depth of defect not indicated Sensitivity decreases with increase in scattered radiation
Thermal (thermochromic paint, liquid crystals)	Lack of bond Hot spots Heat transfer Isotherms Temperature ranges	Brazed joints Adhesive-bonded joints Metallic platings or coatings Electrical assemblies Temperature monitoring	Very low initial cost Can be readily applied to surfaces which may be difficult to inspect by other methods No special operator skills	Thin-walled surfaces only Critical time–temperature relationship Image retentivity affected by humidity Reference standards required
Ultrasonic	Internal defects and variations; cracks, lack of fusion, porosity, inclusions, delaminations, lack of bond, texturing Thickness or velocity Poisson's ratio, elastic modulus	Wrought metals Welds Brazed joints Adhesive-bonded joints Nonmetallics In-service parts	Most sensitive to cracks Test results known immediately Automating and permanent record capability Portable High penetration capability	Couplant required Small, thin, complex parts may be difficult to check Reference standards required Trained operators for manual inspection

Destructive Testing

Test Sample Locations

Samples are cut from the full circumference of the section of pipe to be tested. Drawing shows only samples from Quadrants 3 & 4

Two Side Bends
Two Root Bends
Two Tensiles
Two Nick-Breaks
Two Face Bends
Two Side Bends
Charpy Impact
Crack Opening Displacement

Location of Test Section

Section removed with weld from pipeline

Quad #1

Quad #4

Quad #3

Quad #2

Side Bend

Force

Face

Root Bend

Force

Root

Charpy Impact

(Sample is machined, V-notched on weld face and struck with a sharp impact force.)

Impact Force

Face

Machined V-notch

Crack Opening Displacement

(Sample is machined, cut with saw and bent with a constant force)

Force

Saw cut .006" width

Force

Face

Neck-Break

Saw cuts

Force

Tensile

Force

Face Bend

Force

Face

Microhardness

Force applied to 23 points

Charpy V-notch impact tests
Test temperatures, minus 20 and minus 50 degrees Fahrenheit.

Microhardness survey test
Load: 10 kilograms
Indentations: 6 in base metal, 3 in weld and 14 in heat affected zone.

Chemical analysis:
(Elements measured) carbon, manganese, phosphorus, sulfur, silicon, nickel, molybdenum, chronium, vanadium, and copper.

Figure 15.32. The types of destructive tests that were used on the Alaskan pipe line. (*Courtesy Alyeska.*)

591

general a weld has acceptable integrity if the cross-sectional hardness does not exceed R_c 32. This hardness would be typical of U.S. Steel's T-1 as quenched-and-tempered HSLA steel. A maximum of R_c 40 is generally permissible when the welds are loaded in compression. When the hardness exceeds R_c 40, chances of cracking increase greatly.

Guided Bend Test. The guided bend test is made with the aid of a jig as shown in Fig. 15-33 and in accordance with ASTM specifications. The specimens are machined from welded plates, the thickness of which must be within the capacity of the jig. The specimen is placed across the opening of the lower half of the jig or die. The hydraulically operated plunger forces the specimen into the die. To fulfill the requirements of this test, the specimens must bend to 180° with no cracks greater than ⅛ in. (3.17 mm) in any dimension appearing on any surface. Face bends are made when the face of the weld is placed in tension (outside of the bend), and root tests are made when the root is placed in tension

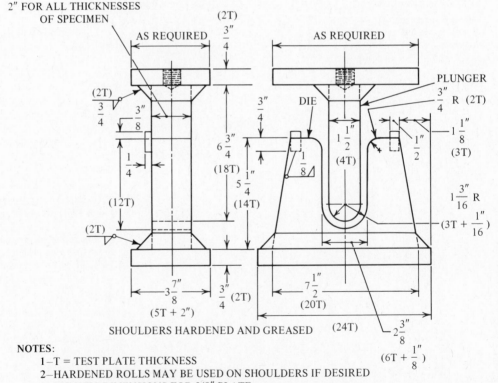

NOTES:
1—T = TEST PLATE THICKNESS
2—HARDENED ROLLS MAY BE USED ON SHOULDERS IF DESIRED
3—SPECIFIC DIMENSIONS FOR 3/8″ PLATE

Figure 15.33. Guided bend testing jig.

(Fig. 15-34). The guided bend test is especially useful for checking the quality of the weld and base metal as well as the degree of penetration and fusion of the weld.

Nick Break Test.   The nick break test is used to determine if a weld has internal defects such as slag inclusions, gas pockets, poor fusion, and/or oxidized or burned metal. The specimen is obtained from a welded butt joint either by machining or by cutting with an oxyacetylene torch. Each edge of the weld at the joint is slotted by means of a saw cut through the center as shown in Fig. 15-35. The piece thus prepared is bridged across two steel blocks and struck with a heavy hammer until the section of weld between the slots fractures. The

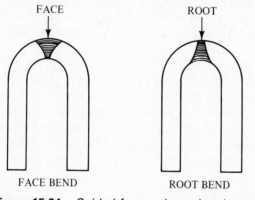

Figure 15.34.   Guided face and root bend tests.

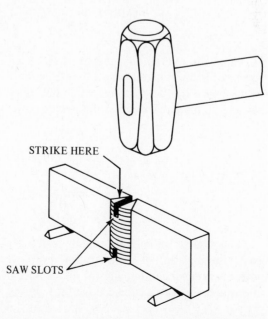

Figure 15.35.   Nick break test.

exposed weld should be completely fused and free of slag inclusions. The size of any gas pocket must not be greater than $\frac{1}{16}$ in. (1.60 mm) across the longest dimension and the number of gas pockets or pores per square inch should not exceed six.

ECONOMICS OF WELDING

The economics of welding begins with the design and follows through to the final inspection of the finished product. It is at the design stage where the type of joint and the method of assembly are pretty well determined. Some feedback may come from production regarding modifications and desirable design changes. Thus the designer should be well acquainted with the wide variety of joining methods available and the comparative advantages and disadvantages of each for the applications at hand.

Economics is also very much in the hands of the production engineer. For example, the design may call for a welded assembly but the exact process may be left to production. The production engineer must weigh the cost of the joint preparation and decide whether assembly will be manual, semiautomatic, or fully automated. As an example, if welding is to be the method of assembly, a cost comparison of automatic versus manual welding can be estimated from standard data. The labor cost of manual arc welding may be calculated using the following formula:

$$\text{Labor cost/ft} = \frac{\text{Hourly rate}}{(\text{Travel speed in./min}) \times \dfrac{(60 \text{ min/hr})}{(12 \text{ in./ft})} \times \text{Duty cycle}}$$

To obtain the overall cost the overhead, electrodes, and power must be added. Overhead is often taken at two times the direct labor rate. The cost of electrodes will be based on the cost per pound and the number of feet of a given size weld that can be made. To obtain the power cost the following formula may be used:

$$\text{Power cost/ft} = \frac{(A) \times (V) \times (\text{cost/KWH}) \times \dfrac{(1 \text{ KW})}{(1000 \text{ W})}}{(\text{min}) \times (\text{ipm}) \times (60 \text{ min}) \times \dfrac{(1 \text{ ft})}{(12 \text{ in.})}}$$

The higher initial investment cost for automatic equipment can result in greater savings only if the operating time is favorable. Usually this is a problem of production scheduling and the availability of versatile tooling to keep the setup time down and to minimize downtime during loading and unloading.

Questions and Problems

15-1. Shown in Fig. P15.1 is a joint requiring two 4-in. fillet welds. Assume the tensile shear load is 50 ksi with occasional shock loads. Are the welds adequate if they are made with E7010 electrodes?

15-2. (a) A butt weld is made in a HSLA steel. The material is 2 in. thick and 4 in. wide. What is the normal stress on the joint if it is placed in tension at 20 ksi?

    (b) If the butt weld of (a) has a $\frac{1}{2}$-in. thick plate, 3 in. wide, placed on the top and bottom of it and welded along both ends of the plate only, what will be the normal stress on these plates?

15-3. As shown in Fig. P15.2, plate B is to be welded to plate A. Upon completion, the weldment must resist 38-ksi tensile shear at the axis of the plates as indicated by the arrows. The following information is provided. (1) Plate B is low-carbon steel $\frac{3}{8}$ in. $\times$ 3 in. $\times$ 15 in. with a tensile strength of 45,000 psi. (2) Plate A has the same tensile strength and is $\frac{3}{8}$ in. $\times$ 4 in. $\times$ 4 in. (3) An E7016 electrode is prescribed.

    How many inches of fillet weld are required on each side of plate B to guarantee the proper strength?

15-4. A mild steel tank of $\frac{1}{2}$-in. wall thickness with a 10-in. dia. and 3 ft long has a welded cap on each end that comes down on the sides 1 in. as shown in Fig. P15.3. A longitudinal seam weld is used on the cylinder. All welds are full-penetrating butt weld and made with an E6010 electrode. What would be the safe allowable load for each seam?

15-5. Is the cost per pound of deposited weld metal a good way to estimate costs? Explain.

15-6. A 3-in. (7.62-cm) dia. hole is made in $\frac{5}{8}$-in. (15.875-mm) mild steel plate that is to be welded to another plate by a $\frac{1}{2}$-in. (12.70-mm) fillet weld around the inside of the hole. What load will this weld support in shear if an E6010 electrode is used?

15-7. A manufacturer of earth-moving equipment uses two different steels in a welded axle shaft assembly. The axle is made of 1045 steel and the gear out of 1020 steel. If this assembly is to be put under higher loads would it be feasible to change to a 4640 steel for the shaft? The composition of 4640

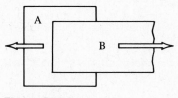

Figure P15.1.

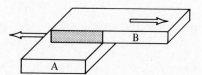

Figure P15.2.

Figure P15.3.

is as follows: C 0.38–0.43, Mn 0.60–0.80, P 0.040, S 0.040, Si 0.40–0.35, Ni 0.65–2.00, Mo 0.20–0.30.

15-8. A welded structural steel member is to be fabricated by making a Tee joint from pieces of plate $\frac{1}{2}$ in. $\times$ 8 in. $\times$ 10 ft long and the other 4 in. $\times$ 10 ft long. The 4-in. wide plate is to be mounted vertically on the center of the 8-in. wide plate and welded with a $\frac{5}{16}$-in. intermittent fillet weld on both sides. The welds are to be 2 in. long with a center distance of 5 in. Make an isometric sketch of the weld and the appropriate welding symbol.

15-9. Sketch the welding symbol that would be used if a butt joint is to be made in $\frac{1}{2}$-in. thick plates in the field with complete penetration and the face of the weld to be ground flush.

15-10. Sketch a symbol for a resistance spot weld $\frac{1}{8}$-in. dia., 6 welds at a pitch of 1 in.

15-11. Show by sketch the joint preparation that should be made for $\frac{3}{4}$-in. thick plate being joined to a vertical column. The $\frac{3}{4}$-in. plate can be welded from both sides.

15-12. What is the effective throat size of a $\frac{1}{2}$-in. fillet weld?

15-13. What is the effective throat of a fillet weld having a $\frac{1}{2}$-in. vertical leg and $\frac{5}{8}$-in. horizontal leg?

15-14. What is the difference between a backing bar and a backing strip?

15-15. (a) When is it not necessary to calculate the strength of butt welds?

   (b) What are the special conditions involved?

15-16. (a) What is the main cause of angular distortion in welding?

   (b) How can it be minimized?

15-17. How can the hardness of a high-strength steel be best preserved through the welding process?

15-18. What would the strain in inches per inch be for a high-strength steel that has a yield point of 85,000 psi?

15-19. Two 2-in. thick plates of 1040 steel 8 ft long are to be joined by butt welding in the flat position. State the following: welding joint design (sketch), welding process, and preheat if any.

15-20. Assume two pieces of 1050 steel, 0.500 in. thick, 6 ft long, and 20 in. wide, are to be butt welded. State the following:

   (a) recommended preheat and interpass temperature.

   (b) 60 v and 300 amp are used with a GMA process. The travel speed is 100 ipm. What would be the heat input in joules per mm?

15-21. Which NDT inspection method would be best for detecting the improper assembly of metal and nonmetal parts?

15-22. What would the recommended minimum weld size be for a lap weld made of two $\frac{1}{2}$-in. thick plates?

15-23. The flat end of a 4-in. dia. rod is welded to a $\frac{3}{4}$-in. thick plate. If a $\frac{1}{4}$-in. fillet weld is used, will the rod be able to withstand a torquing moment of 3000 psi?

Case Study

The axle drive assembly shown in Fig. 15CS.1 is made out of two steels, 1045 for the shaft and 1020 for the plate. The welds are large, 0.500-in. (13-mm) fillet on one side and 0.500-in. (13-mm) bevel groove weld ground flush on the opposite side. These welds were made by the GMAW process to keep the heat input down. The welds proved unacceptable due to lack of penetration and cracking. Also the HAZ on the 1045 side was harder than R_c 50.

State what you would recommend as to

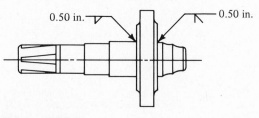

0.50 in. ⟍   0.50 in.

Figure 15CS.1.

welding process and other changes you would make.

Bibliography

Blodgett, Omer W. "New AISC-AWS Allowables Enable Cost Reductions and the Use of Advanced Technologies in the Design of Structure and Weldments." *Welding Journal* 49 (August 1970): 619–638.

Brosilow, Rosalie, ed. "Weldability—What It Means to Fabrications." *Welding Design and Fabrication* 51 (June 1978): 74–80.

Cary, Howard B. *Modern Welding Technology.* Englewood Cliffs, N.J.: Prentice-Hall, 1979.

Cummins, R. W. "What Designers Should Know About Welded Structures." *Welding Design and Fabrication* 51 (July 1978): 86–89.

Editor. "Cold Cracking in Welds in ASLA Steels." *Welding Design and Fabrication* (February 1979).

Giachino, J. W.; W. Weeks; and G. S. Johnson. *Welding Technology.* Chicago: American Technical Society, 1973.

Lindberg, R. A., and N. R. Braton. *Welding and Other Joining Processes.* Boston: Allyn and Bacon, 1976.

Schaffer, George, ed. "Nondestructive Testing." *American Machinist Special Report 692* 120 (November 1976): 86–110.

Thielsch, Helmut. *The Sense and Nonsense of Weld Defects.* Lake Zurich, Ill.: Monticello Books, 1967.

Welding Inspection. Miami: American Welding Society, 1980.

CHAPTER SIXTEEN

Adhesive Bonding

Adhesives were formerly thought of, especially by engineers, as a joining method to be used for noncritical applications. The aircraft industry led the way for the initial boost in the development of chemical fastening systems. Structural bonding of parts is now being extensively used for automotive, office, and industrial equipment.

An example of the use of adhesives in the auto industry is that of a custom-wheel specialist. A structural adhesive was used to join the die-cast aluminum hub and spoke center to the steel rims. The bond had to withstand both high and low extremes of temperature and a great amount of stress, especially shear stress. Two tests were used by the company to estimate the performance of the bond. A press was used to apply pressure evenly to the hub-and-spoke center in order to determine the pressure required to break the bond from the rim. The minimum pressure required was 42,000 lb. The bond withstood 70,000 lb. The second test was a rotary fatigue test. This test indicated the casting would break before the bond.

Structural adhesives have been used for quite a number of years to bond brake shoes to the brake drum. The bond develops strengths of 10,000 lb in shear compared to 2500-lb shear stress developed by the rivets formerly used.

CLASSIFICATION OF STRUCTURAL ADHESIVES

There are many ways of classifying adhesives. A broad classification is that of *nonstructural* and *structural*. In the first category are household-type glues of vegetable or animal base. Typical materials involved are casein, rosin, shellac, and asphalt. They are characterized by poor moisture resistance, but generally good resistance to heat and chemicals within their range of usage.

Structural adhesives are those that are capable of sustaining load-bearing joints for a long period of time.

Structural Adhesives

Structural adhesives are composite systems containing several components that are available in various forms, including liquids, pastes, solids, pellets, cartridges, tapes, and films.

The major factors in selection of an adhesive for a specific application include the influence of the materials (adherends) being joined, joint design, operational requirements (strength, temperature variations, and other environmental conditions), and whether disassembly will be required for inspection, maintenance, or replacement.

The most widely used structural (higher-bond strength) adhesives today are epoxies, acrylics, anaerobics, cyanoacrylate, urethanes, and hot melts.

Epoxies.   Epoxy resins, a reaction product of acetone and phenol, are the most popular of the adhesives used today. Epoxies may be cured (crosslinked) by the addition of a catalyst, resulting in a room-temperature cure, or by subjection to an elevated temperature. Crosslinking between polymer chains is shown in Fig. 16-1.

The two-part, room-temperature curing types are satisfactory for many applications, but higher strengths and better heat resistance can be obtained with the heat-cured type. Epoxies bond well to most surfaces, even with slight oil films; however, for maximum strength (up to 10,000 psi in tensile shear) (700 MPa) surfaces should be cleaned and specially prepared. Although the tensile shear strength is good, the bond has low flexibility and poor impact strength. Little or no setting pressure is required but only contact pressure upon application.

Modified Epoxies.   Since thermosetting-plastic adhesives exhibit "brittle" characteristics when cured, modifications have been incorporated that include thermoplastics and rubber. Some popular modifications are epoxy-nylon, epoxy-polysulfide, epoxy-phenolic, and epoxy-nitrile.

The modified epoxies are excellent for metal-to-metal bonding and bonding to concrete. One of their advantages is that they consist of 100% solids, so there is no problem of solvent evaporation and shrinkage when joining impervious surfaces. Another advantage is their ability to *wet* metal, glass, and cementitious surfaces readily. Wetting refers to the ability of a liquid to spread over a solid surface evenly without voids. For example, water on an oily surface stays in

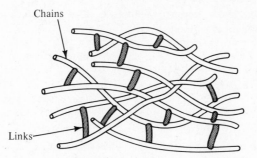

Chains

Links

Figure 16.1. Crosslinking between the polymer chains adds strength, heat, and solvent resistance. Most structural adhesives are cross-linked. (*Courtesy* American Machinist.)

Figure 16.2. An example of poor wetting. Air entrapment between the adhesive and adherent provides place for corrosion to start.

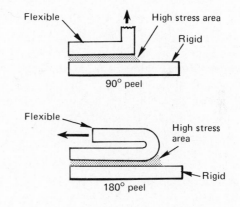

Flexible   High stress area   Rigid   90° peel

Flexible   High stress area   Rigid   180° peel

Figure 16.3. Peel strength is measured in the ability of the adhesive to withstand separation at the edge of the joint.

droplets, but on a clean, dry surface it will spread out uniformly. Poor wetting action is shown in Fig. 16-2.

In addition to being more flexible, modified epoxies show an improvement in *peel strength*. Peel strength refers to the ability to withstand separation at the edge of the joint (Fig. 16-3).

All of epoxy's basic virtues—high adhesion, tensile strength, rigidity, creep resistance, easy curing, and heat resistance—can be enhanced by additives. Some modifications involve tradeoffs, but others do not. It is possible to formulate an epoxy-based adhesive that combines high peel and shear strengths, temperature resistance of over 500°F (260°C), low creep, and low coefficient of thermal expansion. New formulations are constantly being created to meet spe-

Table 16-1. Typical adhesive strengths. (*Courtesy* American Machinist.)

	Shear (psi)	Peel (pli)*
Epoxy amine	4000	5
Epoxy nylon	6000	60
Epoxy phenolic	4000	10
Epoxy polysulfide	3000	50
Thermoset urethane	3500	70
Modified acrylic	4000	15
Structural anaerobic	3000	30
Cyanoacrylate	4000	10
Hot melt	300	20
Vinyl phenolic	4000	50
Nitrile phenolic	4000	60

* Peel strength/linear inch.

cial needs, thus rapidly outdating the handbooks. Table 16-1 lists the shear and peel strengths of common adhesives, including several modified epoxy resins.

Urethane. Like epoxy, urethane adhesives are available in one- or two-part systems, as solid hot melts, liquids or pastes, or as solvent systems. The basic urethane is a thermoplastic material and is often used in that form. A variety of means are available for crosslinking, making it a fairly rigid thermoset adhesive.

The distinguishing feature between urethane or polyurethane and epoxy is its basic flexibility and high peel strength. Although it may take hours or days to cure completely, most formulations cure to touch in a few minutes, allowing urethane-bonded parts to be handled for assembly and packaging.

Urethane is usually cured at room temperature. Elevated temperature cures work with some systems, but the temperatures are much lower than for epoxy and the operating temperatures are also lower, being about 150°F (66°C).

Urethanes may be applied to both surfaces and after a specified time are brought together when the adhesive is tacky enough to hold the parts together. Alternatively, the two parts may be allowed to dry and then heated before bringing them together.

In general terms, urethane is preferable to epoxy when greater elongation or a shorter cure time is needed, or if the joint is to be subjected to prolonged vibration or other cyclic stress.

Acrylics. The acrylic family, which includes sheet materials like Plexiglas, ranges from low-temperature-tolerant, runny adhesives that cure by evaporation to quick-cure thread-locking anaerobics to a new subfamily of structural adhesives that can be applied to oily surfaces and still generate shear strengths as large as 4000 psi.

Until the late 1960s, the thermosetting acrylics had limited usefulness as a metal-to-metal structural adhesive largely because they did not adhere well. About then, however, a new formulation became available that made the acrylics more tolerant of dirty surfaces than any other structural adhesive. An agent in either the primer or accelerator penetrates or reacts with surface contamination, especially oils, enabling the adhesive to bond with the underlying metal. The new adhesive developed by the mid 1970s is referred to as a "second-generation" or "reactive" adhesive. By chemical manipulation, the adhesive actually incorporates the oil into the structure of the cured adhesive.

The new acrylics are challenging epoxies for bonding supremacy, offering flexible bonds with superior peel and impact resistance.

Acrylics are two-part adhesives, but are not mixed for most applications. The resin is applied to one surface and the accelerator to the other. The parts can be set aside for weeks with no detrimental effects, and once they are mated, handling strength is typically achieved in a few minutes. Curing can be done at room temperature. Acrylics can now handle many of the tasks originally reserved for epoxies.

Cyanoacrylates. The prime advantage of these adhesives is their rapid cure, usually within seconds. Cure is affected by contact with moisture or alkali. Almost any surface has sufficient moisture to affect a cure.

Cyanoacrylates are costly, but the amount used is small since a very thin glue line is required to cure properly. The cure is too fast for many applications. Cyanoacrylates are used to a limited extent in surgery where their ability to cure almost instantaneously and give a "chemical" suture is extremely useful. Often the cyanoacrylates can be used to retain components in position while other operations such as welding are carried out. In this way expensive jigs or difficult clamping can be avoided.

Cyanoacrylates have high tensile strength (up to 5000 psi) (3520 kg/mm^2) and low peel and impact strengths. Like other adhesives, cyanoacrylates are undergoing development that will improve their peel and impact strength without harming their tensile strength.

The adhesive is thin and watery in its basic form, which allows it to wick into tight spaces by capillary action but limits its gap-filling ability to about 0.005 in. (0.13 mm). Accelerator–primers increase the practical gap size to approximately 0.03 in. (0.76 mm). Temperature tolerances range from about 160°F to 475°F (71–245°C).

Application can be by manual or by automated equipment as long as the adhesive is not constrained within the application equipment for more than a couple seconds. This problem has been taken into account in several commonly available dispensing units. Solvents such as gasoline, alcohol, and light oils have practically no effect on bond strength.

Anaerobics. Anaerobic sealants are so called because they harden satisfactorily only in the absence of atmospheric oxygen. Hence they must be stored in the presence of air. When in the liquid form, they are capable of flowing into the most minute crevice and thus the whole of free space between assembled

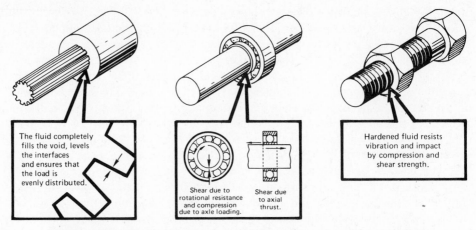

Figure 16.4. Four common applications for anaerobic materials involve compression loads, shear and compression forces, and actual shear followed by destructive crushing. (*Courtesy* Adhesives Age.)

components is readily and completely filled with fluid. After the fluid has cured and formed a tough wedge, the assembly is sealed because there is virtually no shrinkage and therefore no passage for gases or liquids.

The initial problem of what to keep anaerobics in was solved by low-density polyethylene containers that are porous to oxygen and other gases and by designing application equipment that avoids entrapping the material to confined spaces.

For the past two decades, anaerobics were associated with their use for locking threaded and other close-fitting assemblies, as shown in Fig. 16-4, and were popularly known by the trade name Locktite. At the present time, however, anaerobics have been developed for bonding flat surfaces as well.

The properties of the original anaerobics were low viscosity, low peel and tensile strength, but exceptional torque-shear strength.

Anaerobics progressed from a specialized retaining material to a structural adhesive after being combined with urethane to counteract brittleness and its correlative low peel and impact strengths. These "second-generation" anaerobics have tensile shear strengths up to 10,000 psi (7000 kg/mm²), peel strength up to 60 pli (10 kg/cm), cure time that can be adjusted from 10 sec to 5 months, long-term temperature tolerance of 400°F (204°C), and impact strengths up to 40 ft lb (552 kg cm).

Hot Melts. Hot-melt adhesives were initially used in packaging and furniture manufacture and have traditionally been regarded as nonstructural. Newer formulations offer sufficient strength for low-stress product assembly application. Hot melts may be formulated to make flexible or rigid bonds. They are usually applied with a suitable gun and are fast-setting, reaching 80% of their bond strength within seconds of application. They are used to bond both permeable and impermeable surfaces. Hot melts are insensitive to moisture and most solvents but they soften repeatedly at elevated temperatures of about 200 to 300°F (93–149°C).

SURFACE PREPARATION

Most everyone knows that a joint has to be clean in order to have a good adhesive bond (except for acrylics). Why some surfaces bond easily and others are practically impossible to join has been the subject of many theories; in fact, there are enough theories to keep graduate students in engineering occupied for years.

For this presentation only a few basic concepts are important. First, the adhesive must be able to wet the substrate to achieve a good bond. To wet the adherent (material to be joined), the adhesive must have a lower surface energy than the adherend. Since a clean surface has high surface energy, it can be wet easily. This also explains why materials of low surface energy, such as Teflon, are hard to bond adhesively.

Surface roughness is also important, for there is a mechanical entanglement between a rough surface and an adhesive applied to it that locks the two mechanically. There may also be an increase in electrical effects because a rough surface has an increased area.

Cleaning processes may be classified as mechanical, passive chemical, and active chemical.

Mechanical. Common examples of mechanical cleaning are sanding, abrasive scrubbing, wire brushing, grit blasting, grinding, and scraping.

Passive Chemical. Passive chemical treatment of a surface consists of removing soil from the surface by chemical means without chemically altering the parent material. This may be done by vapor degreasing, solvent washing, alkaline and detergent cleaning, or ultrasonic cleaning.

Active Chemical. An active chemical treatment alters the surface physically and chemically to increase its free-energy level and make it receptive to adhesion. This may be done with acids, caustic and sodium etchants, anodic films, or chemical films.

If high strengths are not required, the surface treatment may be either a mechanical or passive chemical treatment. The oxide film on aluminum, for example, may be removed by wiping with methyl ethyl ketone (EK) or trichloroethylene.

For some metals, particularly aluminum, an active chemical treatment is needed to give maximum strength. For aluminum, this consists of a vapor degrease with trichloroethylene, followed by a rinse, then a sulphuric acid chromic etch, and then a rinse and forced-air drying.

Generally, carbon steels offer maximum bonding strength by mechanical treatment, such as a dry-abrasive blast of 80-grit aluminum oxide followed by flushing with trichloroethylene and then a rinse. For detailed accounts of surface preparation, see references at the end of the chapter.

Because of the likelihood of contamination and oxide formation, it is desirable to use the prepared materials as soon after treatment as possible. If storage is necessary, the metal should be kept in airtight containers. Care must be taken so that surfaces that have been etched are not touched. Handlers must wear clean cotton gloves.

ADHESIVE APPLICATION AND CURE

Application.   Adhesives can be applied in a variety of ways. Epoxies, for example, can be formulated for spraying, brushing, dipping, roll-coating, dusting with dry powder, extruding, and trowelling.

Tape-type adhesives have become quite popular because they eliminate the need for mixing and the application will have a known, uniform *glue-line thickness*. Glue-line thickness refers to the amount of adhesive that remains after the pressure is applied and cured. For example, to achieve an ultimate glue-line thickness of 1 to 3 mils, anywhere from 5 to 15 mils of 20% solid, wet-type adhesive must be applied.

If the adhesive is applied in solution, time must be allowed for evaporation of the solvent before the two members can be brought together. Solvent reduces the viscosity of the adhesive, allowing it to penetrate the pores of the substrate. The amount of penetration into the substrate is not important as long as good wetting of the surface has occurred. In fact, too much penetration of the adhesive into the substrate may be detrimental since it may result in glue-line starvation.

Rough surfaces must have enough adhesive to fill in the small depressions plus enough to achieve the desired glue-line thickness. The gap between the two surfaces should not exceed a few thousandths of an inch.

An increasingly popular technique for both adhesive and sealant application is silk screening or stenciling.

In silk screening, a fine nylon or stainless-steel mesh is the base to which a stencil is applied. Adhesive is smeared across the end of the screen and a rubber squeegee is dragged across the stencil, pushing the adhesive through the screen. Silk screening can be used with most liquid adhesives but is especially neat and effective with anaerobics since the adhesive will not cure on the screen as long as it is exposed to air.

The screening process can be automated and is capable of controlling deposition thickness down to 0.001 in. (0.03 mm).

Aircraft manufacturers use large quantities of adhesive in the form of films, either unsupported or backed with a scrim of fiber. Thickness is ensured by this approach, and because the film can be cut before or after it is applied to the adherend surface, shape is ensured. The films are simply laid between the two adherend surfaces. The films are usually epoxy, phenolic, or an alloy mixture and are cured by heat. Often they are refrigerated or frozen to prevent them from curing before being applied. Sandwiching different resin systems together in a single film permits combinations of strength and toughness that are not available in any other form.

Cure.   Adhesives may be cured by solvent release, crosslinking caused by heat and pressure, or a catalyst. Some adhesives, such as one-part epoxies, contain latent catalysts that are activated by heat and crosslinking results with only contact pressure. There are no volatile solvents to be driven off. Phenolics, on the other hand, do have volatile solvents and need high pressure during cure to ensure against solvent entrapment. The time required varies according to the heat, pressure, and catalyst used.

Some adhesives, known as *dry-film* adhesives, may be applied several days before assembly. The surfaces are then activated by one of two methods.

1. The surface is rubbed lightly with a solvent or a liquid adhesive of the same type.
2. The adhesive is heated, usually by infrared lamps, to the reactivation temperature.

Hot-melt adhesives are copolymers of polyethylene with polyvinyl acetate, polypropylene nylon, or polyester. Hot melts bond instantly. They set by contacting a surface, cooling, and solidifying. Hot-melt systems are often automatic, applying the adhesive from a pressurized tank through hoses to the applicator. They have limited toughness and heat resistance and are not used where high structural strengths are required.

ADHESIVE-JOINT DESIGN

The strength of an adhesive joint is dependent on the joint geometry, material properties of the constituents (discussed previously), and the type of loading the joint is subjected to. Typical adhesive strengths are shown in Table 16-1.

Joint Geometry.   As with brazing and soldering, joint strength for adhesives is based on having an adequate area over which the stress may be distributed. As a consequence, the principal joint design is a lap type. When a lap joint is subjected to tensile shear, the ends have a tendency to deflect. Hence the ends of the lap experience the maximum stresses and the center of the bond experiences the minimum stress as shown in Fig. 16-5. When the ends of the joint are tapered as shown in Case 3 of Fig. 16-5, the tensile stress in the adherends and the shear stress in the adhesive are constant, thus giving a stress concentration of unity. Both Case 2 and Case 3 require relating the material properties of the two adherends to the taper ratios. However, Case 3 is almost impossible to manufacture since the thickness of the adherend must diminish to zero at the joint ends.

Joint Loading.   Adhesive joints are stressed primarily in tension and shear, as shown in Fig. 16-6. The stresses produce shear, peel, and cleavage. Lap shear strengths are directly proportional to the width of the overlap, but the unit strength decreases with length. The optimum shear strength of the bonded joint

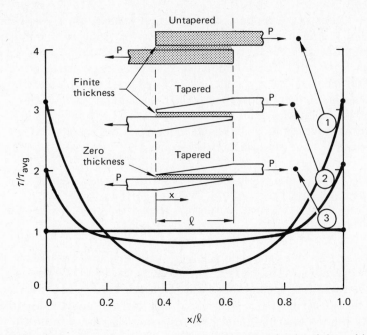

Figure 16.5. The change in stress distribution of a lap joint with geometry of the joint.

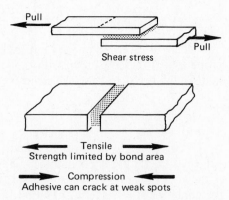

Figure 16.6. Adhesive joints are stressed primarily in shear and tension.

is largely dependent upon the shear modulus of the adhesive and its optimum thickness. The thickness may vary from 0.002 in. (0.0051 mm) for high-modulus adhesives to 0.006 in. (0.0152 mm) for low-modulus material. Failure occurs when stresses overcome the cohesive forces of the joint.

Adhesive lap joints made out of thin aluminum strips and subjected to peel tests have shown that failure need not be in the adhesive. Nylon-epoxy, for example, has sufficiently high peel strength to cause failure in the substrate rather than the adhesive bond. A standard ASTM peel-test specimen is shown in Fig. 16-7.

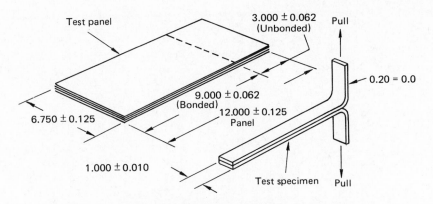

ASTM D 1876 (dimensions, inches)

Figure 16.7. A standard ASTM peel-test specimen.

Adhesive joints may be tested for flexibility or bend qualities by wrapping the completed joint over a mandrel. In general, the joint should be designed to:

1. Reduce bending of the joint for a given load by stiffening the adherends or decreasing the thickness of the adherends, particularly toward the edges to allow the joint to have more flexibility.
2. Decrease the length of overlap required.

ADHESIVES AND TEMPERATURE

Adhesives are constantly being subjected to increasingly hostile environments. Consequently, formulations have been made that maintain good strength at both cryogenic and elevated temperatures. One approach toward withstanding elevated temperatures has been to add poor heat conductors to the adhesive, such as carbon fibers, cork composites, and cast silicon elastomers. There are relatively few adhesives available that will perform well above 350°F (166°C) (see Table 16-2). A high softening temperature is not the only requirement for a successful high-temperature adhesive. The material must be able to stand up for a reasonable amount of time under attack by agents that oxidize and degrade the resin.

Most high-temperature adhesives, which consist of crosslinked networks of large molecules, have no melting points. Their limiting factor is strength reduction due to oxidation. Oxidation starts a progressive breaking (scission) of the molecular bonds in the long chains. This scission results in loss of strength and elongation leading to failure.

Oxidation and degradation of an adhesive is affected by any catalytic reaction between the adhesive and the substrates. As an example, under identical conditions, an adhesive used for bonding stainless steel may oxidize more when bonding aluminum but an adhesive with a different chemical makeup may show exactly the opposite effect (Fig. 16-8).

Table 16-2. Typical properties of resinous high-temperature adhesives. (*Courtesy* Machine Design.)

Adhesive	Cure Condition			Maximum Service Temperature (°F)		Lap-Shear Tensile Strength† (psi)	Environmental Recommendations
	Temperature (°F)	Time (min)	Clamping Pressure (psi)	Short Term*	Long Term		
Epoxy	75–350	120–10	0	500	400	2500	Resists moisture, brine, petroleum fuels, and oils; degrades slowly in air at maximum service temperature
Epoxy-phenolic	350	—	15–50	1200	500	2000	Similar to epoxy, but degrades faster in air at maximum service temperature
Polyimide	500	60	15–100	1000	600	2200	Resists oxidation at >600°F better than other adhesives
Polybenzimidazole	600	60	—	1000	500	1900	Oxidation resistance at high temperature almost equal to that of polyimide
Phenolic	300	30	100	—	600	2000	Degrades rapidly at high temperatures
Cyanoacrylate	75	2	0	325	285	1160	Degraded by UV, moisture, and alkali materials
Polysulfone	700	5	80	—	300	2200	Resists acids, alkalis; dissolved by polar organic solvents and aromatic hydrocarbons; degraded by moisture
Silicone	300	5	0	700	500	38	Resists most oils, acids, and bases; nonoxidizing; dissolved by solvents such as xylene and toluene
Fluorosilicone	70	120 hr	0	700	500	—	Particularly resistant to fuels and oils

* Less than 2 min.
† At maximum long-term service temperature.

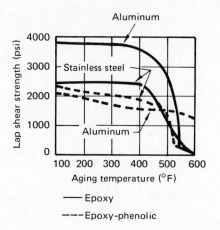

Figure 16.8. A comparison of the strength of epoxy and epoxy-phenolic adhesive bonding aluminum to aluminum and stainless steel to stainless steel. (*Courtesy* Machine Design.)

Newer polyamide-imide adhesives are able to retain considerable strength at temperatures as high as 650°F (343°C), even after 1000 hours continuous exposure in air. This is a significant accomplishment when compared to solders that melt at 370–450°F (188–232°C) and organic chemicals that usually burn and catch fire at much lower temperatures. Putting it another way, two pieces of steel could be adhesively bonded with this material and then immersed in a pot of molten lead at 621°F (327°C). They could remain there for several hours and they would not lose their bond strength.

Advantages and Limitations

There are many advantages that make adhesive bonding attractive in the metal-working industry. The fact that industry is now spending over 100 million dollars per year for adhesives attests to the benefits brought to manufacturing. Some of the main advantages in using adhesives for metal fabrication are:

1. The ability to join unlike materials.
2. The ability to join like or unlike materials without the stress concentrations usually associated with other joining methods such as spot welds, rivets, or bolts.
3. Adhesives do not conduct electricity, and therefore electrolytic corrosion is not set up in joints between dissimilar metals.
4. Flexible adhesives can absorb shock and vibration, increasing the fatigue life. In many cases, adhesive-bonded metal joints have up to ten times more fatigue life than riveted joints.
5. Bonded joints present a smooth, often improved, appearance.
6. Dissimilar thicknesses can be joined, e.g., foils can be joined to heavy-gage plate.
7. Less joint cleaning is required after it is made.
8. Adhesives often permit extensive design simplifications.
9. Large areas can be bonded in a relatively short time.
10. Adhesives provide sealing action in addition to bonding.
11. The elimination of fastener holes allows lighter-gage material to be used. For example, a 0.020-in. bonded-aluminum joint has been used to replace

a 0.051-in. riveted-aluminum joint and still maintains equal or better mechanical properties.

The limitations involved in adhesive joining are:

1. Adhesives are more subject to deterioration by environmental conditions than is metal bonding.
2. Adhesive joints are difficult to inspect once assembled.
3. Adhesives have less strength than some other joining methods. The poor resistance to peeling may require the usage of additional fasteners at stress points.
4. Adhesive properties tend to degrade with time.
5. Elaborate jigs and fixtures may be needed to apply heat and pressure, depending upon the bonding cycle.
6. Adhesives are not suited to elevated temperatures, only a few are reliable over 600°F (316°C).
7. Most adhesives have a limited shelf life.

Questions and Problems

16-1. What is meant by a structural adhesive?

16-2. What are the two types of cure used with epoxy resins?

16-3. How have epoxy resins been improved to increase their useful range in adhesives?

16-4. How did *anaerobic* adhesives obtain that name?

16-5. (a) What would be the difference in peel strength for an epoxy nylon joint 3 in. wide and 2 in. deep, and the same joint made with a thermoset urethane?

(b) Difference in shear strength?

16-6. Why have the acrylics suddenly become competitive with epoxies?

16-7. Compare the shear strength of lap-joint samples made from strips of 16-gage aluminum and stainless steel 2 in. wide at 300°F. The lap length is $\frac{1}{2}$ in. made with epoxy and epoxy phenolic.

16-8. In what respect are the cyanoacrylates competitive with or better than an epoxy?

16-9. Why is an ordinary lap joint not a good design for adhesives?

16-10. Which adhesive should be used if good shear strength and particularly good peel strength is required?

16-11. Is it mainly because the adhesives soften due to heat that the joint weakens in high-temperature adhesives?

16-12. Which adhesive can stand prolonged temperatures at 650°F?

16-13. What are four advantages adhesives have over riveting?

16-14. What is a newer technique used to produce a very thin coating of an anaerobic adhesive to a patterned surface?

Case Study

The public service commission was faced with the problem of replacing some old gas pipelines and adding new lines. They wanted to have a fuel gas line that would last 30 years and remain leakproof at pressures of 500 to 700 psi. They had a guarantee from an aluminum pipe manufacturer that their product would last 50 years or more. The only uncertainty was how long the pipe joints would last. The joint shown in Fig. CS16.1 was normally used; however, the commission felt that there was inadequate information concerning its service life.

State what recommendations you would

make regarding (a) surface preparation of both the aluminum coupler and the pipe ends; (b) adhesive used and why; (c) any special considerations that might affect the joint; and whether or not (d) a straight epoxy adhesive, if it were used, would exceed the bursting pressure of a 4-in. dia. 6061-T6 aluminum pipe of 4280 psi?; (e) plan a preliminary test program that gives some practical assurance that the pipeline joints are going to withstand actual operating and environmental conditions.

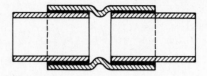

Swaged from each end toward center

Figure CS16.1.

Bibliography

Aronson, Robert B., ed. "Adhesives Are Getting Stronger in Many Ways." *Machine Design* 51 (February 8, 1979): 54–60.

Bruno, E. J., ed. *Adhesives in Modern Manufacturing.* Dearborn, Mich.: Society of Manufacturing Engineers, 1970.

Hegland, Donald E., ed. "Adhesives Prescription for Profits." *Production Engineering* (March 1979).

Huntress, E. A., ed. "Adhesive Assembly." *American Machinist Special Report 716* 123 (October 1979): 145–160.

Rantz, L. E. "Adhesive Bonding: Surface Preparation." Paper Presented at World of Adhesives. Canton, Mass.: Emerson & Cuming, Division of W. R. Grace & Co., 1978.

Schneberger, G. L. "Adhesive Bonding–Basic Concepts." *SME Paper AD79-349.* Dearborn, Mich.: Society of Manufacturing Engineers, 1979.

Zalucha, D. J. "High Performance Structural Bonding of Unprepared Metals." *SME Paper AD78-975.* Dearborn, Mich.: Society of Manufacturing Engineers, 1978.

CHAPTER SEVENTEEN

Automation

"Automation" is a relatively new word, coined in 1935 by D. S. Harder, then vice president in charge of manufacturing for the Ford Motor Company. The word, short for "automatization," implies doing something automatically or without human assistance. This may refer to a single operation or an entire continuous process.

In some of the previous chapters, certain operations were described as manual, semiautomatic, or automatic. These terms refer to hand operation, an operation with some human assistance, and an operation without human assistance, respectively.

Automatic control systems are not a product of this generation. James Watt, as early as 1788, conceived of an automatic speed control for his steam engine in the form of a *fly-ball governor*. The two fly balls, acting by centrifugal force, would extend as the rpm of the engine increased. This, in turn, caused a motion that reduced the opening at the intake valve. With less steam intake, the speed reduced and the balls moved closer to the control shaft, thus reopening the intake valve.

Another example of early automation that is still common in many shops is the fully automatic screw machine. The inventor, Christopher Spencer, had worked on manually operated turret lathes. He realized that by attaching a cam shaft, strip cams, levers, and segment gears for actuating the turret, collet, and cross slide, it could be made to operate automatically. Thus the automatic screw machine was born. Surprisingly enough, Spencer did not realize the import of his invention but was content to let the machine produce screws and other similar items rather than to develop the machine itself.

About the same time as Spencer's automatic screw machine was built in the United States, the Swiss were working out similar ideas for an automatic screw machine. There seems to be some reluctance to give any one man credit for the invention of what has come to be known as the Swiss automatic screw machine. However, the contributions of a brilliant young engineer named Bechler cannot be ignored. His concept of moving the material axially as it was being cut provided a means of generating an infinite variety of shapes with a minimum of tooling. The machine, entirely cam-operated, was fully automatic. One man could look after five or six of these machines, primarily to see that they did not run out of material and that the parts produced were within tolerance limits.

SEQUENCING CONTROLS

An essential element of automation is being able to make a series of events take place at the desired time and in the desired sequence. These controls are either mechanical or electrical. Mechanical controls are usually cams, valves, or specially constructed mechanisms. Electrical controls are of a wide variety—such as limit switches, control relays, timer relays, and solenoids.

The automatic screw machine is an example of one of the first machines that was based on making a series of events occur as planned or essentially in sequence. Controls used to accomplish this are mechanical (cams), electrical, or combinations (electromechanical). The power required is usually pneumatic, hydraulic, electrical, or combinations of these types.

Sequencing Control of a Pneumatic Circuit.   It is intended in this brief presentation to show only some basic concepts of automatic controls. Because of their relative simplicity, compactness, versatility, low initial cost, and wide acceptance, some basic principles of pneumatic circuits will be presented.

Figure 17-1 shows a simple sequential, or automatic-cycling, pneumatic circuit. To start it the operator pushes button A, which energizes solenoid A. This shifts the two-position valve to the right. One can visualize the action by mentally sliding the symbol of the two crossed arrows to the square on the right. In this position, the line pressure is able to flow into the left side of the cylinder, extending the piston until it closes the normally open limit switch, (1LS). Closing this switch energizes solenoid B, which shifts the two-position valve back to its original position. Line pressure now flows into the right-hand end of the cylinder, causing the piston to shift to the left, completing one cycle. To make an automatic reciprocating circuit, a limit switch (2LS) can be added to the left side of the cylinder, which would be used to activate solenoid A.

A variation of this circuit is one that incorporates speed control, as shown in Fig. 17-2. In this circuit the line pressure is shown as passing through the

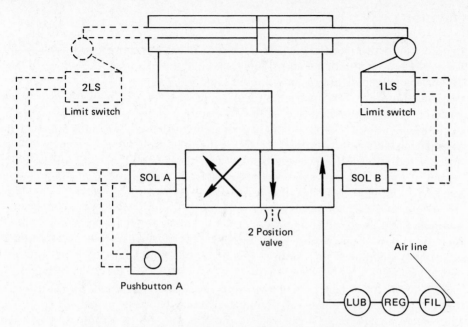

Figure 17.1. An automatic-cycling pneumatic circuit.

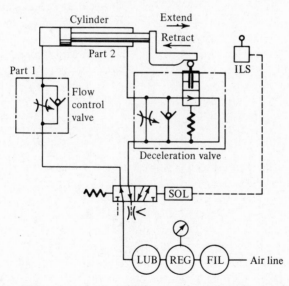

Figure 17.2. A typical multiple-speed pneumatic circuit.

two-position valve into the back end of the cylinder. As shown, the *flow-control valve* offers no restriction to the flow of air since the bypass to the right is open. The piston extends at full speed until the cam on the end of it strikes a secondary two-position valve. The normally open air passage is now closed and it must go through the restricted orifice of the deceleration valve. As the piston extends to the right, a limit switch (1 LS) is activated causing the solenoid attached to the main valve spool to shift to the left. Line pressure can now go through the unrestricted passage to the right-hand end of the cylinder. The piston will not retract at full speed, however, since the air from the back side of the cylinder must pass through the restricting orifice of the flow-control valve.

These two circuits serve to show how the combination of pneumatic (or hydraulic) and electrical components can be combined to achieve relatively simple automation. Compressed air is usually used where economy of the initial investment, simplicity of control, and ease of maintenance are of primary consideration.

Although the two circuits shown were pneumatic, essentially the same symbols are used for hydraulics. Hydraulic circuits have the advantage of being able to produce higher forces with smaller components. As an example, a 2000-lb

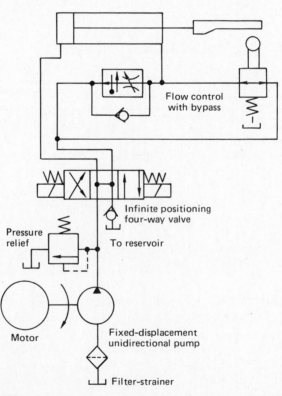

Figure 17.3.   A hydraulic circuit similar in function to the pneumatic circuit shown in Fig. 17.2.

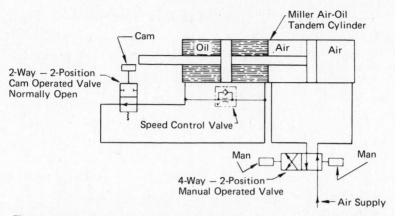

Figure 17.4. A combination air-hydraulic circuit. (*Courtesy Miller Fluid Power Division, Flick-Reedy Corp.*)

force can be exerted by a 1.5-in. dia. hydraulic cylinder, whereas a 5-in. dia. pneumatic cylinder is barely adequate. This assumes the normal line pressure of 80 to 100 psi for pneumatics and 2000 psi for hydraulics. However, it is not uncommon to have hydraulic pressures operating in the range of 3000 to 5000 psi.

Figure 17-3 shows a hydraulic circuit that is very similar to the pneumatic circuit shown in Fig. 17-2. You will note that a motor and pump are shown, as well as a return line to the reservoir rather than to the exhaust. Hydraulic fluids have the advantage of being virtually incompressible, whereas pneumatic action may be somewhat spongy under heavy load conditions. Pneumatic valves, on the other hand, have a shorter response time and will, for a given size, shift from three to four times as fast as a solenoid-operated hydraulic valve.

In some cases it is advantageous to combine air and hydraulic circuits into one system. This is usually done where a smooth, even movement is required, as in machine-tool feeds. The hydraulic circuit is considered "captive." That is, it is built into the machine and air is supplied as shown in Fig. 17-4.

AUTOMATIC POSITION CONTROL AND MEASUREMENT

In addition to sequencing, automation is concerned with accurate position control. This implies the measurement of a location relative to some reference point. As was shown in Fig. 17-1, a signal was sent from the limit switch to the two-position valve when the piston had advanced the required distance. Position control may be sensed mechanically, pneumatically, optically, electrically, etc. Of these methods, the most popular is electrical due to its flexibility, small size, and low power requirement. After limit switches, the next most-used electrical method of sensing position is with transducers.

LENGTH

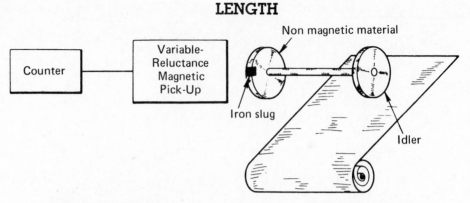

Figure 17.5. The measurement of material length with a variable-reluctance transducer. (*Courtesy Baird Atomic, Inc.*)

Transducers.   Transducers may be defined as any device that converts one form of energy into another. For example, thermocouples transform heat energy into electrical energy and a loud speaker converts electrical energy into mechanical motion. Transducers are used extensively in automation because they can be used to sense a physical quantity and convert it into an electrical value for measurement.

An example of the use of a transducer to aid in length measurement is shown in Fig. 17-5. Two wheels are firmly attached to a common axle. The idler wheel rides on the material to be measured and must have a high coefficient of surface friction. The sensing wheel is usually plastic and contains an iron slug imbedded in its periphery. As the material is moved, the idler rotates and causes the sensing wheel to rotate. Each time the iron slug passes a variable-reluctance magnetic pickup, a pulse is initiated and registered on the counter. Proper sizing of the idler and sensing wheels can make each pulse represent inches, feet, meters, etc. Thus, not only length can be measured but velocity as well. Another method often used to measure velocity of a shaft is by coupling it to a generator. The voltages produced can be calibrated directly in terms of rpm.

Potentiometers, or resistive transducers, are particularly good for measuring low-speed rotation. As the shaft of the potentiometer rotates, it varies the resistance in the circuit that may be translated into a given control signal. For example, the cam shown in Fig. 17-6 represents a mechanical signal that is picked up through linkage by the potentiometer, which in turn produces a variable electrical signal.

Velocity measurements can be made by the use of photocells, as shown in Fig. 17-7. Two photocells are set up with individual light sources a fixed distance apart. When the beams are interrupted by a passing object, each photocell emits a pulse. With the spacing of light beams known, the velocity of the moving object can be computed. The time interval can be measured with an electronic counter or shown on the screen by an oscilloscope.

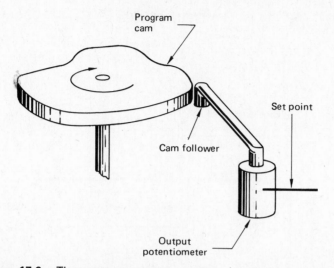

Figure 17.6. The cam represents a program (mechanical type) that is picked up through linkage by the potentiometer, converting into a variable electrical signal.

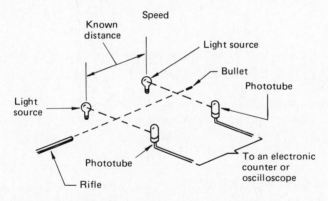

Figure 17.7. Photoelectric transducers are positioned to measure the velocity of a bullet. (*Courtesy* Machine Design.)

The measurement of quantity is often an important part of automation. Parts may be counted automatically by capacitive, magnetic, or photoelectric transducers. Counting with capacitive and magnetic units requires that the items be metallic, such as tin cans. Photoelectric transducers can be used with either metallic or nonmetallic materials. Figure 17-8 shows a photoelectric counting operation used in packaging. Containers on a conveyor belt are placed so that the products being discharged from a hopper interrupt a light beam to a photocell. The interrupted signal is sent to a preset counting unit. When the required number of products have been deposited in the container, the preset counter

COUNTING

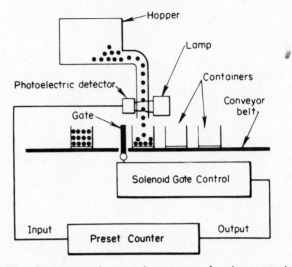

Figure 17.8. An automatic counting system for the control of the number of items fed into each container. (*Courtesy Baird Atomic, Inc.*)

activates the device controlling the gate and permits the filled container to move along the conveyor. An empty container moves into position and the cycle is repeated.

This brief presentation of transducers serves only to introduce them and show their importance in automation. For a more complete discussion refer to the bibliography at the end of the chapter.

AUTOMATIC ASSEMBLY MACHINES

Automatic assembly of manufactured components has received a great deal of attention in the last decade. Estimates show that 55% of the total production man-hours in major industries are devoted to this area. The major elements of a semiautomatic assembly line consist of workpiece orientation, part transfer, and part placement.

Workpiece Orientation

Many devices have been developed to automatically orient and move manufactured components. Best known among those used for small-part orientation is the vibratory hopper (Fig. 17-9). The bowl is made to vibrate in a twisting motion by means of leaf-type springs and electromagnets. The motion causes the parts

Figure 17.9. This four-track vibratory bowl is used to orient this gear and pinion assembly at the rate of 300 parts per minute. (*Courtesy Moorfeed Corporation.*)

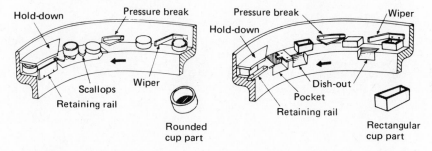

Figure 17.10. Examples of round and rectangular piece-part orientation on the track of a vibratory bowl.

to climb the ramp on the inside of the outer edge of the bowl in a series of tiny hops. Orienting devices such as *wipers, dish-outs, pockets,* and *hold-downs* are incorporated along the track (Fig. 17-10) to reject parts that have not assumed the proper position. The track leading out of the bowl can be used to turn the parts to any desired position. Feeds of 10 to 15 fpm are considered normal. This does not mean that all parts will be oriented correctly at this rate. A part takes several positions as it travels along. The desired feedrate is set at the number of correctly oriented parts required per minute. More than one track can be used at one time, as was shown in Fig. 17-9.

Part Placement

After each work station, parts must be transported to the next station and placed in an exact location for the next operation. It is important to have the placement done consistently, without damage to the part or loss of control, and at relatively high rates of speed. Placement mechanisms must be simple in design and ruggedly made to provide reasonably good service.

Basic types of parts-placement devices include feed tracks with escapement mechanisms, pick-and-place mechanisms, and various types of industrial robots.

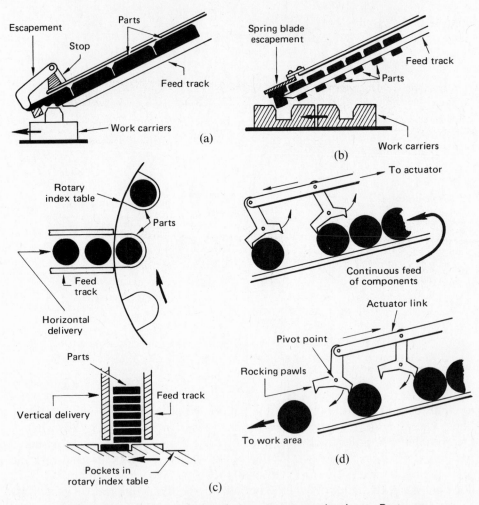

Figure 17.11. Various types of escapement mechanisms. Parts are held in place by a rocker arm (a) or a spring (b) or are simply allowed to drop (c). The ratchet-type escapement (d) involves the use of a linear motion. (*Courtesy* Automation.)

Feed Tracks and Escapements.   Feed tracks are usually gravity-powered and are used to bring the part from the point of orientation or prior work station to the next operation.

As the part moves down the feed track to the end, it is released by an *escapement* mechanism. An escapement may be defined as a mechanical device that allows a predetermined number of parts to be released from a feeding system on demand, as shown in Fig. 17-11. Usually the escapement loads the part directly into the work-holding fixture. Direct loading is inexpensive and reasonably reliable (at slow speeds) and for some parts it is the best way to affect final placement. Direct placement may also have some disadvantages such as: (1) clearance problems between the escapement and the workhead or the fixture, (2) the force of gravity may not be sufficient for proper seating, and (3) loss of control of the part can cause misfeeds and/or jam-ups in the machine and possible damage to the part.

Pick-and-Place Mechanisms.   Because of the possible problems cited for escapement mechanisms, parts are often positioned in the work-holding fixture by more elaborate methods, as shown in Fig. 17-12. Some of these are commercially available and others are made up for in-plant use.

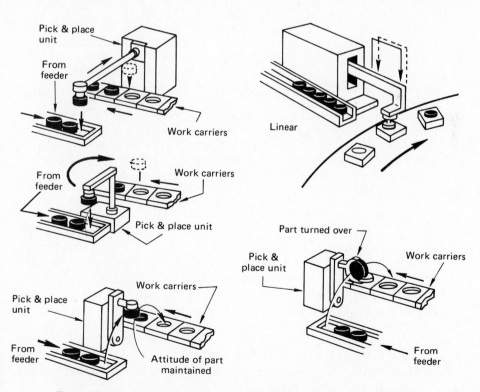

Figure 17.12.   Various types of automatic pick-and-place systems. (*Courtesy* Automation.)

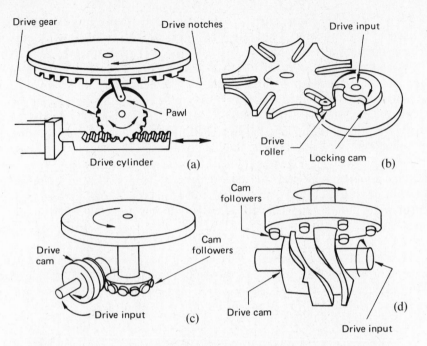

Figure 17.13. Four basic types of rotary-drive systems. The ratchet and pawl system (a) is suitable for light-duty and low-speed applications. The power source is usually a pneumatic or hydraulic cylinder. The Geneva mechanism (b) is suitable for larger jobs. During the dwell time, the locking cam holds the driven member for positioning accuracy. Cam mechanisms are the most consistent and accurate indexing mechanisms with high torque capacity, smooth acceleration, and longer wear life. They can be furnished with 3 to 48 stops. The cam drive shown at (c) applies power to the cam followers, which are mounted with their axes perpendicular to the table spindle. The cam followers shown at (d) are driven by a small-diameter drive cam in a compact unit. (*Courtesy* Automation.)

Synchronous or Intermittent Motion. Figure 17-13 shows four basic types of rotary motion used extensively in automation equipment. They all provide for a positive rotary motion and a dwell time. The cycle time can be made to vary by changing the speed on the drive input, but the relationship between index and dwell is synchronous or fixed. For example, in the Geneva mechanism shown there will be six stops or indexing motions for a complete revolution. The time ratio is based on 120° index and 240° dwell. Each indexing motion will require 20° and each dwell time 40°. Figure 17-14 shows a dial-index machine that is fed by three vibratory hoppers.

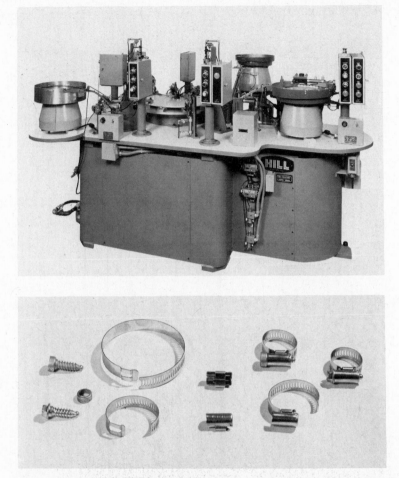

Figure 17.14. A dial-type indexing table equipped with a vibratory-bowl feeder. In this illustration hose clamps are being assembled at the rate of 1800 per hour. (*Courtesy Hill-Rockford Co.*)

AUTOMATED ASSEMBLY

Basic System Concepts

The term automated assembly generally has the connotation of assemblies being made entirely automatically; however, this is not usually realized even though it is the ultimate aim. Many assembly systems have been developed; a general system concept is shown schematically in Fig. 17-15 as single station, synchronous (indexing), nonsynchronous (power and free accumulative), and continuous.

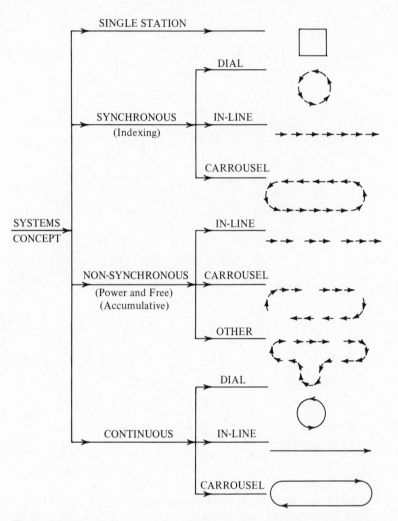

Figure 17.15. Basic concepts of automated assembly lines. (*Courtesy Society of Manufacturing Engineers.*)

Single Station. The single work station may be used to perform one or more functions. Two or more parts of an assembly are joined together or set in a fixture. Parts orientors and selectors are used to feed the station.

The dial-index table may range in capacity from 4 to 32 stations. Each station may be tooled for one or more operations. The physical size may range from 6 in. (152.40 mm) to 200 in. (500 cm). Speeds may range up to 3600 parts/ hr or more if multiple tooling is used. The dial-indexing machine is often tied into an in-line system to provide subassemblies.

Synchronous In-Line Indexing. The in-line systems consist of a series of pallets, or pallets and fixtures, attached to chain belts or moved by fingers from

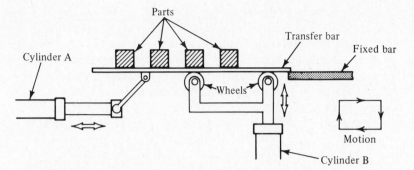

Figure 17.16. A basic lift-and-carry transfer mechanism. The "walking beam" is similar but contains more linkages.

one work station to another in a straight-line motion. All the fixtures are moved at the same time over the same distance. Indexing intervals may be controlled by timers or on a "demand-type basis," indexing only when all stations have completed their individual functions.

In-line machines with fixtures usually return the fixtures by the conveyor line below the assembly area. If the part serves as its own fixture, it may not be necessary to return any part as in some "lift-and-carry" or "walking-beam" systems as shown schematically in Fig. 17-16.

Advantages and Limitations.   More operations are possible on the in-line machine than are usually done on a dial-type machine. The concept of having parts and assemblies come in one end and out the other provides a good work-flow pattern.

Because of the accumulated inefficiency of many stations, the in-line system is usually limited to 15–20 working stations that are fed by hoppers or pick-and-place robots.

The whole line must be paced by the slowest operation. Sometimes it is possible to break up a slow operation into two or more stations. Speeds up to 1500 assemblies (3000 small parts with multiple tooling) are possible.

The size of the parts is not limited to small parts or assemblies. Parts such as car bodies are assembled on the synchronous line with approximately 600 spot welds made by 20 robots.

Synchronous Carrousel.   The carrousel consists of a series of holding fixtures attached to a chain or belt that moves the pallets from one station to another in a rectangular path. All parts are indexed at the same time and for the same distance on either a timed or demand basis.

The pallets and fixtures, which may be in either a vertical or horizontal plane, are all in use since a return line is not needed.

Nonsynchronous Motion.   In this system pallets are free to move at a pre-determined rate but the overall machine cycle rate is dependent upon the speed with which the individual operations are performed. This system is sometimes referred to as *power-and-free* and is usually used in semiautomatic machine operations, that is, where some manual contact is involved. As an example of the nonsynchronous system, parts may be assembled at most stations with 2 sec of dwell time and a manual spring-insertion step at another station might require 6 or 8 sec. While the rest of the machine can operate at a faster cycle rate, the work-holding carrier must stop at the manual station for the slower nonsynchronous operation, as shown in Fig. 17-17(a). This type of operation may be speeded up by varying the number of manual stations. Stations may be added or subtracted with relative ease as shown schematically at (b).

Advantages and Limitations.   The nonsynchronous line allows more flexibility in using single tools on relatively fast operations and multiple tooling on slower operations.

Electric motors are attached to each pallet in the "power-and-free system." The motors receive low-voltage current from buss bars below the pallet and are turned on and off by cam action. An alternative arrangement is to have clutches on each pallet with a lock-and-unlock sprocket engagement on a continuously running chain below the pallet.

The accumulative characteristics of the independent pallet travel permit "averaging out" the variations that occur in manual operations without penalizing the automatic operations.

Virtually an unlimited number of operations can be tied together subject only to the drive torque of the machines.

It is possible to add manual standby stations for any or all of the machines in the line so that as one machine goes down another can be quickly set in place. Adding stations becomes much more practical if a programmable controller is used. Programmable controllers are discussed in the next chapter.

Manufacturers build nonsynchronous lines so that they are completely modular and work stations can be taken out or added as needed. The configurations to which the line can be built are almost limitless, such as T, U, or L.

The nonsynchronous assembly machine is much higher in initial cost than the synchronous line. Control elements are more complex and extensive.

Several fixtures should be provided for each work station in order to take advantage of the accumulative characteristics of the machine if the need arises.

Production rates are usually limited to 600 to 900 parts/hr.

Pallets are quite expensive and have high maintenance costs.

Continuous Motion.   This method, by definition, is the transportation of piece parts and/or work-holding fixtures at a constant or variable speed without interruption. This type of system requires that the work stations move with the work-holding fixtures. Continuous motion transfer is widely accepted in such operations as wave soldering, bottling, packaging, printing or marking, and surface coating. Continuous systems can operate at relatively high speeds since there are no stop-and-start, inertia, or acceleration problems.

Continuous motion systems, either rotary or linear, are the easiest to design, simplest to operate, and least expensive. However, unless the continuous motion

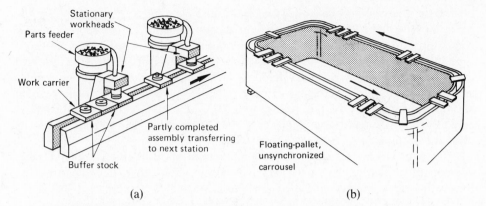

(a)   (b)

Figure 17.17. In the nonsynchronous system, pallets are transferred by means of a continuously moving belt or chain. When pallets arrive at a work station, a clutch or latch mechanism operates to permit the pallet to remain stationary. Once the operation is performed, the clutch is released automatically or manually and the pallet moves to the next work station (a). Pallets may be added or taken off from the nonsynchronous carrier (b) with relative ease. (*Courtesy* Automation.)

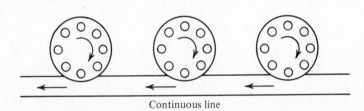

Continuous line

Figure 17.18. In this continuous line system, dial-index machines are used to wash, fill, and cap bottles. A basic standby indexing machine may be needed for replacement of a malfunctioning unit.

is being used for materials handling only, or for simple assembly operations, work must be performed "on the fly." Although this is not difficult if the work is done manually, automatic work stations, which must match with the constant-motion transportation system, can be troublesome and expensive. In terms of fully automatic, high-speed assembly, metal removal, or many other processing applications, continuous motion is not acceptable.

Figure 17-18 shows an arrangement whereby individual items are diverted from a continuous line for various operations and returned to it without interfering with the forward line movement.

Advantages and Limitations of Automatic Assembly. The advantages of automation are fairly obvious—such as reduction in cost per part, increased productivity, a more consistent product, and removal of operators from hazardous operations. However, the larger and more complex the system becomes, the more difficult it becomes to stay advantageous. For an in-line sequence or a rotary closed-loop sequence, the more operations or value-added series, the more rapidly the efficiency falls off. Each station contributes its own inefficiency to the overall inefficiency of the system. As an example, if each station on a nine-station machine has an efficiency factor of 99.1%, which is high for any given station, the efficiency of the system is 0.991^9, or 92.1%, which may be unacceptably low. Given a large enough number of stations, this type of system is theoretically unable to produce finished parts.

Two approaches to offset the series function falloff in efficiency have been developed; one is known as the demand system. The other, developed by the Swanson-Erie Company, is the SynchroBank Assembly System.

Demand System. The demand system shown schematically in Fig. 17-19 is a typical series-parallel assembly sequence. This design is predicated upon the ability of the supply modules to provide perfect subassemblies to the demand module. To accomplish this, the supply module operates at a faster pace than the cadence of the demand module, which may be of the in-line closed-loop design similar to the carousel, shown in Fig. 17-17(b). The supply modules will periodically produce defective or incomplete subassemblies, which are then rejected. However, the increased speed of the supply modules means that their buffer storages will always be full and waiting for part pickup by the demand modules.

The SynchroBank System. Figure 17-20 illustrates three assembly modules used in this system. Each module is a synchronous carousel assembly system.

In this three-module system, operations indicated in the drawing are not necessarily all produced in the individual modules, although they may be. In some instances they are shunted out of the modules and into the dial-index machines (not shown) for multiple assembly operations. Of special importance are the buffer banks shown between the modules. Since Module I operates at a faster speed than Module II—and II is faster than III—the flow of parts is

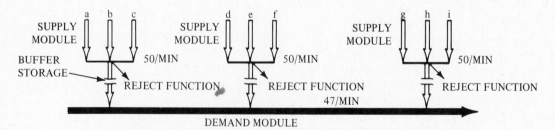

Figure 17.19. A schematic of a value-added series-parallel assembly sequence sometimes referred to as the supply and demand system or simply the demand system. The design is predicated on the ability of the supply modules to provide perfect subassemblies to the demand module. (*Courtesy* Manufacturing Engineering.)

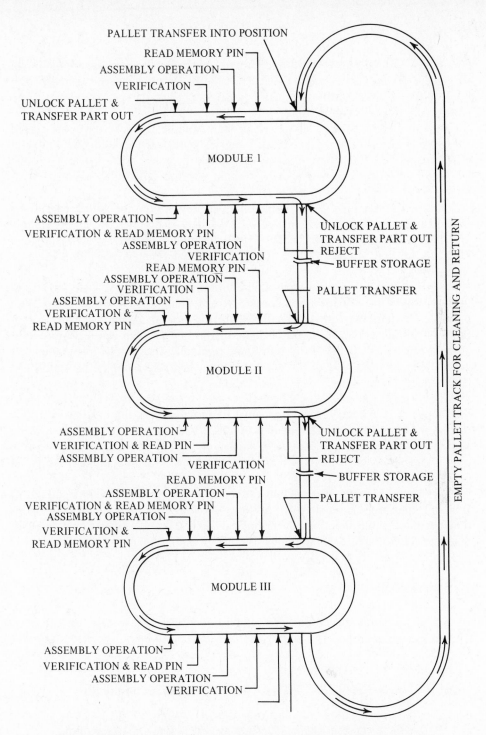

PALLET TRANSFER INTO POSITION
READ MEMORY PIN
ASSEMBLY OPERATION
VERIFICATION
UNLOCK PALLET &
TRANSFER PART OUT

MODULE 1

ASSEMBLY OPERATION
VERIFICATION & READ MEMORY PIN
ASSEMBLY OPERATION
VERIFICATION
READ MEMORY PIN
ASSEMBLY OPERATION
VERIFICATION
ASSEMBLY OPERATION
VERIFICATION &
READ MEMORY PIN

UNLOCK PALLET &
TRANSFER PART OUT
REJECT
BUFFER STORAGE

PALLET TRANSFER

MODULE II

ASSEMBLY OPERATION
VERIFICATION & READ PIN
ASSEMBLY OPERATION
VERIFICATION
READ MEMORY PIN
ASSEMBLY OPERATION
VERIFICATION & READ MEMORY PIN
ASSEMBLY OPERATION
VERIFICATION &
READ MEMORY PIN

UNLOCK PALLET &
TRANSFER PART OUT
REJECT
BUFFER STORAGE

PALLET TRANSFER

MODULE III

ASSEMBLY OPERATION
VERIFICATION & READ PIN
ASSEMBLY OPERATION
VERIFICATION

EMPTY PALLET TRACK FOR CLEANING AND RETURN

Figure 17.20. A schematic of a three-module SynchroBank system used in automatic assembly. (*Courtesy* Manufacturing Engineering.)

continuous with the excess loads contained in the buffer storages. The buffer storages consist of an elevator that carries the pallet to an overhead track. The track moves it into a buffer (series of spiral storage conveyors) that carries the pallets to a downstream module.

As the palletized part advances through Module II, more parts are added to the assembly. In some cases parts or subassemblies are added by off-line dial-index machines. Each part must pass inspection or leave the system through a reject position. If the part passes, it again goes into the buffer bank.

Finished Assembly.   After completing the series of value-added operations in the third module, the part leaves as a finished assembly. The empty pallets are unloaded into a line that carries them through a wash conveyor and then back to the first storage station for reloading and reentry into the system at Module I.

Thus, linear transportation machines, or carousels, offer a wide latitude of design. Since the action is not confined to movements around a circle, large parts can be handled with greater flexibility for tooling. As was shown, two or more basic linear machines can be joined to provide a high-efficiency process. With the automatic control systems now available, principally programmable controllers, as discussed in the next chapter, can make complex assemblies at the lowest possible cost.

The disadvantage of the linear-type machine is the initial cost. If close tolerances are necessary, each part must be mounted in a fixture that is in turn mounted on a pallet. Each pallet is coded to move in or out of designated stations. Once it is in the station, it must be precisely positioned with locating pins. Some nonsynchronous systems are designed with individually energized pallets.

Rotary-table indexing machines are usually limited to relatively small parts. Indexing time must be geared to the longest operation. Accessibility for maintenance and trouble shooting is very limited. As the number of stations increase, dial size must increase, which adds to starting and stopping problems.

INDUSTRIAL ROBOTS

Industrial robots as discussed here refer to general-purpose manipulators used for tool changing and parts handling that are capable of being synchronized with the machine tools with which they are associated. These robots are relatively sophisticated point-to-point transfer robots with seven "degrees of freedom" that are programmed by numerical control (NC) or computer numerical control (CNC). The seven degrees of freedom include a wide variety of "arm" and "finger" movements that can be programmed. Figure 17-21 shows two types of robots: (a) is the Versatron made by American Machine and Foundry, and (b) is the Unimate made by Unimation. Figure 17-22 illustrates examples of arm and finger movement capabilities.

The most effective type of gripper has three "fingers." It can easily pick

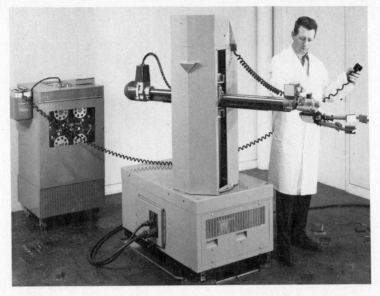

(a)

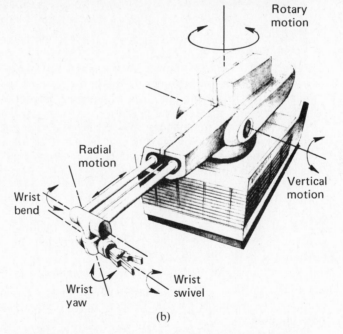

(b)

Figure 17.21. Examples of two types of industrial robots: the Versatron, capable of positioning parts on a point-to-point program or by continuous path (a); and the Unimate, which also has extensive programming capabilities (b). (*Courtesy AMF Incorporated.*)

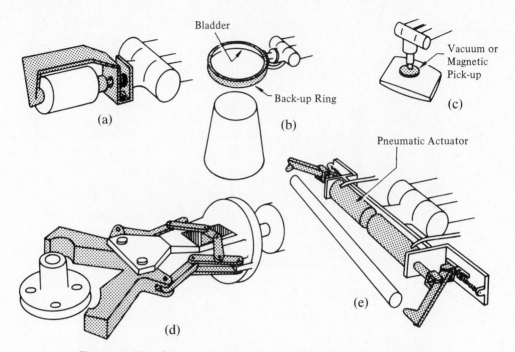

Figure 17.22. Some examples of various types of clamping devices developed for the Unimate robot. The gripper can be lowered onto the part by a wrist action (a). The expandable bladder (b) is useful for flexible and tapered walls. Many parts can be handled by a simple vacuum or magnetic pickup (c). If the part is not always in the same spot, a wide-opening hand device is used (d). Long, slender parts may require special clamping techniques such as fingers that enter the tube (e). (*Courtesy* Automation.)

up either rectangular or cylindrical workpieces from a maximum variety of angles. It also provides a firmer grip than the two-fingered grippers used in a wide variety of applications.

Some robots are built with electromagnetic memory drums or floppy disc memories that can store several hundred sequential commands. The various types of memories available for both robots and NC equipment are discussed in the next chapter.)

Robot Evolution.   The first-generation robots were fixed-sequence devices. Once they had been set and programmed to perform a given set of operations they could perform the sequence tirelessly and reliably. In order to do a different job, however, the robot had to be reprogrammed.

Since the sequence of operations is preprogrammed and fixed, the workpiece must be precisely in the right place at the right time. This means that parts must be accurately fixtured and they must be stationary. The latest of the first-generation robots augmented with computer control have greatly increased flexibility. They are able to alternate through a variety of fixed sequences, depending on which part arrives at the work station.

The second-generation robots are, as of this writing, already in the development stage. In order to adapt to a continually changing environment, the robot must be provided with sensory inputs—of which the most useful is sight. If a robot can see where a part is, it can figure out how to pick it up. A sense of touch can also be very useful in many assembly operations. Hearing would be useful, for it would enable a robot to respond directly to operator commands.

The sense of vision has received the most attention. When a person hears a word, he is immediately connected with a picture or a "model" of the part. As an example, if a person hears the word "connecting rod," a mental picture appears in his mind. Or if the term is unfamiliar and the person is shown a picture of a connecting rod and then shown a real rod, he could reach out and pick it up.

For a robot to obtain vision the following items are needed:

A device similar to a TV camera to serve as the eye.

Vision programs, along with models to interpret the data from the camera in order to know what the part is and where it is located.

Control information telling the robot what to do with the part.

All of the above actions must be done at production speeds, which usually means a few seconds per part.

An early second-generation robot at General Motors research and development center has the ability to pick out a specific unoriented part from a moving conveyor and put the part where it is needed (Fig. 17-23).

In this system (see Fig. 17.24), a light source mounted at an angle illuminates the conveyor. When a part creates a shadow, the camera picks up an image. The camera scans only along a single line. With each sweep, the camera sends the computer 100 data points. Each point is converted to a number corresponding to a gray-scale value ranging from 9 for white and 0 for black. After 100 sweeps, a 10,000-digit image is formed. When the digits are converted to gray values, the shape of the part is identified. To enhance the image and make it easier for the computer to identify the part orientation, the edges are clearly defined. This is done by a computer program that responds to contrasts (large differences in gray and white value numbers).

This system, consisting of an electronic camera and a minicomputer, can be acquired for less than $10,000, which is one-tenth the cost of comparable hardware when robotic research was started. With rapid changes in electronic technology, the cost of the hardware is expected to decrease.

Research is now being directed toward processing visual data optically rather than digitally in order to reduce the amount of data with which the computer has to cope. One approach is the use of a laser light image of the part that filters out everything except the essential information.

Robots have been given a sense of feel through the use of infrared light, a development of the Bureau of Standards. The light source beams its rays

Figure 17.23. A robot equipped with "sight" and a computer brain is able to select parts from a moving conveyor and place them where needed. (*Courtesy General Motors.*)

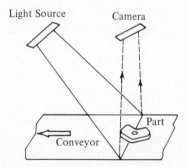

Figure 17.24. A light source mounted at an angle illuminates the conveyor. When a part creates a shadow, the camera picks up the image. The camera scans along a single line at a time. With each sweep, the camera sends the computer 100 data points that correspond to gray-scale values from 9 to 0. A computer program is then used to define the edges of the part. (*Courtesy General Motors Research Center.*)

through a fiber optics bundle to each finger of the robot gripper. When the proper proximity has been reached, the intensity of the light reflected back from the surface of the object approached is measured by a matching fiber optics bundle. The bundle is connected to a photosensitive cell that triggers a command to the robot. Infrared scanners are now being used in full production in forging, heat-treating, and quenching operations.

Pneumatic pressure sensors that detect an absence of vacuum buildup are being used in conjunction with grippers equipped with vacuum cups to signal the absence of the last workpiece in a pallet stack. They then send a signal terminating the robot program and triggering the advance on the next loaded pallet on a conveyor, shuttle, or gravity loader.

Advantages.   Robots are excellent for dangerous, boring, highly repetitive, and demeaning jobs. The advantages of the robot over the conventional transfer systems are lower first cost, faster procurement, greater uptime, higher productivity, and easier maintenance. Judging from today's developmental progress, within a few years robots should be able to respond to human voice commands. The response will result in a desired sequence of actions previously established in the memory.

Robots usually bring about an increase in productivity. In such tasks as unloading, die castings, inspection, and trimming, a 50% increase in production and a 300% increase in capital equipment utilization have been realized. In addition, the robot does not require break times or holidays.

Robots now on the market can position a tool or workpiece to a range within ±0.030 in. (±1.00 mm), depending on the degree of sophistication and cost. Robots with all-electric servomechanisms offer maximum accuracy and repeatability with minimum maintenance and adjustment.

Disadvantages.   Existing robots are bulky, and therefore require a large amount of floor space. Robots are usually actuated by servovalves that dissipate a lot of energy. Energy of 5 to 10 hp is needed to equal that of a human being expending about 1/2 hp. As mentioned, the all-electric servomotor drives are more efficient than electro-hydraulic drives.

Since robots will be working in close proximity to human beings, they must be made safe either by voice command or sensing devices.

Economic Considerations.   The decision of whether or not a robot should be used should be based on a favorable economic investigation. Several factors should make up this study: (1) availability of other methods, including suitable special-purpose automation; (2) number of shifts; (3) duration of job; (4) labor cost; (5) installation cost; (6) cost of auxiliary equipment such as other parts orienters, conveyors, or inspection equipment; (7) cost of service and operation versus special-purpose automation; and (8) availability of funds for investment in equipment. A basic equation for determining the payback period for a robot installation is:

$$P = \frac{I}{L - M}$$

where: P = payback period in years
I = total investment for robot and accessories
L = annual labor savings
M = annual maintenance expense

Typical inputs may be: I = $30,000, L = $18,000, M = $2000 for one shift and $3000 for two shifts. Thus, substituting in the equation for a one-shift operation:

$$P = \frac{\$30,000}{\$18,000 - \$2,000} = \frac{\$30,000}{\$16,000} = 1.8 \text{ years payback period}$$

A basic rule-of-thumb is as follows: if a robot can replace two machine operators in a one-shift operation or one person per shift in a two-shift operation, the robot will pay for itself in 1 to 2 years.

The growing use of robots in industry was signified by the first International Symposium on Industrial Robots held in 1975. In addition, standards for the development of robots are now organized under the American National Standards Institute (ANSI). These standards emphasize safety factors and guidelines for both producers and users.

TRANSFER-LINE AUTOMATION

The specialized transfer-line machine probably comes nearer the average conception of automation than any other system in the metalworking industry. It is on these specially built machines that hundreds of operations are performed on rough-cast engine blocks and many other complicated high-volume production parts. The line is arranged in sections of identical machines that are fed by a conveyor. Parts feeding along the conveyor are automatically deflected into loading chutes for each machine section that is empty at that particular time. Thus the operation is not held up because of difficulty with any one machine nor is progress based on the longest cycle time. As the sophistication of the line increases, so must the complexity of safeguards that must be incorporated into it. Parts placement must be kept within close tolerances and circuits must be designed so that trouble lights will immediately light up upon the detection of part absence, parts out of position, parts out of dimension, etc. Several types of transducers are usually incorporated into the line that monitors the line on the basis of light, sound, temperature, proximity, and torque. The signals received may lead to one or more of the following: (1) machine stoppage to apply corrective action, (2) a process slowdown to permit corrective action, (3) input into the memory system for subsequent disposition, (4) an ink or paint marker to identify the part in error, or (5) an automatic corrective process.

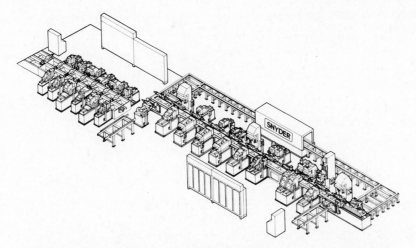

Figure 17.25. System for machining 10 different rear-axle housings combines a palletized transfer line and a free-transfer, lift-and-carry line. (*Courtesy Snyder Corp.*)

AUTOMATION TRENDS

Because 50 to 75% of the manufacturing in the machine tool industry is of the batch type, automation has taken on a relatively new concept variously called flexible manufacturing systems (FMS) or integrated manufacturing systems (IMS).

FMS is defined as a series of machining and associated work stations (cleaning, inspection, etc.) linked by a hierarchical common control to provide automatic production of a family of workpieces through a sequence of tasks. Figure 17-25 illustrates an FMS system used to machine 10 different rear-axle housings automatically.

FMS is very similar in concept to what is better known as group technology, discussed in Chapter 19. FMS involves the use of NC machines and CNC and is sometimes referred to as the "unmanned factory" discussed in the next chapter.

Questions and Problems

17-1. What are the essential components needed to make a simple automatic pneumatic circuit?

17-2. What function does a flow-control valve serve in a pneumatic or hydraulic circuit?

17-3. What are some advantages of a pneumatic circuit over a hydraulic circuit?

17-4. Could a turbine generator in a power station be classified as a transducer? Why or why not?

17-5. What would the actual index and dwell time be for a five-station Geneva drive driven at 1 rpm if the time ratio is 108° index and 252° dwell?

17-6. What are some disadvantages of rotary-table indexing machines?

17-7. Why is the rotary table often used for automation despite its disadvantages?

17-8. What type of automation system is best for operations that vary in length?

17-9. What are some elements in the vibratory hopper that make it possible to orient parts automatically?

17-10. Company ABC decided to buy three robots at a total cost of $70,000. Each robot will be used on two shifts. Preliminary studies show that each robot can do the work of an unskilled worker who is paid $5.00/hr (including fringe benefits). The plant runs 40 hrs/wk per shift but shuts down two weeks for vacation and eight holidays. How long would it take to recover the investment? Yearly maintenance = $200./machine.

17-11. Sketch (fill in) the pneumatic sequential clamping circuit (Fig. P17.1) to show the following features:

(a) Clamping is done first.

(b) The drill advances at rapid traverse and then slows to the drilling feed and retracts rapidly when the drilling is complete.

(c) Unclamping.

17-12. Why can a large automatic assembly machine be self-defeating?

17-13. How is the demand system able to overcome the series-function decrease in overall efficiency?

17-14. How does the SynchroBank system overcome the series-function inefficiency?

17-15. Distinguish between a first- and second-generation robot.

17-16. Why are robots comparatively easy to use?

17-17. About how long does it take for a robot to amortize itself?

17-18. Why are transfer lines of limited usefulness?

17-19. What is being used to replace the regular mass-production transfer line?

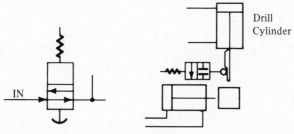

Figure P17.1.

Bibliography

Boothroyd, G., and A. H. Redford. *Mechanized Assembly.* Maidenhead, Berkshire, U.K.: McGraw-Hill House, 1968.

Dallas, Daniel B., ed. "Systems Engineering in Advanced Assembly." *Manufacturing Engineering* 83 (October 1979): 54–57.

Winship, John, ed. "Robots in Metalworking." *American Machinist* 119 (November 1975): 87–110.

Editor. "90% Uptime from Unmanned Machining Systems." *Production* (June 1979).

Gettelman, Ken., ed. "Programming by Part Description." *Modern Machine Shop* 52 (July 1980): 99–103.

Gettelman, Ken, ed. "Robots Drill and Profile." *Modern Machine Shop* 52 (April 1980): 129–136.

Harrison, H., and J. Bollinger. *Introduction to Automatic Controls.* Scranton, Pa.: International Textbook Co., 1969.

Schneider, R. T., ed. "Versatile Hydraulic Robot Handles Tough Tasks." *Hydraulics and Pneumatics* (July 1979).

Smith, Bradford M. "New Dimensions in Automation Technology." *SME Technical Paper MS79-381,* Dearborn, Mich.: Society of Manufacturing Engineers, 1979.

Smith, T. C. "NC's Stake in the Productivity Race." *NC Commline* 7, 5 (September–October 1978).

Treer, Kenneth R. *Automated Assembly.* Dearborn, Mich.: Society of Manufacturing Engineers, 1979.

CHAPTER EIGHTEEN

Numerical Control

A relatively short time ago machines were operated by craftsmen who determined many variables such as speeds, feeds, depth of cut, etc. Now much of this work is being assigned to computers and machines that are numerically controlled. The word numerical is defined as "the expression of something by numbers." Control is defined as "the exercise of directing, guiding, or restraining power over something." By combining the two definitions we find that numerical control consists of directing, guiding, or restraining power over something by the use of numbers. Since numbers are merely symbols and letters are used as well, numerical control could well be called symbolic control.

Numerical control has become commonplace in the world of machine tools, opening up new doors for more advanced concepts. Today, entire preplanned and programmed design and production functions can be processed, stored, and finally executed with the aid of a computer. These functions are termed CNC (computer numerical control), CAD (computer-aided design), CAM (computer-aided manufacturing), and CIM (computer-integrated manufacturing).

Numerical control (NC) is considered to be more than an improved control for machine tools that provides higher productivity. It is considered by many leading scientists to be the first significant step in what is referred to as the second industrial revolution. The first industrial revolution was ushered in by

James Watt's steam engine. Human muscle power was replaced and extended into all types of industrial equipment. The second revolution is an extension of man's mind. NC, with the aid of the computer, takes a symbolic input and restructures it to an output that extends man's concepts and thinking into creative, productive results.

The birth of NC may be traced to the late 1940s. At that time John Parsons and his associates were manufacturing helicopter blades. They found the specifications, contours, and weights of the airfoil very critical and demanding. Struggling with the contour of the rotor blade, they devised an incrementally moving dimensional inspection system. The gathered dimensional data was fed into an early model computer that contained the pure mathematical formula for the airfoil. The inspection data were compared to the basic formula. The idea worked so well that Mr. Parsons concluded that if a computer could tell what had happened after a part was made, why not use the same data to control a machine tool to make the part. A concept is far from reality, of course. However, in 1949 the U.S. Air Force, which recognized the value of the technique, awarded

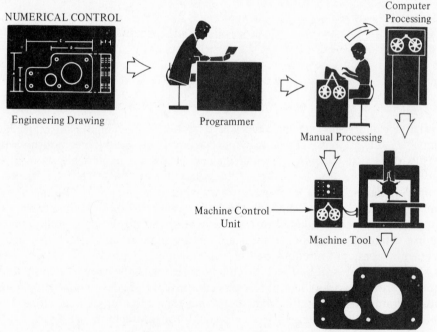

Figure 18.1. The operating efficiency of an NC machine is largely dependent on the programmer. He interprets the engineering drawing and writes the program that contains all the basic instructions for the machine control unit to follow. Some of the simpler programs are processed manually, whereas the more complex programs require computer processing.

a development contract to the Massachusetts Institute of Technology and by 1952 the first numerically controlled machine tool was operational.

Numbers, letters, and symbols gathered together and logically organized to direct a machine tool for a specific task are called an *NC program*. The concept is revolutionary in that the control of a machine has been moved away from operator skill and intuition to an entirely written program that can be interpreted by the machine control unit to perform all the required movements that produce a finished part (Fig. 18-1). Not only has NC replaced human skills and intuition, but it has made it possible to generate complex machine motions with a degree of precision that is far beyond any human capability. If a surface can be mathematically defined, NC can control the machining of it. Because of the higher productivity achieved by NC, it is believed that 75% of all mechanical parts not mass produced could be made by NC machine tools by the year 2000.

NC OPERATION

As with anything to be manufactured, the concept must be formalized into a design. The part programmer studies the drawing and conceptually visualizes all the machine motions that will be required to make the part. He then proceeds to document all these motions in a logical order on a programming manuscript. All the steps must be logically conceived and clearly stated so that the machine control unit (MCU) can follow the coded commands. Most MCUs today have a basic computer design.

The programmed codes that the MCU can read may be one of the following four media: perforated or "punched" tape, magnetic tape, signals directly from computer logic or computer peripherals, such as coated-metal discs or thin plastic "floppy discs," or voice command.

Tape Reading.   The predominant method of communicating with the machine is by using punched tape 1 in. wide with eight channels running the length of the tape. The holes are punched in a predetermined pattern according to the number, letter, or symbol represented. The reader senses the pattern by one of several methods, the most common being that of a light beam passing through the tape holes to complete an electric circuit with photoelectric cells. The MCU converts the input signals from the tape to output signals that can be utilized by the power elements actually running the machine tool.

The signal from the tape goes into data storage. The storage unit has two parts, active and buffer storage, respectively. Active storage emits the signals for the operation being performed, while buffer storage holds the command just taken off the tape. When the active storage command is completed, the command in buffer storage moves into active storage. A continuous, smooth operation provided by buffer storage is essential for contour-type cutting. With commands already electronically integrated into the machine's buffer storage, the input for tool motion goes immediately from one command to another.

OPEN- AND CLOSED-LOOP CONTROL SYSTEMS

As the NC servomotors position a milling machine table, the pen of a drafting machine moves over the paper, or the torch of a flame-cutting machine glides over the metal, doubt may develop as to the accuracy of these various movements.

Designers of NC machine systems cope with this doubt in two ways. Some simply put faith in the system and trust the drive motors to follow the electronic commands precisely, which is called *open loop*.

Alternatively, a checking system may be built into the controller to constantly keep track of what the machine is doing, which is called *closed loop*.

Open-Loop System. The open-loop NC system (Fig. 18-2) is simpler and less costly than the closed-loop system. Open-loop systems are based on the stepping motor. Stepping motors can be electric, hydraulic, or a combination of electric and hydraulic. Each pulse of electric current from the control system exerts a given torque on the motor, causing it to rotate a given amount. If the motor is overloaded, it will stall. To prevent stalling, the prevailing theory in open-loop design is to "overpower" the system to make sure that there is adequate torque behind the stepper to meet all contingencies. If a maximum load of 1000 oz in. is expected, the design may specify a motor rated at 2000 oz in. of driving torque or even higher. Electrohydraulic stepping motors develop torques to well over 100 lb ft at pulse rates to over 16,000 pps (pulses per second).

Of course, a value judgment must be made about the bearings, leadscrew, and other components involved. Instead of inflicting structural damage to the machine, it may be better to let the machine stall and destroy the workpiece and fixture.

Another method of building assurance into the open-loop system is to have the machine return to a zero reference periodically. If the registers are off, the operator will immediately be alerted so that corrections can be made.

The open-loop system is most often used on large machines that have been *retrofitted*. Retrofitting can be done much more easily with the open-loop system than with the closed-loop system. Here retrofitting refers to equiping an ordinary machine with NC capability.

Closed-Loop Systems. There are two types of closed-loop systems, direct and indirect. In the direct feedback system, the tape command is read and then the digital positioning system compares the position command number read from the tape with the stored-position feedback number (Fig. 18-3). An error detector produces a difference signal by subtraction of these two numbers, and the magnitude and sign of the difference are used to command the axis-positioning servodrive to rotate in the direction that reduces the error to zero. For convenience and reliability in processing, all numbers are converted to binary form. Shown in the diagram is a box representing "bit" storage. A bit is a binary digit or

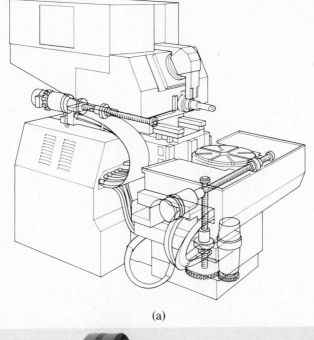

(a)

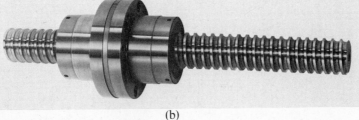

(b)

Figure 18.2.   (a) An open-loop 3-axis control system. Each axis is driven independently with a high-powered dc servomotor. Each drive transmits motion through a precision preloaded ball screw (b). (*Courtesy Cincinnati Milacron.*)

the smallest unit of data in a digital computer, such as 1 or 0, on or off. The bits of the command position information are introduced one at a time to the bit storage device. The bits then appear in the same order at the output as a series of pulses for the pulse generator. Since each pulse is equivalent to a unit of distance, each time the machine moves enough to cause the pulse generator to change state, a bit is added (or subtracted) from the circulating feedback number. The numbers or bits are continuously circulated and stored in the memory until the next bit is processed. The memory is then reset to wait for the next pulse. When the error decreases to zero, the servodrive motor stops. Provisions are made so that any backlash (free motion) that may occur in changing the direction of the leadscrew is accounted for.

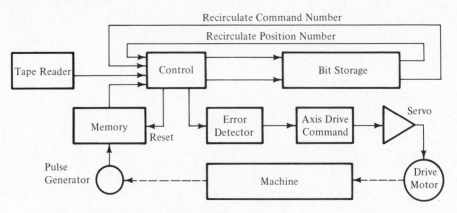

Figure 18.3.   Simplified block diagram showing closed-loop feedback. Control used on NC machines.

Indirect Feedback System.   The most common feedback system for NC today is indirect. Rather than measuring position directly, this system typically utilizes the rotary position of the leadscrew or pinion. The linear travel is inferred from the rotary displacement. This system retains some of the same elements that the direct feedback system utilizes but the cost as well as the accuracy is less.

Direct feedback has become synonomous with high cost and high accuracy and is generally used only on the largest precision machines.

Perhaps the only place that the closed-loop system is absolutely required from a technical standpoint is where adaptive features must be built into the control.

ADAPTIVE CONTROL

Adaptive control of a machine tool may be compared to a perfect machinist, one who intuitively senses every cutting condition and immediately adjusts the proper controls to produce the optimum cutting performance at all times (Fig. 18-4).

Several types of adaptive control systems are available for machine tools but are generally limited to sensing spindle horsepower and changing programmed feedrates. More sophisticated systems have the ability to sense changing stock conditions, lack of stock, work hardness, cutter loads, and cutter wear.

With adaptive control, the programmer need not be concerned with feeds and speeds. The machine is programmed for optimum performance, and therefore the feedrate and RPM will automatically be adjusted to compensate for tool wear and any changes in the workpiece material or size.

In one example of adaptive control the normally programmed feedrate for a milling operation was 5 ipm. Adaptive control programmed the feedrate as

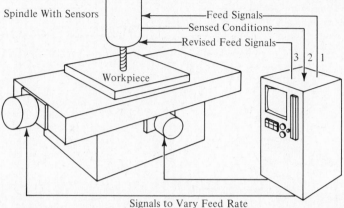

Spindle With Sensors

Feed Signals
Sensed Conditions
Revised Feed Signals

Workpiece

3 2 1

Signals to Vary Feed Rate

Figure 18.4. Adaptive control utilizes a closed-loop system to sense such conditions as heat, torque, vibration, pressure, and deflection as generated by the cutting action of the tool. When sensed, the information is fed back to the MCU where speeds and feeds can be changed as necessary to realize optimum cutting conditions.

high as 20 ipm with an average of 11.4 ipm. The increase in feedrate attained overall was 128% of what would have normally been used.

Obviously adaptive control is not simple. Without computers it is doubtful if a complete adaptive control system could be developed. In addition to reduced machining time (approximately 50%), closer tolerances can be expected as well as a minimum of scrap.

THE COORDINATE SYSTEM

Although the cartesian coordinate system was conceived more than 300 years ago by a French mathematician, Descartes, no system has been found that surpasses it as a basis for NC control. Rectangular coordinates provide an excellent means of locating an NC tool with respect to the workpiece in either two or three dimensions. The three primary motions for an NC machine are given as X, Y, and Z. The primary X motion is normally parallel to the longest dimension of the machine table. The Y motion is normally parallel to the shortest dimension of the machine table and the Z motion is the movement that advances and retracts the spindle [Fig. 18-5(a)]. Letters A, B, and C designate angular and circular motions as shown at (b). Shown at (c) is the type of work that can be done on a five-axis, continuous-path control machine.

The diagram shown in Fig. 18-6 is used to represent the table of a simple drilling machine with the workpiece located in the upper right quadrant. Quite often the entire table will be designated as the upper right quadrant so that no minus dimensions are used. As shown in the diagram, a hole may be drilled at

MACHINE ELEMENT MOVEMENTS
X Axis — Table side to side
Y Axis — Table front to rear
Z Axis — Spindle vertically into work

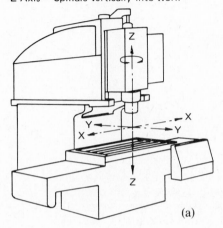

(a)

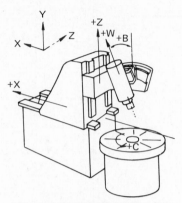

MILLING MACHINE PROFILING
AND CONTOURING

(b)

Figure 18.5. A sketch showing the basic machine axes (a). Angular and circular motions about axis Z are designated by letters A, B, and C, as shown at (b). Shown at (c) is the type of work that can be done with a 5-axis machine. An end mill is being used to generate a contour, or a circular form. It may also be used to make elaborate die cavities or airfoil contours. These operations are referred to as profiling. During profiling the cutter and workpiece move about several or all axes simultaneously.

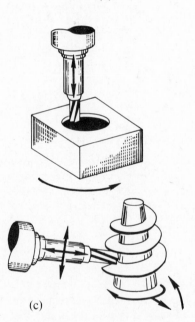

(c)

the +X3, +Y4 location or, if a slot is to be milled as represented by the broken line, the center point of the cutting tool is sent from the point of origin to +X3, +Y4. The slot width is determined by the cutting-tool diameter.

Considerable work is done on machines that have only two axes under NC control. In this case, the depth of holes and slots is determined by limit-switch settings made by the operator when the part is placed on the machine. More advanced NC involves the Z coordinate. Positive values for all three axes are in the upper right quadrant, as shown in Fig. 18-7. The tool always points in the −Z direction. The coordinate system described assumes the tool movement

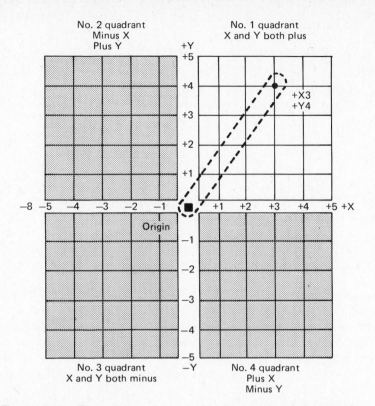

No. 2 quadrant
Minus X
Plus Y

No. 1 quadrant
X and Y both plus

No. 3 quadrant
X and Y both minus

No. 4 quadrant
Plus X
Minus Y

+X3
+Y4

Origin

Figure 18.6. The origin point may be the corner of the machine table or it may be relocated at a point on the workpiece itself. If a hole is to be drilled at the X3, Y4 location or a slot is cut as shown by the broken line, the MCU will be directed for three units of X motion and four units of Y motion. Although there is a tendency to seek setups that allow programming in the all-plus quadrant, programmers find no difficulty for any quadrant.

Figure 18.7. Three-dimensional NC involves the use of the Z coordinate. On most machines the Z measurement begins when the tool spindle is fully retracted. Positive values are shown above the point of origin and minus values below.

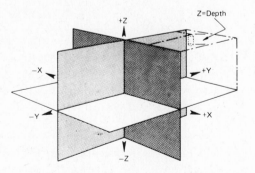

with respect to the work. If the tool is stationary and the table moves the signs will be reversed, e.g., if the table on a vertical milling machine moves to the operator's right, it will be in the $-X$ direction. In actual practice (bearing in mind the $+$ and $-$ signs just stated) one pretends the tool moves rather than the table.

POSITIONING

Point to Point.   Normally positioning refers to directing a tool to a specific location on a workpiece for operations such as drilling, boring, reaming, and punching. This type of positioning to an exact coordinate location is called *point to point* (PTP). It makes no restriction as to the cutter path in arriving at the desired location.

Continuous Path.   As the name implies, the cutting tool is in constant contact with the workpiece as the coordinate movements take place. An example of continuous-path programming is profile milling, which is sometimes referred to as sculpture milling. Contouring also applies to lathe work, sawing, flame cutting, continuous-path welding, and grinding.

POSITION PROGRAMMING

Fixed Zero.   The programmer starts out with an engineering drawing visualizing all the necessary machining sequences needed to produce the part. These steps are written down in a manuscript from which a program tape can be developed. An important first step is to establish the reference point. Invariably a workpiece is loaded on the machine so that a flat surface is parallel to the table and a straight side will be parallel to machine axis X or Y. The point where the X, Y, and Z axes cross is called the *zero point*. The zero point may be either *fixed* or *floating*. The fixed-zero machine is usually made so that the zero location is at the lower left-hand corner of the X–Y coordinate. On some systems the fixed zero may be moved, by means of switches or dials, to anywhere within the full travel range of the machine. Ordinarily, most fixed-zero machines are made so that all machining will be done in quadrant 1 with only a small (0.050 in.) zero shift for adjustment after the part is loaded in place.

Floating Zero.   A floating zero allows the operator to place the zero at any convenient location without regard to the initial fixed zero point. The operator merely moves the table to the desired location by means of jog buttons. Jog buttons make it possible for the operator to "manually" move the table to the desired location at feedrates set on the dial. When the table is at the exact desired location, the operator depresses a button that establishes this point as the zero point. The part programmer usually determines where the zero should

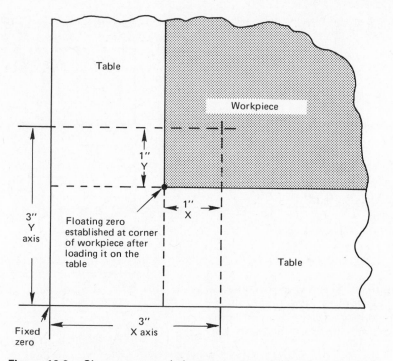

Figure 18.8. Since most workpieces are mounted on the center of the table, the machine zero does not match the workpiece zero. In the example shown, the edges of the workpiece are 2 in. from the bottom and side of the table. The center of the hole to be drilled is 1 in. from the bottom and side of the workpiece. If the program is for a fixed zero, both X and Y axes will be programmed as 3 in. If the machine has a "floating zero" or "zero shift," the origin point will probably be set at the corner of the workpiece and X and Y dimensions will be programmed as 1 in.

be and passes this information on to the operator. Figure 18-8 shows a comparison between a fixed and floating zero.

Absolute and Incremental Dimensioning. Dimensions on an NC drawing may be either absolute or incremental in nature. Absolute dimensions always start from a fixed-zero reference point, as shown in Fig. 18-9(a). All incremental locations are given in terms of distance and direction from the immediately preceding point, as shown in Fig. 18-9(b). The direction of motion reverses itself when the magnitude of numbers decrease. Thus in absolute dimensioning with increasing numbers 2, 5, 7 causes motion in one direction, while decreasing numbers 7, 5, 2 will reverse the direction. This method eliminates the use of algebraic signs to indicate direction as long as the motions occur in the first quadrant.

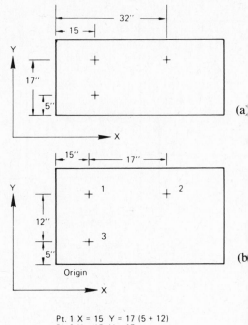

Figure 18.9. Absolute measurements are given from the zero origin (a). Incremental locations are given in terms of distance from the immediately preceding point (b).

Pt. 1 X = 15 Y = 17 (5 + 12)
Pt. 2 X = 17 Y = 17
Pt. 3 X = 15 Y = 5

INTERPOLATION

Interpolation is the method of getting from one programmed point to another so that the final workpiece is a satisfactory approximation of the programmed design. Interpolation must be a function of any continuous-path contouring system and is the main distinguishing feature that separates contouring controls from the more simple positioning controls. There are four interpolation methods or means of connecting defined coordinate points to generate angular or curved shapes: linear, circular, parabolic, and cubic.

Linear Interpolation.   Linear interpolation is found on nearly all continuous-path machines. Figure 18-10 shows a slope-line motion as accomplished by PTP programming. The control will follow path AB because the ratio Y/X = 1, the feedrate in both axes is the same. At point B, the control interpolates to give an additional increment in the X direction, bringing the slope nearer that of the theoretical. The same interpolation would occur at point D, etc.

An arc can be approximated by a series of steps as shown in Fig. 18-11. The interpolation is similar to that of making a slope except that the lines are now tangential to an arc. The theoretical radius of the cutter, the center dashed line, is given as additional input data along with the tolerance. Because of the large number of short lines, the outline appears to be analog in nature. Numerical control, however, is not analog but digital. Therefore, to approximate an arc or a circle a sufficient number of points have to be described within the framework

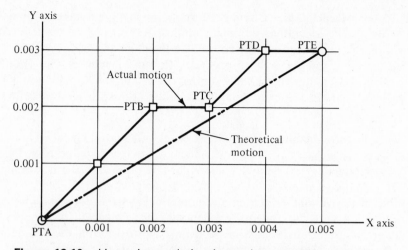

Figure 18.10. Linear interpolation is used to move the tool in an angular direction from one point to another.

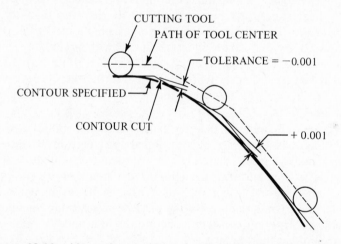

Figure 18.11. Linear interpolation used to define an arc. The x-y coordinate positions are given to the computer along with the cutter diameter and desired tolerance. The computer will then calculate the cutter path.

of rectangular coordinates. It can easily be seen that the length of the program tape is in direct proportion to the number of coordinate positions needed to define the arc. In some sculptoring operations the length of line may be only 0.0002 in. The standard procedure is to have a supply of tapes that have been programmed by the computer for various-size quadrants of a circle made for either clockwise or counterclockwise motion. These tapes are then copied into the regular part program when needed, or they may be taken directly from the computer memory.

Circular Interpolation.   Circular interpolation is a great aid to programming where many cuts are either complete circles or portions of circles. All the information that is needed are the coordinate locations of the center, the radius, and the direction the cutter is to proceed. In most machines, the circular interpolator is a computer that breaks up the circle into small (usually 0.0002 in.) straight lines and automatically computes enough of them to describe the circle.

Parabolic Interpolation.   Parabolic interpolation finds its greatest application in free-form designs such as mold work and automotive die sculpturing. Parabolic interpolation goes between three nonstraight-line positions in a movement that is either a complete parabola or a portion of one (Fig. 18-12).

With the growth of CNC (computer numerical control), the parabolic method of interpolation with its complex mathematics has fallen out of general favor. The lower computer costs have made it possible to rely on the simpler linear interpolation, which keeps costs within a more acceptable range.

Cubic Interpolation.   Cubic interpolation is applicable to the making of large "free-form" sheet metal dies as characterized by the auto industry. By using a cubic (third-degree curve) interpolation, it is possible to generate sophisticated cutter paths necessary to machine sheet metal dies with a relatively small number of input data points. The cubic interpolation, which is done with a larger-than-average minicomputer memory, will not only describe the curves, but also will smoothly blend one curve segment to the next without discontinuities or demarcation points.

It should be remembered that circular, parabolic, cubic, and any other interpolation methods that may be developed are specialized forms of linear interpolation that have been worked out by the control unit developers. Their sole purpose is to eliminate the need for a vast multitude of coordinate axes points in the actual workpiece program. With one of the higher interpolation forms, it is necessary only to describe the particular curve and the computational capacity of the control unit will then generate the multitude of points and direct the cutting tool along that path.

The Cutter Path.   Assuming that a continuous-path program is to be made manually, the programmer will have to consider feedrates and the tool path. If a cutter, for example, is 1 in. in dia. and is used for profiling, it will have to be offset as shown in Fig. 18-13. Note that the offset must always be made perpendicular to the surface being machined, regardless of the direction that surface faces. Determining offsets not parallel to a machine axis will usually involve trigonometric calculations.

Most MCUs have cutter-compensation capabilities to take into consideration the variable diameter of cutting tools or tools that were programmed with one diameter in mind and have been changed or have been resharpened to a smaller diameter.

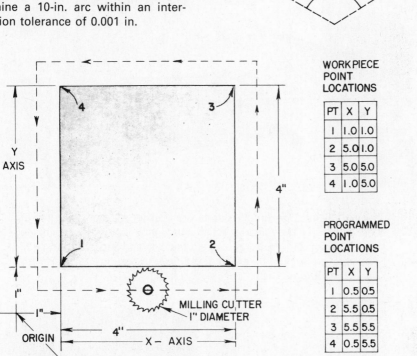

Figure 18.12. The program for a parabolic span-type control unit requires enough spans to stay within an acceptable tolerance as indicated by T. Thirty-degree spans are sufficient to machine a 10-in. arc within an interpolation tolerance of 0.001 in.

WORKPIECE POINT LOCATIONS

PT	X	Y
1	1.0	1.0
2	5.0	1.0
3	5.0	5.0
4	1.0	5.0

PROGRAMMED POINT LOCATIONS

PT	X	Y
1	0.5	0.5
2	5.5	0.5
3	5.5	5.5
4	0.5	5.5

Figure 18.13. A simple program showing the difference between the workpiece location points and the cutter location points.

PROGRAMMING BY SCANNING AND DIGITIZING

Programming may be done directly from a drawing, model, pattern, or template by digitizing or *scanning*. A simple two-dimensional part that needs nothing more than positioning operations such as sheet-metal punching or hole drilling can be digitized (given X-Y coordinate locations) from a fully dimensioned drawing. The drawing is placed on the table (Fig. 18-14) and an optical reticle or other suitable viewing device connected to an arm or arms is placed over the location points. Transducers are made to identify the location and, upon a signal, transfer it to either a tape punch or other suitable programming equipment. Most digitizing equipment also has auxiliary input to allow the establishment of reference points or the addition of data such as position refinement.

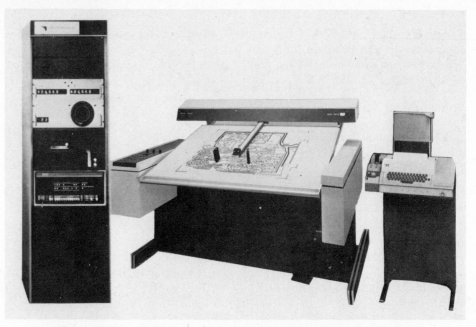

Figure 18.14. Coordinate points are picked up from the drawing and digitized. The unit in which the digital readouts are located also contains buffer storage, scale factors, and programming aids. The tape is punched by a unit at the right of the digitizer. (*Courtesy Data Technology, Inc.*)

Figure 18.15. Scanning provides simplified programming of a nonmathematical configuration. Data obtained through tracer movements are converted into punched tape by a minicomputer. (*Courtesy Cincinnati Milacron.*)

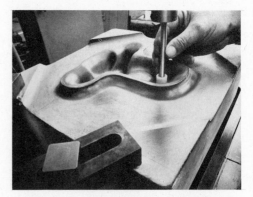

A scanner enables an operator to program complex free-form shapes by manually moving a tracing finger over the contour of a model or premachined part (Fig. 18-15). Overall accuracy of the scanning system from model to tape is ±0.005 in.

Many digitizing and scanning units have the capability to "edit" or revise the basic data gathered. For example, the drawing or model may not be as accurate as required for the finished product. It may be possible to obtain the basic data and then refine it with an overriding corrective input. The design of a surface, for example, may call for an arc but the model may not be near the required tolerance. After the rough data have been scanned, the computer can be called on to smooth the data to a perfect arc.

EDITING

No matter how good the programmer or how flawless the operator who operates the tape preparation unit or data transmissions device, there will still be errors. Several editing systems have been developed to check a program before it is committed to the machine tool loaded with a workpiece.

The units shown in Fig. 18-16 make up both a tape preparation system and an editing system. Each unit may be considered a "stand-alone" programming resource or all of them may all be interconnected so that the plotter, tape punch, tape reader, video display, and floppy disc memory storage may be called on as necessary to make or edit a program. Communication with a remote computer is also possible through the teleprinter. The line between tape preparation units, editing, and computational resources is becoming less sharp, particularly as microprocessor technology makes the combination of such functions less costly. The functions of each of the units shown in Fig. 18-16 will be discussed briefly.

Keyboard or Telewriter.   The keyboard operator copies the information from the program manuscript to prepare the punched tape. As the tape is punched,

Figure 18.16.   The floor editor consists of a keyboard, printout, plotter, tape reader, video display, and tape punch. (*Courtesy White Sundstrand Machine Tool Inc.*)

G72 Stock removal in facing   G73 Pattern repeating   G76 Thread cutting

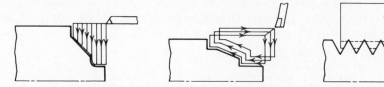

Figure 18.17.   Examples of "canned" cycles that can be called from the computer memory. (*Courtesy General Numeric.*)

a proof sheet is printed that serves as a record of the program tape and as a ready method of editing.

Tape Reader.   The tape reader is used to check the punched tape or to reproduce additional tapes as needed.

Plotter.   The punched tape, or data direct from the computer storage unit or a floppy disc-type memory device, may be used to direct the movements of the plotter. The plotter consists of a table on which a piece of paper is mounted. A driving mechanism with two axes carries pens over the surface to show the cutting path of the tool. Since the paper has two axes, there is often a provision made for selecting the two axes that will be portrayed, XY, XZ, or YZ. Some plotters also provide an isometric view. The plotter will quickly disclose several types of errors, such as cutter paths or misplaced decimal points.

Video Display or CRT.   The cathode-ray tube also provides a video display of tool paths in three standard planes plus isometric views. The screen may also be used to display the printed program.

Floppy Discs.   Floppy discs are thin, flat discs similar to 45-rpm records that are used as memory storage devices. These will be discussed in more detail in the computer section of this chapter.

Canned Cycles.   Canned cycles are repetitive operations that can be put in the computer memory and called up when needed, as shown in Fig. 18-17. Often, however, the cycle may not fit the program exactly and requires modification. The modification can be done in the computer memory, checked on the CRT or plotter, and when satisfactory, a new tape can be punched and a hard copy printed out on the teleprinter.

TAPE FORMAT

Despite the advantages now available of being able to go directly from the keyboard to the computer in the MCU, perforated tape is still used extensively.

Two tape systems were developed, EIA (Electronics Industries Association) and ASCII (American Standard Code for Information Interchange). ASCII is becoming known as ANSCII (American National Standards Code for Information Interchange). An example of the EIA tape is shown in Fig. 18-18. Most of the newer MCUs can handle either EIA or ASCII coding.

A tape feature that has become virtually universally accepted is the BCD (binary coded decimal) system of individual digit coding.

Binary-Coded Decimal System. Binary is based on two digits, 0 and 1. Channels 1, 2, 3, and 4, or each hole across the width of the tape, have, respectively, the binary values of 1, 2, 4, and 8, which are the exponential equivalents of 2^0, 2^1, 2^2, 2^3. Any hole punched in channel 1 will have the value of 2^0, or numerical 1. Channel 2 will be 2^1 or 2, channel 3 will be 2^2 or 4, and channel 4 will be 2^3 or 8. Zero normally would be indicated by an absence of a punch, but since there is greater reliability in reading a punch for 0, channel 6 is assigned to represent 0. These five channels can be used to represent any digit (single number) of numerical data. For example, 3 is represented by holes in channels 1 and 2, and 7 is represented by holes in channels 1, 2, and 4. Channel 5 is used for *parity check*. The parity system is to make sure there is an odd number of holes in each row (EIA standard). The parity check reduces the possibility of error, as in the case of holes being punched but the paper remaining in place. The tape punch automatically adds the punch to make an odd number of holes. The control equipment has circuitry that will detect an even number of holes anytime during data reading, stopping the reader and lighting a "read error" light when detection occurs.

The last two channels are assigned, respectively, an X code and an EB code. The X code is used in combination with alphabet codes to designate miscellaneous functions. The EOB, or simply EB, is the *end-of-block* channel. It signifies the end of all pertinent data for one particular action. This data group could be compared to a sentence. It also provides for easy visual examination of the tape. Figure 18-18 shows a section of EIA-standard BCD-punched tape.

Letter codes are punched in channels 6 and 7 in combination with other channels. Alphabetic punches are equal to numeric positions in the alphabet. For example, A is 1, B is 2, etc. In addition, letters A through I are identified by having both channels 6 and 7 punched. Thus A is 1, 6, and 7.

The next group of nine letters, J through R, are always punched in channel 7 plus the required numbers. Thus J is 1 and 7, with 5 for parity check.

The last group has only eight letters, S through Z. In this case, each letter is numbered from 2 through 9 and channel 6 is the common code punch for this group. An example of a tape command is shown in Fig. 18-19. The command is to "move on X-axis 5.25 in. and on Y-axis 10.5 in." The manuscript, which is typed while the tape is being made, would appear as X05250, Y10500. This is known as a block of data. It contains all the instructions required for one command. Other pieces of information that should be included are feedrates, spindle speeds, etc. It is assumed that the decimal is between the second and third digits.

TAPE PUNCH	8 EL	7 X	6 O	5 CH	4 8	•	3 4	2 2	1 1
0			○			•			
1						•			○
2						•		○	
3				○		•		○	○
4						•	○		
5				○		•	○		○
6				○		•	○	○	
7						•	○	○	○
8					○	•			
9				○	○	•			○
a		○	○			•			○
b		○	○			•		○	
c		○	○	○		•		○	○
d		○	○			•	○		
e		○	○	○		•	○		○
f		○	○	○		•	○	○	
g		○	○			•	○	○	○
h		○	○		○	•			
i		○	○	○	○	•			○
j		○		○		•			○
k		○		○		•		○	
l		○				•		○	○
m		○		○		•	○		
n		○				•	○		○
o		○				•	○	○	
p		○		○		•	○	○	○
q		○		○	○	•			
r		○			○	•			○
s			○	○		•		○	
t			○			•		○	○
u			○	○		•	○		
v			○			•	○		○
w			○			•	○	○	
x			○	○		•	○	○	○
y			○	○	○	•			
z			○		○	•			○
. (PERIOD)		○	○		○	•	○	○	
, (COMMA)		○	○		○	•	○		○
/		○	○			•			○
+ (PLUS)		○	○		○	•			
− (MINUS)		○				•			
SPACE				○		•			
DELETE		○	○	○	○	•	○	○	○
CARR. RET. OR END OF BLOCK	○					•			
BACK SPACE				○	○	•	○		
TAB			○	○	○	•	○	○	
END OF RECORD					○	•		○	○
LEADER						•			
BLANK TAPE						•			
UPPER CASE		○	○	○	○	•	○		
LOWER CASE		○	○	○	○	•		○	

Figure 18.18. The punch tape is shown in the EIA standard RS 350. It has an odd parity (odd number of holes in each line). (*Courtesy* Modern Machine Shop.)

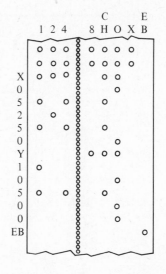

Figure 18.19. An example of tape command for X, 5.25 in., Y, 10.5 in., using the Binary Coded Decimal system. The first six channels carry values of 1, 2, 4, 8, CH, 0, respectively. The CH or parity check channel is utilized to make certain that an odd number of holes appear across the tape.

MACHINING CENTERS

The trend in NC machines is the type of machine in which a number of operations can be done in a single setup, which is called a "machining center." A machining center (Fig. 18-20) is built to do a number of operations such as milling, drilling, boring, facing, spotting, counterboring, threading, and tapping in a single setup on several surfaces of a workpiece.

A numerically controlled machining center must have the following characteristics:

At least three linear axes of positioning control, X, Y, and Z.

Index table to permit access to more than one side of the workpiece automatically.

Automatic tool changer, a tool magazine to hold a large selection of tools, and a tool change arm that selects and places the tools into the machine spindle.

Ability to select spindle speeds and feedrates, and to bring tools to the workpiece automatically on command.

Miscellaneous functions such as coolants off/on, part clamping, program start alerts, peck drilling, and other automatic functions.

The cutting tool is automatically selected from the storage magazine. The number of tools in the magazine has no theoretical limit but is typically 24, 48, or 60.

Machining centers have X, Y, and Z motions within a rectangular coordinate description of locations. However, more complex machining centers are capable of many other axis movements. These motions include tilt and swivel of the spindle head and column as well as rotation and even tilt of the work-holding table.

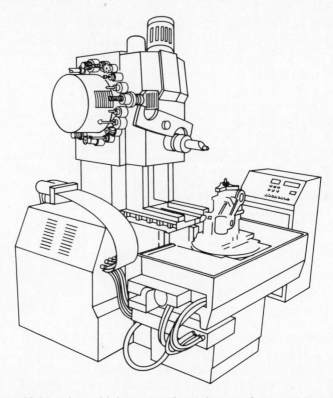

Figure 18.20. A machining center is made to perform a wide variety of operations in five sides of a workpiece. The workpiece table is made to index and different tools are automatically set in and out of the spindle.

The machining center offers the highest potential for NC utilization and versatility. Its multiaxis capability has made it practical in many cases to combine several components into a single part that is machined from casting, forging, plate, or bar stock. Design configurations that were previously impossible can be achieved if enough axes of motion can be simultaneously controlled to develop the configuration.

Most often preset tooling for machining centers comes from the toolroom, but sometimes machine operators have both the time and ability to make up the preset tools needed for their location. Many shops maintain a file of tool-setting drawings for use by both programmers and toolroom employees. These drawings describe the cutter, identify it by a tool number, and establish a setting length.

Tooling for machining centers has wide ramifications. For example, a hole drilled through the wall of a casting has no critical depth setting, although a bore may have critical dimensions for both depth and diameter, even to the extent that the hole is checked and the adjustable boring head reset based on the actual metal removal.

Even with the finest tool presetting, the conditions of the actual cut may indicate a need for final adjustment with the control unit offset. By allowing the machine operator to make the final offset adjustment, it is possible to compensate for conditions such as tool deflection or indexing of tool inserts.

Automatic tool changers require only a few seconds to take one tool out (Fig. 18-20) and replace it with another tool. The tools in the automatic tool changer are either positioned in their numerical machining sequence or separately identified and coded to respond to specific tape commands. The latter system enables the tools to be placed randomly in the tool magazine.

COMPUTERS IN NC AND MANUFACTURING

Computers are becoming universally recognized as the most powerful and effective tool for improving productivity in industry, which is the single most important concern of every manufacturing manager.

The most readily discernible computer applications in manufacturing have been in the area of NC machines and tooling. From its inception, NC has been involved with computer assistance for program development.

Computers excel in three areas of manufacture: collecting information, reaching a decision (based on either well-established rules and principles or guidelines gained from experience), and issuing an order. A computer can be taught the rules or made to remember given experiences.

COMPUTER INPUT AND OUTPUT (I/O)

Input.   There are many ways of getting information in and out of a computer, some of which are shown in Fig. 18-21. Computer input may be a standard tabulating card prepared on a keypunch or even on another computer. As mentioned before, one of the most used methods in NC is the perforated tape. Magnetic tape is used but is usually the result of another computer effort. Input may also be from an almost endless number of specially designed sensors that record phenomena and convert it to binary signals that are computer compatible. The most recent input form is by voice command.

Output.   The output from the computer may be in the form of tabulating cards, perforated tape, printout sheets, magnetic tape, or even direct pulses to another computer. Processed information may be transferred to direct storage that can later be accessed and put back in the computer for further processing. Most often used for this purpose are the random access disc packs (discussed later under memories).

Output may also be in the form of signals that are amplified so that they can be used for all types of industrial applications, such as opening and closing valves in industrial process equipment or operating the servomechanisms of machine tools.

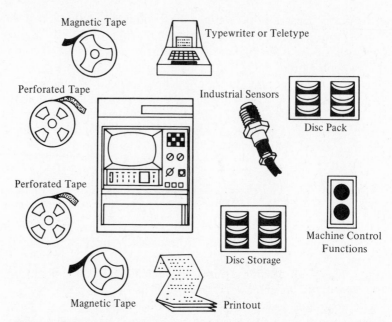

Figure 18.21. There are many forms of computer input ranging from the slower keyboard to the faster tapes. Once the data are placed in storage, they may be edited and checked either in the printout or on the video screen. When correct, they may be transmitted to the computer for full processing.

The method of input and output is dependent on the construction feature of the computer as, for example, perforated tape cannot be put through a punched card reader and magnetic tapes can be read only by magnetic tape readers.

COMPUTER SYSTEMS

Computer systems fall into three reasonably well-defined categories: mainframe, mini, and microprocessor.

Mainframe.   The distinguishing feature of a mainframe in a computer system is that it can partition its memory or logic, therefore being able to service more than one input or output device simultaneously. Thus a mainframe computer may be functioning for an NC programmer, a design engineer, and a company accountant all at the same time. Not all mainframe computers are equipped with a partitioned memory, although they can be. Sometimes the mainframe is defined as the portion of a computer exclusive of the input and output devices and other peripherals.

Minicomputers.   Minicomputers are generally considered to be small machines with a word size of 8 to 18 bits, and with a basic memory of 4000 (4K) words that is often expandable to as much as 32K, or in some cases even larger.

A "bit" is the charged or discharged binary notation usually accomplished by means of magnetic film or ferrite cores. A minicomputer may have a memory of 4000 or 8000 words, whereas a "medium-sized" computer may have up to approximately 64,000 words and a larger computer will jump to about 128,000 to over a million units. The term "4000 or 8000-word computer" is seldom if ever used. The terminology that has grown up with the computer is "4K," "8K," "16K," etc., which is a quick way of stating "kilo" or thousand.

The minicomputer has become very popular in recent years because it resembles the mainframe computer but is much lower in cost. Like the mainframe, it has a central memory or logic section with appropriate input and output devices. While it may seem to work simultaneously with several different devices such as keyboards, card readers, printouts, and display terminals, it actually services each of them one at a time. It switches back and forth from one source of information to another. It is slower than the mainframe, but if jobs can be queued or if a slower operation is acceptable, it does an excellent job.

Because of the many roles a minicomputer is asked to play, it is characterized as having I/O flexibility. The true general-purpose minicomputers are available with a teletypewriter terminal. For some machines this is the only form of I/O available. For others there are many possibilities as shown in Fig. 18-21. An example of the use of these I/O devices may be as follows: A handwritten manuscript is typed on a keyboard. Since the computer memory will not permit storing the long NC tapes for editing and testing, the memory can be augmented with a magnetic tape cassette reader/recorder unit. A punched tape reader may also be supplied for reading tapes to be altered. A cathode-ray tube (CRT) would be convenient to read programming results immediately so that any erroneous data could be erased and replaced with new data at the keyboard. A second magnetic tape will be required to store the corrected version of the tape. Testing of the NC tape would be possible if the computer were instructed through the program to read the information from the magnetic tape and output it directly to the MCU. A tape punch and line printer would be needed to produce permanent copies of the NC program after corrections.

If the peripherals shown in Fig. 18-22 are available, the operator decides what functions are required and uses coded words such as "PUN" to communicate via the keyboard to the computer. PUN means punch a copy of the NC program that was last recorded on magnetic tape. "DIS" would mean display on the CRT the program that was last recorded on magnetic tape. "COP" would mean copy the information from the punched tape onto magnetic tape recorder No. 1. Other instructions may also be necessary, as in the case of "DIS." The beginning and ending points of the part of the tape to be read should be given.

The minicomputer may also communicate with a larger computer at a remote location. This would then give the minicomputer the capabilities of the large computer to allow writing very powerful NC programs such as APT (automatic programmed tool).

Microprocessor. The trend toward ever smaller computers has resulted in the microprocessor. The microprocessor is often referred to as "a computer on a chip." However, this is erroneous since every computer should have a CPU (central processing unit), a memory, and an I/O capability.

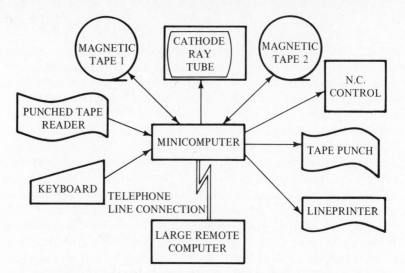

Figure 18.22. A minicomputer with peripherals can be the heart of an NC control system that provides maximum flexibility.

Figure 18.23. Shown at the top is a 40-pin microprocessor integrated circuit chip. Shown below is an integrated circuit that has had part of the top removed to expose the chip in the center. As can be seen, the bulk of the circuit is made up of getting leads out from the chip. This particular chip is made to handle 256 8-bit words.

When the CPU is on a single chip, it is a microprocessor, and a computer based on a single chip is called a *microcomputer*. A 40-pin microprocessor and cutaway of a semiconductor memory unit are shown in Fig. 18-23.

Because of their low cost, microprocessors present an extremely attractive way to incorporate the advantages of a computer in an NC control system. Some modern computer numerical control (CNC) systems include several microprocessors in parallel to increase the power and speed of computation. Other systems combine minicomputers and microprocessors.

The basic element of all three types of computers is the "word," which is composed of a number of "bits." The word is the largest number of bits that can be manipulated as an entity as, for example, the command X03500 would be a six-bit word.

Sometimes computer specialists work with another group of bits called "bytes." This is a somewhat flexible term referring to a sequence of bits operated on as a unit, a subset of a word. Many microcomputers use four-bit words, although eight-bit words are most common. Minicomputers use mostly 12- or 16-bit words.

COMPUTER CONTROL

The three main parts of a computer are the memory unit, the control unit, and arithmetic logic unit (Fig. 18-24).

Computer Memory Storage Devices. Computer memory refers to an area where data or given programs are stored. The speed and capacity at which information can be channeled in or out largely determines what type of storage system is used. Magnetic type storage devices consist of the magnetic core, discs, magnetic tape, and floppy discs. Solid-state storage consists of semiconductor and bubble memories.

Magnetic Core. The ferromagnetic rings that make up a computer memory system are shown in Fig. 18-25. As electrical current is sent through the wires,

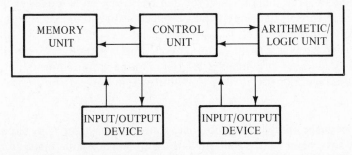

Figure 18.24.   The main parts of a computer are the control unit, memory unit, and arithmetic logic unit.

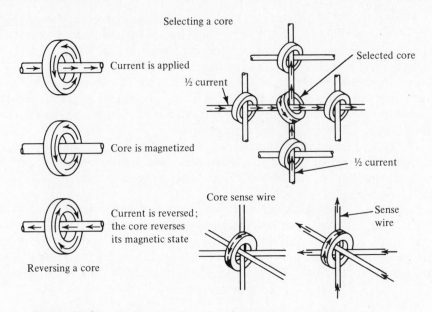

Figure 18.25. The ferromagnetic rings that make up the core of a computer memory.

the cores become magnetized. The direction of the current determines the polarity of the cores. Reversing the direction of current reverses the polarity of the magnetization. This ability to change polarity makes the core suitable as a binary storage element; its two magnetic states can be used to represent the binary digits 0 or 1, which are used for storing computer information. Because any specific location of storage must be instantly accessible, the cores are arranged so that any combination of ones (1s) or zeros (0s) representing a character can be written magnetically or read back when needed. In the grid work, two wires run through each core at right angles. If half the current needed to magnetize a core is sent through each of two wires, only the core at the intersection of the energized wires is magnetized. No other core in the grid is affected. By this principle, a large number of cores can be strung on a screen of wires, and yet any single core in the screen can be selected for storage as reading without affecting any other core.

The major drawback to magnetic-core memories is their cost, which includes quite a bit of auxiliary circuitry needed to get at the stored information. For this reason, core memories are always limited in size in comparison to other storage devices.

Ferromagnetic Discs.   Magnetic disc storage consists of thin, circular metal plates that have been coated on both sides with magnetizable material. Data are stored and retrieved by read-write heads positioned over the face of the disc.

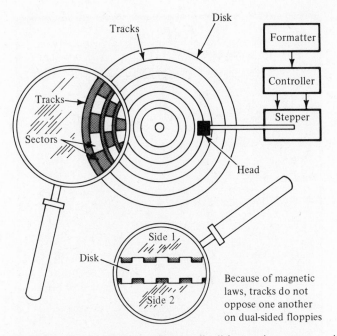

Figure 18.26. The discette or "floppy disc" is used to store coded information that can easily be recorded or retrieved.

Flexible Discs. One of the more recent memory units has been the flexible or "floppy disc." It was introduced as a low-cost minisized random access memory (RAM). The floppy disc has become a popular data input medium for NC machines, replacing punched tape. In appearance, the disc or discette is very much like a thin 45-rpm record. It has a ferromagnetic coating so that data can be recorded as magnetic pulses arranged in concentric circles and pie-shaped sectors on either or both sides (Fig. 18-26).

Semiconductor Memories. Semiconductors or large-scale integrated circuits (LSI) are able to store information in the form of simple electronic charges on chips of silicon. A typical memory chip is less than a quarter of an inch square and can contain 16,000 bits of information (Fig. 18-27). Stored information on the individual chip is accessed by accompanying circuitry located on the same silicon chip. These wonders spawned the microcomputer. The main drawback of semiconductors is that they are volatile; however, small backup batteries can be used to ensure memory retention in case of power failure.

Newer Memory Technologies. For many years computer builders had a limited range of options for trading off memory cost and speed. This limitation was caused by "access-time-gap" (time required to move data into and out of memory), which ranged between very fast but expensive semiconductor mem-

Figure 18.27.   An enlarged view of a single chip used as a memory device that can store 16,000 bits of information. The actual physical size of this chip is approximately ¼ × ¼ in. (*Courtesy Intel.*)

ories and lower cost magnetic discs and tapes. (Semiconductor memories have an access time of less than 1 microsec; magnetic memories have access times of about 5×10^3 microsec.) Two newer technologies developed by Bell Telephone Laboratories seem destined to fill this vacancy; these are CCDs (charged coupled devices) and magnetic-bubble memory.

CCD.   The CCD is a memory structure formed by growing a silicon-dioxide insulating layer on the face of a silicon substrate whose charges can be controlled by applying appropriate bias voltage to the substrate.

CCDs are predicted to double their memory capacity each year to a projected 4-M bit memory by 1986. The present access time is expected to be reduced from the present 400 microsec by 75%.

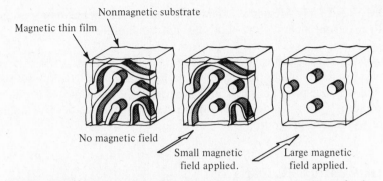

Nonmagnetic substrate

Magnetic thin film

No magnetic field

Small magnetic field applied.

Large magnetic field applied.

Figure 18.28. Magnetic bubbles are formed on a garnet wafer by applying a magnet field. The applied field expands the magnetic regions of the same polarization as the field and shrinks regions of opposite polarity. Sufficiently large fields shrink opposite polarity regions into circular cylinders called magnetic bubbles. (*Courtesy Machine Design.*)

Magnetic-bubble Memory (MBM). Although fabricated by the same process as CCD devices, magnetic-bubble memory devices are not semiconductors. Thin films of magnetic garnet are grown on a nonmagnetic garnet substrate that has randomly distributed serpentine-shaped magnetic domains. When a magnetic bias field is applied perpendicular to the thin films by two permanent magnets placed on each side, the domains shrink into microscopic "bubbles" (Fig. 18-28). The bubbles are cylindrical magnetic domains of fixed volume whose polarization is opposite to that of the film. The presence or absence of a bubble at a given position in a loop corresponds to the logic bit, 1 or 0.

Permanent magnets maintain the bubbles when memory power is lost. Data can remain intact for at least 100 years after power loss. This makes MBM ideal for mass storage of information in portable computer subsystems or for any other application requiring 300-k bit to 3-M bit memory storage capacity.

A 1-M bit device on a chip about 1 cm² has been developed by Rockwell International. Both Rockwell and Texas Instruments have 256-k bit MBMs that are commercially available.

Computer Memory Types. There are three main types of computer memory: R/W, ROM, and EPROM.

R/W (Read/Write). Because of the nature of the operation the content of magnetic-core memories is destroyed during the normal reading cycle; therefore the data must be replaced or rewritten into the memory after it has been read. R/W memories are able to read (sense the information) or "write" data into the memory. The major limitation of R/W memories is that they must have a constant power source or they will "forget."

ROM (Read Only Memory).   The read only memory has data permanently built into it during manufacture that may not be altered.

EAROMS (Electrically Alterable Read Only Memories).   These memories are similar to ROM memories except that programs may be rewritten many times. The original program must first be erased by exposure to an ultraviolet light through a quartz window. It would take about 10 years of ordinary light exposure to erase a program.

ADVANTAGES OF COMPUTER CONTROL

1. The computer eliminates some of the mechanics of program execution. If 100 parts are to be machined on a hard-wired unit, it would mean that the tape would have to be read 100 times. Reading from the computer memory will not involve any mechanical movement; only electronics will be involved.
2. Workpiece programs can be quickly optimized via editing facilities and the computer memory.
3. Perhaps the most significant advantage of computer use is the ability to expand into the overall concept of computer-aided manufacture (CAM).
4. Experience has shown that machines operated by dedicated computers, as in CNC (computer numerical control), require less overhaul and maintenance than the older hard-wired variety.
5. Computers on the factory floor also serve as an instant communications network for such information as parts completed, parts in progress, parts waiting, etc.

SOFTWARE AND POST PROCESSING

Software.   Software refers to a collection of programs, routines, and documents associated with the computer. Programs are compiled by the programmer as the programmer writes a specific set of instructions to machine a workpiece (Fig. 18-29).

Post Processing.   The computer is not equipped to handle the instructions as written by the programmer. The basic intelligence required to change the program into computer language is usually termed a "post processor." As an example, if the programmer wishes to program an arc, the processor must have an arc generation capability after being given the appropriate coordinates and radius.

Every bit of information that makes up the processor language occupies some of the available logic space in the computer. If the processor requires 10K of words and the computer has a 16K capacity, then only 6K words of computer memory will be available for programming. There is now a trend to use a small computer for processing a workpiece program while all processor language storage is kept in an auxiliary piece of equipment, such as a disc pack.

PROCESSOR LANGUAGES

The large number of processor languages now available can be confusing to those who are uninitiated; however, some help is given by the fact that they can be divided into three broad groups; generalized, specific, and parametric.

Generalized Processor Language.   Generalized processor languages have one thing in common: they must be post processed in order to adapt their output to the specific machine tool or control unit combination that will be employed to produce the workpiece.

In the early days of NC each aerospace company tried writing its own processing language. They soon found this was too time consuming; therefore the Aerospace Industries Association (AIA) pooled all of their talent to broaden the scope of the oldest and most powerful NC processor language—APT (automatic programming of tools). APT is a syntax of English—like words that have specific and precise meanings. All of the words are limited to six letters and include such terms as fedrat (feedrate), from, point, plane, golft (go left), gorgt (go right), circle, past, etc. Some are descriptions of geometric shapes or patterns and others are of cutting tool movements. The part programmer must intimately know the vocabulary of the processor language being used.

Specific Processor Language.   Specific processor languages are designed for a specific type of machine tool or situation. There are languages specifically designed for drills, lathes, milling machines, sheet metal punching machines, and inspection and drafting machines. The main advantages are that they require smaller computers, less programming skill, and may not require post processing if the processing is designed for a specific machine or control unit combination.

Parametric Languages.   Parametric languages, also called parts generator, are limited to one-of-a-kind part or parts made on a specific type of machine tool. An example of an application of this type of language would be for cutting cams, where a great many coordinate locations are required to describe the shape within acceptable tolerances. Although limited in scope, these languages are extremely easy to learn, are often developed by users, and require only a small computer. There are other part generator processors where the user merely describes the outline of the finished workpiece and the language will then enable the computer to generate all the machine instructions to produce the part. This approach is frequently used with lathe work.

A newer version of parametric programming involves a keyboard terminal, minicomputer or microprocessor, and other peripherals such as a tape punch and a plotting unit. The parameters capable of being programmed usually involve geometric elements such as lines, circles, arcs, points, etc. Programming is done in answer to a series of questions that appear on the CRT or print unit. By answering the questions, the programmer actually creates a program that will be used to make the desired workpiece.

FINISHED PART:

SIDE	LENGTH	REF. NO.	DIAMETER	ANGLE	RADIUS A	RADIUS B	DEPTH	FLAT	GROOVING	THREADING	TYPE	PITCH	RH/LH	B	R	T
S	L 2.478	N	D 3.000	A	RA	RB	UD	UF	GS	TS		P				
S	L 2.978	N	D 2.000	A	RA .5	RB	UD	UF	GS	TS		P				
S	L 3.231	N	D 2	A	RA	RB	UD	UF	GS	TS		P				
S	L 3.938	N	D 1	A	RA -.5	RB	UD	UF	GS	TS		P				
S	L 3.939	N	D 1	A	RA	RB	UD	UF	GS	TS		P				
S	L 4.064	N	D 1	A	RA	RB	UD .065	UF .125	GS 400	TS		P				
S	L 4.5	N	D 1	A	RA .031	RB	UD .065	UF .125	GS 400	TS		P				
S	L 5.013	N	D 1	A	RA .050	RB	UD .065	UF .200	GS 400	TS		P				
S	L 5.875	N	D 1.0	A	RA	RB	UD	UF	GS	TS 300	N 3	P 12	R H		R	
S	L 5.938	N	D	A -45	RA	RB	UD	UF	GS	TS		P				
S	L 5.938	N	D .7174	A	RA	RB	UD	UF	GS	TS		P				
S	L	N	D .6874	A -135	RA	RB	UD	UF	GS	TS		P				
S	L 5.438	N	D .6874	A	RA	RB	UD	UF	GS	TS		P				
S	L 5.438	N	D .5	A	RA	RB	UD	UF	GS	TS		P				
S	L 4.438	N	D .5	A	RA	RB	UD	UF	GS	TS		P			R	
S	L	N	D	A	RA	RB	UD	UF	GS	TS		P				
S	L	N	D	A	RA	RB	UD	UF	GS	TS		P				
S	L	N	D	A	RA	RB	UD	UF	GS	TS		P				
S	L	N	D	A	RA	RB	UD	UF	GS	TS		P				
S	L	N	D	A	RA	RB	UD	UF	GS	TS		P				
S	L	N	D	A	RA	RB	UD	UF	GS	TS		P				
S	L	N	D	A	RA	RB	UD	UF	GS	TS		P				
S	L	N	D	A	RA	RB	UD	UF	GS	TS		P				
S	L	N	D	A	RA	RB	UD	UF	GS	TS		P				

,END ;   CLFILE , PRINT   ; ;   CLFILE , PLOT   CLFILE , SAVE ;   SCALE ,   FLIP ;

SHELDON MACHINE CO., INC. *Builders of precision machine tools*   4258 N. KNOX AVENUE CHICAGO, ILLINOIS 60641

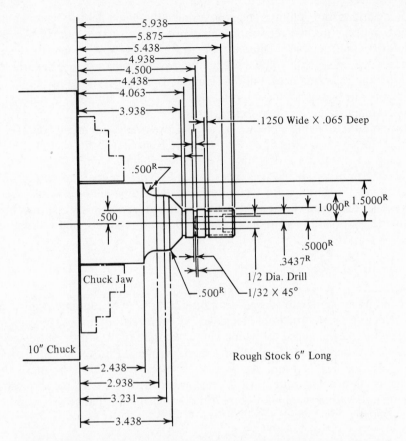

Figure 18.29. An example of a part drawing dimensioned for NC and the corresponding program (opposite page).

Voice Programming

A speech processor has been developed that converts a programmer's analog voice signal into the digital language of a computer, which in turn permits part programs to be generated.

The programmer can enter data by voice directly from a part drawing. The voice commands are processed, formatted, and re-presented to the programmer on the CRT screen for verification. Afterward the commands are entered into the NC processor. A prompter then requests the next bit of information, and eventually a complete part tape is punched.

The operator wears a close-talking noise-canceling microphone. From the microphone output, a wave spectrum analyzer derives features from these signals to indicate the overall spectrum shape. Combinations and measurements of these values are then processed in hardware to produce a set of 32 significant acoustic features.

A total of 512 bits of information are required to store a feature map or reference pattern for each utterance. In training the machine, the system extracts a feature matrix for each repetition of a word. A consistent matrix of features are stored in the memory. Thus an operator "trains" the system to recognize and accept the operator's voice by repeating each command word several times. Thereafter the operator need only be identified prior to a program input.

The operator uses common terms such as "turn," "thread," "mill line A," etc., plus dimensional and coordinate numbers.

In addition to being able to check the program statements on the CRT screen, a plotter shows the operator the exact coutours of the part programmed and a printout produces a description of the machining steps.

A tape can be produced five times faster by voice numerical control (VNC) than even by the large-scale computers used by time-sharing firms. Another advantage is that direct labor overhead is reduced because VNC draws heavily on machine shop skills and reduces the need for special computer skills.

EVALUATION OF PROCESSOR LANGUAGES

One of the most significant decisions a purchaser of NC equipment must make is the selection of a processor language. A hierarchy of processor languages developed by the Industrial Development Division of the Institute of Science and Technology at the University of Michigan are shown in Fig. 18-30. Studies show that if there are five or more NC machines in the plant, there is more than a 90% chance that a computer will be needed to develop workpiece programs. Even with two to four machines, there is more than a 50% chance that a computer will be needed.

Figure 18.30. A hierarchy of computer languages. At the top of the list are those used specifically for NC. These languages require the most computer translation effort and thus utilize more of the computer logic. They are also the easiest for the programmer to use. At the bottom of the scale are the simple binary notations that require the least computer logic but are almost impossible for a programmer to use without a computer.

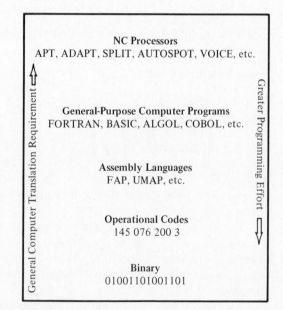

There are three broad areas to consider in choosing the best NC processor language:

1. The geometric capability of the language. If more than one type of machine is used will the language be able to work on all of them?
2. The method in which the language is made available. Is it limited to time sharing? Is it available only on an in-house computer? Does the user buy the language or only the rights to use it? How often is the language updated? Can the user make changes? These kinds of considerations are applicable to the post processor as well.
3. Programming tryout. Once satisfactory answers have been obtained for the first two points, the user should select a number of representative workpieces for which the language will be used. The parts should be programmed and run the same as production parts so that the actual use of the language can be tested in the user's environment. A few errors may be programmed in to check the processor's language capability at error diagnosis. This will provide an opportunity to observe the method of handling them.

COMPUTER CONTROL OF NC

Computer controls range from massive systems with hierarchies of smaller computers to interpret and disseminate the logic from the larger computer to the individual machines. The large central computer may have enough capacity to

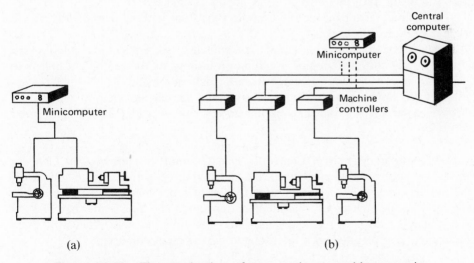

(a)   (b)

Figure 18.31. The production of two or three machines can be controlled directly by a minicomputer, as shown at (a). In larger, more elaborate systems programmable machine controllers are used for each machine and these may in turn be interfaced with a minicomputer or, in extremely large systems, a central computer.

handle several hundred machines. The second level may handle a dozen different machine tools and a small unit that handles only one or two machines (Fig. 18-31). In such a system, there is usually a remote communications unit at each machine tool so that conversation may be carried on with the central computer. The large systems are of course powerful and expensive and represent the frontiers of manufacturing technology, computer numerical control (CNC), and direct numerical control (DNC).

CNC AND DNC

CNC. The distinguishing feature of CNC is a "dedicated" computer, usually a minicomputer, to one specific machine tool or perhaps to two or three simple machines.

The computer is not used for workpiece program generation. Instead, a workpiece program that has been written and is "stored" in the computer and "read" as punched tape would be read in the execution of a program to machine a workpiece.

CNC units look very much like standard hard-wired NC units except the CNC unit has a complete computer module or microprocessor as part of its working system. Programming is also quite similar, by tape, or directly from the keyboard to the microprocessor. It is possible to "read in" a length of programmed tape that may average 100 to 200 ft long, which is often enough for one or more complete programs. Once in the memory, the tape will not have to be run again until the workpiece geometry is changed.

Editing may also be done right on the shop floor without even involving the original workpiece program.

Some CNC executive programs (permanently stored in the memory) are written so that they will compute common routines, as for example a bolt hole circle or a pocket milling sequence from a single command.

As CNC applications continue to grow and gain acceptance, machines will become easier to program and more memory can be added to the soft-wired controls.

DNC (Direct Numerical Control). DNC is another step beyond CNC in which a number of CNC machines are connected directly to a remote computer. The number may be small or as large as 100 or more machines. The program tape is fed to the control via a communications line. The tape reader is bypassed or eliminated, and as a result the machines can operate more efficiently since commands have already been interpolated by the main computer.

PROGRAMMABLE CONTROLLERS (PC)

Every machine tool operation or process system has some type of control built into it. For simple operations, a control panel with a set of switches will tell the system what to do and when to do it. As systems become more complex,

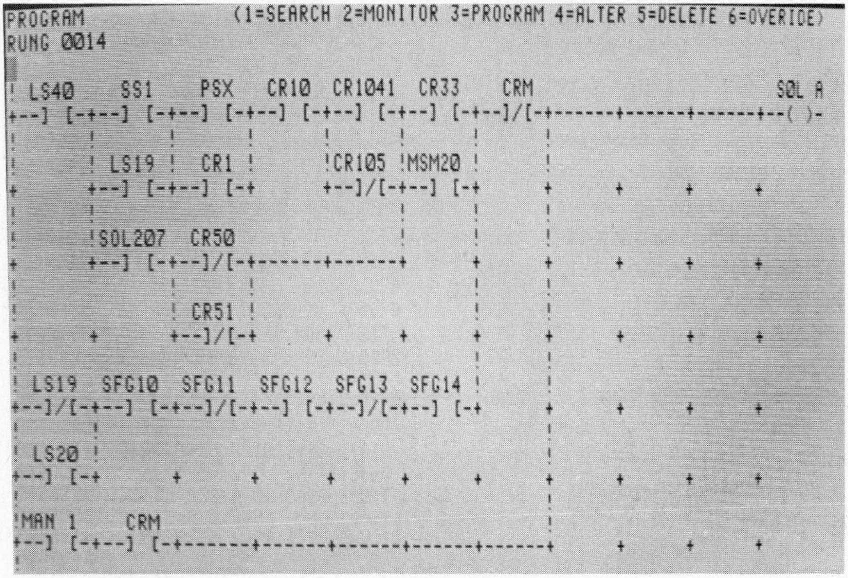

```
PROGRAM            (1=SEARCH 2=MONITOR 3=PROGRAM 4=ALTER 5=DELETE 6=OVERIDE)
RUNG 0014

! LS40    SS1    PSX   CR10  CR1041 CR33    CRM                       SOL A
+--] [-+--] [-+--] [-+--] [-+--] [-+--] [-+--]/[-+-----+------+------+--( )-
!         !       !       !              !       !       !
!         ! LS19 ! CR1 !        !CR105 !MSM20 !               !
!         +--] [-+--] [-+       +--]/[-+--] [-+      +       +       +
!         !               !                    !               !
!        !SOL207 CR50             !        !               !
+         +--] [-+--]/[-+------+------+       +       +       +       +
!         !       !                    !               !
!         ! CR51 !                            !               !
!         +       +--]/[-+      +       +       +       +       +       +
!         !                                   !               !
! LS19 SFG10 SFG11 SFG12 SFG13 SFG14 !       !
+--]/[-+--] [-+--]/[-+--] [-+--]/[-+--] [-+      +       +       +
!                                         !               !
! LS20 !                                  !               !
+--] [-+      +       +       +       +       +       +       +       +
!                                         !
!MAN 1    CRM                             !
+--] [-+--] [-+------+------+------+------+------+      +       +       +
```

Figure 18.32. A programmable controller ladder diagram on CRT display. (*Courtesy Square D Company.*)

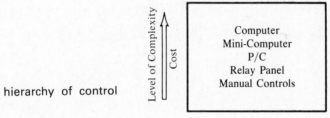

Figure 18.33. A hierarchy of control costs.

automatic controls in the sense of being preprogrammed are needed and may include motor-driven cams, relay banks, solid-state circuits, punched cards, and tapes. These devices are generally referred to as being *hard wired* since they must be physically changed when a change in function is required.

The programmable controller offers the advantage of being able to store an easily modifiable and readable set of controls. A typical program consists of a series of commands, responses to error conditions, checkpoints, and timing restraints set up on a ladder diagram basis (Fig. 18-32).

Minicomputers and microprocessors can do the same thing but they are more expensive and more complex. The computer's potential is also wasted since the demands do not usually approach the capability of the computer. A hierarchy of control system costs is shown in Fig. 18-33.

Essentially a PC is a low-cost, easily modifiable control device. The programmable control is built on Boolean algebra logic. Certain predetermined logic

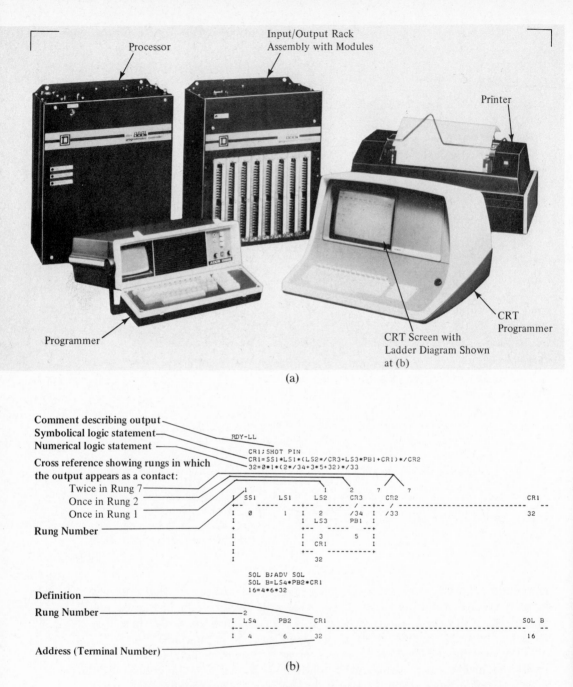

Figure 18.34. A programmable controller (a), and an enlargement of the CRT display and explanation (b). (*Courtesy Square D Company.*)

684

outputs will be triggered by certain inputs. Generally speaking PCs cannot perform calculations but can handle sequential or Boolean logic. Data may be entered either manually through a keyboard that is connected to a CRT or through numerically coded input devices such as punched tape or cards (Fig. 18-34). An enlargement of the CRT ladder diagram display is shown in Fig. 18-34(b).

After the program has been entered, it is stored in the "memory section." Here the program gives specific information to the processor about what normal operating conditions are and what to do to correct discrepancies.

The processor within the PC compares the input signal with the program stored in the memory section sequentially (one parameter at a time as directed by the clock).

Two basic memory types are available, R/W and ROM, as discussed earlier in the computer section of this chapter. Repeated changeability is also available with EAROMS (electrically alterable programmable ROMS).

If any adjustment is required when the signal is fed back into the PC, the control section begins making corrections. Simultaneously, the processor sends correction signals back to the memory section, which in turn is connected to a line printer or CRT so that the operator can immediately see the status of operation.

Advantages of a PC

1. Solid-state devices have about 99% of their problems occur in the first 70 hours of operation, whereas conventional (relay) system failures increase with time.
2. PCs are more rugged and also easier to repair than minicomputers. Repairs can be made by in-house electricians.
3. Most PCs can be reprogrammed on the spot in minutes. This flexibility makes PCs especially useful if product lines change frequently.
4. PCs lend themselves to centralized control. Time formerly taken up by computers can be turned over to the PC. If desired, the PC can act as a computer interface.

CAD/CAM

Computer technology has risen in an exponential curve. Only a generation ago computers were not really needed for NC control. Except for relatively few five-axis machines, NC was point to point, which was easily accomplished with manual programming.

With the advent of two-dimensional profiling, however, a need arose for numerous calculations to generate the curved profile. At that point the computer was looked upon as a useful tool in manufacturing engineering. After some development, it was found that punched or magnetic tape could be dispensed with entirely in favor of computer programs that controlled the machines directly. Then came the minicomputers, which when operating with the central processor brought about the first real concept of the automatic factory.

All of these developments were important steps in the advance of technology. The computer is no longer viewed as a remote piece of hardware used primarily in accounting and engineering departments, but instead as an important tool to be used increasingly in CAD (computer-aided design) and CAM.

Today CAD and CAM have become closely linked for complete processing from design to production. The design engineer can now create the desired design on a CRT screen that is linked to a computerized graphics plotter. The same system can be used to determine the viability of the design by analyzing stresses and deflections that will take place under loading conditions. The functioning of mechanical linkages may also be checked. After the design is complete, the production group can draw on the information developed in CAD for computer-aided manufacture. Thus the unified CAD–CAM team has been developed to bring the product from the design concept through manufacture.

CAD Functions

Initially CAD systems were primarily automated drafting stations in which computer plotters produced engineering drawings. The systems were later linked to graphic display terminals, which allowed the user to communicate with the computer in pictures instead of raw columns of numbers. Now advanced systems have added analytical capabilities. CAD functions may be grouped into four major categories: geometric modeling, engineering analysis, kinematics, and automated drafting.

Geometric Modeling.   Traditionally, engineers do the sketching and draftsmen the drafting. Engineers then discover needed changes or improvements and draftsmen draft what is required. Moving a bracket 2 in. to the left causes an entire chain reaction throughout a complete set of drawings.

Now the computer has opened up a new way to meet design engineering needs with speed and reliability in products ranging from high-performance aircraft to complex, integrated circuits. The designer can sketch directly on the screen of a graphics terminal. Lines and contours are defined by pressing keys and positioning a light pen, and the computer displays what is expressed.

Curve fitting, or reducing the design to a set of control equations, is done at the terminal, eliminating coding, card punching, and repeated computer batch runs; then CAD converts the preliminary design to a dimensional drawing with auxiliary views.

If the drawing reveals a problem of compatibility with a subassembly, the pieces can be moved around with the light pen (Fig. 18-35). The computer then converts this pictorial representation to a mathematical model stored in the computer data base for later use. The model may be used for other CAD functions or it may be recalled and refined by the engineer at any point in the design process.

Geometric modeling is often considered to be the most important feature of a CAD system because so many other design functions in the system depend heavily on the model. The geometric model may be used to create a finite element model of a structure for stress analysis, as will be discussed later. The model

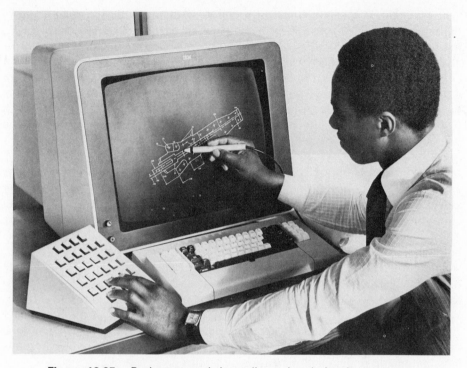

Figure 18.35. Both two- and three-dimensional sketches are produced on the CRT screen. Light pens may be used to interact directly with the image on the screen to place designated symbols or "draw" designated lines. (*Courtesy IBM*)

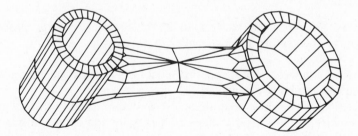

Figure 18.36. Most geometric modeling is done by wire frames as shown. Some CAD/CAM systems create processing tapes directly from the model. The program is then verified at the terminal by computer simulation.

may also serve as an input for automated drafting to produce engineering drawings. If CAD/CAM systems are interfaced, the geometric model can be used to create NC tapes for fabricating the part.

Most modeling is done with a wire frame that represents the part shape with interconnected line elements, as shown in Fig. 18-36.

Analysis.   After the geometric modeling is done, the CAD system allows the designer to move directly to analysis. With simple keyboard instructions, the computer is asked to calculate weight, volume, surface area, moment of inertia, or center of gravity of a part. However, the most powerful method of analyzing a part is by the finite element method. In this technique the structure is broken down into a network of simple elements that the computer uses to determine stresses, deflections, and other structural characteristics (Fig. 18-37). Such analysis requires the tremendous computational power of a mainframe computer.

Kinematics.   Many CAD systems have kinematic features for plotting or animating the motion of simple, hinged parts such as doors or cranks. Such analysis can ensure that moving components do not impact on other parts of the structure. The design of an automobile hood linkage, for example, requires only a few minutes of interaction with a computer, whereas hundreds of hours can be required for manual design of the same part.

Experts believe that in the future computers and programs will be available to completely synthesize a complex mechanical system based solely on its intended function.

Drafting.   With automated drafting, detailed engineering drawings may be produced automatically from a data base. In addition, most drafting systems have automatic scaling and dimensioning features.

A typical CAD system with automated drafting can generate up to six views on the screen. Any design change made on one view is automatically added to the others.

CAM Functions

CAM functions center around four main areas: numerical control, process planning, robotics, and factory management.

Numerical Control.   As stated previously NC uses coded information to control machine tool movements.

The computer can now generate an NC program directly from a geometric model of a part. Automatic capabilities are generally limited to highly symmetric geometrics and other specialized shapes.

Process Planning.   Process planning is involved with the detailed sequence of production steps from start to finish. Essentially, the process plan describes the state of the workpiece at each work station. The use of computers as an aid to process planning is comparatively recent and has led to a rebirth of what is known as group technology (GT). Group technology is based on organizing all similar parts into families to allow standardization of fabricating steps. Just how this is done is explained in detail in the next chapter.

Currently under development is a process planning system that is able to produce process plans directly from the geometric model data base with almost

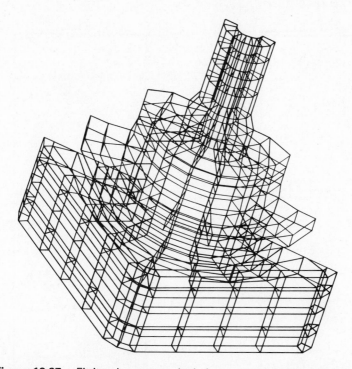

Figure 18.37. Finite element analysis is greatly aided by computer graphics. Models such as this created with the UNISTRUC system from Control Data Corporation are represented by a network of line elements that the computer converts to a mathematical description for a subsequent analysis. (*Courtesy* Machine Design.)

no human assistance. In this system, the process planner would review the impact from the design engineer via computer communication and then enter this input into the CAM system, which would generate a complete set of process plans automatically.

Robotics.   Many advances are being made to integrate robotics into CAM. One of these efforts is the U.S. Air Force Integrated Computer-Aided Manufacturing (ICAM) project, where the goal is to organize every step of manufacturing around computer automation. As part of this program, a robot is used to drill sheet-metal aircraft parts. The robot selects drill bits from a tool rack, drills a set of holes to 0.005 in. tolerance, and machines the perimeter of any one of 250 types of parts. Production rates are four times faster than conventional manual fabrication.

Factory Management.   This portion of CAM ties together the other areas to coordinate operations of an entire factory. The management system relies heavily on group technology with its families of similar parts. Computers also perform various management tasks such as inventory control and scheduling in

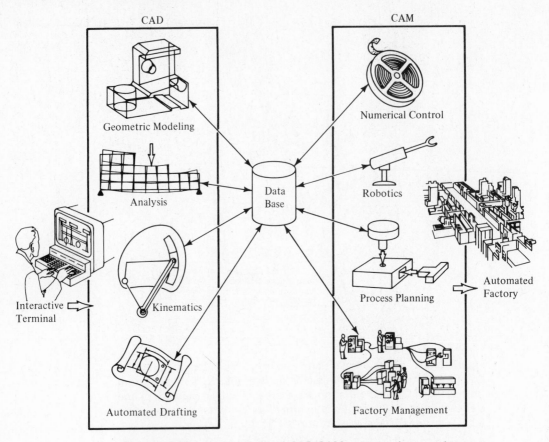

Figure 18.38. In an idealized CAD/CAM system, the user interacts with a computer via a graphics terminal where the part may be designed from start to finish. A three-dimensional model may be constructed from which an analysis and kinematic study can be made. With CAM, the user creates NC instructions for machine tools, produces process plans for complete manufacture and assembly. (*Courtesy* Machine Design.)

systems that have come to be known as material requirements planning (MRP) systems.

A diagram (Fig. 18.38) shows how all the CAD/CAM functions are tied together by a common data base.

THE AUTOMATIC FACTORY

The automatic factory has been the dream of those concerned with progressive manufacturing methods for many years. This goal has not yet been achieved but lightly manned factories are now in existence. Enough castings are manually loaded on pallets during one shift to run two shifts. The castings are moved in

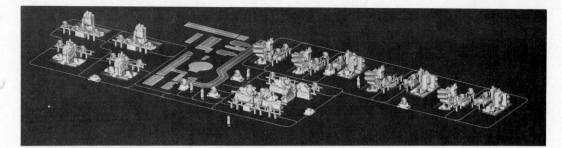

Figure 18.39. Rough castings enter the integrated manufacturing system of ten NC machines where 400 metal-cutting operations are performed on 17 different parts. Virtually the entire spectrum of machining operations is covered in this line. (*Courtesy Cincinnati Milacron.*)

and out of machining stations automatically. The process is monitored by TV cameras, which may have the receiver at home or at the plant.

Although not considered an automatic manufacturing operation, the system shown in Fig. 18-39 comes close to it. In this CAM-type operation, rough castings enter one end of the line of 10 NC machines and are completed 148 ft away. The system shown can handle 17 different parts and perform over 400 machining operations.

CAM operations of this type have the versatility of general purpose machines combined with efficiency close to that of a dedicated transfer line. As discussed previously, a transfer line is a high production automatic type of machine, but it is made for only one particular part such as automobile engine blocks.

A typical computer manufacturing system (CMS) with six NC machines may have 6 to 10 different parts in process at the same time and the batch size for each part may be from one to several hundred.

CMSs are very expensive, sometimes costing several million dollars for the equipment alone; therefore their effective operation is of prime importance.

Both NC and the computer have improved tremendously over the last two decades. The computer has been applied to many design and manufacturing problems that go far beyond that of generating programs for NC machine tools. It is now being realized that the computer, not numerical control, is the key to the new manufacturing revolution.

Problems

18-1. What would be the coordinates of a point in quadrant 3 that is 2.125 units from the x axis and 3.250 units from the y axis?

18-2. A point having a plus sign for the x coordinate could be in either of which two quadrants?

18-3. The coordinates of a point are $x = +4.250$, $y = 3.500$. A second point lies 6 units directly to the right of this point and down in the $-y$ direction a distance of 4.750 units. What are the coordinates for the second point?

18-4. Suppose the *xy* plane is 3.000 in. above the work table. How many inches above the work table would a point be that has co-ordinates *x* = − 10.000, *y* = +7.750; *z* = − 2.125?

18-5. (a) Considering the *xy* plane only, point *A* is located at coordinates *x* = +8.000; *y* = +4.000. Point *B* is located at co-ordinates *x* = −2.000; *y* = −4.000. What are the incremental *x* and *y* dis-tances when moving from point *A* to point *B*?
    (b) Show points *A* and *B* on coordinate graph paper, label each axis.

18-6. How would the following X-Y coordinate numbers appear on tape?
    (a) *x* = +6.250,
    (b) *y* = − 12.500,
    (c) *x* = −.005,
    (d) *y* = +0.10.

18-7. Refer to Fig. P18.1 and identify the coor-dinate points. (Note + signs need not be shown.)

18-8. How can machine design users be reason-ably certain that the NC cutting tool is fol-lowing the part program when using an open-loop system?

18-9. What type of interpolation is applicable for sheet-metal dies?

18-10. When is it imperative to have a closed-loop system?

18-11. Give an example of when canned cycles would be useful.

18-12. How is the magnetic-bubble memory formed?

18-13. Why do MBMs have great potential for solid-state devices?

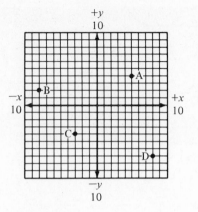

Figure P18.1.

18-14. What is the advantage of an EAROM mem-ory over an ROM type?

18-15. What factors make a machining center versatile?

18-16. What programming language would be used for cutting surfaces where many coordinate locations are required?

18-17. How has CAD changed engineering design?

18-18. Compare the access time gap of CCDs and semiconductors.

18-19. What is the function of a post processor?

18-20. What is the advantage of having a specific processor language?

18-21. What is the main difference between a transfer line and FMS?

18-22. What is meant by software?

Case Study 1

The manufacturing engineer must decide which machine to use for a gear box cover plate. Standard times show the part could be made by conventional machining in 6.985 hr and by an NC machine in 3.16 hr. In order to machine the part in 3.16 hr on the NC machine, an index table would be required at a cost of $2000.00. It was decided that the cost of the table should be amortized over the cost of the 100 parts. The tooling cost for the conventional machines was $1000.00.

The manufacturing engineer constructed a break-even chart to find how many parts would have to be made before the extra tooling cost for the NC machine would be paid for. Show, by constructing a chart, where the break-even point occurs. The cost per hour on the NC machine is $25.00 and $15.00 on conventional machines.

Case Study 2

Company XYZ had been producing jet engines since they were first developed. The engine entails many small groups of complex parts, most of them precisely machined to close tolerances. Thus most of the plant functioned as separate job shops on a massive scale.

The company decided that to run these shops more efficiently a good communications system was needed—that is, letting each machine operator know what needs to be done and making sure he has everything he needs to do it.

Numerical control has already optimized the metal-cutting operation once the part is in the machine. DNC has already optimized the method of getting the NC program to the machine.

The task remained of using CAM to optimize the rest of the process. At present a typical machine tool is in the production cycle about 45% of the time. The long-range plan of CAM is to increase the production cycle time.

The first part to be tackled was a rotating part because the operation encompassed about 100 NC machines.

Plan a system using DNC, machine monitoring, machine diagnostics, quality control, and other elements that will provide better communication to the job shops and will increase production.

Show, with a schematic sketch, how the machines and computer communication system could best be organized in order to provide constant input to each machine as well as a feedback monitoring system. Computers available are mainframe (on a shared basis with the office), master minicomputer, DNC minicomputer, and microcomputers.

Bibliography

Barash, M. M. "The Future of Numerical Controls." *Mechanical Engineering* (September 1979).

Childs, James L. *Principles of Numerical Control.* New York: Industrial Press, Inc., 1979.

Cox, Gerald. "A Revolution in Computer Memories: Storing Data in Magnetic Bubbles." *Machine Design* 51 (March 8, 1979): 60–65.

Delarosa, J. G.; J. Carlos Goti; and Susan M. Holic. "Computer Controlled Manufacturing." *Machine Design* 44 (September 7, 1972): 126–131.

Gettelman, M. M., ed. "Fundamentals of NC/CAM." *1980 NC Guidebook, Modern Machine Shop.*

Hegland, Donald E. "The Many Faces of CAD/CAM." *Production Engineering* (June 1979).

Krouse, J. K., ed. "CAD/CAM: Bridging the Gap from Design to Production." *Machine Design* 52 (June 12, 1980): 117–125.

Moorhead, J., ed. *Numerical Control Fundamentals.* Dearborn, Mich.: Society of Manufacturing Engineers, 1980.

——. *Numerical Control Applications.* Dearborn, Mich.: Society of Manufacturing Engineers, 1980.

Poisson, Normand. "Buying a Programmable Controller: Look for Flexibility." *Automation* 21 (April 1974): 68–74.

Schaffer, George, ed. "Computers in Manufacturing." *Special Report 703, American Machinist* 122 (April 1978): 115–130.

Schaffer, George, ed. "Machining by the Numbers." *Special Report 711, American Machinist* 123 (May 1979): 131–158.

Teshchler, Lee, ed. "Stripping the Mystery

from Microcomputers.'' *Machine Design* 49 (December 8, 1977): 161–170.

Zurica, Tom, ed. *A Numerical Control Glossary.* Wheeling, Ill.: Numeridex, Inc., 1979.

CHAPTER NINETEEN

Group Technology

Mass production is discussed so widely that it is assumed that most manufacturing is carried on that way. Nothing could be further from the truth, however. The Department of Labor estimates that 75% of all metalworking in the United States is done in lot sizes of 50 or less. There are many compelling reasons for these small lot sizes despite the obvious disadvantages. A typical job shop is a random type of business and at first impression it appears chaotic. Every job looks different, the batch sizes vary, and machine loading appears to be guesswork, with delivery times uncertain. Although management is aware of these problems, most job shop managements cannot afford the human resources required for proper process planning for the typically large variety of parts manufactured.

Time is frequently unavailable or else it would be too time consuming to calculate plans for each part; thus plans consist of no more than routing sheets merely indicating the sequence of manufacturing stations or processes to be followed.

With inadequate process planning, it follows that pricing cannot be adequately administered.

Failure of overburdened planners to recognize identical parts, or parts very similar to those previously manufactured, leads to unnecessary process planning when existing plans or their revisions should be used. Rather than standardize an optimum plan, each planner has a tendency to "reinvent the wheel," with

Figure 19.1.   Parts families of similar size and shape that have most of their machining operations in common. (*Courtesy* Manufacturing Engineering.)

the result that two different manufacturing costs may go out for the same or very similar part. One approach to making order out of this chaos is to apply group technology and the associated computer-assisted manufacturing.

Group technology (GT) is a philosophy of manufacturing based on the premise that similarities exist in design and manufacture of discrete parts. Based on this premise, parts are organized into families that require similar manufacturing processing and as a result some benefits of mass production can be achieved in job shop operations.

A part family may be defined as a collection of related parts that are similar. The parts are related by geometric shape and/or size and require similar machining operations. Alternately, they may be dissimilar in shape but related by having all or common machining operations. Figure 19-1 shows an example of a part family.

Total implementation of GT requires three basic steps; formation of parts families, cellular manufacturing organization (machine groups that match parts families), and coding systems. The coding system, now computerized, classifies parts by families that can be quickly identified by key numbers and letters.

The classification of parts into families is not new. At the turn of the century, F. W. Taylor, well known for his work in metal cutting and time study, introduced a classification system similar to that used in GT today. Evidence that the Russians used GT is in a book by S. P. Mitrofanov in 1959, *Scientific Principles of Group Technology*. Many European countries also used the concept, but the work of Professor H. Opitz of Aachen Technical University, West Germany, expanded the concept to include machines, product design, and plant layout.

Edward G. Brisch, an expatriate from Poland, did much to develop the concept in England. His designs for GT have been used in many companies in England and the United States. In Japan, GT has been developed jointly by industry and government (The Japan Society for the Promotion of Machine Tools).

Although GT is usually associated with small lot or batch production, it should be pointed out that the concept can also be applied to mass production if fixed automation (e.g., transfer line machines) is used.

PLANT LAYOUT

Traditionally, metalworking manufacturing plants have used a *functional layout* rather than a *line layout*. The merits of these two plant layouts will be discussed briefly along with *cellular layout* as used in group technology.

Functional Layout.   Functional layout is based on many specialized departments or sections such as the lathe section or grinding department. Each department is manned by persons who specialize in a number of operations for a type of limited work. This type of layout is popular because operators can become skilled in limited repetitive work; work loads can be assigned more flexibly to machines; and both men and machines can be kept "busy" (which is the most compelling reason from management's standpoint).

Figure 19-2 shows a schematic of a functional plant layout contrasted with a cellular layout, discussed later. The workpieces take a nearly unpredictable path through the plant in order to reach all the necessary processing and inspection locations. Production planning and control for a functional layout can become quite complicated.

Disadvantages of Functional Layout

1. There is considerable lost motion in moving parts from one machine department to another according to the process requirements.
2. Individual machines are loaded with an "economical order quantity" greater than the immediate needs in an attempt to optimize machine operations in which setup times are relatively long compared to cyclic rates. It can easily take from 30 to 60 days to complete a part run with as much as 95% of the time being consumed by waiting.
3. Individual machines are assigned to produce parts, not because the parts are needed now, but to keep the machines and workers busy for accounting purposes.
4. Excessive inventories of raw materials and in-process work cause:
   a. Storage and equipment problems.
   b. Excessive in-plant movement and handling not related to immediate needs.
   c. Extensive problems of planning and control in attempting to achieve coordination and to avoid losses.
   d. Low morale among production personnel due to uncertainty and confusion

A. COMPLICATED MATERIAL FLOW SYSTEM (Functional Layout)

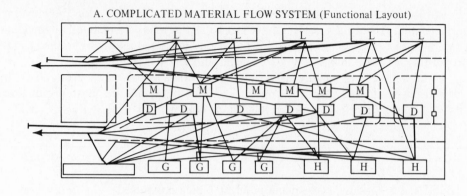

B. SIMPLE MATERIAL FLOW SYSTEM (Work Cell Layout)

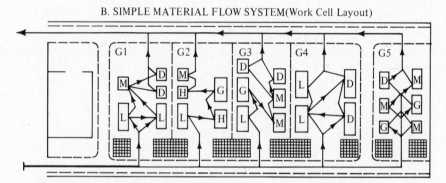

Figure 19.2. A functional layout versus a cellular layout. (*Courtesy MICLASS–TNO and Organization for Industrial Research, Inc.*)

in carrying out their tasks. Occasionally parts simply disappear in the manufacturing maze and delivery schedules slide down the drain.
  e. Pressure on management to buy more equipment so that delivery schedules can be met, even though in casual observance many idle machines can be easily spotted.
5. Parts continue to proliferate since it is often impossible to trace down a print of a given part. Therefore parts are "reinvented."

Functional Layout and Economic Lot Size.   Functional layout is based on the premise that parts can be routed through the plant in economic lot sizes. An economic lot size is based on production volume, machining cycle time per component, and inventory costs.

The normal procedure for determining the economic lot size is to balance the setup change cost with the inventory carrying cost. A simple formula often used is:

$$Q = \frac{S}{(T)(K)}$$

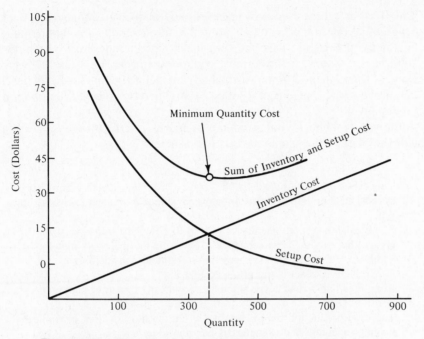

Figure 19.3.   The relationship of setup cost to inventory cost.

where: Q = production quantity
           S = setup time in minutes
           T = standard machining time in minutes
           K = efficiency factor (usually 10%)

The minimum economical quantity is established at the point where the setup cost and inventory cost are equal (Fig. 19-3). For any quantity less than the minimum, the cost curve climbs quickly. For any quantity greater than the minimum, the cost increase is not as rapid. The manufacturing engineer must always be aware of how production and corresponding inventories will influence the competitive status of the company.

Several factors that weigh heavily in favor of a small lot size are:

1. Product features can be changed more easily to take advantage of new developments.
2. Defects can be designed out more quickly.
3. There is less inventory in process.
4. Lead times are reduced, making it advantageous for gaining new contracts.
5. Changes in schedules can be effected at short notice.

Generally, the smaller the batch size, the more sophisticated will be the equipment needed to keep setup times to a minimum. An example is quick change group tooling, in which only a minimum of time is needed to check the fit and alignment of locating pins, rest pads, etc.

Line Layout.   Line layout of machines and auxiliary equipment is the simplest type of layout in that materials are fed in at one end and components leave the line at the other end. This does not mean that the line must be straight. It may take many shapes, with the most common being L, U, S, and O, or combinations of these shapes. The principal advantage of the line layout is the flow-through processing arrangement as often exemplified by powered or gravity parts conveyors.

Although the line layout produces a more orderly progression of parts through the plant, it has some serious deficiencies:

1. The "pace" is determined by the slowest machine, hence cutting speed may be deliberately reduced on some machines to reduce excessive idle time.
2. A breakdown of one machine may lead to complete stoppage of the machines that follow in the sequence. Some standby machines may be kept available, although this increases the investment considerably.
3. Supervision is general and not specialized.
4. Long flow lines create large amounts of work in process or high inventory cost.
5. The total production time is relatively long because of time consumed in waiting (queuing) and in materials handling.
6. Owing to the diversity of jobs in specialized departments, higher grades of skill are required.

Cellular Layout.   The layout used in group technology is somewhat similar to a line layout. However, it overcomes most of the problems mentioned by grouping machines on the basis of manufacturing parts families. All parts requiring similar machines and tooling are processed in a sequence that increases the number of parts per setup.

The machines are organized into cells or, as it is often called, *cellular manufacture*. The overall organization is still that of a line but a given component is completed before it leaves the cell.

Each cell is self-contained and self-regulating. Each member of the cell staff is capable of using several machines or processes so that there are usually fewer operators than machines. The cell usually is made up of about 10 machines and a fewer number of operators who organize themselves to manufacture a particular family of parts.

Benefits and Costs of Cellular Layout.   The personnel advantages are:

1. Having each cell responsible for its own quality control and work movement.
2. Development of supportive work group.
3. Less stress than that caused by the usual interdepartmental friction.
4. A higher morale since each group is aware of its contribution to the overall production.
5. Reduced absenteeism and turnover.

Production Advantages:

1. Reduction in setup time (about 60%).
2. Increased productivity.
3. Reduced work in progress, because of reduced cycle times.
4. Reduced finished inventory, because of smaller batch sizes.
5. Reduced scrap, because of greater operator familiarity.
6. Lower cost and better status information on each part more readily available with less paperwork.
7. Improved management costing, with costs more closely tied to components produced.

Costs of Adopting Cellular Layout:

1. Grouping of parts into families and deciding on the appropriate cell structure.
2. Planning the new layout.
3. Moving existing machinery and installing new machines as required.
4. Redesign of tooling.
5. Design and installation of materials handling equipment within the cells.
6. Modifying production control procedures and paperwork systems.
7. Developing a new operator pay structure.
8. Education of workforce and management.

CLASSIFICATION AND CODING SYSTEMS

As stated previously, group technology is based on families of parts. In order to recognize the similarities of parts in a manufacturing operation, various classification and code systems are used. Three basic methods are used to form parts families. The classification is done by:

1. Design features.
2. Production features.
3. Production flow analysis.

Design Features.   Sorting out parts by looking at their common design features and then placing them into family groups may be the quickest way of defining parts families. This method suffers from the fact that the families may not represent the best machining advantages and may require a high level of skill to obtain the desired results.

Production Features.   Component classification is based on a complete analysis of all part drawings. Family groups will be made based on part size range, geometry, and machining operations. Although this is the preferred method of establishing family groups, it is also the most time consuming.

The drawings and associated production data must be collected together, coded according to a classification system, and then sorted into code numbers. A side benefit of this process is that it builds on design and process standardization.

Production Flow Analysis (PFA).   Existing data for production methods are used to group operations into families. Parts with common operations and routes are grouped and identified as a manufacturing family. An advantage of this method is that family groups can be made without a coding system, since part families are established from operation sheets and route cards instead of part drawings.

Steps used in organizing a PFA system are:

Data collection
Sorting the process routings
PFA chart building
Analysis

Data Collection.   The first step in production flow analysis is to decide on the scope of the study, that is, if all parts made by the plant are to be included. Once the number of parts to be included is decided, the data needed for analysis are gathered. This will include part numbers and machine routing sequences for each part.

Sorting of Process Routings.   The second step in production flow analysis is to sort all the routing sheets into separate groups according to the sequence of operations they represent. If the population is large (thousands of parts) it will be necessary to code the information from the first step onto computer cards. A sorting system can then be set up that will sort the routing cards into groups or "packs" that have the same process routings. Each pack is given a separate identification number.

Process Flow Chart Making.   The information from each pack of cards is made into a PFA matrix type of chart for ease of examination. The machine code numbers represent one axis and the routing code number the other axis. By Xing in the intersection points, certain patterns can be identified.

Analysis.   The analysis of the process flow charts is the most demanding step in the PFA operation and the most crucial. The patterns that developed in the chart grids can be rearranged so that similar packs can be brought closer to their respective routings.

Of course, not all parts will fit well into the process routings of the revised charts. These parts will best be handled in a separate, more functional type of routing.

Advantages and Limitations of PFA.   The process flow analysis system of classification produces the most thorough classification of parts. Once made, it will be useful in eliminating considerable duplication of parts and machining operations.

The PFA method has the disadvantage of relying on existing production data and routing methods, which may not be complete and may entail bad practices that will be perpetuated. There may be ten part numbers and five different process plans for essentially one part. It may take more time to compare these designs and processes than it would to simply create a new part from scratch.

Coding Systems.   Coding is the function of assigning alphabetical or alphanumeric identification to parts drawings. The code numbers identify part geometry, processing information, size, and materials, etc.

Companies often establish their own coding system or may have a system made specially for them by an outside consultant. Detailed information often contained in the code may be divided into design and engineering data and manufacturing information.

Design and Engineering Code Information.

Main shape	Assembly
Material	Prototype building
Major and minor dimensions	Records
Tolerances	

Manufacturing Code Information.

Major and minor operations	Production time
Cutting tools and gages	Surface finish
Jigs and fixtures	Inspection
Lot sizes	Production control
Data processing	Quality control
Setup time	

There are two main coding systems: hierarchical and chain types.

Hierarchical Code.   In the hierarchical code the information represented by each subsequent digit is dependent on the preceding digit, and the code is assigned by a step-by-step examination of part characteristics. This method is often referred to as the "Brisch" system (see bibliography at the end of the chapter).

Chain Type Code.   The chain type of code has a fixed digit significance where a given digit value always represents the same feature. Codes will often combine both systems, hierarchical codes being used for the primary numbers and the chain code for secondary numbers. The number of digits in a code may run from 5 to 30 with a general average of about 8 or 10 digits. A balance must be maintained between the amount of information needed and the number of digit columns required to provide it. Each additional column generated increases the handling problem.

Combined Codes.   Examples of the combined code (hierarchical and chain) are the Opitz, CODE,[1] and MICLASS[2] systems.

THE OPITZ CLASSIFICATION SYSTEM

The Opitz parts classification system was developed by Professor Opitz of the University of Aachen in West Germany (see bibliography at the end of the chapter). The basic code consists of nine digits, which are intended to convey

[1] CODE—a system of classification produced by Manufacturing Data Systems Incorporated, Ann Arbor, Mich.

[2] MICLASS—a code system produced by TNO, Waltham, Mass.

1st Digit

Component class		
0	rotational parts	L/D ≤ 0.5
1	rotational parts	0.5 < L/D < 3
2	rotational parts	L/D ≥ 3
3	rotational parts	
4	rotational parts	
5	rotational parts	
6	non-rotational parts	
7	non-rotational parts	
8	non-rotational parts	
9	non-rotational parts	

2nd Digit

External shape, external shape elements			
0			smooth, no shape elements
1	stepped to one end		no shape elements
2	stepped to one end	or smooth	thread
3	stepped to one end	or smooth	functional groove
4	Stepped to both ends		no shape elements
5	Stepped to both ends		thread
6	Stepped to both ends		functional groove
7			functional cone
8			operating thread
9			all others

3rd Digit

Interhal shape, internal shape elements		
0		no hole, no breakthrough
1	smooth or stepped to one end	no shape elements
2	smooth or stepped to one end	thread
3	smooth or stepped to one end	functional groove
4	stepped to both ends	no shape elements
5	stepped to both ends	thread
6	stepped to both ends	functional groove
7		functional cone
8		operating thread
9		all others

4th Digit

Surface machining	
0	no surface machining
1	surface plane and/for curved in one direction, external
2	external plane surface related by graduation around a circle
3	external groove and/or slot
4	external spline (polygon)
5	external plane surface and/or slot, external spline
6	internal plane surface and/or slot
7	internal spline (polygon)
8	internal and external polygon, groove and/or slot
9	all others

5th Digit

Auxiliary holes and gear teeth		
0	no gear teeth	no auxillary hole
1	no gear teeth	axial, not on pitch circle diameter
2	no gear teeth	axial on pitch circle diameter
3	no gear teeth	radial, not on pitch circle diameter
4	no gear teeth	axial and/or radial and/or other direction
5	no gear teeth	axial and/or radial on PCD and/or other directions
6	with gear teeth	spur gear teeth
7	with gear teeth	bevel gear teeth
8	with gear teeth	other gear teeth
9	with gear teeth	all others

Main Code

Figure 19.4. The first five digits of the Opitz "form code."

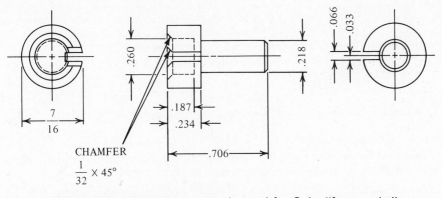

Figure 19.5.   Workpiece example used for Opitz "form code."

both design and manufacturing data. The code may be extended by adding four more digits.

The first five digits, 12345, are called the "form code" and are used to describe the primary design attributes. The first digit divides all parts into rotational and nonrotational, and also describes the general shape of the part. Figure 19-4 shows the coding system used for the first five digits.

> **Example:**   Figure 19-5 shows an example of a part used to illustrate the use of the Opitz "form code."
>
> The overall length/diameter ratio, $L/D = 1.61$, which makes the first digit = 1. The part is stepped on one end with a functional groove so that the second digit of the code is 3. The third digit is 1 because it has a smooth hole on one end. The fourth and fifth digits are both 0, since no surface machining is required and there are no auxiliary holes or gear teeth required. The whole Opitz form code for the part shown would then be 13100. To complete the supplementary code the sixth through the ninth digits with data on dimensions, material, starting workpiece shape, and accuracy would have to be added.

THE CODE SYSTEM

The CODE system uses a base of 16 (hexadecimal) instead of the common base 10 numbering system. This allows for six additional variables within the framework of each of the digits assigned. The CODE system numbers are identified as follows:

1st digit   The part is turned.

2nd digit   The basic shape is turned but may be a cylinder, multiple diameter concave, convex, conical, variable, etc.

3rd digit   Other primary operations such as drilling or boring a hole concentric with the outside diameter. This digit also indicates if the hole is blind, through, or multiple diameter and threaded or not.

4th, 5th
and
6th digits
    } These three digits are concerned with secondary operation holes other than those in digit 3 and internal and external grooves, threads, slots, flats, etc.

7th digit   The largest outside diameter or section.

8th digit   The overall length

Example: CODE is applied to a typical part as shown in Fig. 19-6.

1. The basic shape is cylindrical so it is classified in the first digit as 1.

2. The second classification has concentric parts or diameters. A check through the chart shows the second digit to be a 3, multiconvex cylinder.

3. The third digit is concerned only with a center hole description. Since this part is solid, the third digit is 1.

4. The fourth digit determines whether or not there are holes other than the center hole. In this particular part, there are four. Therefore the fourth digit is 8.

5. The fifth digit concerns grooves or outside diameter threads. This part is threaded on one end so the fifth digit becomes 8.

6. The sixth digit is concerned with the design of flats, slots, protrusions from the main shape, and concentric variations, or any of 11 possible combinations of them. This part contains concentric variations, slots, flats, or combinations of "1," "4," and "8." The combination of these descriptions is a "D," which is used as the sixth digit of the code number.

7. The last two digits of this eight-digit code are used to address the overall dimensions. The seventh digit is determined by locating the maximum outside diameter in the proper size range. In this case, the major outside diameter is 1.50 in. and is located in the range of diameters greater than 1.2 in. and less than 2 in. Therefore the code number 7 is assigned to the seventh digit.

8. The eighth number classifies the overall length of the part. This part is 5.7 in. long and it falls within a 4.4 to 7.2 in. range. A descriptor number 5 completes the eight-digit code number. This completes the first step in putting a coding or classification system to work.

The second step of classification is to record the code number on the source data collection card, also shown in Fig. 19-6 for input to the computer. First the eight-digit CODE number is entered, followed by the part name or description. Ordinarily this name will match the name on the drawing. The material identification may be listed by standard code or company standard.

Space is provided for dimensional information. Dimension 1 is given as the maximum external diameter or section, and dimension 2 is the minimum external diameter or section. Dimension 3 is the minimum internal diameter or section, and dimension 4 is the maximum overall length. These overall dimensions provide specific information for quick sorting and retrieval of parts.

Other information may be added to this card, such as heat treat specifications, tolerance class, or surface coating. The card shown provides for a metric flag or a dual code identifier for those few parts that need to be classified in more than one manner.

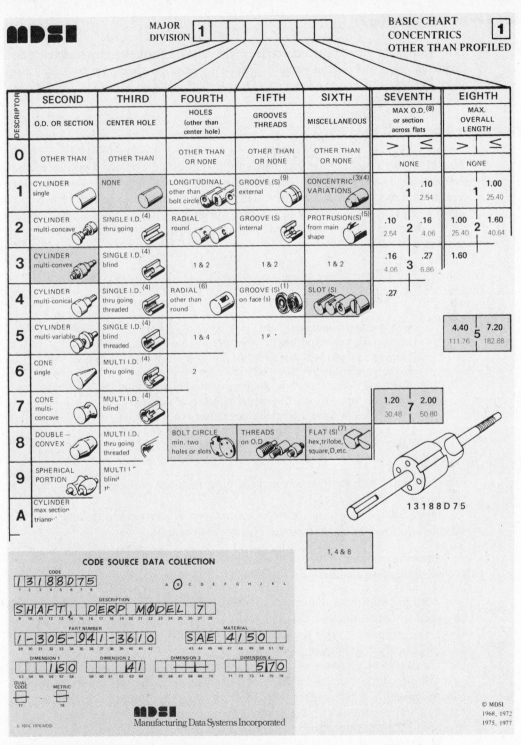

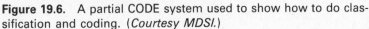

Figure 19.6. A partial CODE system used to show how to do classification and coding. (*Courtesy MDSI.*)

New parts should also be coded by reference to the family and the basic sequence of operations identified. The new operation sheet can be easily produced by making minor changes to the existing methods.

THE MICLASS SYSTEM

The MICLASS system uses 12 digits to describe and classify workpiece characteristics. The coding may be done manually but it is usually done by conversational mode with a computer. The number of questions asked by the computer varies with the complexity of the workpiece ranging from a minimum of 7 for a simple part to about 20 for a complex part. No special training in computer language is required.

> **Example:** A partial example is given for the part shown in Fig. 19-7. The answers are shown as underlined.
>
> Is it a rotary component? <u>Yes</u>
> Largest diameter and length? <u>196, 87</u>
> Does the rotary part deviate? <u>No</u>
> Is the axis of motion threaded? <u>No</u>
> Must any eccentric holing, planing, or slotting operations be done? <u>No</u>
> Must the top sides or outer form be turned? <u>Yes</u>
> Has the outer form a special groove(s) or cone? <u>No</u>
> Are the outside diameters increasing from both ends? <u>Yes</u>
> Must the inner form be turned? <u>Yes</u>

Questions will also be asked regarding material, tolerances, and surface roughness.

Once all the questions have been answered, the computer will assign a classification number to the part. The printout will be:

Drawing number = abc
Classification number = 1330   4021   2104

With the information given and the classification number, the computer can be asked to obtain numbers of similar drawings or related parts that are already in the company files. The similarity may be all-inclusive or it may involve only a single element of the part.

The retrieval system may also be used for gathering other information, as for example all the "shafts" in the company files or all the workshop operations used to produce a given piece. Data can be requested regarding machine tool loadings and production costs. With this information, short- and long-term investment plans can be made.

The MICLASS code information is used to form more efficient work cells or families of machines for workpiece families.

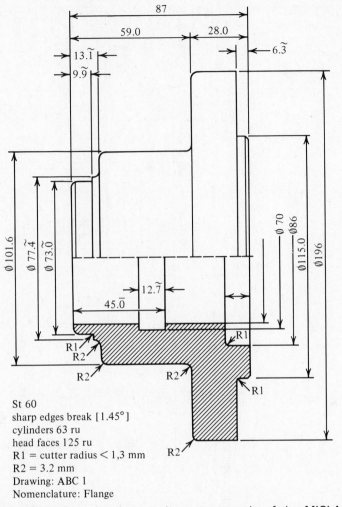

St 60
sharp edges break [1.45°]
cylinders 63 ru
head faces 125 ru
R1 = cutter radius < 1,3 mm
R2 = 3.2 mm
Drawing: ABC 1
Nomenclature: Flange

Figure 19.7. Part drawing used as an example of the MICLASS coding system. (*Courtesy TNO and Organization for Industrial Research, Inc.*)

CODING BENEFITS

In addition to the main benefits of a better flow of products through the plant, coding provides for a better utilization of tooling, ease of checking validity of cost estimates, group purchasing, check of tooling duplication, setup changes, and design reviews.

Validity of Cost Estimates. Cost estimates of coded parts can quickly and easily be checked. Any discrepancy in the cost of a part in relation to other similar parts will stand out and can quickly be investigated as to the cause. Table 19-1 shows a CODE printout sheet of seven similar parts. One can see

Table 19-1. Computer printout of family code information.

"Code No." 13D0956C	Matl. 1045	Fin	Ht 00	O.D. .812	O.D. .438	I.D. .313		Length 10.578	Size 329	Cost 8.2341	No. 9876523
13D0957E	1050	A		1.600	.750	.438		14.358	61	11.0078	6956333
13D0957E	1045			1.600	.750	.438		14.358	44	12.3363	30163859
13D0957E	1050	B		1.600	.750	.438	B1	14.358	24	15.5471	10034876
13D0957E	1050			1.600	.750	.438		14.358	5	7.6463	21023756
13D0957E	1050			1.600	.750	.438	B2	15.848	6	9.8730	45628019
13D0957E	1050			1.600	.750	.438		15.848	92	9.6871	40628737
13D0957E	1050			1.600	.750	.438		15.848	73	14.6165	5003131
13D0957E	1050			1.600	.750	.438		15.848	5	19.7045	7396211
13D0957E	4140	C		1.625	.997	.438		15.865	14	8.7327	21035687
13D0957E	4140			1.625	.997	.438		15.865	27	15.5839	11286598
13D0957F	1045	D		1.625	.985	.438		16.015	5	52.7780	27913621
13D0957F	1045			1.625	.985	.438		16.015	20	17.8950	47603001
13D0957F	1045			1.625	.985	.438		16.015	6	38.7534	39368598
13D0957F	1045			1.625	.985	.438		16.015	115	5.7089	13507765

ID	Code	Group					Sub		Count		Value
13D0957F	4140		1.625		.997	.438		17.365	120	7.5679	36975106
13D0957F	1045		1.625		.997	.438		17.365	34	5.2199	7834590
13D0957F	4140		1.625		.997	.438		17.365	101	8.2131	9234501
13D0957F	1045	E	1.625		.985	.438	E1	17.515	75	8.3187	13706948
13D0957F	1045		1.625		.985	.438		17.515	40	13.2586	14713526
13D0957F	1045		1.625		.985	.438		17.515	10	32.4476	47329873
13D0957F	1045		1.625		.985	.438		17.515	26	17.1325	50016713
13D0957F	1045		1.625		.985	.438		17.515	14	28.7932	50173721
13D0957F	1045		1.625		.985	.438	E2	18.015	19	21.4583	30659875
13D0957F	1045		1.625		.985	.438		18.015	20	16.6572	30264781
13D0957F	1045		1.625		.985	.438		18.015	12	31.9997	26145697
13D0957F	1045		1.625		.985	.438		18.015	6	39.0012	10301002
13D0957F	4140		1.625		.997	.438		19.365	87	15.1214	21203697
13D0957F	1045	F	1.625		.985	.438		19.515	120	5.5562	8340021
13D0957F	1045		1.625		.985	.438		19.515	115	5.4136	9569572
13D0957F	1045		1.625		.985	.438		19.515	5	28.7397	40078561
13D0957F	1045		1.625		.985	.438		19.515	10	13.7008	84182111
13D0958F	4120		2.125	00	.969	.438		18.797	2000	9.6878	4076336
13D0958F	1140	G	2.125	00	.969	.438	G1	18.797	25	27.7252	10102593
13D0958F	1140		2.125	00	.969	.438		18.797	100	23.6501	73965111
13D0958F	1140		2.125	00	.969	.438	G2	20.859	1000	15.4715	6474372
13D0958F	1140		2.125	00	.969	.438		20.859	1100	13.2603	72964344

at a glance that items 5 and 8, reading from the top of the list, have a large difference in standard cost. The first impression is that the lot size is responsible; however, the cost of six or five items should be very close. Thus further investigation will be needed.

Also cost estimates can be quickly made for new parts merely by adding or subtracting the cost of additional features to the standard cost of the existing physical characteristics. Only those parts that are completely different need to be estimated from scratch.

Purchasing.   The raw material requirements may be more easily arrived at and forecasts made based on the coding data. The seventh digit of the eight-digit CODE specifies the maximum size of the material required. Table 19-2 shows the relationship between numbers with a "6" or a "7" in the seventh column. By multiplying the lot size by the maximum length, plus a cutoff allowance, the total bar stock requirements for parts with a "6" can be made as:

<u>6297 bars of 1045, 11.5 ft long</u>

CODE 13DO957F, Tables 19-2 and 19-3, represents 21 parts with a total lot size of 1612 pieces, of which 1304 parts are made from 1045 steel and 308 from 4140 steel. There is only one change in diameter and that is in the minimum O.D. from 0.985 to 0.987 in. There are five changes in overall length. The overall quantity of each type of steel can be quickly calculated.

Tooling Duplication.   A review of the printout, Table 19-3, raises several questions about duplicate tooling, such as: (a) How much duplicate tooling is being used? (b) Is some of the tooling of better design and more efficient? (c) Is the cost difference between part 84182111 with a standard cost of 13.7008 and part 47329873 with a standard cost of 32.4476 realistic, or is it caused by inadequate tooling?

Setup and Methods.   If after the tooling for the two parts mentioned in the preceding paragraph has been checked and found to be satisfactory, the machining setup and processing should be checked for optimization. Again, are the methods used to produce part 84182111 more efficient than those used to produce part 47329873? Or are the optimum methods used to produce part 7834590? Other questions that arise: Can the standard cost be reduced by planning production based on minimum setup changes? How many parts can be eliminated with a resultant increase in lot size and reduction in cost? For example, can part 7834590 made of 4140 steel be combined with part 36975106 and part 9234501 to make one lot size of 255 parts?

Design Changes.   By use of the code system and the memory of a computer, all existing designs can be made quickly available for review. The designer will be able to see in a matter of minutes the numbers of all similar parts. He can then ask for the most pertinent designs to be shown on a video screen. With

Table 19-2. Computer printout of family code information.

"Code No."	Matl.	Fin	Ht	Maximum O.D.	Smallest O.D.	Smallest I.D.	Maximum Length	Lot Size	Standard Cost	Part No.
13D0936C	1045		00	.812	.438	.313	10.578	8	24.4225	10002693
13D0936C	1045		00	.812	.438	.313	10.578	25	9.7762	21069377
13D0936C	1045		00	.812	.438	.313	11.078	2000	1.1969	7825986
13D0936C	1045		00	.812	.438	.313	11.141	864	2.3189	4206524
13D0936D	1045		00	.812	.438	.313	12.078	249	1.9061	6910234
13D0936D	1045		00	.812	.438	.313	12.453	2822	2.2180	10011539
13D0956C	1045		00	.812	.438	.313	10.578	329	8.2341	9876523
13D0957E	1045			1.600	.750	.438	14.358	44	12.3363	30163859
13D0957F	1045			1.625	.985	.438	16.015	5	52.7780	27913621
13D0957F	1045			1.625	.985	.438	16.015	20	17.8950	47603001
13D0957F	1045			1.625	.985	.438	16.015	6	38.7534	39368598
13D0957F	1045			1.625	.985	.438	16.015	115	5.7089	13507765
13D0957F	1045			1.625	.997	.438	17.365	34	5.2199	7834590
13D0957F	1045			1.625	.985	.438	17.515	75	8.3187	13706948
13D0957F	1045			1.625	.985	.438	17.515	40	13.2586	14713526
13D0957F	1045			1.625	.985	.438	17.515	10	32.4476	47329873
13D0957F	1045			1.625	.985	.438	17.515	26	17.1325	50016713
13D0957F	1045			1.625	.985	.438	17.515	14	28.7932	50173721
13D0957F	1045			1.625	.985	.438	18.015	19	21.4583	30659875
13D0957F	1045			1.625	.985	.438	18.015	20	16.6572	30264781
13D0957F	1045			1.625	.985	.438	18.015	12	31.9997	26145697
13D0957F	1045			1.625	.985	.438	18.015	6	39.0012	10301002
13D0957F	1045			1.625	.985	.438	19.515	120	5.5562	8340021
13D0957F	1045			1.625	.985	.438	19.515	115	5.4136	9569572
13D0957F	1045			1.625	.985	.438	19.515	5	28.7397	40078561
13D0957F	1045			1.625	.985	.438	19.515	10	13.7008	84182111

Table 19-3. Computer printout of family code information.

"Code No."	Matl.	Fin	Ht	Maximum O.D.	Smallest O.D.	Smallest I.D.	Maximum Length	Lot Size	Standard Cost	Part No.
13D0957F	1045			1.625	.985	.438	16.015	5	52.7780	27913621
13D0957F	1045			1.625	.985	.438	16.015	20	17.8950	47603001
13D0957F	1045			1.625	.985	.438	16.015	6	38.7534	39368598
13D0957F	1045			1.625	.985	.438	16.015	115	5.7089	13507765
13D0957F	4140			1.625	.997	.438	17.365	120	7.5679	36975106
13D0957F	1045			1.625	.997	.438	17.365	34	5.2199	7834590
13D0957F	4140			1.625	.997	.438	17.365	101	8.2131	9234501
13D0957F	1045			1.625	.985	.438	17.515	75	8.3187	13706948
13D0957F	1045			1.625	.985	.438	17.515	40	13.2586	14713526
13D0957F	1045			1.625	.985	.438	17.515	10	32.4476	47329873
13D0957F	1045			1.625	.985	.438	17.515	26	17.1325	50016713
13D0957F	1045			1.625	.985	.438	17.515	14	28.7932	50173721
13D0957F	1045			1.625	.985	.438	18.015	19	21.4583	30659875
13D0957F	1045			1.625	.985	.438	18.015	20	16.6572	30264781
13D0957F	1045			1.625	.985	.438	18.015	12	31.9997	26145697
13D0957F	1045			1.625	.985	.438	18.015	6	39.0012	10301002
13D0957F	4140			1.625	.997	.438	19.365	87	15.1214	21203697
13D0957F	1045			1.625	.985	.438	19.515	120	5.5562	8340021
13D0957F	1045			1.625	.985	.438	19.515	115	5.4136	9569572
13D0957F	1045			1.625	.985	.438	19.515	5	28.7397	40078561
13D0957F	1045			1.625	.985	.438	19.515	10	13.7008	84182111

this information, drawings for new designs can be made with the assurance that previous design efforts will not be duplicated.

The designer can also see from the code information the possibility of salvaging obsolete parts by having the overall dimensions, the number of parts, and the type of material immediately available.

ADVANTAGES OF GROUP TECHNOLOGY FOR GROUP SCHEDULING

1. Reduction of setup and idle time. The setup time for each job is reduced because several jobs are grouped and processed in sequence. In practice, special tooling and fixtures can be effectively employed to reduce setup and idle time. As an example, a drill jig can be made so that it can be adapted to accommodate all parts of a family by adjustments or special inserts.

2. Optimizing Job Sequence (JS) and Group Sequence (GS). It is often a problem to select both an optimal group sequence and a job sequence in each cell. This two-phase scheduling technique reduces the computation effort if effective theorems and algorithms are used. In GT the task is simplified since the jobs can be processed into several reasonable groups with an effective classification and coding system.

3. Flow Shop Scheduling. With GT the job flow is expected to be easily handled so that "flow shop scheduling" can be used rather than the complicated "job shop scheduling," as is done in an ordinary functional plant layout. Production can be controlled by a day-to-day loading of groups of machines. Also smaller batch sizes give greater flexibility to schedule changes and expediting is easier.

4. Quality Control (QC). Because of the low batch quantities, the operators must be sufficiently skilled to set up the machines and produce an acceptable first piece within the stipulated setup change time. Only input raw material and output finished components from each group need to be checked by a separate QC inspector on a sampling basis.

5. Maintenance. Since machines are restricted to individual groups, maintenance priorities depend on the criticalness of the machines in each group. The need for maintenance is therefore increased compared to machines grouped by function. It is assumed that sufficient time is allowed within the group to carry out a regular preventive maintenance program to minimize stoppages caused by machine failures.

6. Tool room. Because of the more intimate knowledge of the nature of production sequencing, tooling repairs can be anticipated and attended to in time. Smaller batch sizes limit the wear and tear on tooling from reaching chronic levels before receiving attention for repairs. Also modifications of tooling needed for new designs can be carried out quickly without causing stock obsolescence.

7. Plant Layout and Material Handling. Because of the negligible in-process inventory, machines can be spaced in a compact manner which also helps in carrying out multiple operations. Small batch quantities allow simple handling equipment such as crates, tote boxes, trolleys, and gravity rollers

are used for heavier parts. It is typical to see forklift trucks being used only for transporting material between stores and input or output stages of shops in a grouped machining arrangement. Homogeneity of parts makes for more compact handling of parts possible. Also with an emphasis on reduced setup change times it is necessary to provide storage space for tooling near the machines that will be used for the entire product line. The layout is planned to be flexible so that as the product changes rearrangement of machine groups can be done with a minimum of disruption. Therefore wiring should be supplied from overhead bus bars and permanent foundations should be avoided. If they are unavoidable, they should at least be minimized.

8. Personnel. In the prevailing functional plant, there is often boredom among workers caused by working mechanically on a single machine for years. Also an unsafe feeling develops when inventory starts to pile up. A sense of distraction develops from seeing a heterogeneous grouping of personnel without any apparent sense of purpose.

Where group technology has been introduced (usually a three-year process), there is a very noticeable lack of the conditions mentioned in the previous paragraph. Group technology, because it is more dynamic and places more emphasis on the small group and individual, prevents these morale problems from occurring. It provides some of the most sought-after contributions to job satisfaction through more worker involvement in decisionmaking, personalized work relationships, variety in tasks, freedom to determine methods and workplace layout, and meaningful pay incentives.

PROBLEMS ASSOCIATED WITH GT

1. Work Load Requirements. A steady work load is required to assure worker satisfaction since work-in-process is relatively low.
2. Implementation. Implementation requires extensive reorganization for plants now organized on a functional plan. Machines must be moved to cellular groupings. This not only requires time and capital investment but also a temporary slowdown in production.
3. Reorganization. Reorganization affects most all departments, including engineering, accounting, management, and the shop employees.
4. Redesign of Parts. Some redesign of the products will be required to fit standardization and modularization.
5. Machine Usage. Since GT typically involves fewer people than machines, some machines will have a lower utilization level.
6. Supervision. Supervision of manufacturing cells may be more difficult than that of the functional layout. More knowledge of the total manufacturing process is required in each cell.
7. Group Dependency. Each group must be made to realize how dependent they are on every other group. Cooperation of all groups is absolutely essential. All employees must accept this concept for their own best interests and that of the plant.

GROUP TECHNOLOGY AND NC

NC and Batch Production.   Batch production cannot normally take advantage of automated sophisticated equipment because it requires too much capital investment for the rate of return. The advantages of automated equipment are usually outweighed by the problems of production planning, tooling, poor utilization, and long setup times. This is not true when NC machines are applied to GT. NC machines find their best usage in batch type of production, 10 to 1,000 parts. NC machining centers are readily adaptable to a wide range of production with a minimum of setup time. Their versatility makes it possible to reduce the number of machines and personnel required in each group. Machining centers can be used coincidentally as submachine groups or subcells within the group, which practically eliminates queuing and greatly simplifies layout.

Master Family Planning.   One of the important applications of GT is the software development for NC called part-family programming or master family planning (MFP). MFP is an NC program system that groups common or similar program elements into a single master computer program. This master computer program, or preprocessor, is a permanent base from which NC tapes can be prepared for any part in the family. Therefore part-family programming can be used to increase the productivity of the costly NC operations by saving programming time, manpower, and tape proveout time.

This is accomplished by making a composite part to represent each family as shown in Fig. 19-8. The composite possesses all the shape characteristics and machining of the part family. If the tooling is set for the composite component, any part that fits within the envelope of that shape can be machined without changing tools. This greatly simplifies tape preparation. The programmer need only refer to the composite geometry already in the processor memory and identify each line, radius, curve, taper, etc. Each horizontal line is called HL1,

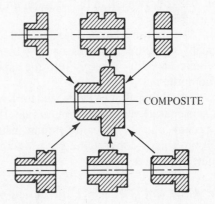

COMPOSITE

Figure 19.8.   An example of a composite part for a family group.

HL2, etc., starting at the top, and each vertical line, starting from the center, is VL1 etc. Then chamfers and taper lines and radii are TL1 and R1, etc.

Logic may also be incorporated in MFP so that not only can any size part of a given family be run from the composite, but also changes can be made creating different configurations.

MFP makes use of the computer for setup, preset operator, and inspector instructions. The instruction given on the printout varies with each family of parts and type of NC machine. As an example, on a vertical turret lathe the tool selector will print out the listing of each tool as coded in the tool book starting with the longest bar, then describing the turning tools and their position on the turret. Usually the carbide grade is also given, such as C2, C5, or ceramics if used. This is the information that the presetter and the machine operator will use. Similarly, printout instructions are given to the inspector. Thus the information clarifies the task for the setup men, and for the operators, foremen, and inspectors.

Building Block Tapes.   Closely associated with GT and MFP is a technique of building tape libraries of standard operations much like that of modular or building block construction used in commercial buildings. At first programmers made a library of many standard operations so that new tape programs could be prepared with a minimum of time and effort. These prepared short tapes were referred to as building block or "canned macros." As the programmer looked over the blueprint, he could see where he could splice in the building blocks and reap the time-saving benefits. Good as the system was there were problems, such as efficient storage and retrieval systems. Also once the tape was located it had to be spliced into the new program, which was cumbersome and time consuming. Most troublesome was the fact that the canned macros were often almost but not quite right in one or two dimensions. These slight modifications required time to reprogram.

Several companies introduced a method of combining the building blocks through a tape "editor." The programmer was now able to modify and manipulate the building blocks with ease and to build new programs via the computer. Again this had a drawback in that it required expensive computer time, even on a time-sharing basis.

At this time Numeridex, a well-known name in numerical control, found that by using microprocessors it was not only possible but also quite practical to build an editor directly into the tape preparation systems without the use of a computer.

By adding floppy discs, video screen, and memory capacities to the basic tape preparation system, programmers can not only implement the building block approach but also it can be done instantly without having to punch a tape or call up a hard copy of the program until the programmer is confident that it is right.

CAD and GT.   Once the design, materials, and manufacturing codes have been assigned, the computer will be able to search the records for similar or identical parts to see whether or not a new design is actually needed. If a new design is needed, it would be possible for the computer to aid in the design, based on previous decisions, and to utilize the complete code to generate the processing documents, prepare the NC tapes, and issue schedules, orders, and bills of materials. This approaches the goal of the automated factory, which is now beginning to emerge, as discussed in Chapter 18.

Problems

19-1. Three machine parts are shown, MA-100, MA-101, and MA-106 (see Figs. P19.1a, P19.1b, and P19.1c).

    (a) Using the partial CODE sheet shown in Fig. 19-6, develop the eight-digit code for each part.

    (b) What would the economic lot size be for straddle milling operations for part MA-100 if a milling fixture is used that

holds 10 parts? Twenty minutes are allowed for setting up the milling fixture.

19-2. Using the Opitz system, what would the first digit of the code be for part MA-100?

19-3. What is the main difference between functional layout and cellular layout?

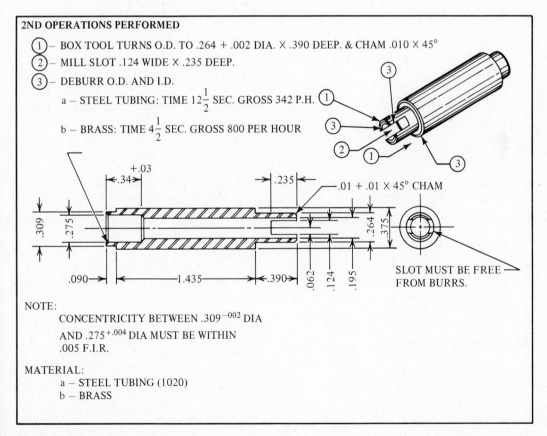

2ND OPERATIONS PERFORMED

①– BOX TOOL TURNS O.D. TO .264 + .002 DIA. × .390 DEEP. & CHAM .010 × 45°

②– MILL SLOT .124 WIDE × .235 DEEP.

③– DEBURR O.D. AND I.D.

    a – STEEL TUBING: TIME $12\frac{1}{2}$ SEC. GROSS 342 P.H. ①

    b – BRASS: TIME $4\frac{1}{2}$ SEC. GROSS 800 PER HOUR

.01 + .01 × 45° CHAM

SLOT MUST BE FREE FROM BURRS.

NOTE:

    CONCENTRICITY BETWEEN $.309^{-.002}$ DIA

    AND $.275^{+.004}$ DIA MUST BE WITHIN

    .005 F.I.R.

MATERIAL:

    a – STEEL TUBING (1020)

    b – BRASS

Figure P19.1a. (MA—100)

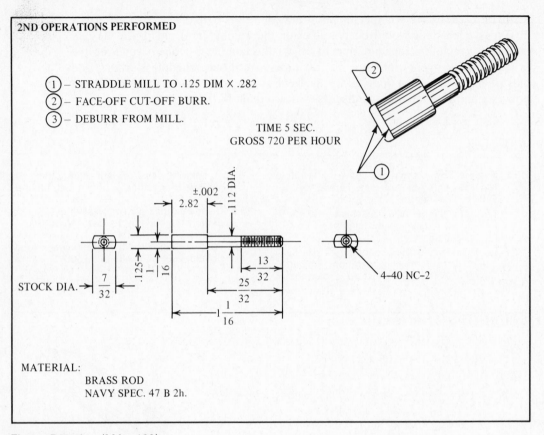

2ND OPERATIONS PERFORMED

① – STRADDLE MILL TO .125 DIM × .282
② – FACE-OFF CUT-OFF BURR.
③ – DEBURR FROM MILL.

TIME 5 SEC.
GROSS 720 PER HOUR

±.002
2.82

.112 DIA.

STOCK DIA.→

$\frac{7}{32}$

.125

$\frac{1}{16}$

$\frac{13}{32}$

$\frac{25}{32}$

$1\frac{1}{16}$

4–40 NC–2

MATERIAL:
    BRASS ROD
    NAVY SPEC. 47 B 2h.

Figure P19.1b. (MA—100)

19-4. Give several reasons why most manufacturing is done on the batch basis.

19-5. (a) Briefly state the underlying philosophy of GT.
(b) What are the three main sources of savings due to GT?

19-6. What is the requirement of GT in respect to work load?

19-7. In what way are GT and NC related?

19-8. What is the effect of MFP?

19-9. What is a distinguishing feature used in the MICLASS coding system?

19-10. What would the second and third digits be for Part MA-101 using the Opitz code?

19-11. (a) What is the main advantage of the PFA system of classifying parts?
(b) What is the main disadvantage of the PFA system?

19-12. How does a hierarchical code system differ from a chain code system?

19-13. Is the Opitz code chain or hierarchical?

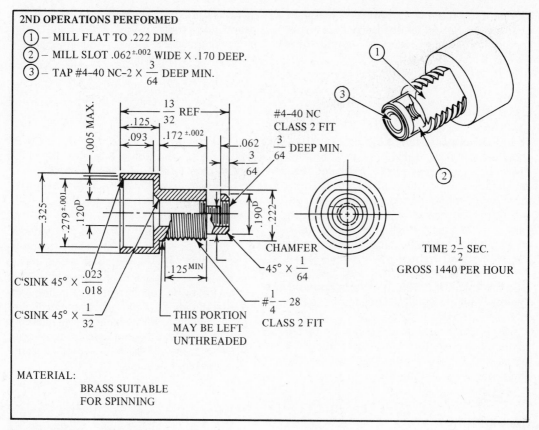

2ND OPERATIONS PERFORMED

① – MILL FLAT TO .222 DIM.

② – MILL SLOT .062$^{±.002}$ WIDE × .170 DEEP.

③ – TAP #4–40 NC–2 × $\frac{3}{64}$ DEEP MIN.

TIME 2$\frac{1}{2}$ SEC.

GROSS 1440 PER HOUR

MATERIAL:

BRASS SUITABLE
FOR SPINNING

Figure P19.1c. (MA—106)

Case Study 1

In response to growing demands for increased production at one of the typewriter assembly plants, IBM followed the standard procedure of lengthening the assembly lines. Eventually each line was more than 230 feet long and employed more than 70 workers. Each worker put an average of three minutes work into each typewriter.

The typical typewriter has 2500 parts.

Besides 18 standard models there are 25 specials, with more than a hundred type heads and keyboards for different languages.

The production line is fast but the error rates are high, and the plant had to spend 12% of all man-hours in overtime to repair defective typewriters. Labor turnover is 30% per year.

Propose a solution for this problem.

Case Study 2

A manufacturer of electronic components has a plant that is organized in a traditional way. Different kinds of products are produced in

a single work sequence moving from one functional department to the next for assembly, welding, etc. Each department has a foreman and there is a production manager over the whole operation. Delays are widespread and overdue orders account for fully a third of the work load.

Make a plan that could help correct the production problem. Your answer should include suggestions as to management, personnel, and their duties.

Case Study 3

The ASO company started out as a small job shop on the West Coast and grew rapidly. However, a study made in 1972 showed their actual chip-cutting time to be 15%. Government contracts warranted adding 30 NC machining centers increasing production considerably and the chip-cutting time rose to 30%.

The plant produces missile system components ranging in size from $\frac{1}{2}$ in. cube-shaped parts to parts nearly 6 ft in dia. Lot sizes typically range from 20 to several hundred. The machining generally ranges from 10 min to 2 hr floor-to-floor time.

Equipment now used in addition to the 30 machining centers consists of about 20 lathes and 6 special machines such as riveting machines and jig borers.

You have been asked to come up with a plan that would potentially raise the chip-cutting time to 50% on existing machines.

A preliminary investigation reveals that during the 70% non-chip-cutting time, the following conditions existed:

a. More than half this time was labeled "no operator" or "awaiting operator action" (AOA).

b. The remainder of the nonproduction time was spent on setup, loading and unloading, and *miscellaneous machining*. Miscellaneous machining turned out to be a catchall term covering mostly shop layout problems and material handling.

Bibliography

Periodicals

Bobrowicz, V. F. "CODE: Group Technology Classification System." *Proceedings of CAMI Coding and Classification Workshop.* Arlington, Texas, 1975, pp. 91–104.

Brisch, E. G., and R. S. Geoghidan. "Simplification and Standardisation for Automation." *Journal of the Institution of Production Engineers,* 36 (August 1957): 571–582.

Ham, Inyong. "Introduction to Group Technology." *Society of Manufacturing Engineers Technical Report MMR76-03.* Dearborn, Mich., 1976.

Houtzeel, Alexander. "The Many Faces of Group Technology." *American Machinist* 123 (January 1979): 115–129.

MiClass News. "Current Developments in Group Technology." TNO Organization for Industrial Research, Waltham, Mass.

Wagle, R. P. "Group Technology as a Concept in Manufacturing Planning." MDSI (June 7, 1979).

Books

Burbridge, John L. *The Introduction of Group Technology.* New York: John Wiley, 1975.

DeVries, Marvin F.; Susan M. Harvey; and Vijay Tipnis. *Group Technology: An Overview and Bibliography.* Cincinnati: Machinability Data Center, August 1976.

Gallagher, Colin C., and W. A. Knight. *Group Technology.* London: Butterworth, 1973.

Groover, M. P. *Automation, Production Systems and Computer-Aided Manufacturing.* Englewood Cliffs, N.J.: Prentice-Hall, 1980.

Opitz, H. *A Classification System to Describe Workpieces.* Oxford: Pergamon Press, 1970.

CHAPTER TWENTY

Nontraditional Material Removal Processes

Nontraditional metal removal refers to processes that do not develop the conventional chip. These processes used to be called "chipless machining." The growth of newer metal removal processes has been accelerated in recent years because of the need to cut more exotic materials, often in the hardened condition and intricate designs. At first the processes were largely developed and used by the aerospace industries but are now prevalent in the auto and appliance industries and in tool and die work. Cuts made by the newer processes are essentially stress free.

There are more than 30 newer processes that can be broadly divided into four main classifications. The processes discussed in this text will be those most used, as follows:

Chemical
    Photochemical Milling (PCM)
Electrical
    Electric Discharge Machining (EDM)
    Electrochemical Machining (ECM)
    Electrochemical grinding (ECG)
Mechanical
    Abrasive Jet Machining (AJM)

Thermal——
    Laser Beam Machining (LBM)
    Electron Beam Machining (EBM)

PHOTOCHEMICAL MACHINING

Although PCM is not exactly new, it probably rates as one of the newer metal-machining methods. It is basically an etching technique that has been developed to compete on a production line basis. The steps involved are preparation of

(a)

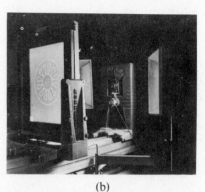

(b)

(c)

Figure 20.1. Steps in the process of photochemical machining. Preparation of the drawing (a). The image is transferred to the metal (b). Spray etching the metal to leave only the finished part (c). (*Courtesy Chemcut Corp.*)

the part drawing, coating the metal with a photo resist, baking the resist, image printing, etching, and resist removal.

The drawing, as shown in Fig. 20-1(a), is very carefully and accurately made since the finished part can be no better than the artwork from which it originated. Drawings of very small parts are sometimes made 100 times actual size to improve accuracy and are then photographically reduced on film to the required size.

Before the photo resist is placed on the metal, it is thoroughly cleaned, usually with the aid of solvents. When the resist is completely dry, it is baked at about 180°F (82°C) for several minutes. The master negative is then placed against the resist (b). A powerful light source is used to expose the resist through the negative. Development of the resist leaves a chemical resistant image on the metal in the shape of the part.

The metal is then spray etched (c) to dissolve all unprotected areas and leave only the finished part, which is washed in a solvent and flushed with water to remove the resist.

For greater accuracy, particularly with thicker metals, two master negatives are made. One is a mirror image of the first and is exposed on the back side of the metal so the etchant can act on both sides of the metal at the same time.

Advantages

1. Eliminates die costs.
2. Excellent for thin parts (Fig. 20-2).
3. Provides for quicker delivery.
4. Produces stress-free parts.
5. Produces burr-free parts.
6. Virtually any metal can be used.
7. Has perfect repeatability.
8. Makes short runs feasible.
9. Makes design changes easy.
10. Both large and small parts can be made.
11. Produces close-tolerance parts.

Disadvantages

1. If close accuracy is required, etching all the way through is limited to metal thicknesses of about 0.0625 in.
2. Etchant vapors are very corrosive and must be isolated from other plant equipment.
3. Production yields are relatively low. Part size, thickness of metal, and quantity required determine the break-even point. The metal-removal rates and tolerances are shown in Table 20-1.
4. Because of the inherent undercut, there is a minimum-size limit for slots and holes and other piercings. For example, the smallest hole that could be pierced in 0.010-in. brass while still providing a near vertical wall would be 0.007 in. in dia. The progressive undercutting action is shown schematically in Fig. 20-3. An etch factor of 3 : 1 means that for every 0.003 in. of etched depth, 0.001 in. of undercut will occur. The following limitations on hole size are characteristic: $0.7t$ for copper alloys, $1.0t$ for steel alloys, and $1.4t$ for aluminum alloys.

Figure 20.2. Representative parts that have been photochemically machined. Parts may range in thickness from 0.0005-in. (0.0013-mm) foils to 0.125-in. (3.17-mm) thicknesses. (*Courtesy Chemcut Corporation.*)

Table 20-1. Removal rates and tolerances for chemical milling. (*Courtesy ASTM*)

Material	Depth of Etch per min. (in.)	Total Depth of Cut (in.)	Milling Tolerance (in.)
Aluminum	.001	up to .020	± .001
		.021 to .060	± .002
Magnesium	.0013	up to .020	± .001
		.021 to .060	± .002
		greater than .060	± .003
Stainless steel and heat-resistant alloys	.005	up to .020	± .001
		.021 to .060	± .002
		greater than .060	± .003
Titanium alloys	.005	up to .020	± .002
		.021 to .060	± .0035
		greater than .060	± .005

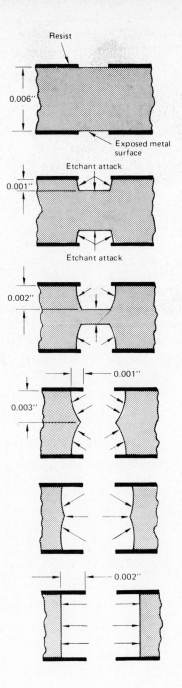

Figure 20.3. The undercutting action of chemical machining explains why there is a lower limit on the size of small holes and slots that can be made.

ELECTRIC-DISCHARGE MACHINING

Perhaps you have observed an arc caused by an accidental short circuit and noticed the pitting that occurred on the surface of the shorted material. This is the principle that operates in EDM machining. In EDM the spark is pulsed at frequencies ranging from a few hundred to several hundred thousand kilohertz and the workpiece and electrode are separated by a dielectric, as shown schematically in Fig. 20-4. The dielectric is usually a low-viscosity hydrocarbon or mineral oil. As the tool and the workpiece are brought together, the gap between them is maintained by a servomechanism comparing actual gap voltage to a preset reference voltage.

The gap is raised until the dielectric barrier is ruptured, which usually occurs at a distance of 0.001 in. (0.03 mm) or less and at about 70 volts. Technically, the dielectric is ionized to form a column or path between the workpiece and the tool so that a surge of current takes place as the spark is produced, as shown schematically in Fig. 20-5.

The spark discharge is contained in a microscopically small area, but the force and temperatures resulting are impressive. Temperatures around 10,000°C

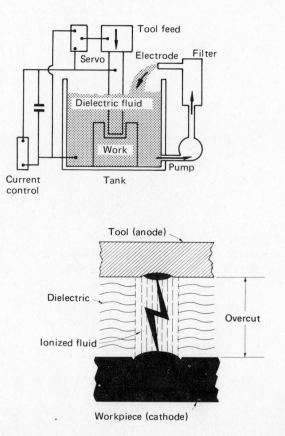

Figure 20.4. Schematic of EDM.

Figure 20.5. When electrical energy ionizes a portion of the dielectric fluid, a spark from the tool erodes a tiny crater on both the tool and workpiece. (*Courtesy* American Machinist.)

and pressures many thousand times greater than atmospheric are created, all in less than one microsecond for each spark. Thus a minute part of the workpiece is vaporized. The metal is expelled as the column of ionized dielectric vapor collapses. The tiny particles are cooled into spheres and are swept away from the machining gap by the flow of dielectric fluid.

The wear of the electrode is relative to the heat and ion flow that attack it. Therefore, the higher the melting temperature of the electrode, the less it wears. For this reason, high-density carbon is one of the best overall electrode materials. However, the fine carbon dust produced while machining is quite objectionable and must be drawn off with an adequate vacuum system.

An example of a blanking die produced by EDM is shown in Fig. 20-6. In this case the electrode is placed on an arbor and is started from the back side of the die. Since the hole produced is slightly tapered, it provides a positive taper for the die opening when it is turned right side up. The clearance between the punch and the die can be regulated but may be as small as 0.001 in. (0.03 mm). The cost of producing a graphite electrode is very small compared to having a broach made to produce the same hole. In this case, the machining time for three holes on a $\frac{5}{8}$ in. (15.87 cm) thick die block was 47 min. Although the actual machining time is relatively slow, other operations that are often done by hand, such as providing positive taper and controlling the surface finish and clearance, are taken care of all at the same time.

Figure 20-7 shows a forging die being cut by the EDM process using a graphite electrode.

Advantages.   EDM has many advantages that stem from three basic facts:

Figure 20.6.   A blanking die produced by EDM. Shown at the right is the graphite tool used to machine the serrated holes. (*Courtesy South Bend Lathe Company.*)

Figure 20.7.   An example of an EDM machine being used to cut a connecting-rod forging die cavity in a solid die block. The electrode (top part) is machined from graphite. Fast cutting speeds and feeds can be used and still provide a good finish. The EDM process removes between 4 and 5 in.^{3}/hr at 150–200 amp. The tolerance is ± 0.002 to 0.003 in. (±0.05–0.08 mm). (*Courtesy Cincinnati Milacron.*)

1. The hardness of the workpiece is not a factor. As long as the material can conduct current, it can be machined.
2. Any shape that can be produced in a tool can be reproduced in the workpiece. Complicated tooling may be made up in segments and fastened together with an adhesive such as Eastman 910.
3. The absence of almost all mechanical force makes it possible to machine the most fragile components without distortion. A 0.002 in. (0.05 cm) diameter hole, for example, can be produced in a small, delicate part using a very fine wire as a tool.

Disadvantages:

1. Tool wear requires stepped tooling or redressing of tools for deep holes.
2. EDM leaves a recast layer at the surface of the cut. This may be an advantage or disadvantage. Where undesirable, it may be removed by a light finishing cut or by polishing. Depending upon the current density used, the recast layer may be from 0.0002 to 0.005 in. (0.005 to 0.13 mm) in thickness.
3. EDM is slow when compared to conventional methods. Whenever possible, the cavities are roughed out prior to heat treatment and then finished by EDM after heat treatment.

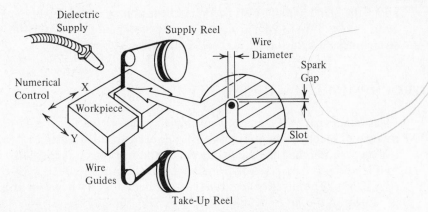

Figure 20.8. A schematic of traveling-wire EDM. The profile of the cut is determined by the NC command to the X and Y axis. (*Courtesy* Modern Machine Shop.)

Traveling Wire EDM. (TW–EDM). The significant difference between conventional EDM and TW–EDM is that the electrode is a wire that travels through the workpiece from a feed reel to a takeup reel, as shown schematically in Fig. 20-8.

The wire is often brass or copper although it may be tungsten or molybdenum. It is usually 0.008 in. (0.20 mm) in dia. but it may be as small as 0.003 in. (0.08 mm) or as large as 0.012 in. (2.59 mm). By coupling the traveling wire concept with NC, it is possible to have the machine run up to 50 hrs unattended while constantly cutting intricate shapes in hardened steel or carbide.

About the only stop or attention required in the machining operation is what the programmer calls a "glue stop." A glue stop is used to prevent a profiled section from falling against the wire as it is parted from the parent material. Before the cut section falls out, the cut is stopped and a drop of glue is placed on it along with a strap to bridge the gap over the cut. After the profiling is complete, the strap can be used to lift the cut section out.

By using CNC control and proper voltage and frequency, the amount of overcut can be accurately controlled and the computer can be told whether it should be on the inside or the outside of the programmed line. Overcuts can be controlled to within 0.0002 in. (0.013 mm) on small parts. A machine setting to angle either the workpiece or wire can be used to generate a clearance angle when making die openings.

Deionized water is the dielectric used for the TW–EDM. A small stream is directed at the wire or workpiece interface to flush the chips from the cut as well as to serve as a dielectric.

Cutting speeds vary with wire size, workpiece thickness, and applied current. An average figure is about 1.5 in. (38.10 mm) per hour in D-2 die steel. Cutting speed is not proportional to workpiece thickness because only a limited amount of current can be transmitted by a wire. Generally thick or thin pieces require about the same amount of time. Workpieces up to 6 in. (152.40 mm) thick have been cut by TW–EDM; however, in order to maintain a 0.0002 in. (0.013 mm) straightness a light 0.0002 in. (0.013 mm) finishing cut is made.

ELECTROCHEMICAL MACHINING

Electrochemical machining is a machining process that may be described as the reverse of electroplating. High-density direct current is passed through an electrolyte solution that fills the gap between the workpiece (anode) and the shaped tool (cathode). The electrolyte must be considered, in effect, a part of the tool (Fig. 20-9). The electrochemical reaction deplates the metal of the workpiece as shown in the more detailed schematic (b).

The Process.   The electrode is brought close to the workpiece with a working gap that ranges from 0.003 in. to 0.030 in. (0.08 to 0.76 mm) with 0.010 in. (0.25 mm) being average. The current density (50 to 1000 amp/sq in.) is the chief factor in setting feedrates and smoothness. The highly conductive electrolytes are aqueous solutions of inorganic salts such as NaCl, KCl, $NaNO_3$, or mixtures with proprietary additives operating at 100 to 120°F (38 to 49°C). The electrolyte is pumped through the machining gap at pressures often as high as 300 psi (20 kg/cm^2).

The machining gap is maintained by a servo control mechanism. Any contact of the tool with the workpiece will result in arcing, with damage to both members. On the other hand, if the gap becomes large (0.030 in; 0.76 mm), more electrolyte will flow, offering increased resistance for the current with a subsequent reduction in stock removal. In the application of Ohm's Law, $E = IR$, or as transposed, $I = E/R$. We can see that if the resistance (gap between the electrode and work) is small, a higher current density will result. However, when the gap gets too small, hydrogen bubbles tend to crowd together and increase the resistance across the gap, thus reducing the current flow. Hence the maximum penetration rate is based on a number of variables but is primarily a function of a precisely controlled machining gap.

Surface Finish and Accuracy.   Surface finish is dependent on the alloy being machined and the current density. Areas finished on the frontal surface of the cathode where the current densities are high generally have fine finishes. Finishes of 20 microinches are not uncommon, and on high-grade materials such as nickel-chromium alloys it may be as little as 2–5 microinches. Bearing races have been finished in production quantities by ECM to within 6 microinches.

The overcut per side of the cathode is about 0.005 in. (0.13 mm) and repeatability is good to within 0.0015 in. (0.04 mm).

Advantages.   The advantages of ECM are similar to those given for EDM with the following additions:

1. The tool does not wear. Once the tool is developed, it can be used indefinitely.

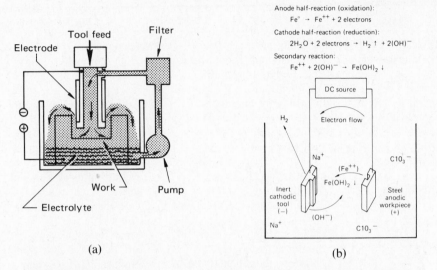

Anode half-reaction (oxidation):

$Fe^\circ \rightarrow Fe^{++} + 2$ electrons

Cathode half-reaction (reduction):

$2H_2O + 2$ electrons $\rightarrow H_2 \uparrow + 2(OH)^-$

Secondary reaction:

$Fe^{++} + 2(OH)^- \rightarrow Fe(OH)_2 \downarrow$

(a)

(b)

Figure 20.9. A schematic of the ECM process (a). A detail of the chemical process (b).

Figure 20.10. A typical ECM machine showing electrolyte work enclosure and control unit. Machines normally range in size from 500 to 5000 amp in steps of 500 amp. (*Courtesy Chemform, Division of KMS Industries, Inc.*)

2. There are no thermal or mechanical stresses on the workpiece.
3. Faster stock removal and better surface finish can be obtained. As a rule of thumb for estimating the metal removal rate, ECM is considered to have the ability to remove 0.10 cu in. (1.64 cm^3) of metal per 1000 amp used or 1 cu in./min (16.4 cm^3/min) for 10,000 amp as compared to up to 10 cu in./hr (164 cm^3/hr) for EDM. The surface finish range is generally from 4 to 50 microinches.

Disadvantages

1. The basic cost of the equipment is several times that of EDM. A representative ECM machine and control unit are shown in Fig. 20-10.
2. Rigid fixturing is required to withstand the high electrolyte flow rates.
3. The tool is more difficult to make than for EDM since it must be insulated

Table 20-2. A comparison of EDM and ECM process characteristics. (*Courtesy Chemform, Division of KMS Industries, Inc.*)

Data	EDM	ECM
Typical applications	Machining of: 1) Dies (stamping, cold heading, forging, injection molding) 2) Carbide forming tools 3) Tungsten parts 4) Burr-free parts 5) Odd-shaped holes and cavities 6) Small diameter deep holes 7) High strength and high hardness materials 8) Narrow slots (0.002″–0.012″ width) 9) Honeycomb cores and assemblies and other fragile parts	Machining of: 1) High temperature alloy forgings 2) Turbine wheels with integral blades 3) Jet engine blade airfoils 4) Jet engine blade cooling holes 5) Deburring of all kinds of parts 6) Odd-shaped holes and cavities 7) Small deep holes 8) Honeycomb cores and assemblies and other fragile parts 9) High strength high hardness materials
Tolerances	Practical: ±0.0005″ Possible: ±0.0001″	Practical: ±.002″ Possible: ±0.0005″
Surface	Finish is affected by rate of removal 0.010 in.^{3}/hr–30 rms 0.5 in.^{3}/hr–200 rms 3.0 in.^{3}/hr–400 rms Heat affected zone 0.0001″–0.005″	4–50 rms can be attained. No heat affected surface is produced. No burrs. Must guard against selective etching in remote areas exposed to electrolyte
Practical removal rates	Approximately 0.00025 in.^{3}/amp./min. or up to 10 in.^{3}/hr	0.1 in.^{3}/min./1000 amp is often used for rough approx. or 1 in.^{3}/min. for 10,000 amp.

Machining characteristics	Process produces taper and overcut (finished cavity minus original electrode size) in workpiece as well as corner radii
	Typical values:
	Taper 0.001–0.005 in./in./side
	Overcut 0.002–0.005 in./side
	Min. corner radius 0.001–0.20 in.
	Electrode wear ratio:
	  Metallic electrodes 3/1
	  Carbon 5–70/1
	Corner wear ratio: 2/1
		Process does not introduce machining stresses.
		Virtually no electrode (tool) wear occurs. Process may produce overcut, taper, and corner radii. Tools subject to damage by arcing if process malfunctions

to maintain the correct conductive paths to the workpiece. The tool may also serve as a device for introducing the electrolyte into the machining gap and must then have a center hole. However, this is not true if the tool is connected to the negative terminal and becomes the cathode.

4. The most common electrolyte, sodium chloride, is corrosive to the equipment, tooling, and workpiece.

A comparison of the process characteristics of EDM and ECM is given in Table 20-2.

ELECTROCHEMICAL GRINDING

Electrochemical grinding is the removal of metal by a combination of electrochemical decomposition and the action of diamond abrasive particles contained in the grinding wheel. The electric current flows from the negatively charged abrasive wheel to the positively charged workpiece through an electrolyte (saline solution), as shown schematically in Fig. 20-11. The resulting electrochemical oxidation produces a soft oxide film on the workpiece surface. The grinding wheel then "wipes away" this oxide film. Since the oxides are very much softer than the parent metal, there is very little grinding-wheel wear.

The tool looks like an ordinary grinding wheel, and it is, except that the bonding material is mixed with a conductor such as pure copper or carbon. The abrasive particles which are nonconductive tend to maintain a gap between the

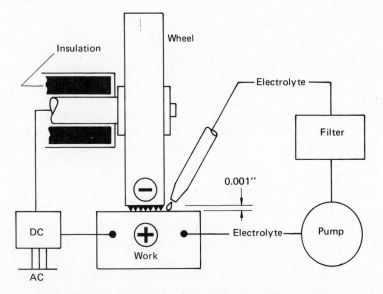

Figure 20.11. Schematic of electrochemical grinding.

metal-bonding agent and the work. Applying Faraday's Law, the metal-removal rate is almost directly proportional to current density.

Form Grinding.   The wheels may be dressed to facilitate form grinding with the aid of a single-point diamond dressing tool, a metal-bonded diamond dressing wheel, or a crushed-formed vitrified grinding wheel in a plunge grinding-wheel operation.

Power Supply.   The power used for electrochemical grinding varies in size up to 1000 amp and normally has a dc voltage output which ranges from 4 to 15 v.

Advantages

1. Higher, about 80% faster, metal-removal rates are obtained with ECG than conventional grinding of hard materials such as carbides, germanium, cobalt, cast alloys, etc. On hardened alloy steels, the metal removal rate is about 0.100 cu in./min/1000 amp (1.64 cm³)/min/1000 amp). On carbides, it ranges down to 0.060 cu in./min/1000 amp (1.0 cm³/min/1000 amp).
2. Wheel wear is drastically reduced. Depending upon feed rate, size of the wheel, and other process variables, the wheel life can be increased by a factor of 10 on most ECG applications.
3. No heat is generated, so there is no danger of burning or heat distortion.
4. Only a small amount of mechanical force is applied to the workpiece, so the possibility of burrs or distortion is also eliminated.

Disadvantages

1. As with ECM, the salt solution is corrosive. The machine should be washed down periodically to remove salt deposits.
2. The electrolyte should be changed about once a week.
3. Initial cost of the equipment is high when equipped with large power supplies.
4. Copper-resin–bonded diamond wheels may be dressed to simple forms only. Intricate forms with multiple radius tangents and very deep forms require single-layer, plated-metal, bonded diamond wheels.

ABRASIVE JET MACHINING (AJM)

Abrasive jet machining may be simply described as a fine stream of abrasive particles propelled through a nozzle by a high-pressure gas for the purpose of cutting a wide variety of materials. The gas used to obtain the high pressure may be carbon dioxide, nitrogen, or filtered compressed air. The nozzle through which the abrasive particles are propelled is usually made of tungsten carbide or synthetic sapphire. A typical AJM unit is shown in Fig. 20-12(a) and a sketch of the process at (b).

The jet can be used to make fine (0.005 in.; 0.13 mm) line cuts or a wide spray to "frost" large areas. The width of the cut is dependent on how far the

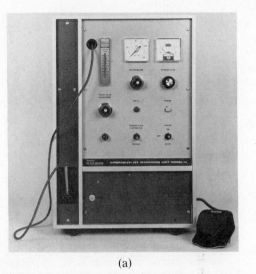

(a)

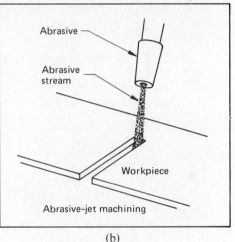

Abrasive

Abrasive
stream

Workpiece

Abrasive-jet machining

(b)

Figure 20.12. A typical abrasive-jet machining unit. (*Courtesy S. S. White Industrial Products.*)

abrasive must travel before it impinges on the workpiece. As the jet leaves the nozzle opening, which may be either round or rectangular, it begins to diverge at about $\frac{1}{16}$ in. (1.58 cm) from the opening. When the nozzle is kept relatively close to the work, the cut will be straight; at greater distances, the cut will become triangular in cross section. In materials that are too thick for a "straight" cut or where the nozzle distance is too great, a straight side is possible simply by angling the jet. Straight cuts have been made in steel to a depth of 0.060 in. (1.52 mm), and in glass to depths up to 0.250 in. (6.35 mm). Currently the narrowest cut possible is 0.005 in. (0.13 mm), made with a 0.003 × 0.060-in. (0.08 × 1.52-mm) rectangular nozzle opening held at a distance of $\frac{1}{32}$ in. (0.793 m) from the work.

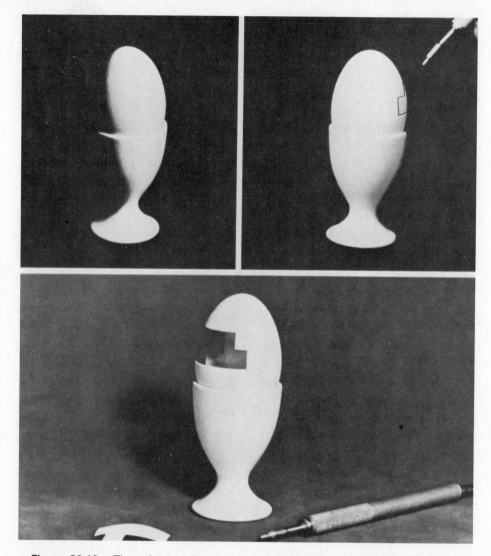

Figure 20.13. There is very little impact to AJM as shown by the ability to cut an eggshell. (*Courtesy S. S. White Industrial Products.*)

Cutting Variables.   The cutting rate can be altered by several variables such as flow rate, distance of nozzle to the work, gas pressure, and type of abrasive used. The maximum flow of abrasive is about 10 gpm (37.8 lpm). Gas pressure is usually 75 psi (6.07 kg/cm^2), but it may vary from 1 to 125 psi (0.070 to 8.6 kg/cm^2).

The abrasive powders used range from about 10 to 50 microns. The hardness may vary from a soft sodium bicarbonate to a hard silicon carbide. Other commonly used abrasives are aluminum oxide, dolomite, and ground nutshells.

Applications.   Etching and surface preparation are common applications of AJM. The process has been successfully used in cleaning corrosion and other contaminants from such diverse parts as electronic components to priceless leather artifacts. A common use is in trimming resisters of hybrid power-amplifier circuits for a specified value within a tolerance of ±0.5%. The entire operation is automatic. The time for trimming each resistor is 0.5 sec.

Advantages

1. The abrasive jet can be used to cut any material. Even diamonds have been cut, using diamond dust as the abrasive.
2. Because of the low velocity of the abrasive particles, there is virtually no impact and very thin brittle materials can be cut without danger of breaking. This is shown by the example of cutting an eggshell (Fig. 20-13).
3. Virtually no heat is generated in the workpiece.
4. The process is safe. The operator can pass his hand under the jet without injury.

Disadvantages

1. Because of the very small stream of abrasive particles, the material removal rate is low. On plate glass, for example, the removal rate is about 0.001 cu in./min (0.03 mm^3/min), or the equivalent of a "slot" 0.020 × 0.010 × 5 in (0.57 × 0.25 × 127 mm).
2. The abrasive powders cannot be reused since the points and edges get worn down. However, the cost of most abrasives is relatively low.
3. Because of its nature, AJM usually requires some type of dust-collecting system. Most machines are used with an enclosure to which a vacuum hose has been attached.

LASER-BEAM MACHINING

The principles of the laser were discussed in Chapter 14 in connection with its use in welding. It may also be used in special applications for machining. The two principal types of lasers are often referred to as "glass and gas." Glass refers to the solid-state lasers: ruby, Nd : YAG, Nd : glass, etc. The gas lasers are of the CO_2 type.

The continuous-wave (CW) gas laser as used in metal cutting is relatively new. Bell Labs received the patent on it in 1965. This laser had a power of several milliwatts. Since then the power of the CO_2 laser has grown astronomically to almost 100 kilowatts.

The laser is probably the most versatile of all nontraditional machining processes. It can be used to cut and drill virtually any material—hard, soft, brittle, or ductile, insulator or conductor. As an example, lasers can be used to slice 0.5 in. (12.70 mm) acrylic plastic, drill 0.001 to 0.04 in. (0.03 to 1.02 mm) holes in tungsten sheet, cut 0.4 in. (10.16 mm) thick steel (with oxygen assist), and drill 0.01 in. (0.25 mm) holes in 0.025 in. (0.63 mm) thick ceramic.

Table 20-3. Laser cutting rates for metals.* (*Courtesy* Modern Machine Shop.)

Metal	Thickness, inches	Power, watts	Cutting rate ipm
Mild steel	.252	500	20
Mild steel	.051	500	142
Stainless steel	.012	500	146
Stainless steel	.252	500	20
Low alloy steel	.024	850	23
Tin plate	.020	500	240

* Note that these cutting rates are with a power output less than 1.2 kw. A standard metal-cutting system would cut 400 series. 0.250-in. thick martensitic stainless steel at approximately 40 in./min.

Some manufacturers are offering lasers combined with conventional CNC punch presses. The conventional hard tooling is for the regular jobs and the laser can be quickly programmed to turn out the odd-shaped and prototype parts made of hard-to-machine materials. Table 20-3 shows the expected cutting rates for lasers in steel. Each axis of the laser has a rigid frame weldment that allows the laser to cut within an accuracy of ± 0.005 in. (0.13 mm) per foot of travel.

Table 20-4 shows the advantages of the laser cutting process.

Disadvantages

1. The laser beam can be dangerous if it is not used carefully. The visible laser beam produced by the ruby can penetrate the cornea and do retinal damage. The CO_2 laser, on the other hand, is in the infrared region and cannot be seen. Special glasses are necessary for the ruby laser but regular safety glasses will protect against the CO_2 laser.
2. Production laser machines are costly, ranging from about $18,000 to $100,000 for CO_2 units. Ruby lasers are considerably below this range.
3. The laser depends on thermal properties of the material being cut. When exposed to the laser beam, some materials char and bubble, as in the case of fibreglass-reinforced structures. In general, a material's boiling and vaporization points determine how suitable it is for laser machining. The higher the vaporization point, the more difficult it is to laser machine. Also, the closer the material's boiling and vaporization points, the harder it is to machine.

Applications. In recent tests of a $CWCO_2$ laser with an oxygen-gas jet assist (Fig. 13-26) 0.5 cm (0.195 in.) thick stainless steel sheet was cut at the rate of 75 cm/min (30 ipm). Plywood of about the same thickness was cut at the rate of 520 cm/min (195 ipm) with an argon-gas jet assist. The same laser system can be easily changed from cutting metals to cutting flamable materials by substituting an inert gas for a reactive gas in the jet-assist apparatus.

At the present time, relatively little of this new technology has been exploited. It is expected that, as the technology expands, there will be CO_2 lasers "on-line" as a production tool in a variety of industries.

Table 20-4. Laser cutting advantages. (*Courtesy* Modern Machine Shop.)

Product	More economic prototypes or specials Low total heat input and self-quenching No part distortion caused by excessive heat or tool- ing force Sharp corners cut because of small laser beam spot size Internal cuts easily made
Tooling	Reduce necessity for tool storage and retrieval Exceptionally long life; laser never grows dull No conventional tooling or machine setup required No repair, sharpening or repairing of broken tools A noncontact process, it requires minimal work- holding fixtures
Production environment	Operators require less skill for total turnkey system Material savings through narrow kerf, typically 0.005 to 0.012 in. Simplify operations on materials that are expensive and difficult to process, such as refractories and composites Operates in normal manufacturing environment; fast cutoff speeds Easily adaptable to all types of control systems; re- duce inventory and work-in-process

ELECTRON-BEAM MACHINING (EBM)

The electron beam is usually associated with welding and was discussed in Chapter 14; however, it may also be used in metal removal.

EBM may be described as machining by sublimation. A stream of electrons is used to bombard a piece of metal until it "perforates" a hole in it. A stream of electrons, with densities typically of 10^5 to 10^6 W/cm^2 can generate dozens of holes from 0.004 to 0.040 in. (0.10 to 1.02 mm) in microseconds. The process is not affected by the type of material being drilled or by its hardness.

The drilling capability of the beam varies with the material condition and the hole depth required. As an example, a 0.004 in. (0.10 mm) hole can be made in 0.001 in. (0.03 mm) thick sheet to a 0.040 in. (1.02 mm) dia. hole in a 0.250 in. (6.35 mm) thick material. Generally deeper holes are larger in diameter.

EBM is performed in a vacuum to avoid dispersion of the beam by collision of the electrons with gas molecules. Machining time estimates must take into acocount the time required to pump down the working chamber each time a workpiece (or fixture of many workpieces) is loaded.

EB drilling systems are usually controlled by a digital computer to take advantage of the extremely high-speed drilling capability of the process. The computer controls the beam current as a function of the workpiece position and time, and triggers the energy pulses. The workpiece moves almost continuously under the beam, and its position is recorded by the computer.

Metal is removed by a combination of fusion and evaporation. The vapor pressure causes the liquid material to eject. As in EB welding, a white-hot capillary column of molten material forms in the workpiece under the beam, vapor pressure builds up, and the material is ejected in a sudden burst.

Advantages. EBM is particularly well suited to high-speed perforation of multiple holes, relatively small diameter holes in thick material, tapered holes, noncircular holes, engraving of metals and ceramics, and engraving vapor-deposited layers. An example of engraving would be making a miniature electronic hybrid circuit. Copper or other conducting material is vapor deposited on a ceramic substrate. An EB trace 40 micrometers wide is used to engrave the circuit. Circuits can be made at 5 m/sec or faster.

EB is fast because it is entirely nonmechanical. It is electromagnetically generated, focused, and guided. The EB travels at two-thirds the speed of light and is free from mechanical errors and is not slowed by moving mechanical guides or lenses.

Disadvantages. EB machining is done in a vacuum, which adds to the expense and setup time.

A computer is needed to provide three-dimensional simultaneous movement of the focal point in combination with a moving workpiece. Computer control is also required to produce a hole of the desired shape based on 50 to 1000 cycles per second.

Because EB is a thermal metal-removal process, thin parts may distort. When possible, it is better to have more small holes with less heat than a few large ones.

The melting and evaporation of the material in EB holes usually makes them have a slight bell or trumpet shape.

Problems

20-1. A hardened alloy-steel block is to be ground by the electrochemical process. The area to be machined is 1×2 in. and it is to be cut to a depth of 0.060 in.

(a) How long will it take to machine this surface assuming 1000 amp will be used?

(b) Assume a 300 amp machine will be used.

(c) Assuming a 0.010 cu in./min/100 amp metal-removal rate, what will the feedrate be?

(d) What is the time required to make the total cut?

(e) Determine the feedrate from the proportion: time for cut/60 = depth of cut/feedrate.

20-2. A carbide tool with a surface area of 0.5×1 in. is to be cut to a depth of 0.0625 in. Assume 900 amp will be used to ECG this surface. Power supplies are available in capacities of 300, 600, 1000, and 1500 amp.

(a) What power supply will be needed?

(b) What is the volume of metal to be removed?

20-3. What is the minimum hole size that can be photochemically machined in $\frac{1}{32}$ in. thick copper, $\frac{1}{16}$ in. thick aluminum, and 0.010 in. thick brass?

20-4. If 2 in. dia. brass circles are to be photochemically machined from $\frac{1}{16}$ in. thick material, how much larger should the drawing be made? Assume an enlargement of 5 : 1.

20-5. A material that has an etch factor of 3 : 1 is to have several slots cut into it. If the material is $\frac{1}{16}$ in. thick and the slots are to be $\frac{1}{16}$ in. wide, what width should they be drawn? Assume a 10 : 1 enlargement.

20-6. Shown in Fig. P20.1 is a stamping that can be made in various ways.
   (a) What tolerance could be expected if the part were photochemically machined?
   (b) What size would the electrode punches be for the holes if the die cavity were to be made by EDM?
   (c) How long would it take to etch out one piece or a sheet of parts by photochemical machining using the double-sided etching method?
   (d) If just ten of the parts shown in Fig. P20.1 were to be made, how would you propose they be made?

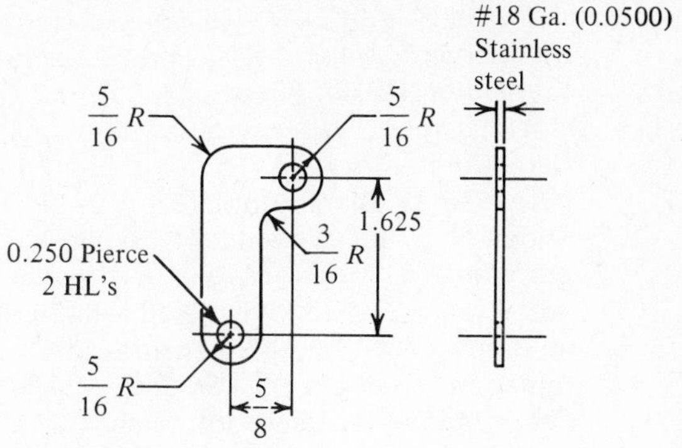

20-7. Assume a forging die must have 15 cu in. of metal removed and have a 100 microinch finish or better.
   (a) What methods could be used to produce the die if it is to be machined in the hardened condition?
   (b) How long would the machining take?

20-8. A 1 × 1 in. square is to be removed from 0.25 in. thick plate glass. Approximately how long would it take if AJM were used? Assume the cut will be 0.020 in. wide.

20-9. (a) Assume 100 parts, as shown in Fig. P20-1, can be cut out of one sheet of mild steel. A CNC program will be used to guide the laser beam. About how long

#18 Ga. (0.0500) Stainless steel

Figure P20.1.

would it take to cut all the parts? Assume 0.125 in. spacing between the parts. Include the two holes.
   (b) How long would it take to cut 100 of these parts out of 0.252 in. thick stainless steel?
   (c) How long would it take to cut the same parts out of 0.200 in. thick plywood if an oxygen-gas jet assist were used?

20-10. Just how does the electron beam remove metal?

20-11. What makes the EB so well suited to perforating multiple holes in metal?

20-12. Why are some punch presses being fitted with lasers for contour cutting in some punch presses but not EB?

Bibliography

Bellows, Guy. "Nontraditional Machining: Where Does It Stand?" Four part series, *Modern Machine Shop* 49 (April 1976 through July 1976).

Boothroyd, G., and A. H. Redford. *Mechanized Assembly.* McGraw-Hill House, Maidenhead, Berkshire, England, 1968.

Dallas, Daniel B., ed. "Systems Engineering in Advanced Assembly." *Manufacturing Engineering* 83 (October 1979): 54–57.

Drew, John. "Electron Beams Tackle Tough Machining Jobs." *Machine Design* 48 (February 26, 1976): 94–98.

Editor. "90% Uptime from Unmanned Machining Systems." *Production* (June 1979).

Gettelman, Ken., ed. "Programmability and Sensing: Robot's New Dimensions." *Modern Machine Shop* (June 1979).

——. "Robots Drill and Profile." *Modern Machine Shop* (April 1980).

——. "Untying Man from Machine." *Modern Machine Shop* (August 1979).

Harrison, H., and J. Bollinger. *Introduction to Automatic Controls.* Scranton, Pa.: International Textbook Co., 1969.

Marshall, Harry E. "EDM—No Longer Nontraditional." *Modern Machine Shop* 15 (September 1978): 90–97.

McDonald, James. "When Ordinary Methods Can't Cut It—Try AJM." *Cutting Tool Engineering* 31 (March/April 1979): 8–11.

Schneider, R. T., ed. "Versatile Hydraulic Robot Handles Tough Tasks." *Hydraulics and Pneumatics* 32 (July 1979): 55–58.

Smith, Bradford M. "New Dimensions in Automation Technology." *SME Technical Paper MS79-381,* Dearborn, Mich., 1979.

Smith, T. C. "NC's Stake in the Productivity Race." *NC Commline* 7 (September/October 1978).

Springborn, R. K., ed. *Non-traditional Machining Processes.* Dearborn, Mich.: Society of Manufacturing Engineers, 1967.

Treer, Kenneth R. *Automated Assembly.* Dearborn, Mich.: Society of Manufacturing Engineers, 1979.

Winship, John, ed. "Should You Buy a Robot?" *American Machinist* 119 (November 1975): 105–108.

CHAPTER TWENTY-ONE

Measurement— Quality Control— Product Liability

The progress of measurement has played a large part in man's scientific advancement. Early attempts at standardization of length measurements were based on the human body. The width of a finger was termed a digit, and the cubit was the length of the forearm from the end of the elbow to the tip of the longest finger. These measurements were in use at the time of the construction of the Khufu pyramid (4750 BC) and are mentioned in connection with the building of Noah's Ark.

Metric Measurement. From simple beginnings, man progressed gradually toward standardization of basic measurements. A very significant advance was made at the Convention of the Meter held in Paris in 1875. At that time, the meter was defined as one ten-millionth of the distance from the north pole to the equator, measured on the meridian quadrant that passes through Paris. The metric system was worked out with related factors of 10 so that conversions could be handled by simple shifting of decimal points.

In 1960, 36 countries, including the United States, participated in the 11th

General Conference of Weights and Measures, meeting in Paris. At that time, the modernized International System of Units was adopted. It may be abbreviated SI from its French name, *Systéme International d'Unités*.

The SI meter is now defined in terms of wavelength of light, 1 650 763.73 wavelengths of krypton light.

A new metrology lab announced that it had the capability of measuring one-tenth of a microinch. To understand how small this measurement is, the following example is given: A 1 in. dia. 10 in. long quartz bar (the most perfect elastic material known) is placed on simple supports at either end. For every gram load at the center, that point will sag by about 1 microinch.

The International Organization for Standardization (ISO) is responsible for

Table 21-1. International system of units.

Physical Quantity	Unit	Symbol
Base Units		
length	meter	m
mass	kilogram	kg
time	second	s
electric current	ampere	A
thermodynamic temperature	kelvin	K
luminous intensity	candela	cd
amount of substance	mole	mol
Derived Units		
area	square meter	m^2
volume	cubic meter	m^3
speed	meter/second	m/s
acceleration	meter/second squared	m/s^2
density	kilogram/cubic meter	kg/m^3

Combination Derived Units

Physical Quantity	Name of Unit	Symbol	Derived Units	Combination Base Units
force	newton	N		$m \cdot kg \cdot s^{-2}$
pressure & stress	pascal	Pa	N/m^2	$m^{-1} \cdot kg \cdot s^{-2}$
energy	joule	J	$N \cdot m$	$m^2 \cdot kg \cdot s^{-2}$
power	watt	W	J/s	$m^2 \cdot kg \cdot s^{-3}$
conductance	siemens	S	A/V	$m^{-2} \cdot kg^{-1} \cdot s^3 \cdot A^2$

Supplementary Units

Physical Quantity	Unit	Symbol
plane angle	radian	rad
solid angle	steradian	sr

international development of industrial and commercial standards in almost every field of technology. It is supported by 55 nations, including the United States.

The SI system is built up from three kinds of units: base units, derived units, and supplementary units, as shown in Table 21-1.

An essential feature of SI is the systematic use of prefixes to designate decimal multiples and decimal fractions of base units. The most commonly used SI prefixes are shown in Table 21-2.

Tables of conversion factors, permitting conversion from English to metric units, are widely available and an abbreviated example is shown in Table 21-3 and in Appendix H. Some useful approximations are: a meter is about 10% longer than a yard; a liter has 5% more volume than a quart; to get kilograms, subtract one-tenth of the number of pounds and divide by 2; to calculate kilometers (or km/h), add half of the miles (mph); a Btu is about one kilojoule; and water freezes at 0°C and boils at 100°C.

Not standard at the present time, but widely used, is the system of using a comma to separate a decimal fraction rather than a decimal point. Thus $2\frac{3}{4}$ is written 2,75. Also, a blank space is used rather than a comma to group every three digits of a number. Thus a million is written as 1 000 000.

The Development of Useable Standards

The use of the earth's meridian (or some specified fraction of it) was fine in theory, but it was difficult to translate into everyday needs. Even with the definition of the meter as the distance between two engraved lines on a bar of a platinum-iridium alloy at zero degrees centigrade (now Celsius), it was impractical for manufacturers to use.

Gage Blocks. Toward the close of the nineteenth century, Carl Johansson of Sweden assumed the task of providing measurement standards that would be

Table 21-2. Commonly used SI prefixes.

Prefix	Factor	Symbol
tera	10^{12}	T
giga	10^{9}	G
mega	10^{6}	M
kilo	10^{3}	k
hecto	10^{2}	h
deka	10^{1}	da
deci	10^{-1}	d
centi	10^{-2}	c
milli	10^{-3}	m
micro	10^{-6}	μ
nano	10^{-9}	n
pico	10^{-12}	p
femto	10^{-15}	f

Table 21-3. Basic conversion factors.

To Convert from	To	Multiply by
Inches	Millimeters	25.4
Feet	Meters	0.3048
Yards	Meters	0.9144
Miles	Kilometers	1.609
Square inches	Square centimeters	6.4516
Square feet	Square meters	0.0929
Square yards	Square meters	0.836
Acres	Hectares	0.405
Cubic inches	Milliliters	16.387
Cubic feet	Cubic meters	0.0283
Cubic yards	Cubic meters	0.765
Quarts	Liters	0.946
Gallons	Liters	3.784
Ounces	Grams	28.35
Pounds (mass)	Kilograms	0.454
Pounds (force)	Newtons	4.448
Pounds per square inch (psi)	Kilopascals	6.895
Horsepower	Kilowatts	0.746
Btu	Kilojoule	1.055

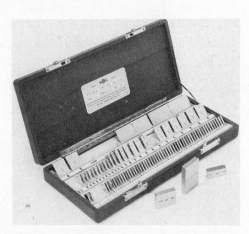

Figure 21.1. A standard 81-block gage set. (*Courtesy DoAll Company.*)

available to industry. They were to duplicate the international standards as nearly as possible. To accomplish this, he painstakingly made a set of steel blocks that were accurate to within a few millionths of an inch. This unheard of accuracy, and how it was achieved, remained a secret with Johansson for many years. Since these steel blocks were so accurate, they could be used to check manufacturers' gages and other measuring instruments. The blocks were made up in sets and are known as "gage blocks." A typical set consists of 81 blocks with sizes ranging from 0.05 in. to 4.0 in., as shown in Fig. 21-1.

Gage Block Use.   Gage blocks maintain accuracy even though several have to be used in combination to produce a specified dimension. By carefully sliding the gaging surface of one block over the gaging surface of another, two blocks can be "wrung" together so tightly they are difficult to separate (Fig. 21-2).

Gage blocks may be used in direct measurement, as in checking the size of a keyway or the distance between milling cutters (Fig. 21-3). They may also be used for indirect measurement, as in checking a snap gage or setting a dial indicator, as shown at Fig. 21-4.

Gage Block Calibration.   We have seen how a useable standard was developed for industry, but how could the accuracy of these blocks be checked consistently within millionths of an inch?

As early as 1827 a French physicist, Jacques Babinet, suggested that the wavelength of light had possibilities as a standard of length. His proposal could not be realized until much later when the concept of interferometry had been developed. In 1892–1893 Professor Michelson of the University of Chicago perfected methods of optical measurement and defined the international meter in terms of a number of light wavelengths. The concept, which is relatively simple, is that when the crests of two light waves coincide, they reinforce each other; when a crest coincides with a trough, the two waves cancel. In the first case the interference is constructive and in the second case it is destructive, hence the alternate light and dark bands.

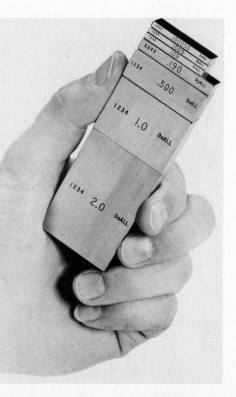

Figure 21.2. Gage blocks that have gaging surfaces wrung together form effective length rods for specific measurements. (*Courtesy DoAll Company.*)

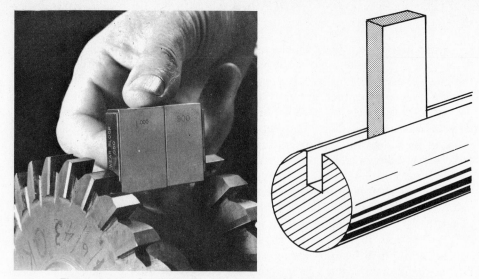

Figure 21.3. Gage blocks may be used for direct measurement as in checking a keyway or measuring the distance between milling cutters. In the latter case, carbide-wear gage blocks are wrung in for protection of the gaging surfaces. (*Courtesy DoAll Company.*)

(a)

(b)

Figure 21.4. The dial indicator is an example of an indirect-measuring tool. The workpiece height is compared to that of a gage block (a). A gage block or a combination of gage blocks can be wrung in to check a snap-gage setting (b).

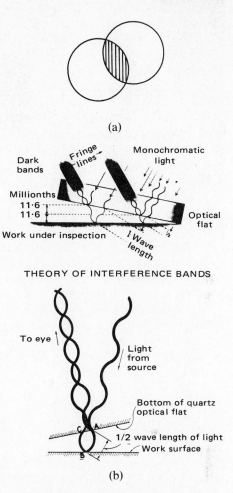

Figure 21.5. Interference bands show between the surfaces of two ceramic-glass gage blocks even under fluorescent lighting (a). The straightness of the lines indicate the flatness and the degree of wring. The theory of interference bands is shown at (b).

A simple interferometer can be made by placing one optically flat glass plate on top of another. Shown in the sketch [Fig. 21-5(a)] are two ceramic-glass gage blocks wrung in at the edge. Shown in Fig. 21-5(b) is a more detailed view of how the fringe lines are formed. The light of a single color, or wavelength, is directed to the upper surface. When this light reaches the lower surface of the upper flat, it is divided into two portions. One portion is reflected upward by the lower surface and the other portion continues downward to be reflected from the upper surface of the lower flat. The two series of reflected rays recombine in the upper flat, where they interfere to form alternate light and dark bands as seen from above. Such interference will occur at every point where the distance from the flat to the work is in odd multiples of one-half the wavelength of the light used. By counting the number of fringes between the edge of the upper plate and its point of contact with the object to be measured, the object's length can be determined. It will be equal to the wavelength of the light used multiplied by one-half the number of dark fringes to the point of contact. The most-used light source in metrology laboratories is helium, which has a wavelength of 23.2 microinches. The krypton-86 isotope is now the basic international standard of length. The meter is defined as being exactly 1,650,763.73 wavelengths of this source, measured in a vacuum. The krypton lamp is generally used only in standardizing laboratories due to the cooling requirement, and because the lamp emits so many different wavelengths (spectral lines) that a fairly elaborate monochromator is required to separate them. It now appears that the helium-neon laser is the best light source for the measurement. It is capable of producing light that is far more monochromatic and more intense than other light sources, and therefore extremely sharp fringe lines are produced. The American National Standards Institute has related one laser line, the neon line, at a wavelength of 6328 angstrom units, to the krypton standard with an accuracy of one part in 100 million, which is close to the limit of accuracy inherent in the krypton wavelength.

Conceivably the laser could extend the limit of interferometric measurement to hundreds or even thousands of kilometers. Such measurements, however, present the difficulty of counting several hundred million interference fringes, a task that could be accomplished only by automatic means.

The flatness of the gage block can be checked by noting the pattern of the fringe lines. If they are straight and parallel, the surface is flat within one-half wavelength of the light used. If the surfaces are not flat, the fringe lines will not be straight but will show the amount of deviation by their contour, as shown in the three sketches of Fig. 21-6.

Precision Measurement Principles

The science of measurement has progressed until "millionths of an inch" is becoming a commonplace term. It is difficult to grasp the true meaning of so fine a measurement. To calibrate a gage block or to make any measurement to within 1 microinch requires considerable precaution and effort since temperature, humidity, ambient air conditions, and other factors must be at specified standards. Thus controlled environment laboratories, or metrology labs as they are often referred to, are necessary to insure precise measurements.

The temperature of standard metrology laboratories is maintained at the international reference level of 68°F (20°C) within ±0.5°F. Owing to differences in linear coefficients of expansion of various materials, even small deviations in temperature can cause significant dimensional differences.

Foreign particles cannot be tolerated in the atmosphere of a standards laboratory due to the adverse effects on the accuracy of measurement and dependable operation of instruments. Clean air is accomplished by careful filtering and by having a higher air pressure in the room than outside of it. Personnel enter through a double air lock.

Humidity is maintained at 45 to 50% to eliminate the possibility of rust, which forms quickly on highly polished surfaces in the presence of moisture.

Precision and Accuracy.   Precision and accuracy, two key concepts in the science of measurement and gaging, are frequently misused. *Precision* may be

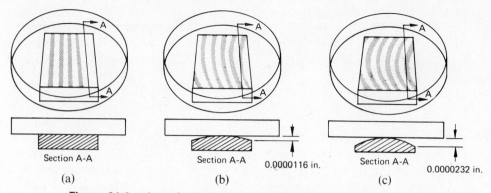

Figure 21.6.   A perfectly flat surface (a) produces uniform straight interference bands when viewed through the optical flat. A slightly convex part (b) produces curved bands. The amount of curvature is one band or 0.0000116 in. Shown at (c) is a curvature of two full bands or 0.0000232 in.

defined as the closeness with which a measurement can be read directly from the measuring instrument. It is the smallest marked increment on the instrument as on a scale 1/64 in. or on a dial as 0.001 in.

Accuracy is a measure of how close a reading is to the true size of the part being measured. This involves temperature, the mechanics of the measuring instrument, the pressure exerted by the instrument, and at times the skill of the user. Thus a measuring instrument may have a precision of 0.001 in. (0.03 mm) and a repeatable accuracy of 0.005 in. (0.13 mm).

A level of accuracy and precision is inherent in every measuring tool. Many of the hand-operated tools in the lower echelons of accuracy depend on operator skill and care. Instruments of high accuracy, however, are protected against operator-initiated errors.

Basically, all measuring devices can be broadly classified into two groups: direct and indirect. Direct-measuring instruments obtain the measurement without the aid of other equipment, as in the case of a micrometer. Indirect-measuring tools require a standard of reference, as in the example of a dial indicator shown in Fig. 21-4.

Measurement can be further divided into five main areas: linear (length, diameter), geometric forms (roundness, flatness), geometric interrelationships (parallelism, squareness, and concentricity), angular dimensions, and surface texture.

Linear Measurement

Direct

Linear measurements may be taken directly with rules (scales), micrometers, vernier scales, and measuring machines.

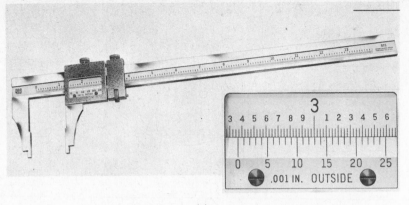

(a)

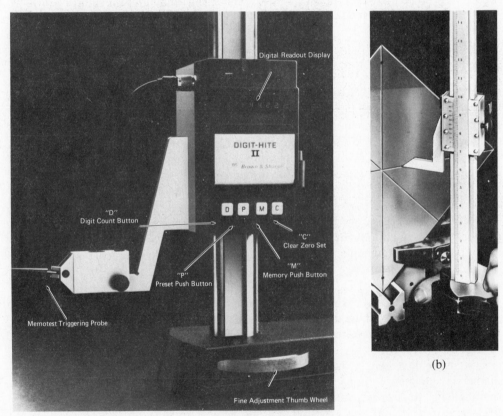

(c)

(b)

Figure 21.7. The vernier scale is used for direct measurement in the vernier caliper (a) and the vernier height gage (b). Shown at (c) is a digital readout height gage. (*Courtesy Brown & Sharpe.*)

Steel Rule.   The steel rule or scale is the most common measuring device. Graduations commonly used are fractions of an inch down to $\frac{1}{32}$ in. Millimeter scales are also available.

Vernier Scales.   Vernier scales improve the accuracy of the scale by an order of magnitude. The vernier scales are used on calipers and height gages shown in Fig. 21-7. The reading on the insert scale is as follows.

> 2.000 in. (the zero on the vernier scale has passed 1 in.)
> + 0.300 hundred-thousandths on the scale
> + 0.050 two twenty-five thousandths lines
> + 0.018 thousandths, where the vernier line coincides with a line on the scale

Total = 2.368 in.

Dial Caliper.   The main disadvantage of the vernier caliper has been the difficulty in reading it. This has now been remedied by the introduction of the dial caliper shown in Fig. 21-8(a) and most recently by the digital readout caliper shown in Fig. 21-8(b). Both of these calipers are very easy to read, within 0.001 in. (0.03 mm). They have four modes of measurement: inside, outside, depth, and step. A T-bar [Fig. 21-8(c)] may be added to facilitate depth measurement. The digital gage features switchable English/Metric, full-floating zero for comparative measurement and memory to retain a reading.

Micrometers.   Micrometers are universal direct-measuring tools. Using the mechanical reduction of a fine-pitched screw thread, the micrometer reduces great rotational motion to slight linear motion. One full rotation on the micrometer thimble or barrel (Fig. 21-9) produces only 0.025 in. of linear travel.

The basic micrometer principle leads to many special-purpose measuring tools such as the thread and depth micrometers shown in Fig. 21-10.

Figure 21-11 shows a laser micrometer. The part to be measured is placed in the path of the scanning laser beam and the measurement shows up instantly on the digital readout. The laser "bench-top" micrometer shown can measure any dimension from 0.00003 to 1.2000 in. (0.08 to 30.48 mm) with resolutions from 0.00001 to 0.0001 in. (0.00025 to 0.0025 mm). Pushbutton controls enable the micrometer to display readings from the memory of the microprocessor and indicate average, minimum, and maximum readings. Set point limits with visual and/or audible alarms are available.

Coordinate Measuring Machines.   Coordinate measuring machines are highly sophisticated direct-reading equipment that can measure on three axes simultaneously (Fig. 21-12). In operation, a suitable probe tip is first selected and inserted into the stylus. To measure center distances between several holes, for example, a tapered plug probe would be used, as shown in Fig. 21-13. The probe

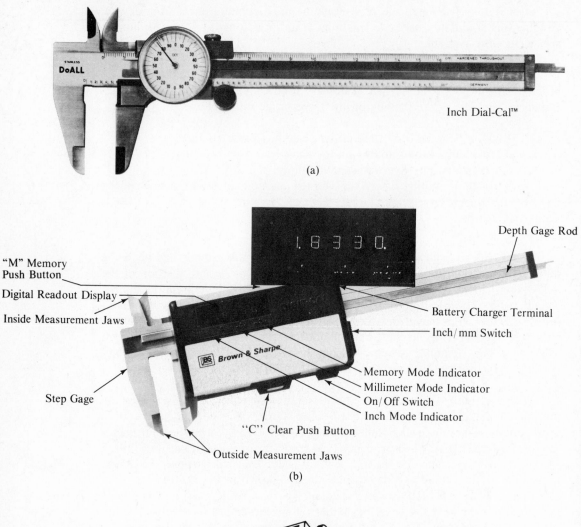

Inch Dial-Cal™

(a)

"M" Memory Push Button

Digital Readout Display

Inside Measurement Jaws

Step Gage

Depth Gage Rod

Battery Charger Terminal

Inch/mm Switch

Memory Mode Indicator

Millimeter Mode Indicator

On/Off Switch

Inch Mode Indicator

"C" Clear Push Button

Outside Measurement Jaws

(b)

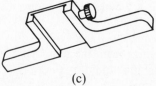

(c)

Figure 21.8. A dial caliper (a), a digital readout caliper (b), T-bar used to facilitate depth measurement (c). (*Courtesy Brown & Sharpe and DoAll Company.*)

is inserted into the first hole and the machine is set to a zero reference. As the probe is moved from one hole to another, movements indicating each position on x, y, or z coordinates are read out on the digital display.

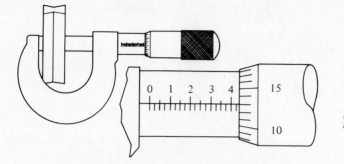

To read any micrometer setting; first, read scribe marks
along hub — this one reads 425 thousandths (.425) inch.
Second, read scribe marks around thimble — this one
reads 13 thousandths (.013) inch. Add reading of
thimble to reading of hub — .013 + .425 = .438.
This micrometer is set at 438 thousandths (.438) inch.

(a)

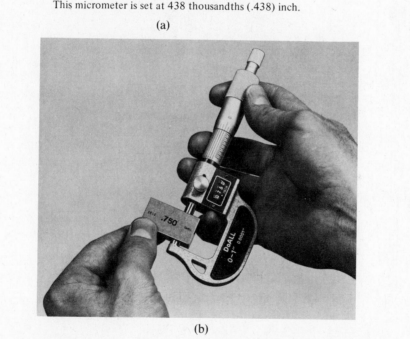

(b)

Figure 21.9.   The micrometer with an enlarged view of reading (a),
a newer digital type micrometer (b). (*Courtesy DoAll Company.*)

A typical measuring machine operates on the moiré fringe-pattern principle.
A ruled grating extends the entire length of travel. An index grating with the
same line pattern is mounted above it at a slight angle. The superimposed pattern
is the familiar moiré fringe [Fig. 21-14(a)]. As the measuring head travels from
point to point, the changing fringe pattern is converted to electrical impulses
through photocells (b).

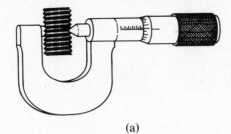

(a)

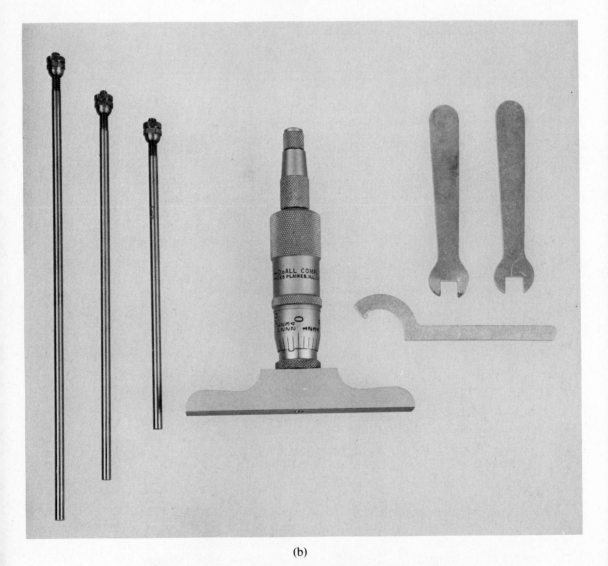

(b)

Figure 21.10. A thread micrometer used to measure the pitch diameter of a thread (a). A depth micrometer complete with interchangeable depth rods and wrenches (b). (*Courtesy L. S. Starrett Co. and DoAll Company.*)

Figure 21.11. A laser micrometer that measures parts instantly with a scanning beam. The part size is given on the digital readout to within 0.0001 in. (0.00025 mm). (*Courtesy Techmet Company.*)

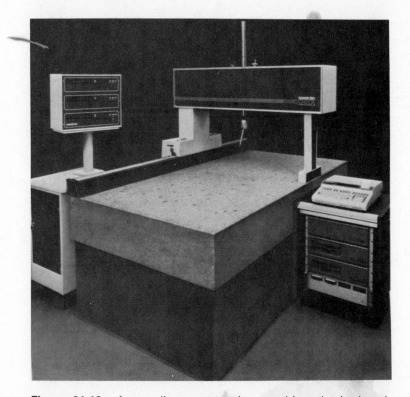

Figure 21.12. A coordinate measuring machine checks location points in three axes to an accuracy of ± 0.00015 in. (0.00375 mm). The probe can be made to move on command from the desk top computer tape or floppy disc memory. A printout provides a permanent record of the measurements. (*Courtesy Hansford Manufacturing Corporation.*)

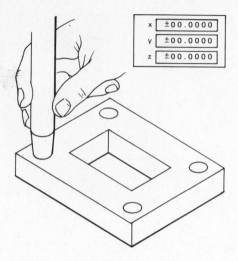

x	±00.0000
y	±00.0000
z	±00.0000

Figure 21.13. A tapered-plug probe is used when measuring the center distances of hole locations.

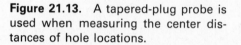

(a)

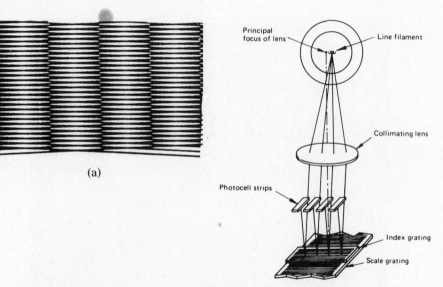

(b)

Figure 21.14. A typical measuring machine operates on the moire fringe pattern, which occurs when one grated plate is placed above another at a slight angle (a). As the measuring head travels from point to point, the changing fringe pattern is converted to electrical pulses (b). (*Courtesy* Machine Design.)

Measuring machines can quickly and easily do jobs that would be long and tedious with less sophisticated equipment. In addition to measuring, they can be used for marking and layout work. The main disadvantage is the high initial cost.

Figure 21.15.   A dial indicator is used to make a comparative measurement between the workpiece and the gage block from the same reference plane. (*Courtesy L. S. Starrett Co.*)

Indirect

Indirect-measuring tools are used to transfer or compare a measurement with a known standard. These *comparators,* as they are referred to, are of four main types: mechanical (dial indicator), electronic, optical, and air.

Dial Indicators.   The dial indicator has a stylus probe that is connected through a precision gear train to the dial pointer. Small linear deflections of the stylus are multiplied into large pointer movements. An example of a comparison measurement is shown in Fig. 21-15. The part surface to be measured is compared to a gage block or a combination of gage blocks that have been wrung in. The gage is set at "0" over the blocks and the workpiece is measured as any deviation from "0".

Comparison measurement frequently requires a good frame of reference. As was shown in Fig. 21-15, the indicator base, the part, and the standard all rest on a precision surface plate for a uniform plane of reference. Any variation in the "flatness" of the surface plate will cause the measurement to be inaccurate. However, granite surface plates usually have guaranteed overall accuracies to within ±0.0002 in. in the larger sizes, ±0.000025 in. in the smaller sizes.

Optical Comparators.   Optical comparators improve accuracy by magnification. When an optical comparator, such as that shown in Fig. 21-16, magnifies an image by 100X, an observed error of 0.10 in. (0.25 mm) is only 0.001 in. (0.03 mm) full scale.

Optical comparators are best suited for silhouette projections, although some models are equipped for surface projection. These machines are ideal for measuring complex outside dimensions as well as tiny intricate grooves, radii, or steps. Frequently templates matching the correct shape are placed on the pro-

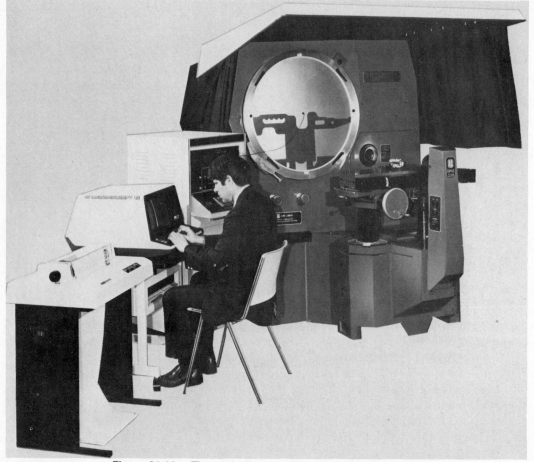

Figure 21.16. The optical comparator magnifies the surface silhouette and is especially useful for checking complicated forms. Shown on the screen is a sensor. As the shadow or light area moves past the sensor aperture, the digital readout displays the edge-to-edge distance the table has traveled. (*Courtesy Jones & Lamson Textron.*)

jection screen as production parts are compared for contours and many dimensions at one time. Direct measurements may also be made with an accuracy of 0.0001 in. (0.003 mm).

Air Gages or Comparators.   Air gages utilize the effect of minute dimensional changes produced by metered air. The air gage consists principally of plugs or rings that are calibrated by checking them with a master or standard. A gentle stream of compressed air escapes between the surface of the standard and the close fitting surfaces of the gage. The differential in air pressure produces

(a)

(b)

Figure 21.17. An air gage (a). Variations in size cause variations in air pressure (b). (*Courtesy Bendix Automation and Measurement Division.*)

a reading that can be adjusted to zero. As more or less air escapes when the workpiece is checked, the float in a glass tube will be correspondingly higher or lower. Figure 21-17 shows a bank of air gages (a) and an example of how the air plug is used to check for waviness (b). Air gages are comparatively easy to use for both inside and outside measurements. They are particularly well suited to checking long, small inside diameters.

Electronic Column Gages. A gage very similar in appearance to the air gage column is the electronic column gage shown in Fig. 21-18. This modular gage operates on a linear voltage differential transformer (LVDT). Up to 15 columns can be coupled together for complex, multidimensional parts. Probes can be arranged in numerous combinations to measure accurately almost any typical dimension such as diameter, radius, or depth. A few of the more common types of measurements that can be made to a resolution of 0.00001 in. (0.0003 mm) at high production rates are shown in Fig. 21-19.

Figure 21.18. The electronic column gage being used to measure a hole diameter. (*Courtesy Valenite.*)

Geometric Forms (Roundness, Flatness, and Angles)

Roundness Measurement. A large number of precision components that must be produced are of a circular nature, such as ball and roller bearing parts, pistons, gear blanks, gyroscope components, hydraulic-valve parts, crankshafts, etc. In the past the relative ease of manufacturing circular parts meant that they could generally be produced to tolerances of thousandths and even tenths with little or no regard to their geometric form. This is no longer true. In working to tolerances of tenths of thousandths or better, parts are often found to be far from a true circular form.

There are several types of out-of-roundness that consist of lobing or bumps about the circumferences. An even number of lobes equally spaced about the circumference results in a variation in the diameter of the part. An odd number

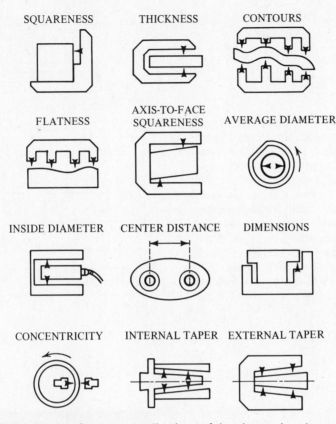

SQUARENESS

THICKNESS

CONTOURS

FLATNESS

AXIS-TO-FACE
SQUARENESS

AVERAGE DIAMETER

INSIDE DIAMETER

CENTER DISTANCE

DIMENSIONS

CONCENTRICITY

INTERNAL TAPER

EXTERNAL TAPER

Figure 21.19.   Common applications of the electronic column gage.
(*Courtesy Valenite.*)

of lobes equally spaced around a circumference can result in an effective constant diameter. A third category of out-of-roundness consists of random lobing or a combination of even and odd lobing conditions. In all of these, the radius of the part will vary from the true center.

Formerly, V-block and bench centers were used to check the roundness geometry of parts, as shown in Fig. 21-20. The V-block is no longer considered accurate since erroneous readings may not be readily detected on parts having an odd number of lobes. The system of bench centers is applicable to only those parts made with center holes. If the center holes on the parts are in good condition, this method can be fairly accurate. However, errors in the center hole and roundness errors may cancel each other.

The preferred method of measuring roundness today is with an electronic trace-type roundness gage, as shown in Fig. 21-21(a). A polar chart is produced (b) that can be used for close examination and comparison. The trace represents a profile in which the radial variation is greatly magnified in order that subtleties of geometrical error can be closely examined.

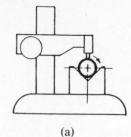

(a)

Figure 21.20. The V-block and bench-center arrangements have only limited use in being able to check roundness geometry accurately. In order to measure the out-of-roundness with a dial indicator and V-block, it is necessary to have the correct angle for the number of lobes present. Thus a three-lobed out-of-round condition requires a 60° included angle V-block and a five-lobed condition requires an 108°. The proper V-block angle is found by the formula 180 + (360/n), where n = the number of lobes.

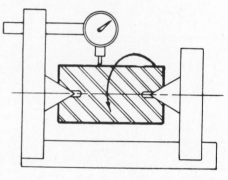

(b)

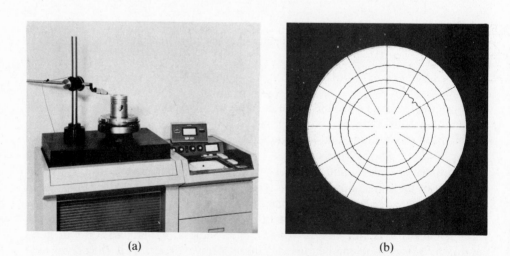

(a)   (b)

Figure 21.21. Roundness measurements are made by centering the workpiece on the rotary table shown at the left. As the part is rotated the gagehead probe follows the surface. The polar recorder, at the right, shows the trace in amplified form on a polar chart. (*Courtesy Bendix Automation and Measurement Division.*)

Figure 21.22. The autocollimator is used to calibrate a surface plate. (*Courtesy Rahn Granite Surface Plate Co.*)

Flatness.   The accuracy of small, flat surfaces (up to 10 in. in dia.) may be checked with optical flats, as discussed in the calibration of gage blocks.

For larger surfaces, a quick rough check can be made by sweeping the surface several times with a straight edge, or more accurately with a specially made knife edge. The light, shining from one side, will leak through the low spots of the surface plate.

Accurate checking of large, flat surfaces such as a planer or milling machine table is done by what is known as *optical tooling*. A line of sight is used somewhat as a surveyor uses a transit or theodolite.

The optical tooling used to check alignment and flat surfaces in machine-tool work is the autocollimator and mirrors, as shown in Fig. 21-22. The fundamental feature of the autocollimator is the perfectly straight optical axis of its small telescope. This straight line, augmented by an integral light source and illuminated reticle, will measure small changes of angle in a distant mirror as an image shift when the image is reflected back in the telescope, as shown in Fig. 21-23. This instrument is the basis of the most common method of calibrating surface plates and large machine-tool tables.

The autocollimator rests at one corner of the plate, its parallel beam of light projecting an image horizontally to a fixed mirror at a second corner. The beam is then reflected to the second mirror located at successive positions in a straight line across the surface plate, either perpendicular or diagonal to the original direction of the beam. The reflected beam from the second mirror retraces its original path according to the contour of the plate. As the image returns, it is observed to be either coincident with its origin or displaced by a measurable value. This measurement, twice the angular change of the second mirror, is caused by a change in slope from its previous position on the plate's contour. A sensitivity of 0.1 arc sec is typical and equivalent to elevations on the plate contour of 0.5 microinches per inch. Special features are available to make automatic photoelectric readings and recordings of the data, or a continuous recording of the trace.

Laser beams can now be used to calibrate surface plates in a fraction of the time required by earlier methods. The laser is also used to calibrate multiaxis machines for positioning accuracy, alignment, and squareness of axes, with a printed or plotted record of each measurement.

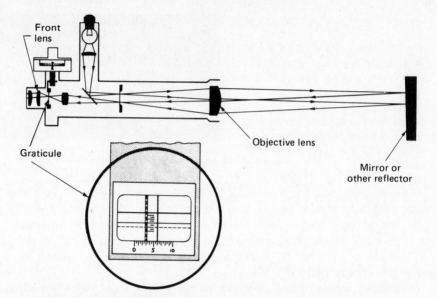

Figure 21.23. A schematic showing the principle of the autocollimator. (*Courtesy Rank Precision Industries, Inc.*)

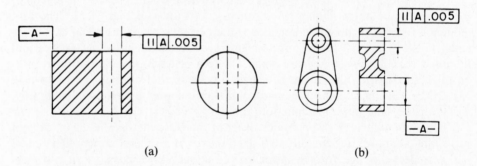

(a)   (b)

Figure 21.24. Examples of geometric tolerancing for parallel axes. The notation in the box calls for the hole to be parallel to datum plane A within 0.005 in. (a). The small hole must be parallel to the large hole (datum plane A) within 0.005 in. (b).

Geometric Interrelationships

Parallelism.   Parallelism may be defined as a condition where a surface line or axis is equidistant at all points from a prescribed datum plane or axis.

Two examples of geometric tolerancing for parallelism are shown in Fig. 21-24. These parts may be checked by what is known as functional gages. As an example, a gage to check the parallelism of the two holes is shown in Fig. 21-25. There are other gage designs that could be used, as well as instrumentation to check the parallelism of the two holes. For example, if the part is not too large, the holes could be fitted with dowel pins and checked for parallelism on

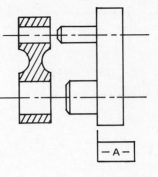

Figure 21.25. An example of a functional gage used to check the parallelism of the part shown. The tolerances for the dowel pins and perpendicularity to datum plane A will have to be closer than that of the part. This gage would also check the center distance of the two holes. One pin is longer to ease assembly.

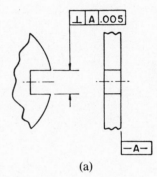

(a)

INTERPRETATION

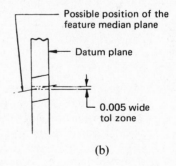

Figure 21.26. An example of a hole axis not being square with the reference or datum plane. The abbreviated notation as used in geometric tolerancing calls for the keyway to be perpendicular to datum plane A within 0.005 in.

(b)

the optical comparator. If the part is large, the dowel pins could be checked for parallelism with the use of a surface plate and a dial indicator or electronic height gage.

Squareness or Perpendicularity. Perpendicularity refers to the condition of surfaces or axes which are exactly 90° from a given datum plane or axis [Fig. 21-26(a)]. Shown at (b) is another example of geometric tolerancing, using the abbreviated notation for perpendicularity. The keyway is to be perpendicular to datum plane *A* within 0.005 in. Again, a functional-type gage can be made by the use of a perpendicular pin in a plate on which the part would be placed,

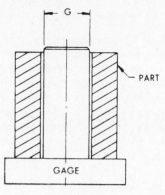

Figure 21.27. A functional-type gage that can be used to check perpendicularity.

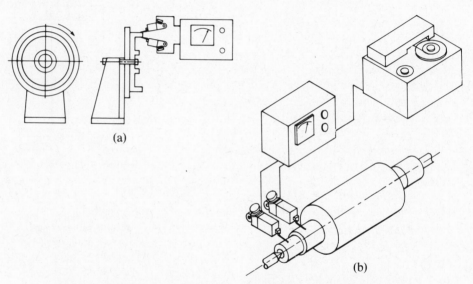

(a)

(b)

Figure 21.28. Concentricity and wall thickness are checked at the same time (a). Roundness and concentricity may be checked right in the production machine to a radial accuracy of 10 millionths of an inch (b). The information may be read directly as shown and it can also be given to a tape recorder, direct printout, or to a computer memory. (*Courtesy Society of Manufacturing Engineers.*)

as shown in Fig. 21-27. The allowance between the gage and the part would have to be less than ±0.005 in.

Concentricity. Concentricity may be defined as a condition in which two or more features (cylinders, cones, spheres, etc.) in any combination have a common axis.

Concentricity and wall thickness can be checked to an accuracy of millionths of an inch using differential techniques, as shown in Fig. 21-28. The two signals given by the gage heads which are set opposite each other (a) can be combined

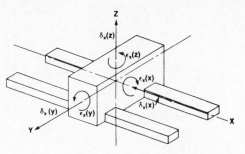

Figure 21.29. The six degrees of freedom of a machine-tool table.

by means of special circuitry to permit input data refinement. For example, when a source of error affects both heads equally, the error can be canceled out. Errors of work support, gage support, temperature, and wear may all be canceled out. High-precision concentric measurements are thus obtainable without high-precision fixturing. Routinely matched heads can cancel an error up to 100 times the desired gaging accuracy. Roundness and concentricity are checked at the same time, as shown at (b). The system is especially useful in checking bearing surfaces concurrently with their manufacture. Adjustments and corrections can be made without removing the part from the machine.

Geometric Relationships and Machine Tools. Perhaps nowhere is there a greater need for geometric relationship accuracy than on machine tools. In the case of drilling, boring, milling, and other similar machines, there are two or three orthogonally related axes. Linear, angular, and straightness interferometers are capable of evaluating five of the six degrees of freedom of a rigid body, roll being the exception (Fig. 21-29). For machines where the workpiece rotates, as in lathes, parallelism between the tool and spindle axes is of fundamental importance. The laser straightness interferometer is able to make both perpendicularity and parallelism measurements, as shown in Fig. 21-30. This is done by comparing two consecutive straightness calibrations of adjacent axes and introducing either a 90° or 180° rotation into the straightness reflector (shown on the lathe cross slide) between traverses.

For perpendicularity, a separate right-angle reference is required, which causes the extended axis of the straightness reflector to be viewed with a 90° deflection. As an alternative, a precision indexing table can be used, which physically rotates the reflector assembly after the first axis has been measured.

Angular Measurement

Direct Angular Measurement. Direct angular measurements are commonly made with a protractor [Fig. 21-31(a)]. For more accurate measurements, the vernier bevel protractor is used, which reads in degrees and minutes (b). Direct angular measurements may also be made in degrees and minutes on the optical comparator or with the autocollimator and a rotating table, as was shown in Figs. 21-16 and 21-24.

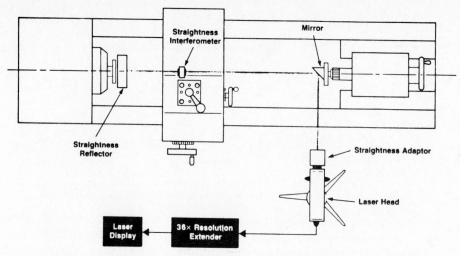

Figure 21.30. A plan view showing how straightness and parallelism are calibrated on the lathe with the use of a laser and straightness interferometer.

(a)

Figure 21.31. Direct angular measurements can be made with the protractor (a). Angular measurements may be made in degrees and minutes with the universal bevel protractor (b). (*Courtesy L. S. Starrett Co.*)

(b)

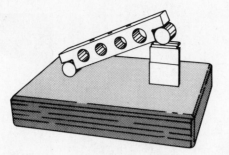

Figure 21.32. A sine bar on a surface plate with gage blocks for precision angular setting. (*Courtesy Brown & Sharpe.*)

SETTING COMPOUND SINE PLATES

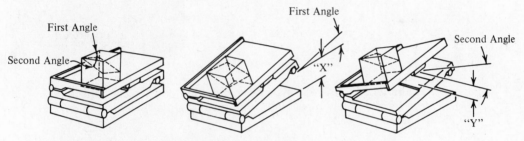

For Setting Two Known Angles at 90° to one another —
Example: — First Angle = 20°   Second Angle = 30°

Figure 21.33. Sine plates are used for layout and inspection of parts where a high degree of accuracy is required. (*Courtesy Brown & Sharpe.*)

Indirect Angular Measurement.   Indirect angular measurement is made with the aid of sine bars and sine tables. A sine bar, as shown in Fig. 21-32(a), is a bar that has a precise cylinder attached at each end. The center distances are made to vary in increments of 5 in. (127 mm) ranging from 5 in. for the smallest sine bars to 20 in. (508 mm) on sine plates and bench centers made with sine-bar bases.

Figure 21-33 shows sine bar tables that are used for accurately positioning of parts for layout work or inspection. Sine bar vises are also available to hold stock for milling or grinding.

Measuring Microscopes

Measuring microscopes are used for direct linear and angular measurement of small parts. The contour of the part as observed through the microscope represents a magnified replica. Although not actually a shadow because it is not projected on a surface, it is commonly called a shadow image or, more correctly, a silhouette. Contours which are partially obstructed from viewing can be measured by special accessories.

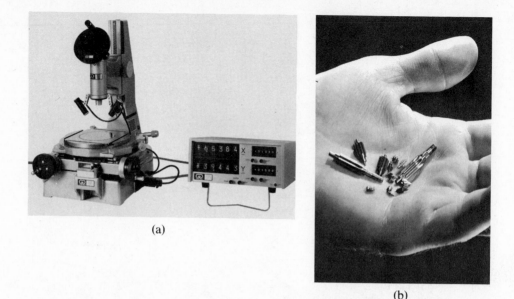

(a)

(b)

Figure 21.34. The toolmaker's microscope and typical parts that can be accurately measured. (*Courtesy Gaertner Scientific Corp.*)

The toolmakers' microscope, as shown in Fig. 21-34, is equipped with micrometer screws which fill the double function of moving the table laterally and horizontally and providing measurements to accuracies of 0.0001 in.

The most common method of measuring is by the use of graticule hair lines. The edge of the part to be measured is lined up with the hair line. At that point, a zero setting can be made on the micrometer dial. The part is then moved until the other edge is lined up with the hair line and the distance noted on the dial. A protractor hair line is used for angular measurements. Parts may be viewed in magnifications of 10X to 100X. Typical applications are shown in Fig. 21-35.

Surface-Texture Measurement

The science of surface measurement has progressed rapidly in recent times. Not too many years ago an appraisal of surface finish was given by rubbing the thumbnail across the surface or by a purely visual examination.

Surface finish is usually specified by the design engineer, and the degree of finish should be carefully related to the end-use component. It is important to remember that the finest finish attainable by a skilled operator is not necessarily the best finish for a particular component. Also, costs increase rapidly as extra operations are added, as shown in Fig. 21-36.

Formerly the designer merely placed a symbol "*f*" on all surfaces to be machined, but in time a much more meaningful standard symbol evolved, as shown in Fig. 21-37. You will note that the symbol has considerable information surrounding it, including roughness, roughness width, roughness width cutoff,

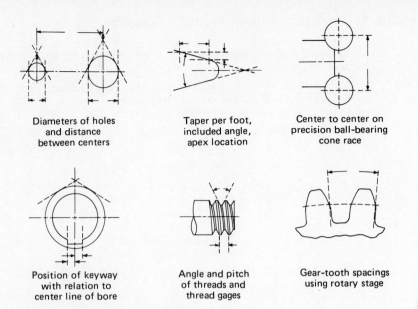

Diameters of holes
and distance
between centers

Taper per foot,
included angle,
apex location

Center to center on
precision ball-bearing
cone race

Position of keyway
with relation to
center line of bore

Angle and pitch
of threads and
thread gages

Gear-tooth spacings
using rotary stage

Figure 21.35. Typical measurements made on a toolmakers' microscope. (*Courtesy Gaetner Scientific Corp.*)

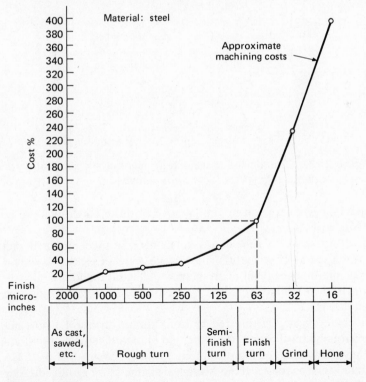

Figure 21.36. There is a proportional increase in cost with improved surface finish up to about 63 microinches, after that there is a rapid rise.

SYMBOLS FOR SPECIFICATIONS AND DRAWINGS

In this example, all values are in inches except R_a values, which are in microinches (millionths of an inch). Metric values (millimeters and micrometers, respectively) are used on metric drawings.

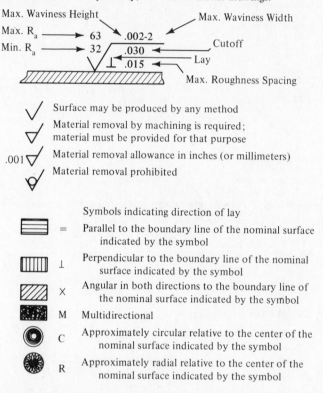

Figure 21.37. Standard surface-texture symbol and illustration of terms. (*Courtesy American Standards Assoc.*)

waviness height, waviness width, and lay. All of these conditions are now covered by the term *surface texture*.

Figure 21-38 shows a schematic sketch of a surface profile depicting the various terms used to describe it. The surface roughness, R_a, is shown on the dial in either microinches or in micrometers. A digital readout is also available. Also shown is the Profilometer used to measure the surface roughness by an amplified stylus trace.

The average surface is referred to by different terms, R_a, AA, and CLA. R_a (average roughness) became the accepted standard term in 1978. AA refers to the arithmetic average of the deviations from the center line X. CLA stands for the center line average. If no surface roughness is specified, the surface produced by the machine is assumed to be satisfactory.

If no cutoff is specified, the 0.030 in. (0.8 mm) instrument setting is assumed. The cutoff should include all surface irregularities and roughness spacings that

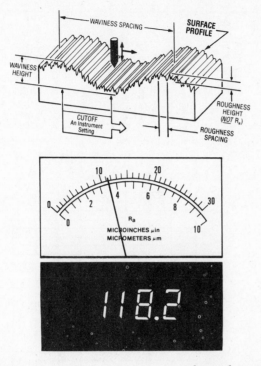

Figure 21.38.   A schematic representation of a surface profile with standard markings. The average surface roughness R_a is shown directly on the dial or digital readout. (*Courtesy Bendix Automation and Measurement Division.*)

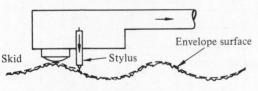

The stylus is sensing variations of the actual surface contour relative to the envelope surface.

(Used for average readings.)

-- Actual surface contour

Figure 21.39.   Conventional surface roughness measurement with skid (shoe) guided stylus traverse.

are of interest. Often machining operations will produce surface irregularities that have spacings that are greater than 0.030 in. (0.8 mm). These irregularities will not be reflected in the measurement.

ANSI B46.1 standard specifies the minimum radius of a single skid as 50 times the cutoff used and 9 times if a double skid is used. (The radius referred to is in the tracing direction.) Figure 21-39 shows a schematic sketch of a stylus and single skid. Standard instrument cutoff settings are 0.003 in., 0.010 in., 0.030 in., 0.100 in. (0.08, 0.25, 0.76, and 2.54 mm).

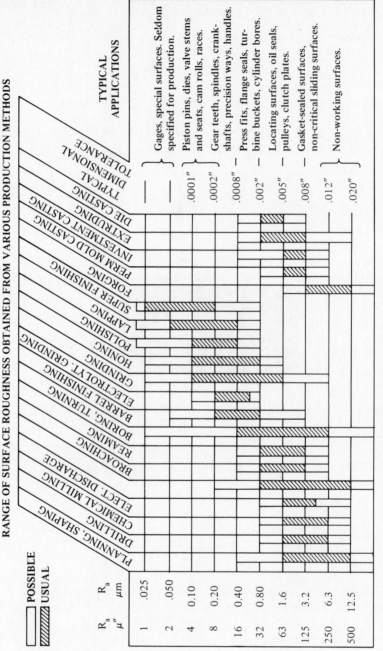

RANGE OF SURFACE ROUGHNESS OBTAINED FROM VARIOUS PRODUCTION METHODS

Figure 21.40. A guide to surface roughness produced by various operations. (*Courtesy American Standards Association.*)

To obtain meaningful measurements, the stroke length in one direction on the part surface must be at least five times the cutoff.

On meter display instruments, the average roughness (R_a) value is the mean reading about which the needle tends to dwell or fluctuate with a small amplitude.

It is essential for the designer to realize what is being specified by the surface-texture symbol and what is not. Flaws are not specified. These are faults due to poor structure of the material or those imposed by handling. A good example of flaws is the pitted surface found on cast iron or hard coated aluminum. Although the roughness reading may be within specifications, the flaws may cause it to be rejected.

All elements of the surface specification are significant, but to overspecify can be as wasteful as to underspecify. The designer must look at the job and decide how the part is to function and how it will be processed before specifying the surface texture.

Design and production engineers must be aware of how each process affects the surface texture. A guide to surface-finish capabilities of various processes given in both micro and metric units is shown in Fig. 21-40.

Gaging

The tools and instruments discussed thus far are largely used for individual part checking, with the exception of the comparators. Comparators such as the dial indicator, electronic type, and air type may be considered *adjustable-type gages* and those to be discussed now are *fixed-type gages*.

Gages in general are used for checking various part dimensions on a quantity basis. They are used as much to *prevent* scrap parts from being made as to check dimensions after the parts are made. In fact, some gages are made to be used right on the machine as the parts are being made. This is known as *in-process* gaging.

The Rule of Ten-To-One.   A general rule in gaging is that a gage must be ten times more accurate than the dimension being checked. For example, a part that must be checked to ±0.005 in. must have a gage that is accurate to ±0.0005 in. (0.0013 mm).

Fixed Gages.   Fixed gages are gages made for one dimension and are not easily altered. They may also be classified as *progressive* or *go/not-go*. These gages may be divided into groups according to the purpose for which they are used.

Plug Gages.   Plug gages are used to check internal diameters. They may be double-ended with a go dimension on one end and a not-go on the other. Plug gages may also be tapered or threaded, as shown in Fig. 21-41.

Automatic Gaging

Automatic gaging provides a means of checking each part as it is being made or immediately afterward. The checking of parts as they are being made is termed

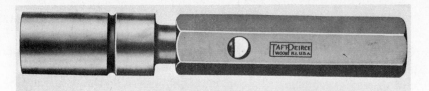

A go—not-go progressive plug gage

A go—not-go double-ended-thread plug gage

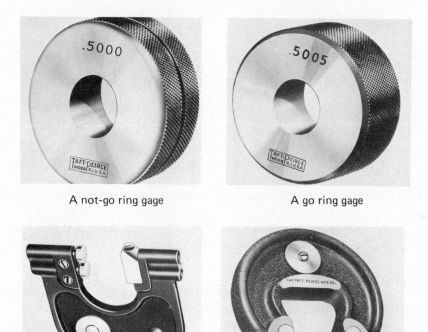

A not-go ring gage   A go ring gage

Figure 21.41. Plug, ring, and snap gages. The snap gage with two anvils is a go—not-go snap gage. (*Courtesy Taft Peirce Manufacturing Co.*)

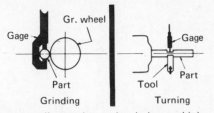

Gage — Gr. wheel

Part

Grinding

Gage — Tool — Part

Turning

"In-process" control or gaging during machining provides the machine or cutting tool with signals to:

• Automatically stop or change the tooling from a roughing to a finishing cut.
• Automatically retract the cutting tool when the part is to finished size.
• Automatically indicate and adjust the tool when the size trend is toward either limit of tolerance.
• Automatically stop the machine after the tools become worn and parts no longer can be machined within limits.

"In-process" control prevents faulty parts from being made by initiating signals for tool correction or replacement before part size is out of control.

Figure 21.42. The "in-process" control gage shown continuously measures the size of the part. The control system produces a feedback signal to stop the machine when the nominal size has been achieved. In grinding, wheel wear is automatically compensated for. (*Courtesy Bendix Automation and Measurement Division.*)

in-process gaging and checking after they are made is termed *postprocess gaging*. It is done primarily by air or electronic means. It may be further classified as contact and noncontact automatic gaging.

Contact In-Process Automatic Gaging. A contact-type automatic gage used on a cylindrical grinder is shown in Fig. 21-42. The size of the part is under continuous control of the gage; when the right size is obtained the grinding wheel retracts. The control circuit can be made to adjust the tool automatically when the size approaches the limit of the tolerance.

Noncontact In-Process Gaging. Historically, inspection has been a bottleneck in the manufacturing cycle. In many instances it takes longer to inspect a part, using traditional techniques, than it does to make it on modern automated equipment. The product liability problem makes sampling a risky business and suggests that 100% inspection is the best hedge against economic disaster in the courts. Many inspection methods depend, to some degree, on visual observations which tend to be subjective and unreliable.

To cope with the burgeoning problem of adequate inspection there is now a quiet revolution in the application of noncontact gaging—an automated means of sensing dimensions or conditions without touching the part as it speeds by. The techniques rely heavily on optical, or more specifically, electro-optical ap-

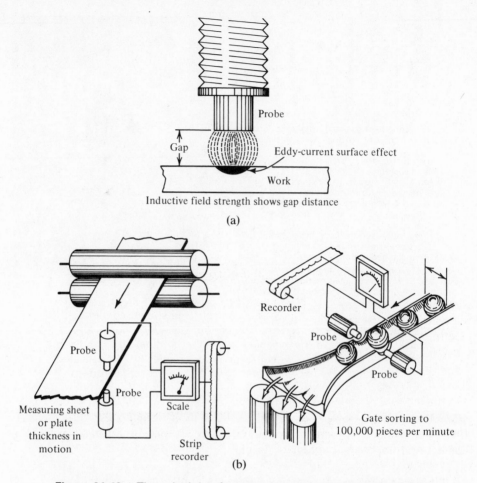

Figure 21.43. The principle of noncontact gaging by the use of the eddy-current principle is shown at (a). Two examples of noncontact eddy-current type gaging are shown at (b). (*Courtesy* American Machinist.)

proaches in which light is the sensing element. But they also use electrical-field effects such as magnetic, eddy current (Fig. 21-43), capacitance, and to a limited extent, radiation and acoustic methods.

Quality assurance, in its ultimate conception, must ensure that every single component and end product produced will meet clearly established criteria—no more, no less. To assure that this occurs, the automated inspection system must keep up with production and have the means to institute corrective actions before the process goes out of tolerance. Noncontact gaging is able to provide this. Data are collected instantly that can classify the part and correct the manufacturing process that may have caused the undesirable deviation in the first place. All this is done automatically and without the need for human intervention.

All optical noncontact inspection systems use some sort of sensing device

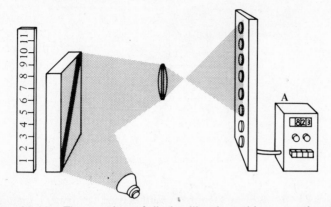

Figure 21.44. The number of diodes illuminated is proportional to the length of the subject. The electronics unit has only to count the illuminated diodes and apply the proportionality factor to display the length of the object. (*Courtesy* Production Engineering.)

that responds to changes in light levels. For the human observer, it is the eye, but in the electro-optical sensing devices it ranges from ordinary photocells to electron-beam-scanning devices such as videcon tubes and, most recently, various self-scanned solid-state detector arrays that generate the signals to be processed by the inspection system "brain," which might be a simple logic circuit or a full-blown computer.

Because the dimensional information is carried by a light beam, the way in which it is coded and decoded is a major factor in determining system accuracy. Therefore the way the measurements are made and the characteristics of the objects to be measured are crucial in deciding which approach is best for a particular application. Three relatively new non-contact-type measuring systems are linear diode array, scanning laser beam, and diffraction.

Linear Diode Array.   A linear diode array consists of photodetector diodes regularly spaced along a straight line. The spacing between the diodes is constant and provides the scale or ruler for measurement. An image of the illuminated object is formed on the linear diode array and the electronics unit counts the number of adjacent, illuminated diodes. A schematic of a linear diode array inspection system is shown in Fig. 21-44.

The length of a part is directly proportional to the number of diodes counted with a correction for the magnification of the image. For example, with an 0.001 in. (0.03 mm) center-to-center diode spacing, and an image with a magnification of one-half, the 20 illuminated diodes correspond to an object length of 0.04 in. (1.02 mm). The accuracy of the measurement depends on the fidelity of the image (which depends on lens performance and image focus) as well as the spacing of the diodes and the uniformity of their response.

When it is necessary to determine only if the length of an object falls within acceptable limits, as few as four diodes may be used. Two of these detect signal

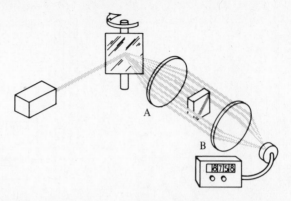

Figure 21.45.   A laser scanning beam is used to measure the linear dimensions of a part by measuring the total time the laser sweeping across the subject at a known rate is cut off. (*Courtesy* Production Engineering.)

presence at one end of the object and the other two diodes determine whether the other end is between the two detectors at the same instant. This system discriminates between objects that are acceptable, too short, or too long. Two additional detectors can be used to alert the operator to trends in the object length before the tolerances are exceeded.

Scanning Laser Beam.   The scanning laser beam employs a rotating mirror to sweep a single beam of laser light across a planer area (Fig. 21-45). The laser beam is reflected from the rotating mirror at its axis of rotation. The axis of rotation coincides with the focal point of the lens. The transmitter consists of a low-power helium or neon laser that provides collimated, parallel beams of light which are reflected off a five-sided rotating mirror and are projected through a special, diffraction-limited, collimating lens (A) to form a constant speed scanning beam. The beam is collimated over the entire measuring field, permitting accurate measurements anywhere in the beam, measuring fast-moving products, and often eliminating the need for fixturing. In addition, up to six objects can be measured simultaneously in the same beam. Linear dimensions are measured by the total time the laser, sweeping across the object at a known rate, is cut off.

Practical scanning systems use low-mass mirrors mounted on rugged galvanometers and are capable of dimensional measurement accuracies of ±0.001 in. (0.03 mm).

Diffraction.   Small dimensions can demand accuracy beyond the capabilities of practical linear diode arrays and laser beam scanning systems. Diffraction effects are well suited for gaging objects like small-diameter wire and also for measuring narrow gaps. Figure 21-46 shows a narrow beam of light incident on a small diameter wire spread, due to diffraction, with the spread inversely proportional to the diameter of the wire.

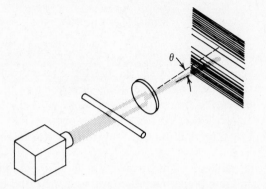

Figure 21.46. The parallel coherent light in the laser beam is diffracted by the wire and the resultant pattern focused on the screen by the lens. At the focal plane of the lens, the angle of the first minimum of light intensity θ is equal to the wavelength of the light divided by the diameter of the wire. (*Courtesy* Production Engineering.)

The diffraction beam can be focused on a screen using a suitable lens, resulting in the typical diffraction pattern of alternate light and dark bands. The linear distance between these bands can be measured using a linear diode array. The measurement will have a direct mathematical relationship to the wire diameter, wavelength of the light used, and focal length of the lens. The latter two are known constants; hence the wire diameter can be calculated directly from the diffraction pattern measurement.

The method is independent of the position of the wire in front of the lens because all parallel light rays incident on a lens at a given angle are focused on one point. Therefore the diffraction pattern is independent of the location of the wire as long as the diffracted light is collected in the lens. The system is just as useful for measuring narrow gaps since the diffraction pattern will be the same.

INSPECTION, QUALITY CONTROL, RELIABILITY, AND PRODUCT LIABILITY

The terms inspection, quality control, and reliability may seem at first to be somewhat synonymous. They are, however, quite different in application. Inspection is concerned with *how well* the physical specifications for a product are being met. Quality control is not as concerned with the physical aspects of the individual part as it is with the prevention of bad parts by use of statistical techniques. Reliability is concerned with the probability of satisfactory operation of the product over a specified time interval in a specified environment. Product liability, a comparatively new term, is concerned with consumer protection.

Inspection

Inspection has two equally important tasks. One of these is to detect errors in workmanship and defects in materials so that the end product will meet specifications; the other is to help minimize future errors. It is the latter that occasionally leads the inspection function to go beyond workmanship and materials and into design specifications. For example, inspection may reveal tolerances that are impractical or materials that are inferior for the intended use.

Inspection Requirements.   The inspection department must have a broad knowledge of the products being manufactured and their end use. The inspector must be familiar with the product design specifications and know the relationship of each component and how it functions in respect to the complete assembly. The inspector must set up a sequence of inspections designating at what time in the manufacturing operations inspection takes place, as well as determining the type of equipment needed to carry out the inspection.

Inspection Procedure.   The manufacturing process is analyzed to establish the best sequence and methods of inspection. Common inspection steps are first-piece, sampling, and final inspection. Inspection procedures may involve various sampling methods.

First-Piece Inspection.   Usually right after a machine has been set up for a new product or a change in dimension the operator will call the inspector to check the first piece before the production run is started. The inspector must have all the specifications, which are usually obtained from the part print. After the first piece has been inspected, critical dimensions are selected thereafter and used as the basis of inspection.

Sampling Plans.   Between first-piece and final inspection, various sampling plans may be set up. A roving inspector may tour the shop inspecting parts at random or he may have certain set intervals and a specified sample size.

Batch Sampling.   Batch sampling is generally done on smaller parts that are removed from the manufacturing floor. Each batch is chosen and sampled according to statistical methods.

Final Inspection.   Inspection of parts and assemblies just before they are prepared for packaging is called final inspection. At this stage much of the inspection may be visual, checking for flaws, defects, and missing or damaged parts. Products such as engines or motors are usually given an operational test. Since all of them are tested, it is considered 100% inspection.

Quality Control

As stated previously, quality control is concerned with the prevention of bad parts. To achieve the desired product standards, agreed upon by engineering and management, quality control sets up sampling plans which assure the process is under control and, at the same time, inspection costs are being kept at an economical level.

Very little was done in the application of statistics to quality control until the beginning of World War II when, faced with the job of buying enormous quantities of war material, the U.S. Government adopted the use of statistical sampling plans. In general, contracts with suppliers indicated that material would be subject to sampling inspection. Failure of several lots to meet specifications as judged by those samples would result in disqualification of that particular manufacturer as a supplier. Individual manufacturers responded by attempting to train people to understand and apply sampling plans within their own industry and the government aided the effort with appropriate courses.

When industries found that their own inspection departments, using these sampling plans, were rejecting high percentages of their own parts, they recognized the desirability of installing "in-process" or "performance" control systems. By this method, they were able to detect and eliminate defective material as soon as possible after it was produced. In addition, the new system improved the possibility of finding and correcting the point in the process that caused the production of defective items. Thus a major tool developed for in-process control, *capability studies* and *control charts*.

Process Capability.   Process capability studies are a quality control technique used for determining how much variation there is in the process and what the average variation is. For example, a lathe is used to turn out parts that have a 1-in. critical diameter with a tolerance of ± 0.005 in. The result of a sampling plan used to inspect the parts is shown in Fig. 21-47. The curve gives the following information:

1. The variation pattern is consistent; that is, the process is in a state of statistical control.
2. The total variation of the lathe is 0.012 in., which is greater than the tolerance (0.010 in.).
3. The lathe is being operated on the high side, 0.003 in. above the nominal dimension.
4. Under current operating conditions, approximately 16% of the product is beyond the upper specification limit and will probably require reworking (determined by computing the area of the normal curve in the out-of-specification segment).

Possibilities for action are:

1. Reset the tool to the nominal dimension. This will reduce the out-of-tolerance product from 16% to 4%. However, the *process capability* is still not acceptable.
2. Recondition the lathe to bring the total variation within the 0.010 in. tolerance.
3. Change the specifications to coincide with the process capability.
4. Transfer the job to more capable equipment.
5. Leave as is and screen defective parts.

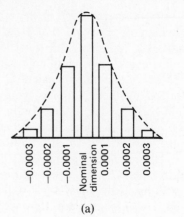

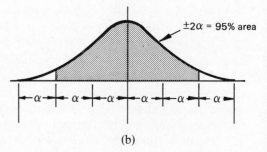

Figure 21.47. (a) A histogram showing normal distribution of acceptable shaft diameters made to a tolerance of 0.001 in. (b) A normal curve with the shaded area representing ±2σ limits, or 95% of the area.

Suppose the first possibility is chosen from the five foregoing alternatives The key factors in determining process capability are the natural tolerance (μ_{na}) and the standard deviation.

The equation for natural tolerance is:

$$\mu_{na} = \sqrt{(\text{tol. of part 1})^2 + (\text{tol. of part 2})^2}$$
$$= \sqrt{0.006^2 + 0.008^2} = 0.010 \text{ in.}$$

The standard deviation may be defined as one-third of the natural tolerance:

$$\sigma = \mu_{na}/3$$
$$= 0.010/3 = 0.0033 \text{ in.}$$

Note: The standard deviation may be somewhat oversimplified since it will be dependent on the sample size. If the sample size is between five and ten, the example given is satisfactory.

After determining the value of the standard deviation σ we know that $\pm 3\sigma$ will include approximately 99.7% of the product and is considered to be the total variation or the process capability. The only precaution necessary is to assure

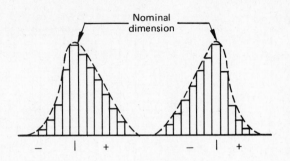

Figure 21.48. Histograms of actual production lots may be skewed either to the right or left. It is important when matching close-tolerance parts by selective assembly that they have similar histograms.

that the process is normally distributed and that it is in a state of statistical control. This can be determined visually from the histogram comparison to the normal curve and from plotting control charts.

In practical mass production, grinding tolerances may be held to 0.001 in. (0.03 mm). However, in hydraulic valves the fit tolerance between the sleeve and the shaft may be 0.0001 in. (0.003 mm). This requires selective assembly, that is, shafts from the regular production lot must be gaged and classified to the nearest "tenth" dimension. Likewise the sleeves. Such classifying work is totally impractical with manual gaging equipment but is entirely feasible on automatic equipment. Automatic gaging equipment is available that can gage and match 1200 assemblies per hour with 100% reliability.

In production, the grinders are set at the nominal dimension and the bulk of the production will cluster at the midpoint in the normal distribution. However, the production may shift or "skew" to the right or left, as shown in Fig. 21-48.

As mentioned previously the lots of shafts and sleeves brought to the automatic gaging machine may be skewed in opposite directions and there will be many parts that will not match. Thus a histogram of each production run is necessary for effective assembly tolerances. In automotive plants where parts are subcontracted, it has become a practice to include histograms of tolerances so that the assemblies can be more effectively matched.

Control Charts.   Control charts are basically a graphic means of presenting in-process data on a particular machine operation. The control chart as first introduced by Western Electric in the 1920s presented a new and powerful concept in that a certain amount of variation in a process, as distinct from the product, must be considered "normal" or expected.

$\bar{X}$-and-R Charts.   One of the most common of the quality control charts is the average-and-range chart or $\bar{X}$ (X-bar) and R chart. For this chart, samples are taken at regular intervals. The size of the sample is usually four or five items. The average value of the critical dimension is plotted for the given time interval on the chart, as shown in Fig. 21-49.

The centerline of the chart is obtained from the grand average of $\bar{X}$. This is usually shown on the chart as a red dashed line. The ranges are also averaged

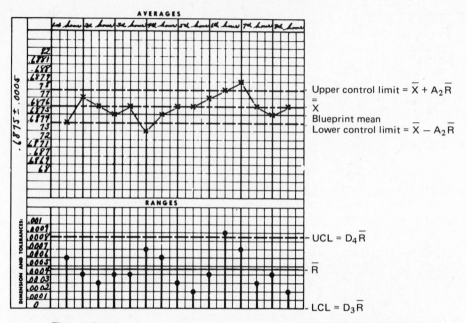

Figure 21.49. An average-and-range chart with sample averages posted each half hour. The range of each sample is shown on the lower half of the chart. (*Courtesy Federal Products Corporation.*)

and a line drawn as shown at $\overline{R}$. The upper and lower control limits can then be calculated by multiplying the average range ($\overline{R}$) by a constant (Table 21-4) which is based on the number of observations in the sample (n).

Control limits can be established after a reasonable number of samples have been taken, say after 20 or 25 samples. Too few samples will not give a true picture.

The $\overline{X}$ chart indicates the level at which a machine is operating and the $\overline{R}$ chart indicates the amount of variation that exists in the machine output. Usually the operator is responsible for controlling the tool setting. Therefore the chart shows his ability to establish and maintain the process average.

The biggest advantage of in-process control is that a malfunctioning machine or an improper tool setting is discovered almost immediately. This reduces scrap and repairs.

Reliability

In more recent years, reliability has become a common term or, simply stated, both the public and industry have become "reliability conscious." The main reason has been the customer's incessant demand for a more reliable product.

Reliability may be defined as the probability of a device giving adequate performance for a given length of time under a specified operating environment. Thus reliability is a property determined by many factors: design, production, operational demand, operator use, maintenance, personnel skills, etc.

Table 21-4. Control chart constants for small samples (*Courtesy* ASTM.)

Number of Observations in Sample, n	Chart for Averages			Factor for Central Line	Chart for Ranges			
	Factors for Control Limits				Factors for Control Limits			
	A_1	A_2	A_3	d_2	D_1	D_2	D_3	D_4
2	2.121	3.759	1.880	1.128	0	3.686	0	3.268
3	1.732	2.394	1.023	1.693	0	4.358	0	2.574
4	1.500	1.880	0.729	2.059	0	4.698	0	2.282
5	1.342	1.596	0.577	2.326	0	4.918	0	2.114

Manufacturers often provide a warranty on their products. Before this can be done, the reliability engineer must determine the probability of having to supply additional equipment or make repairs under the terms of the warranty. Thus some of the personnel in a reliability group should be engineers with statistical backgrounds.

Product Liability

As stated previously, product liability is concerned with consumer safety. An event that happened in 1911 first brought this subject into sharp focus.

A Mr. MacPherson was driving his automobile in upstate New York at a modest speed of 15 mph (24 kmph) when a wheel fell off. The car flipped over and MacPherson ended up in the hospital. He sued Buick for negligence. The case of *MacPherson v. Buick Motor Company* became a landmark case and is still widely quoted as a reference in current legal actions.

In its defense, Buick raised what was then a plausible argument: it had not sold MacPherson anything directly, therefore there was no direct contractual agreement between the company and the car owner. The judge felt otherwise. In the now famous opinion rendered in the New York Court of Appeals, he extended manufacturers' liability to third parties for any product "reasonably certain to place life and limb in danger when negligently made." His decision did leave one vital requirement, the plaintiff must prove that the manufacturer was negligent. This principle has been severely eroded in recent years.

Because proof of negligence is often highly elusive (it is presumed that only manufacturers know all the technical facts), and because products are becoming increasingly sophisticated, more and more courts are finding that plaintiffs need not prove negligence in manufacture. They need only prove that the product had a defect when it left the manufacturer's hands, whether or not that defect was known or could have been prevented.

In 1960 the New Jersey Supreme Court, in a case of *Henderson v. Bloomfield Motors Company,* allowed an injured customer to recover against the manufacturer as well as the dealer. This doctrine is being gradually adopted by courts across the country.

The government's National Commission on Product Safety has encouraged the accreditation of private testing laboratories, control of imported products, safety research and development of test methods and product analysis, and trade regulation rules for those who certify or endorse the safety of consumer products.

In 1972 the Consumer Product Safety Act was signed into law. This was the first real attempt by the government to enforce safety standards by preventive measures. The act has, in effect, forced manufacturers to evaluate their entire product safety organization and all management functions relating to getting a product on the market.

Product Liability Loss Prevention.   From the foregoing discussion of product liability, it would appear that the odds are heavily against a manufacturer unless he takes the utmost precaution to produce safe products. How can this safety be assured without incurring undue cost?

Investigations have revealed that no one group within a manufacturing organization can assume the total responsibility for product liability loss prevention. The errors of omission and/or commission can be traced in many instances to routine or minor decisions made in nearly all areas of a company's operation and at all levels of responsibility. In product liability claims, the magnitude of the error bears little relationship to the magnitude of the potential loss resulting from it. Sometimes the cause is not the result of outright error, but of innocent, well-intentioned, production-line decisions.

These observations indicate that the best approach to the problem is total involvement by all company personnel. A special organization responsible for product liability is not only costly but produces a false sense of assurance to others. However, to be successful a liability loss program does require an organizational mechanism to provide coordination of the effort, followup audits, and a line of accountability to top management. A product liability committee can accomplish these ends.

A product liability loss prevention program can be organized into seven main steps:

1. Education. Emphasis in education is placed on the importance of the program.
2. New product safety review. A special concerted effort is needed to minimize risk associated with new products. Plans for development and end use must be carefully reviewed.
3. Establishment of risk criteria. Some products pose greater risks than others. Products that are rated as "significant risk" must be subjected to more stringent production controls.
4. Control of warranties, advertisements, and technical information. A review by upper management of all warranties, publications, and advertising is needed before release.
5. Complaint and claims procedure. Complaint documents serve the accounting function and are a source of information for quality control. They also serve as a communications link between the manufacturing operation and the market. An early alert by a disgruntled customer often serves to resolve an adverse occurrence before it becomes a dispute.

6. Records retention schedule. The collection of design, production, and sales records is the first critical step in resolving an adverse claim or preparing for a product liability lawsuit. Records cannot be kept indefinitely but decisions as to what should be kept can be made on the expected life of the product.
7. Product liability audit. As mentioned, the key to developing an effective product liability loss prevention program is to obtain the total involvement of each department and not let the task fall on one or two groups closely related to the problem.

In short, a manufacturer's defenses must be built into his product by the effort of all concerned.

Problems

21-1 Shown in Fig. P21.1 are three flat surfaces as seen through optical flats. How much out of flat are the surfaces?

21-2 How could the accuracy of a micrometer be checked to within 0.001 in. at four or five points over its 1 in. length of travel?

21-3 A sine bar fixture is used to check the included angle of a plug gage as shown in Fig. P21.2.

(a) Determine the angle from the information given.

(b) What instruments can be used to check the sine bar plugs to an accuracy of 0.001 in.? The height over the plugs is 1.745 in. and 3.276 in.

21-4 (a) A tapered gib is shown in Fig. P21.3. State three methods that could be used to measure the angle of the gib in degrees and minutes.

(b) What is the taper of the gib block in inches per foot?

21-5 Using a 5-in. sine bar, what is the length of

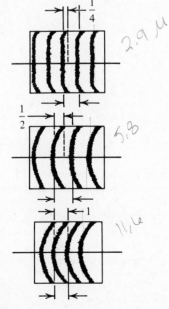

Figure P21.1.

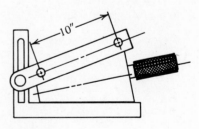

Figure P21.2.

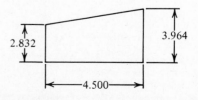

Figure P21.3.

the side *c,* of a right triangle, or hypoteneuse of a setup for measuring tapers?

21-6 What is the percent increase in cost to go from a 63-microinch finish to a 32-microinch finish?

140

21-7 If a closed 1-in. micrometer is opened $4\frac{1}{2}$ turns, what will the reading be?

.1125

21-8 Convert the following to metric measure. Remember the use of the space instead of the comma and the comma instead of the decimal.

 Area
a. 1 in.² to mm²
b. 2 in.² to mm²

 Stress
a. 3000 psi to kPa
b. 30,000 psi to MPa

 Energy
a. 3000 lbf to J
b. .3 lbf to J

 Length, the following to mm, cm, or m— whichever is the most appropriate
a. 8 in.   d. 20 ft
b. 0.008 in.   e. 0.5 ft
c. 2 ft

 Liquid Measure
a. 10 U.S. gal. to liters
b. 10 quarts to liters

21-9 An angle plate is assumed to be 38°50'. You decide to check it with a 10-in. sine bar. What would the height of the gage blocks be for the opposite side?

21-10 Five splines have been cut on a shaft with a root diameter of 4.000 in. (Fig. P21.4). The splines are 1 in. wide. What should distance *F* be?

21-11 Sketch a functional gage that would be used to check the location and perpendicularity of four equally spaced $\frac{1}{2}$ in. square slots cut into a 4 in. dia., $\frac{1}{2}$ in. thick round plate, as shown in Fig. P21.5.

21-12. What instruments would be used to check the part shown in Fig. P21.5 in small quantities, as to:

depth micrometer
vernier caliper

(1) The depth of the slots to within 0.001 in.
(2) The outside diameter to within 0.001 in.

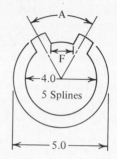

Figure P21.4.

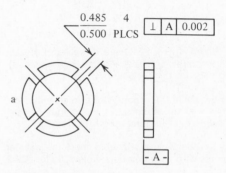

Figure P21.5.

optical comparator
(3) The 90° angular relationship of the slots, to degrees and minutes.
vernier caliper
(4) The minimum diameter to 0.001 in.
electronic roundness gage
(5) The roundness of the part to within 0.0001 in.
gage block
(6) The width of the slots to within 0.0001 in. What instruments would be used to check the following dimensions of the part shown in Fig. P21.5 in large quantities?
go/no go, snap gage
1. The outside diameter to 0.001 in.
2. The minimum diameter to 0.001 in.
plug gage
3. The width of the slots to 0.001 in.

21-13 In constructing an average-and-range chart the grand average ($\bar{x}$) was computed to be 8.502 in. and the range average was 0.008 in. What would the upper and lower control limits be on the $\bar{x}$ chart if the sample size were 3?

21-14 What would the difference in size be between a 3.0-in. steel gage block measured at 68°F (20°C) and at 78°F?

21-15. Two plates, each 2.00 in. ±0.006 in. thick, are assembled on a production line.

(a) What is the total natural tolerance (μ_{na}) of the assembly?

(b) What is the total range or variation of the assembled plates that would be considered satisfactory?

(c) Assume that the actual process, when run, showed a total tolerance of ± 0.003 in. Make two sketches showing how the normal distribution curves would appear, theoretical and actual.

(d) Would the distribution as shown in Fig. P21.6 be satisfactory? Why or why not?

(e) What would the natural tolerance be for the assemblies for the case shown at (d)?

(f) What would the standard deviation be for the case shown at (d)?

21-16. A designer places the following surface finish specification symbol on a given surface:

(a) Is the roughness width cutoff consistent with the roughness width?

(b) What processes would be needed to achieve the surface roughness specified?

Bibliography

ASME. "Measurement of Out of Roundness." *ANSI B89.3.1-1972.* New York: American Society of Mechanical Engineers, 1972.

———. "Surface Texture." *ASA B46.1-1978.* New York: American Society of Mechanical Engineers, 1978.

———. "Temperature and Humidity Environment for Dimensional Measurement." *ANSE B89.6.2-1973* New York: American Society of Mechanical Engineers, 1978.

Amstutz, H. N. "Sources of Error in Surface Texture Measurement." *Technical Paper 1Q76-585.* Dearborn, Mich.: Society of Manufacturing Engineers, 1976.

Axelrod, Norman N. "Gaging with Light." *Production Engineering* (March 1980).

Bryan, James B., and Erwin G. Loewen. "The Limit of Accuracy of Machine Tools." *Paper UCRL-75645,* presented at the International Conference on Production Engineering, Tokyo, April 23, 1974.

Bryan, J. B., and P. Vanherck. "Unification

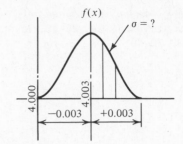

Figure P21.6.

(c) What would the roughness be in metric measurement?

(d) What may be a cause of the waviness?

21-17. In an $\overline{X}$ and R control chart, can the R portion serve as a process capability chart?

21-18. What are three different types of noncontact type measurement?

21-19. Why are histograms often asked to be sent along with shipments of manufactured parts?

21-20. Why is noncontact gaging gaining rapidly in popularity? *100% Inspection*

of Terminology Concerning the Error Motion of Axes of Rotation." *Paper UCRL-76931,* presented at the International Conference on Production Engineering, Freudenstadt, West Germany, August 1975.

Editor. "A New Look at Inspection." *American Machinist* 123 (August 1979): 103–126.

Garcia, G., ed. "Automated Product Evaluation and Regulation at the Assembly Line." *Manufacturing Engineering* 83 (November 1979): 60–65.

Hume, K. J. *Metrology with Autocollimotors.* London: Hilger and Watts, Ltd., 1965.

Kennedy, C. W., and D. E. Andrews. *Inspection and Gaging.* New York: Industrial Press Inc., 1977.

Moore, Wayne R. *Foundations of Mechanical Accuracy.* Bridgeport, Conn.: Moore Special Tool Company, 1970.

Schneider, Eric J. "Roundness Measurement." *Mechanical Engineering* 91, two parts (October 1969): 26–29 (November 1969): 36–40.

Wick, Charles, ed. "Automatic Electro-Optical Inspection." *Manufacturing Engineering* 83 (November 1979): 66–73.

STRESS = Load per unit area $S = \frac{\text{load in lbs}}{\text{unit area}}$ Elasticity = Most engineering materials used under elastic conditions, in material

TENSILE - Pull apart   Ductility   Deformation not permanent

COMPRESSIVE - make shorter   past

Shear - make material layers   Hooke's Law = Degree body bends or stretches out to shape. Direct Proportion force

STRAIN = % change unit length the elongation or   Acting on it

Contraction of Specimen $V = (l_f - l_0 \times 100)/l_0$ Modulus Elasticity = Measure of stiffness of material $E = S/N$

l_f = final length   STEELS Average Modulus Elasticity 30×10^6 STRAIN PROPORTION TO STRESS

l_0 = original length   STEELS Average Modulus Elasticity 30×10^6 STRESS - STRAIN CURVE

TENSILE STRENGTH - Maximum LOAD TENSION material withstand   Proportional Limit = Highest P.T. or

YIELD STRENGTH - STRESS which material exhibits specified limiting permanent Deformation (offset 0.2%)

PERCENT ELONGATION - Increase in length over orig. gage length

DUCTILITY = Ability material withstand plastic deformation without rupture, Probability/Drawability, Brittleness opposite Ductility

COMPRESSION = Measure of extent it deforms under compressive load prior to rupture, Buckling of compression specimen take

BENDING = Place if length-to-diameter ratio greater than 2.5

Outside fibers being placed tension & inner fibers in compression, STRESS on outer fibers depend on Sec Geometry

bea ratio & loading. Decrease Deflection, Section increased on material higher modulus Elasticity. Springback Excavated

TORSION = Torque to member cause it to twist about longitudinal axis   Inbowing

Shear strength = Maximum Load can withstand without rupture when subjected to shearing action

BRINELL = Test based on area indentation a steel or carbide ball 10mm dia; Loads used 3000, 1500, 500 kg. Load applied

for 15 sec on ferrous metals 30 sec non ferrous steels; Impression measured by Brinell microscope; Birinell coarse grained

or nonuniform in structure   Vickers soft VB; V_1, V_2

ROCKWELL = 2 Types Penetrators STEEL BALL & Diamond cone on BRALE used Hard material; Penrelle measures diff depth penetration

between minor & major load; minor load 10kg; soft 100kg B SCALE; HARD 150kg C SCALE LAZER testes is 45kg

MICROHARDNESS TEST = LOADS 1-1000 grams; Insulter 136° diamond Pyramid on Knoop dj; Ron tie & HARD TESTED Knoop T SCALE

Shore Scleroscope = Falls Portable, cause checking hardness; Hardness based height rebound diamond-tipped hammer

Harder material the higher rebound Vickers Cosa A

ANVIL Effect = Thin material checked if sufficient Number Layers Packed together Prevent hammer Penetrating metal

Vickers thin

TENSILE TESTING = Cone closer evaluating fundamental Properties of material use in design any other Test (Both Tension & Compression Test

Specimens usually found, FLAT Size or 0.50 used

IMPACT TEST = MANY machines designed to Absorb impact Load & 2 Test developed Provide measure rupture strength of material

on toughness are Izod & Charpy Tests; 3 Test Performed Average Taking of them

FATIGUE = Failures Start at surface, since stress higher at that PT; common Type Fatigue Test Rotating beam Test

COMMON ALLOYING Elements Used STEEL Carbon, Nickel, Chromium, Nickle

CARBON = .005 - .30% Low; .30 - .60% med & .60 - 1.4% High; Above this P.T. CAST IRONS below Wrought Irons

NICKEL = Increases toughness & resistance to impact   Molybdenum = High Speed Cutting Tools, forged Gears, Crankshafts, turbine rotors

VANADIUM = Provides fine grain structure, own break range temp, various Paro, alloris; forged axles, Shafts turbine rotors

TOOL STEEL = Increases Hardness, resisting Heat; Bod, cutured lattice, Various; used Higher % 18%

& Vanadium resulting steel high Speed Steel on US, reamers lathe

MANGANESE = Secondly to carbon frequency used. High manganese content 13% causes steel strain work harden depleter. 2 - .50 added to

SULFUR = Increases free-cutting action, low carbon; becomes brittle breaks up ferrite in structure helps chips break free of tool   Steel)

PHOSPHOROUS = Impurity in steel; Increase strength & hardness of steel Reduces Ductility   (low carbon .1 / .05 P)

HEAT TREATMENT = Process heating & cooling materials order to obtain certain desires Properties

HARDENING = STEELS with more than 1800 carbon may be Hardening P.T.

HARRY PM

YIELD STRENGTH = $S_y = L/A_0$ ← LOAD / ORIGINAL AREA

TRUE STRESS → STRAIN

$\sigma = L/A_i$, $A_i = $ INSTANTANEOUS AREA

% Elongation = $L_f - L_0/L_0 \times 100$ % RA = $A_0 - A_f/A_0$

Bending stress = $\sigma = my/I$ I = Area moment of inertia (in-4)

M = moment (in-lb) $y = d/2$ shaft Q/A

Shear stress $\tau = P/\pi d t$

P = Punch load in lbs

$d = $ 0.1A in inches

$t = $ specimen thickness inches

$T = 196 \tau J^3$ Hollow = $.19\pi x .86 - .84$

$\delta = $ shear stress outer fibers

$d = $ diameter of shaft

Angle twist θ = $584 \frac{TL}{d^4}$

$G = $ Length shaft

Hollow θ = $584 \frac{TL}{G(d_0^4 - d_i^4)}$

(Carbon equivalent = Ce)

Ce = Ce + 1/4 (%Si + %P)

Ce = total % carbon

Bend Allowance = B.A. = θ/360 · 2π(r + kt) | B.A. = $\frac{\theta}{360} \cdot 2\pi(r + kt)$ θ/360

$R_i = $ inside radius   r = inside radius inches

t = metal thickness   t = metal thickness

$R_i \leq 2t$, then k = .33 R_i inside radius   R inside radius

$R_i > 2t$, then k = 0.5   R radius

— ultimate (max stress) limit

Proportional/Fracture

Proportional/Elastic limit

% Reduction = $\frac{D_0 - D}{D} \times 100$ $D_0 = $ initial Dia $\delta = $ Punch Dia

Extrusion Process: Billets of material Preheated to a Plastic state in Preheat furnace. Billets move mechanically to extrusion Press & loaded into Press. The Ram or Press Pushes billet into container liner & the Ram keep the metal from working, 1 ε Press ram. Ext Press build up of pressure until breaks thru pressure needed. The material is extruded thru Die & out oth side. Press cooled by heat fans mechanical. Puller keeps material from tensiling up. TOL ± .003 inches

Extrusion = metal compressively forced to flow through suitably shaped die to form product reduced cross section

Methods Extrusion = Direct forces metal through Die IN Direct forces the Die into metal

Extrusion Shapes = Solid extrusion include parts rods, bar, the ks. oval Hollow tubing & intricately shapes hollow parts

Solid Extrusion use simple Die for their Production Hollow use die with mandrel or spider or torpedo die

Type of hollow extrusion Plastic coating of wire replace the mandrel with wire. The wire fed through the Extrusion Die & emerge coated with Plastic

Extrusion Dies: made of tool steel. Simple dies made by conventional methods but for intricately shapes Parts 's thin well made by Electrical discharge machining (EDM).

Ext Dies contain no Draft & have allowance for aluminum extrusions 8-20 mutes. Die Dies machined at room temp

becomes plastic/about 8000 F for aluminum   Small Extrusions 8-20 mutes   material extruded in Plastic state, coarse, cold/reg lead, magnesium. Steel some difficulty

Wash out = extruding material gradually erodes away Die material from die cavity

Extrusion = Billet yard → Preheated furnace → Extrusion Press → Runout table → walking beam table → Stretcher → Cut to size → Cut off → Heat Treat

Alloys = substance with metalic properties composed 2 more elements

Non ferrous = No Iron   3 major Iron ores Hematite, Magnetite, Taconite

Ferrous = Iron contains

Common Crystal Structure = FCC - 12 slip planes   HCP - 3 slip planes

BCC - 48   It easier form by Loritile Fracture

Titanium lighter steel Low = Less .30% C

but stronger (aircrafts)   Med = .30% to .60% C

High = .60 to 1.70% C

AISI - SAE 41020

top   1. carbon content

2. Nickel-chromium   6. chromium

3. Nickel   4. molybdenum

2 Nickel   5. chromium

Carbon content

1. carbon steel

2 carbon steel

7. 20% carbon content

6. chromium

7. Nickel-chromium-molybdenum

8. Nickel-chromium-molybdenum

9. silicon-manganese

Compression failure determined by

1 Ductile fracture   2 Barreling   3 cracking

Forgeability = relative ability material formed with-out rupture

Hypereutectic = I-rods with carbon equivalent of over 4.3 - good for thermal shock resistance (ingot + molds)

Hypoeutectic = Higher strength gray I-Rods less 4.3 C

3 main types stainless steel

Formability = ease which metal forced into Permanent change of shape

Punchant forces sheet to have the punch

Drawing = Forming flat sheet metal into hollow shapes means of punch cause metal flow into die cavity

$F = 1.33 S_u W t^2/L$

$S_u = $ ultimate strength

W = width metal bend in inches

L = length span across die inches

$t = $ metal thickness

Harden Stite Austentic ferritic hardenable

Shallow impressions = Patterns on sheet metal on extruded shapes

Coining & stampings = making of metal Product desired shape. requires high pressure.

min = similar metal blank under pressure

STAMPING = Cut wires ? ...

COMPOUND DIES = Two Distinctly seperate operations blanking & forming

Progressive Die = made cut & form Part In successive stages or stations of Die (very high cost used for high production)

TRANSFER Die = used when size & complexity of Part require moved from one Die to another

Roll Forming = change shape of coiled stock into desired contours without altering cross-sectional area

STRETCH Forming = Used produce compound curves in Sheet Stock

STRETCH-WRAP Forming = stretch metal beyond yield P.T. while straight thru wrapping around form block (used only 1 DIE)

Drawing = Oval or bulged bottom

Roll Forging = most simple form - Preheated billet passes between Pair of rolls that deform along it's length

Cross Rolling = Gradually deforming shaft in transverse direction as it passes between Rolls

Up Set Forging = Inserting a blank of a specific length into a stationary Die

Hammer = most widely used forging equipment

Swaging = reduces the cross-sectional area of rods or tubes

Extrusion Forging = called cold-working Process

Radius Line = Die Faces meet, helps determine grain flow & affects Production rate, die cost, & Die Life

Hammer & Press Forgings = Hammer forging done by repeated blows between two platens

Press Forging = utilizes slow squeezing action & relies less on operator skill, more on Proper Design

HERF = Describe method of efficiently applying more energy to work Piece Per unit of time
Be applied to High & & heavy Base materials

$$K = \frac{1}{2} mv^2$$

HUF = High Velocity metal forming

ADV: Large variety forges of formed, exotic & refractory metal, stainless steels
Drayl allowance eliminated - complex parts can be formed in one blow? Repeatability, except ?

DISADV: Tooling Serious limitation, Pool configuration usually limited one-Piece Dies
If several blows are required Process economically questionable.

IT 252 MID-TERM NOTES

Bonville = Entire material cool determines its steel contain 80% T. or more carbon 10. T.carbon = .01% carbon

Annealing = Intermediate-Type structure, forms between Pearlite & martensite cool, forms between Pearlite & martensite at slower cooler rates produce Steam pockets

Tempering = Softens the steel to avoid or minimize, problems of cracking & distortion in the metal is reheated or slower than critical rate

Stress relief Annealing = Heated to temp. close to lower critical temp, followed by any slower rate cooling

Spheroidizing = greater ductility, improves forming qualities ~ Full Annealing = Heating steel 1000℉ left at that temp then cooled very slowly

Flame Hardening = moving oxyacetylene flame over part followed quenching

Dis = Optimum usually low shape of flame of generation ADV: Large Surface machined economically, equip controlled electronically, Precise contact

Induction Heating = Heating metal followed with High frequency alternating magnetic-field fast cooled by surface machinability result chips easily, brittle up

Flame Hardening = HAP, surface with soften, tougher core Induction heating = Heating coil with High frequency alternating magnetic-field

Machinability = Ease which metal be sheared in turning, drilling, milling steam-ing (ASSE) good finish result chips easily, brittle up

Formability = Imelied formed ductility of metal based on crystal structure Semi skilled = partly drook and zero before reaching into mold

Killed = chemically deoxidized lie quickly in mold die cool Solidification still phases crystal-structure

Rimmed = chemically deoxidized lie quickly in mold die cool Semi-killed = partly drook and zero before reaching into mold

High = outer Rim of ingot considered 900 evolution during Solidification Rim'd from liquid state

machinability = Low carbon steels don't machine well because structure mostly ferrite, too soft good Shearing Action

Ultrahigh-strength = provide wear & reduced load

Water Hardening → STEELS = Least expensive ; Not used wenewow

Oil Hardening = USED more depth of Hardening & safety from distortion or cracking Also tolerate less non edge

Shock Resisting = only capable taking great deal of Shock also high fatigue resistance & good wearing qualities

Air Hardening = High Production at this from as get with any used in more wear & pickup resistance

Gray Cast Iron = Appearance large amount of flake graphite on surface

White cast Iron = Tooling cast iron rapidly so carbon cooled form cementite (chilled cast iron)

Magnesium = Meeting chance properties cast iron Small amount may-down one 16 per ton of iron surface

flake graphite fake spheroid shape.

Forging = ADV: Little/no waste Raw material Disadv: High Power Requirements; Expensive Equipment & tooling

metal forming Variables; Interdependent: Variables engineer a designer a control overall situation 3 tool's & dies geometry 2) work piece 3) starting temp = cold work = process occurs ,3 times melting P.T. Warm = .3 to .5 x melting P.T. Hot = .7 x melting P.T.

5) starting temp = cold work = process occurs ,3 times melting P.T. Warm = .3 to .5 x melting P.T. Hot = .7 x melting P.T.

Deform = Design for No control over 1) Forces & Power requirements 2) Exit temp 3) Surface finish & Precision 4) Volume material flow

Ways controlling DEPENDENT Variables 1) Experience 2) Experiments 3) Theory

Presses = Type of Velocity & Process

Forming & Bed Design OBF = Open Back Inclined ; very versatile press

Press = 1) type a) work Deformed 2) method Power transmission A) mechanical - Flywheel B) STEAM - Steam drop hammer C) Hydraulic

Bending = forming material at any angle or radius

Slitting = metal returning to small range then formed by the die

Punching = piece punched is work piece

Blanking = piece blanked out is work piece

Drawing = Processes on forcing flat metal blank into a female conform shape

Shearing = Symbols goods each per

HERF = High energy RATE forming

Forging = Plastic working metal by compressive forces shape part Also oldest process

Roll forming = three roll former 2) Pitch 3) Urethane Rubber

1) Restriking - Hit piece multiple blows
2) Over Bending - over bend angle spring back over's angle
3) Impact forging - Exert O/Reverse

Smith forging
1) open die hammer - Die moves up & Down
2) Drop closed die - Steel by step process (no greatness)
3) upset or Heading

Geometry developed
1) open DIE hammer - Die moves up & Down
2) Drop closed one
3) Upset or Heading

Die Section
1) Forming back down
2) cross-over sectional area parts
3) equalizes Draft

CHAPTER TWENTY-TWO

Protective Surface Treatments

Protective coatings for metals are selected for a number of reasons, such as appearance or customer appeal, cost, corrosion resistance, wear or abrasion resistance, and toughness or resistance to impact. To help ensure lasting qualities of the finish, a thorough preparation, mainly cleaning, of the base metal is necessary.

METAL CLEANING

During manufacture, many types of contaminants are encountered by the product. These usually include cutting oils, greases, waxes, tars, rust preventatives, dirt, heat-treat scale, and drawing and buffing compounds.

Not all contaminants can be removed by any one easy cleaning method. The problem must be met with a knowledge of each of the contaminants involved. A broad classification of cleaner types is alkalines, solvents, and acids.

Alkaline Cleaners

Alkaline cleaners are the most widely used of the three basic types of cleaners because they remove most types of soil and because they can be applied by most any method.

Alkaline cleaners consist mainly of detergents dissolved in water at the time of use. The effectiveness is dependent to a great extent on the detergent or wetting properties of the solution that break the common boundary wherever soil and metal meet and to a lesser extent on the agitation or movement of the bath relative to the parts being cleaned. Common methods of applying alkaline cleaners include electrocleaning, spray cleaning, dip-tank cleaning, and steam cleaning.

Electrocleaning.   Before electroplating, the metal is electrocleaned. This is done by making it the anode in a hot alkaline solution. As the current passes through the metal, thousands of tiny oxygen bubbles are formed on the surface. As these bubbles burst, they perform a scrubbing, scouring, agitating motion over the entire surface of the metal. Thus the dirt can be said to be "unplated" from the metal. This process produces the cleanest surface possible from a conventional alkaline solution.

Spray Cleaning.   Spray-rotary washers are used for small and medium stampings and small machined parts such as nuts and bolts. The alkaline cleaning solution is sprayed on the parts as the drum rotates. After the cleaning spray, a rinsing spray is used. This operation may include some rust inhibitor. The time required by either spray or electrolytic cleaning is usually 0.5 to 1 min at temperatures ranging from 130 to 170°F (54–77°C) for spraying and 160 to 210°F (71–99°C) for the electrolytic process.

Dip-Tank Cleaning.   Parts with light soil can be cleaned by dipping in an alkaline solution. Parts that are too small to put on racks and that are in quantities too large for handling in baskets are frequently cleaned by the use of rotating barrels. A barrel cleaning line should have at least two cleaning tanks. The first stage removes the gross soil. This allows the subsequent stage to operate more effectively in a comparatively unpolluted solution.

Steam Cleaning.   Steam-gun cleaning is used mostly on machinery that is too large for soak tanks and spray-washing machines. The alkaline steam solution used on steel or cast iron surfaces may be strong in caustic soda with additions of orthosilicates.

Ecological Considerations.   An important facet of today's cleaning is disposition of the fluid waste product. Once the soil has been dispersed by the solution, it should reunite away from the work to precipitate out or be skimmed off. The ability of a cleaning solution to perform this function increases its life and produces cleaner work. Most chemical suppliers are now offering biodegradable surfactants (surface-active agents) and phosphate-free cleaners.

Solvent Cleaning

When heavy oils, grease, and dirt are to be removed from the product, either straight or emulsifiable solvents are used.

Table 22-1. How solvents rate in cleaning functionality. (*Courtesy* Metal Progress.)

Solvent	Ability to Dissolve			Surface Tension, dynes/cm or mN/m	Compatibility* with			Chemical Stability
	Oils	Ionic Soils	Water		Metals†	Plastics	Elas-tomers	
Water	Poor	Good	—	73	C	C	C	Good
Acetone	Good	Fair	Good	24	C	X	X	Good
Ketones	Good	Fair	Poor to good	23-25	C	X	X	Good
Alcohols	Good	Good	Good	22-23	C	C	C	Good
Stoddard solvent	Good	Poor	Poor	28	C	C	X	Good
Trichloroethylene	Good	Poor	Poor	29	C	X	X	Fair
Other chlorinated ...... hydrocarbons	Good	Poor	Poor	26-32	C	X	X	Fair
FC-113	Good	Poor	Poor	18	C	C	C	Good
FC-113 azeotropes	Good	Poor to good	Poor to fair	18-21	C	C	C	Good
FC-113 blends	Fair to good	Good	Fair to good	18-21	C	C	C	Good

* C, generally compatible under short-term exposures encountered in cleaning applications; X, high probability of encountering incompatibility in cleaning applications.
† Excludes alkali and alkaline earth metals such as sodium, potassium, and barium in their free metallic form.

The 1977 amendment to the Clean Air Act requires states to revise regulations for solvent metal cleaning as well as other operations involving use of solvents or hydrocarbons. Under this law the regulation of water pollutants can be expected to be more severe. The removal of soil from metal parts with an alkaline water wash requires large quantities of water. The spent water must be treated before disposing of it. In contrast, most solvents can be reclaimed by distillation. This makes the ultimate disposal by the solvent method easier. However, to minimize risk of water pollution, the solvent should be insoluble in water and, to avoid air pollution, photochemically nonreactive. The greatest danger to the worker involves inhalation of excessive concentrations of solvent vapors and solvent flamability.

A complicated factor in cleaning is that products are often made of a wide variety of materials. Unfortunately, there is no one solvent that works well on metals, glass, elastomers, and plastics.

Table 22-1 lists a rating of common solvents, including water, and the relative effectiveness of each. The rating of "poor" indicates the soil has very limited solubility in the solvent, "fair" indicates partial solubility, and "good" indicates the contaminant and solvent are completely miscible.

Detergents are often incorporated into an aqueous cleaner to lower the water's high surface tension and improve the wetting qualities.

Normally, for parts that are heavily contaminated with oils and grease, an emulsifiable solvent is used. Small parts are put in baskets and dipped. Intricate castings are also dipped. Large parts are sprayed as they travel along on a belt-type conveyor. Flat sheet metal parts are sprayed from both sides as they move by on an overhead conveyor.

An alkaline cleaner is used after the emulsifiable solvent, with the same method of application, followed by a clear water rinse and drying.

Cleaning Equipment

Most machines used in cleaning metals are combination types. The monorail degreaser shown in Fig. 22-1 utilizes solvent vapors as the initial step in removing grease. The parts then move to the next station, where they are sprayed with warm solvent. The last station consists of a pure vapor rinse to ensure complete removal of all soil.

Other machines are made to handle parts in baskets (Fig. 22-2). The baskets, suspended from crossbars, can be made to rise and fall alternately or to advance longitudinally in steps from one end of the machine to the other and back again. The baskets may be exposed to vapor rinsing, immersion in solvent, spraying from nozzles, or combinations of these cleaning cycles.

Ultrasonic Cleaning

It is often necessary to remove insoluble particles from hard to reach cavities, indentations, slots, and small holes. Precision parts such as hypodermic needles, automatic transmissions, carburetors, and hydraulic needle valves must be perfectly clean to function properly. One of the most effecient ways of ensuring their cleanliness is to employ an ultrasonic scrubbing action.

The principal parts required for ultrasonic cleaning are shown in Fig. 22-3. The generator produces the high-frequency electrical energy. A transducer changes the electrical energy impulses into high-frequency sound waves that vibrate the cleaning agent and parts to be cleaned.

When sound passes through an elastic medium such as metal, both the sound energy and the medium vibrate at the same rate. When sound passes through a liquid, it, being inelastic, ruptures or *cavitates,* forming small vacuum pockets which almost immediately collapse or *implode.* The rapid action results in scrubbing of a speed and vigor impossible by conventional means.

Ultrasonic cleaning can be very effective in removing scale, rust, and tarnish provided the proper chemical is used. The cleaning agents may be acids, solvents, or detergents. Acid cleaners are now being used for radioactive scale removal, hard water scale, and oxide removal, welding and brazing oxides, heat-treat and forging scale, and plasma-deposited coatings from gas turbine blades.

Many acid cleaners cannot be used directly in the stainless steel ultrasonic tank since they are too corrosive. In this case, an insert tank, usually polypropylene, is used to contain the acid. Water or an alkaline solution is used in the ultrasonic tank to couple the ultrasonic energy to the insert tank.

Although cavitation is dependent on high-frequency energy, there is a point at which it becomes ineffective. That is, a point may be reached in the frequencies per second in which the liquid is incapable of accepting increases in power since the liquid becomes elastic, thus eliminating further transmission of energy. On

Figure 22.1. A conveyorized degreaser used for both vapor and spray-type cleaning. (*Courtesy Detrex Chemical Industries, Inc.*)

Figure 22.2. Automatically operated crossbar containers used in both dip and vapor degreasing applications. (*Courtesy Detrex Chemical Industries, Inc.*)

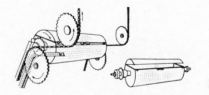

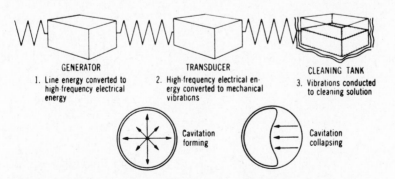

GENERATOR
1. Line energy converted to high-frequency electrical energy

TRANSDUCER
2. High-frequency electrical energy converted to mechanical vibrations

CLEANING TANK
3. Vibrations conducted to cleaning solution

Cavitation forming

Cavitation collapsing

Figure 22.3. Schematic diagram to illustrate the principles of ultrasonic cleaning. (*Courtesy Turco Products, Inc.*)

the other end of the scale, there is a certain threshold level below which cavitation will not occur. For technological reasons, the low-frequency systems (20 to 25 kHz) are more difficult to engineer and manufacture than the high frequencies. However, the lower frequencies produce better cleaning results because of the stronger cavitation field. If the frequency is within the lower audible range, noise becomes a problem. As the frequency approaches 20 kHz or lower, operating noise not only becomes substantial but may exceed maximum safe limits as specified by OSHA or other regulatory measures. For this reason, ultrasonic equipment which operates at a frequency of 40 kHz is increasingly recommended for industrial use even though it is somewhat less efficient.

Mechanical Cleaning

Mechanical cleaning of metal surfaces consists of an abrading action produced by abrasive blasting (wet or dry), power brushing, and tumbling.

Abrasive Blasting. Dry-abrasive blasting equipment consists of a cleaning cabinet, an abrasive reservoir, and an air nozzle. The abrasive is drawn through the hose to join the high-velocity airstream at the nozzle and impinge on the metal surface. The used abrasive slides down the centrally sloped bottom of the cabinet and returns to the reservoir.

The abrasive used may be shot, steel grit, or coarse, sharp sand. Fine sands do not produce a good cutting action. They also present more of a feeding problem.

Abrasive blasting machines are especially good for cleaning stubborn or hard to remove contaminants such as heavy scale, corrosion, or welding slag. They are also useful in providing an excellent surface with "tooth" for protective coatings.

The wet-abrasive blasting machines are quite similar in appearance to the dry machines. The main difference is that the wet machines use an abrasive slurry that is kept in suspension by an agitator. The abrasive is fed to the blasting gun by a centrifugal pump. Once at the gun, the abrasive is forced out at extremely high velocities with compressed air. The abrasives commonly used are silica, quartz, garnet, aluminum oxide, and novaculite (soft type of silicon dioxide).

The abrasive blasting machine is used both for cleaning and finishing metals. The action obtained depends on the size and type of abrasive. The wet abrasive machine is especially suited for finishing. In fact, the manufacturer of one type refers to it as a liquid honing machine. A variety of surfaces can be produced, as shown by the machine in Fig. 22-4.

Power Brushing. Power brushing utilizes both fiber and wire wheels to remove heat-treat scale, weld flux, machining burrs, and other unwanted surface contaminants. Some applications of wire brushing can impart fine microfinishes to metal parts. Several applications of wire brushing are shown in Fig. 22-5.

The principal kinds of power brushes as shown in Fig. 22-6 are wheel

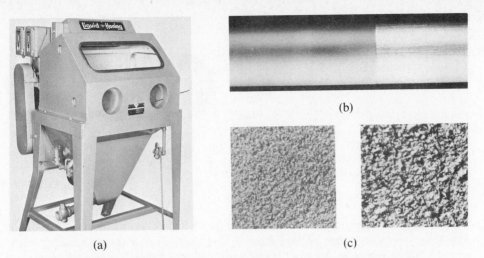

(b)

(a)

(c)

Figure 22.4. The liquid-honing machine (a). A comparison of a conventionally honed (left) and liquid-honed (right) drill rod is shown at (b). Photomicrographs showing the effect of various abrasives on metallic surfaces (c). (*Courtesy Vapor Blast Manufacturing Co.*)

Scale removal

Weld cleaning

Burr removal

Satin finishing

Figure 22.5. Some applications of power brushing. (*Courtesy Osborn Manufacturing Co.*)

Figure 22.6. The main types of power brushes are wheel, cup, end, side action, and wide-faced. (*Courtesy Osborn Manufacturing Co.*)

brushes, cup brushes, end brushes, side-action brushes, and wide-face brushes. The wheel brushes range in size from 1 to 18 in. (45.72 cm) in diameter with wheel widths from ⅛ in. to 48 in. (3.175 to 1216 mm). Fiber brushes consist of natural fibers, including Tampico (treated and untreated), horsehair, and various cord materials. Synthetic fibers are also used. Wire wheels are made of steel, stainless steel, and brass.

The larger-sized brushes are usually operated at 4500 to 6000 sfpm. When used with fine abrasive compounds, they can produce fine microinch finishes on metal surfaces.

Power brushing is also used to blend sharp contours left by grinding so that uniform plating can result. Polishing marks, draw marks, and scratches can also be blended. Even a few seconds of power brushing can reduce microscopic V-notch cracks that go undetected and greatly increase the fatigue life.

PROTECTIVE METAL COATINGS

The variety of protective coatings available today covers a wide range in terms of length of life, ease and difficulty of application, effectiveness of protection, and cost. In order to select the optimum coating system, it is necessary to establish the protective requirements. There is generally a direct relationship between the applied cost of a coating and the degree of duration it will provide. From this standpoint, coatings can be divided into three main types: interim, durable, and "permanent."

Interim Coatings

Interim coatings provide a rust preventative while the parts are in process. Some also provide lubricating qualities to the surface as an aid to further forming operations. In other cases just the opposite is required since adhesives will be used in assembly. The best known and most widely used interim coating process is that of chemical conversion, or simply conversion coatings.

Conversion coatings

Conversion coatings get their name from the fact that certain metals are capable of reacting with various chemical agents in such a way that solid chemical compounds are precipitated on the surface or, more accurately, *with* the surface. The metallic and ionic nature of the metal at the surface is altered by conversion into nonmetallic chemical compounds through mechanisms in which the metal atoms become part of the compounds. The process is thus a transformation of the outer atomic or ionic layers of metal into insoluble chemical compounds of the metal, the result of a precipitation reaction. The best known of the conversion coatings are those employing phosphoric acid.

Phosphate Coatings.   It is said that Roman soldiers stationed in Germany found certain deposits of phosphate-bearing earths which proved effective in removing rust from their armor. Thus, the art of protecting metals by depositing a phosphate coating on them is not new, but the broad scientific base of the process is a more modern achievement.

A phosphating installation requires the removal of all soils. A simple test of cleanliness is to hold a part under cold running water and watch for any appearance of a water break or water separation on the surface. If a water break occurs, the part is not clean and must be reprocessed. Normally, five successive tanks are used for phosphating: cleaning, rinsing, phosphating, rinsing, and a final inhibitive rinsing. Where pickling or other operations are necessary, the number of stages may be as high as nine or more.

Several types of phosphate salts are used: iron phosphate, zinc-iron phosphate, and manganese-iron phosphate.

The iron phosphates provide a light crystal coating for steel. A detergent can be incorporated into the phosphate bath so that cleaning can be carried on simultaneously with the process. The thin phosphate coating under paint is especially good for impact resistance.

The zinc-iron phosphate provides a good base for paint as well as nondrying oils and waxes. It also provides a heavier coating for more extensive outside exposure.

The manganese-iron phosphate provides a heavy black coating that is used on frictional surfaces to prevent galling, scoring, or seizing of parts.

In brief summary, the outstanding characteristics of phosphate coatings are their ability to retain paint and provide lubrication when necessary. Of course the primary purpose is to provide interim corrosion resistance, as shown in Fig. 22-7(a). Machines are available that can accomplish vapor degreasing, phosphatizing, and painting [Fig. 22-7(b)] in a single pass.

Chromate Coatings.   Chromate conversion coatings are used on zinc, cadmium, aluminum, copper, silver, magnesium, and alloys of these metals. The process consists of a simple chemical dip, spray, or brush treatment. The chromate bath is an acidic solution containing hexavalent chromium compounds known as activators or catalysts.

The principal use of chromate coatings is for corrosion resistance. The chromate furnishes a base for painting but it may also be a decorative final finish. Various colors may be added by subsequent dipping in organic-dye baths.

Most chromate treatments are proprietary processes.

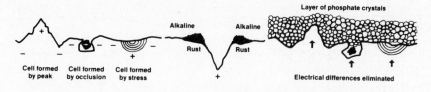

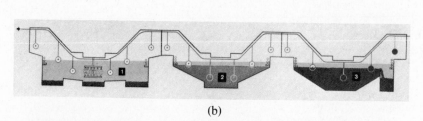

Figure 22.7. Metal surface without phosphate coat (left) is positively and negatively charged. When moisture is present, electrolytic cells form and rust builds up (center). Phosphate coat neutralizes these electrical differences (right) (a). Vapor degreasing takes place at (1), phosphatizing at (2), and dip painting at (3), (b). If baking-type enamels are used, a fourth but shortened baking step can be added. (*Courtesy Oakite Products, Inc.*)

Anodic coatings

Anodized films are produced by electrochemical reaction with the base metal and are another form of conversion coating. Anodizing gets its name from the fact that the base metal becomes the anode instead of the cathode as is the usual case in the electroplating process. The process is used on aluminum and magnesium. An electrolyte capable of yielding oxygen on electrolysis is used. As current is passed, oxygen is liberated at the surface, forming an oxide film. Following the anodic treatment, the oxide film may be sealed with boiling water. This closes the pores and destroys the absorptive characteristics of the coating.

Hard-coat anodizing is much the same process, but with a higher voltage to produce a harder surface. Thus items made from light metal can have a hard-wear resistant surface.

The use of colored anodized aluminum has made rapid progress in recent years, particularly in the giftware and home-appliance field. Color imitations of precious and semiprecious metals such as gold, copper, brass, etc., as well as a wide variety of other colors, are used.

Dying the anodized aluminum is done by dipping in a water solution of dyestuff at a temperature of 150°F (66°C) for about 10 min. The work is then sealed and is resistant to further dying or staining.

Durable Coatings

Durable finishes may also be classified as organic and inorganic finishes. There are several dozen major chemical types of organic coatings that derive from resins such as alkyds, vinyls, and epoxies. The inorganic coatings are porcelain enamels and various plasma-sprayed materials.

Organic Coatings. For the initial selection of an organic coating, all environmental requirements should be clearly defined. Consideration must be given to anticipated exposure (whether indoor or outdoor, marine or tropical), corrosive ambients (such as solvents, chemicals, vapors, or salt spray), and mechanical abuse (such as impact, vibration, abrasion, bending, or elongation).

Inorganic Coatings. Porcelain enamels are mixtures of alumina–borosilicate glass to which coloring oxides and other inorganic materials are added. The powder is spread on the steel sheet and heated up to about 400°F (204°C) to fuse it to the steel sheet. The fusion process forms a tenacious enamel-to-steel bond.

Porcelain enamels are supplied in a wide range of finishes, ranging from the smooth nontextured to the special textures. They have excellent resistance to environmental conditions of fading, mechanical damage, abrasion, and corrosion.

Porcelain enamels are extensively used on all major appliances, plumbing ware, domestic and commercial heating equipment, lighting fixtures, and heat exchangers.

Plasma-sprayed materials consist of borides, beryllides, carbides, and nitrides. These coatings, particularly carbides, have proven to be very wear resistant. In addition to their high hardness they have good chemical stability.

The plasma-spray method used for applying these coating materials furnishes the first controlled temperatures above 20,000°F ever used in industry. The plasma spray process was discussed in Chapter 13. Carbides and nitrides may be plasma sprayed on Inconel, stainless steel, zirconium, and graphite.

Primers

The application of the best coating will be of little value unless the proper primer is used first. Primers provide adequate adhesion and compatability of the topcoat with the surface. They are usually of a different chemical type than the topcoat.

Primers are formulated to provide the most stable bond to the substrate, to smooth out the surface for topcoat applications, or to bring anticorrosive or inhibitive elements such as zinc into contact with the metal. Zinc is known as a sacrificial metal. If a zinc coating on steel is scratched, a small electric current is established which flows atoms from the anode (zinc) depositing it on the cathode, thus sacrificing itself to protect the steel from corrosion. Automotive rust has been drastically reduced since the use of galvanized steel has been used on inside surfaces of doors and fenders.

Application of organic coatings

The choice of coating material is rapidly expanding. The manufacturer may consider liquid, powder, and solid films, or a combination of these types. An automobile, for example, may be first coated with Bonderite, followed by a zinc-rich powder prime on all accessible areas. The powder coating would be partially baked and then the body would be immersed in a cathodic electrocote tank to receive a waterborne material coating on all areas that did not receive a coating from the spray. The coatings will then be final baked in preparation for the topcoat color application.

Why is such a complicated system used? EPA rulings state that the allowable weight of any hydrocarbon solvent must be less than 2 lb (0.9/kg) per gallon per hour. Thus high solids, water-reducible coatings, powder coatings, and chemical coatings are being used more and more.

Chemical Coating.   Chemical coatings can be put on in dip tanks or plating tanks.

Electrostatic Atomization.   Electrostatic spraying utilizes an electrical differential to cause the paint to separate. A disc or bell rotating at 900 to 1800 rpm is used as the atomization device. The paint flows to the periphery of the disc by centrifugal force and levels out. As this action occurs, 120,000 v are applied, resulting in the paint particles taking on an electrical surface charge.

The charge particles seek out the nearest grounded surface and attach themselves to it. This surface of course is the article to be painted, which is at a ground potential of zero voltage. The electrical charge of the particles quickly dissipates upon contact with the surface. Thus, other particles tend to seek out unexposed surfaces, providing a uniform buildup.

Although this method is very good for high-production painting, it is not as smooth as with more conventional methods and sometimes an orange-peel effect is produced. The chief advantage of the process is in paint savings. Properly applied, paint costs can be reduced by as much as 50%.

Air electrostatic spray is another version of electrostatic atomization. In this process, the paint is atomized by the conventional compressed-air method or by the airless technique. As the particles leave the gun, they are directed under an electrode that is charged to about 62,000 v negative (Fig. 22-8). The particles pick up a negative surface charge and proceed toward the workpiece, which is electrically grounded. As the particles approach the work, they rapidly lose velocity and will be almost floating as they reach their destination. The strong negative charge is attracted to the nearest grounded object. Some of the particles may pass the object and the attraction is strong enough to bring them back, depositing paint on the backside of the object. This effect is known as *wraparound* and is one of the biggest advantages of the electrostatic process. Outside corners and edges are usually hard to paint by conventional methods, but are covered very well by electrostatic methods.

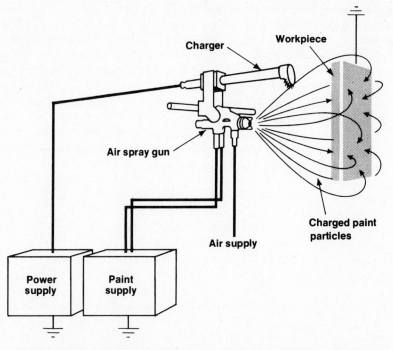

Figure 22.8. Air-electrostatic spray system is designed to handle the increasingly popular water-reducible coatings. The charger creates an electrostatic field between the nozzle and the workpiece, charging passing paint particles. (*Courtesy* American Machinist *and Nordson Corp.*)

Powder coatings

Powder coatings have become popular in recent years because they are 100% solid, thereby affording savings in freight, storage space, and air pollution control equipment. Application efficiency is attainable up to 98%. Thicker coats can be applied in one application, and they do a better job of covering edges and blemishes. They also do not drip or sag.

Powder coatings are either thermoplastic or thermosetting. Common thermoplastic types are acrylic, cellulose nitrate, and vinyls. Thermosetting types are alkyds, epoxies, polyesters, polyimides, and silicones.

Application of powdered coatings

Although a number of techniques exist for applying powder coatings, fluidized-bed and electrostatic-spray methods are used to coat most parts.

The simpler fluid bed dip application has severe limitations in that relatively thick coatings of 6 mils (15 mm) and up are applied. Most parts do not require coatings over 3 mils (0.08 mm) thick. Also coating thicknesses will vary on large nonuniform parts.

Fluidized Bed.   Parts are preheated to 400 to 600°F (200–300°C) and immersed in a fluidized bed of dry plastic powder, or they may be suspended above the fluidized bed as shown in Fig. 22-9(a) in what is called an *electrostatic* fluid bed. In this system, a charging grid is placed across the top of the tank below the top level of the fluidized powder. The bed is fluidized by introducing low pressure air at the bottom of the tank. The air diffuses through a porous membrane where it lifts the powder from the bottom and essentially fluidizes the entire resin powder contents in the tank.

In the electrostatic method, particles are attracted to the part because of unlike charges. This method provides the highest powder transfer efficiency known. The coating can be applied to either a hot or cold part with a high degree

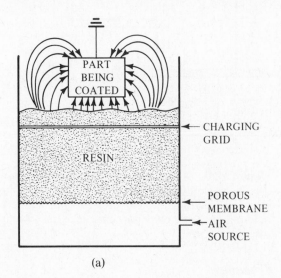

(a)

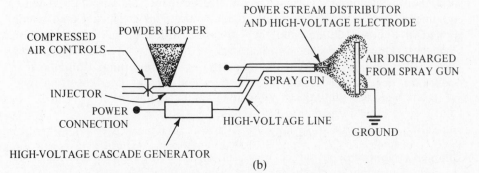

(b)

Figure 22.9.   Schematic representations of the two main methods of applying powder coatings to metals, electrostatic fluid bed (a) and electrostatic spray method (b). (*Courtesy* Manufacturing Engineering and Management.)

of control over the thickness and at a high production rate. However, it has a severe limitation in that only products having heights that would pass within the uniformly dense, charged powder cloud, extending 2 to 4 in. (50 to 100 mm) over the aerated charge, can be coated. Normally 30 to 100 kv are used at a maximum current range of 200 microamp. After coating, the parts are oven cured to provide a final set of the coating. Mil thickness is dependent upon the type of material used, which is almost always thermoplastic, the preheat temperature, and the time in the bed or above it, as in electrostatic application.

The most widely used powders are the epoxies. They are formulated to be tough, highly durable, and highly decorative. They are available in an assortment of colors, glosses, and textures to meet all requirements.

Polyvinylchloride is the most widely used thermoplastic powder. It is relatively low in cost and provides good chemical, abrasive, and impact resistance. It is also the best coating for electrical insulation.

Electrostatic Spray.   In this process, cold parts are sprayed with unheated, electrically charged plastic powder [Fig. 22-9(b)]. This process is especially good for large parts and thin coatings. It is also advantageous for parts that need to be coated only on one side. Coatings as thin as 1 mil can be applied by electrostatic spray.

The major trend in the development of second-generation electrostatic spray equipment for metal powders is the greater use of solid-state components. Currently, electrostatic charging equipment is 100% solid state. Since the powder particles are charged in the spray gun, the strength of the electrostatic field outside the gun is not of primary concern.

Methods for development of the high voltage needed in the electrostatic gun have been improved. The advent of the cascade generator is a significant development in the production of high voltage. By using diodes and capacitors, the cascade generator develops high dc voltage with relatively little power input. This miniaturized, high-voltage generator is located within the gun. The elimination of a high-voltage cable to the gun increases the safety of the system.

In general the surface preparation for powder coating is the same as that for painting. Normally three to five detergent wash and rinse cycles followed by an iron phosphate treatment (to improve powder adhesion) precede the coating operation.

Advantages and Limitations.   One of the main advantages of powder coating compared to conventional finishing techniques is that solvents are not required, thus eliminating the need for air and water pollution controls. Powder coatings can be corrected even after being applied and before curing by simply vacuuming or blowing off the damaged or imperfect coatings. As a result of this simple correction procedure, powder finishing rejects have been cut to as little as 2%.

Powder coating systems are inherently safer than solvent paint systems. A hazard exists only when the concentration of powder suspended in the air exceeds a known limit. Powder manufacturers stipulate the lower explosive limit in g/m^3 or oz/ft^3. Enough air must be provided to avoid this concentration. Automatic limitations on operational current keep it below the combustibility point of the powder–air suspension.

Powder coatings are now being used extensively in the automotive and appliance industry and are increasingly being selected for bicycles, farm machinery, guard rails, highway signs, hospital furniture, lawn mowers, lighting fixtures, office equipment, playground equipment, silos, snowmobiles, toys, and wire goods.

One limitation at present is the colors available—at present there are only five.

The powder coating system is easily automated with all elements on line interlocked, conveyor system, spray guns, powder supply, powder recovery unit, and blending system for recycled and new powders. Sensing devices enable spraying to occur only when a part is in position, and powder spray patterns can be automatically adjusted to accommodate various size parts moving on the same overhead conveyor.

"Permanent" Metal Coatings

In this discussion, metal or inorganic coatings will be considered permanent. Zinc, tin, lead, copper, brass, chromium, aluminum, and stainless steel are the principal metal-coating materials and are applied by a variety of processes, namely, by hot-dipping, plating, metal spraying, vacuum metallizing, and sputtering.

Hot-Dipping

Hot-dipping is used to apply a substantial portion of the total amount of all metal coatings. It consists of first thoroughly cleaning the base metal by acids or other methods, followed by proper fluxing, usually with zinc ammonium chloride, after which the metal is immersed in a bath of the molten coating metal. All of the coatings applied by this method are dependent upon an alloying action between the coating metal and the base metal for the bonding action.

The interface alloy is usually more brittle than the base metal or the coating metal. This may interfere in some forming operations during subsequent manufacturing.

The big advantage is that the costs are relatively low. Hot-dipping may imply a batch process, but high-speed lines are used for coating wire, strip, or continuous sheet coil.

Common applications of the hot-dip process are aluminum, zinc, and tin to steel and lead, and zinc alloys to nonferrous alloys. Zinc coatings on steel are commonly referred to as galvanized coatings.

To ensure an even coating on small parts such as nuts and bolts, after being taken from the bath they are centrifuged until the coating is hard.

Mechanical Galvanizing.   Mechanical galvanizing is a relatively new process for imparting a uniform zinc coating and eliminating many of the problems associated with the traditional hot-dip process. The new process was developed by the 3M company.

Figure 22.10. Mechanical galvanizing provides uniform zinc coating on all surfaces of nut seen at lower right. Nut at upper left was coated by the hot dip process. (*Courtesy Plating Systems Department 3M*).

Parts to be coated are tumbled in a lined barrel at room temperature with water, zinc powder, small glass beads, and a special promotor chemical. The peening action of the beads cold welds the zinc powder onto the parts. Coating thickness is determined by the amount of zinc powder used and the length of time the tumbling action continues.

At this time, mechanical galvanizing is recommended for bulk coating of small thread fasteners, nuts, washers, and other parts used in heavy duty applications involving exposure to atmosphere. Coating thicknesses that can be produced range up to 0.003 in. (0.08 mm).

The advantage of the process is that the coating is consistently uniform on all surfaces, as shown by the example in Fig. 22-10. Adhesion is as good or better than with hot-dip galvanizing. The process is also cleaner for the work area with virtually no fumes. Sludge is easily treated to comply with pollution control regulations.

Electroplating

Electroplating is the most widely used method of applying metallic coatings. The controlled thickness of deposit and uniformity of quality are the method's greatest assets. In practice, coatings of 0.0002 to 0.0003 in. (0.005 to 0.008 mm) are not uncommon.

Surfaces to be plated must be buffed smooth to eliminate scratches and unevenness. The quality of the plated surface depends to a great extent on the preparation of the base metal. The work is then cleaned in suitable cleaning solutions to remove all grease, oxides, buffing compounds, etc. After rinsing, the part is ready for plating.

The four essential elements of a plating process are the cathode, anode, electrolyte, and direct current. Current leaves the anode, which is a bar of plating material, and migrates through the electrolyte (salts of the plating metal in solution) to the cathode, or part to be plated. As the ions are deposited on the cathode, they give up their charge and are deposited as metal on the cathode. The current density largely determines the rate at which the metal is deposited. For example, in chromium plating, good deposits may be obtained at cathode current densities ranging between 0.25 amp/sq in. to 6 or more amps/sq in. The current density in turn is determined by the operating temperatures, the size and shape of the workpiece, the condition and nature of the surface, and other factors.

Chromium will provide a good corrosion-resistant film directly over stainless steel, nickel-silver, or Monel. It should never be used directly on brass or zinc die castings due to the dezincifications that will take place under the coating causing it to lift off at some future date.

Hard chromium, as differentiated from decorative chromium, is applied for functional reasons. If a 0.00005 in. (0.0013 mm) deposit is applied over a hardened surface, then it is a hard film because of its backing. If the same thickness is applied over softer metal, then the chromium will be no harder than the base. If, however, a deposit of 0.0015 in. (0.04 mm) to 0.002 in. (0.05 mm) is applied over the same unhardened surface, it will be very hard.

Chromium plate has a lower coefficient of friction than any other metal and has good antigalling properties as long as it is not used against itself.

Parts to be plated should be designed with generous fillets and radii instead of sharp corners since current concentrations occur at sharp points, resulting in excessive deposits. Plating thicknesses should be taken into consideration when designating design tolerances on mating parts.

Plating of Die Castings.   Most die castings are finished by electroplating. Simple shapes can be plated with about 1.3 mils minimum of copper plus nickel, and 0.01 mil minimum of chromium, in approximately 50 min. Complex shapes require longer times and often special fixturing. In complicated shapes, more

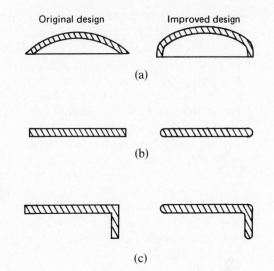

Original design   Improved design

(a)

(b)

(c)

Figure 22.11. Original and improved designs for plating. A convex surface is easy to plate, especially if the edges are rounded (a). Flat surfaces are not as easy to plate. A 0.015-in. crown/in. is recommended to hide undulations caused by uneven buffing (b). Sharp angles and corners are undesirable, all edges should be rounded (c).

metal is plated unnecessarily on projections and other high-current-density areas. Figure 22-11 shows a few sample designs with comments as to the platability of that geometry. In general, flat surfaces should be crowned, edges should be rounded, and inside angles should be given generous radii to improve the platability.

For good corrosion resistance outdoors, the minimum thickness of copper and nickel should be 0.2 and 1.0 mil, respectively, with a 0.03 mil of chromium. For indoor parts, the nickel and chromium can be less, about 0.8 and 0.01 mils, respectively.

Metal spraying

Although metal spraying is a coating process, it is more commonly associated with welding equipment and is discussed in Chapter 13.

Vacuum metallizing

Vacuum metallizing is a process whereby metals and nonmetals are deposited under high vacuum on to prepared surfaces of plastics, metals, glass, paper, textiles, and other materials to produce finishes that are both decorative and functional.

The process consists of placing objects to be metallized in a vacuum chamber with a small amount of selected "plating material." The plating material, depending upon the application, may be aluminum, cadmium, silicon, monoxide, silver, selenium, magnesium fluoride, or a number of other materials. When the required high vacuum has been reached in the chamber, the plating material is heated to its vaporization temperature. The material vaporizes, radiates throughout the chamber, and deposits by condensation on the objects to be metallized.

Objects are held in the chamber on fixtures, the design of which enables either entire surfaces or selective areas to be metallized.

After metallizing, the products are dipped in topcoat lacquer. A wide range of colors can be produced by dipping products in dye solutions, which will color the topcoat. Dyes may also be combined with the topcoat. The topcoat serves to protect the very thin metal film from abrasion and wear. It is standard procedure when metallizing plastics to put a base coat of lacquer on the part before metallizing. A metallized finish is then, in effect, an ultrathin film of aluminum, or other metal, sandwiched between two layers of coatings. Depending upon service requirements, topcoats are air-drying or baking types.

The main advantages of vacuum metallizing are: (1) It provides a method of placing a thin film of metal on a nonconductive material. (2) The deposit may be less than five millionths of an inch, yet it completely conceals the color of the object. (3) It is much less expensive than electroplating. (4) The corrosion resistance of lacquered metal surfaces is high. Salt-spray tests show 500 hours of endurance.

Sputtering

Sputtering is a deposition process wherein a surface, or target, is immersed in an inert-gas plasma and is bombarded by ionized molecules that eject surface atoms. The process is based upon disintegration of the target material under ion bombardment. Atoms broken away from the target material by gas ions deposit on the part (substrate) and form a thin film.

The older and simpler type of sputtering uses a dc diode, where the coating material (target) acts as a cathode and the part to be coated (substrate) acts as an anode. When these electrodes are immersed in an argon atmosphere of 10 to 100 mm and a dc potential of 500 to 5000 v is applied across the electrodes, the gas is ionized and a plasma is generated. Positive ions in the plasma are accelerated toward the target with enough kinetic energy to knock off (sputter) atoms from the target surface. The technique is limited to conductors and certain semiconductor materials. A newer RF (radio frequency) process has eliminated this limitation and now almost any material can be sputtered.

Until recently, the maximum thickness of sputtered coatings was limited to 0.1 micron or less. Because the process was relatively slow and required skilled operators, the process was not used on a large scale. Now, however, sputtering is being used on production lines for such operations as "plating" of plastic automobile grills. The new system imposes RF (radio frequency) on a metal electrode behind the target as shown in Fig. 22-12. The target, which is an insulator, is bonded to the metal electrode, forming a capacitor. The insulator surface in contact with the plasma is alternately bombarded by electrons and positive ions during each RF cycle. When the surface is positive, it attracts electrons; and when it is negative, it attracts ions. Since the electrons in the plasma have a higher mobility than the ions, the electron current to the target is initially much greater than the ion current, causing the target to be coated.

The NASA Lewis Research Center has successfully sputtered cattle bone on metallic prosthetic devices for use in surgical implants as hip-bone replace-

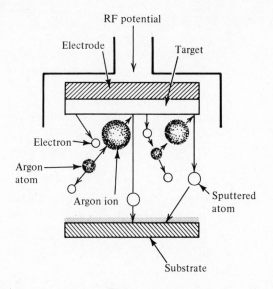

Figure 22.12. Schematic of the sputtering process. (*Courtesy* Machine Design.)

ments. The sputtered bone film promotes bone growth and attachment to living bone.

The widely advertised platinum-edged and chromium-edged razor blades are sputter-coated. Their superior performance is attributed to improved corrosion resistance. To make the process economical, 75,000 blades are strung on a skewer, placed in a vacuum chamber, and sputtered simultaneously. The coating is so thin that the blades do not have to be rehoned after the sputter coating has been applied. In spite of the thinness of the coat, it is pinhole-free and extremely adherent.

Electroforming

Electroforming may be defined as the production or reproduction of parts by electrodeposition on a mandrel or mold that is subsequently separated from the deposit. Occasionally the mandrel may remain in whole or in part as an integral functional element of the electroform. Electroforming is considered an art and its successful practice is largely dependent on the skill and experience of the operator. Although detailed instructions for producing various electroformed objects are available, they do not guarantee the successful production of a part that has not been previously manufactured by this process.

Some sectional views of electroformed parts and mandrels are shown in Fig. 22-13.

Nickel Electroforming.   Most electroformed parts are made of nickel. Studies have been made so that nickel electroforms can now be made with the

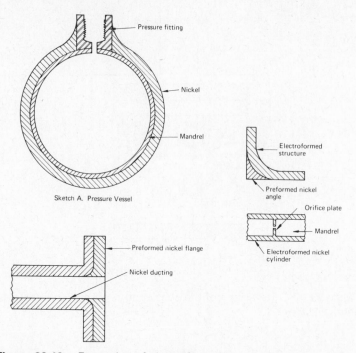

Figure 22.13. Examples of electroformed parts. The electrodeposited nickel may be formed around inserts.

following property ranges: 130 to 250 DPH, 50 to 200 ksi tensile strength, elongation from less than 1% to 37%, and internal stress ranging from compressive to 25 ksi tensile.

One product that is ideally suited to electroforming is that of large, curved metal mirrors. The curved shape, perfect finish, and thin cross section required for precision mirrors can be achieved most economically by the replication of a highly polished glass mandrel. The mandrel may be used either directly in the plating bath (after metallizing the surface) or, more commonly, as a master for casting low-cost, single- or multiple-use replica molds from epoxy resin. The replica is silvered, plated to the desired thickness, and removed to produce a precise metallic reflector.

Advantages.   A number of basic advantages distinguish electroforming from mechanical forming methods. These advantages enable the manufacturer to: (1) form complex, thin-walled small or large shapes (Fig. 22-14 shows a large, 10 ft long, 12 in. dia. thin-walled (0.004 in. thick) rotary screen being formed on a nickel steel mandrel); (2) laminate similar or dissimilar metals without heat or pressure; (3) reproduce surfaces precisely; (4) incorporate inserts and flanges; (5) form integral cavities; (6) enclose and support nonmetallics; (7) form precise openings (Fig. 22-15).

Figure 22.14. A large, thin-walled shape, 10 ft long and 12 in. in diameter is being electroformed to 0.004-in. wall thickness on a stainless steel mandrel. (*Courtesy* Machine and Tool Blue Book.)

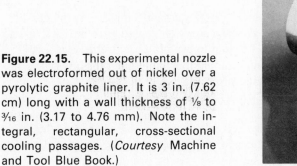

Figure 22.15. This experimental nozzle was electroformed out of nickel over a pyrolytic graphite liner. It is 3 in. (7.62 cm) long with a wall thickness of ⅛ to 3/16 in. (3.17 to 4.76 mm). Note the integral, rectangular, cross-sectional cooling passages. (*Courtesy* Machine and Tool Blue Book.)

Disadvantages.   Parts are limited to materials that can be plated. The mandrel material must be conductive or a nonconductive material such as rubber must be given a very thin conductive coating, usually aluminum or silver, on which the nickel can be electroplated.

Precoated Metals

Although precoated metals have been available for many years, they have become increasingly attractive in recent years, particularly after the passage of the Clean

Air Act of 1970 which created the Environmental Protection Agency. There has been increased monitoring of plant conditions, including paint storage practices, possible fire and explosion hazards, ventilation conditions, and many other factors incident to an in-plant paint operation, all of which come under OSHA. Among the 25 most-cited violations of job safety and health standards are those that include failure to meet industry standards for flammable and combustible liquids and their use, spray finishing, and general fire protection. Thus safety regulations have been among the factors that pushed precoated steel sheet into a more than billion-dollar industry.

Advantages of precoated metal often cited are reduced handling and reduced inventory. Substantially lower insurance rates can often be obtained when a finishing line is eliminated.

A wide variety of coatings are available, including tin, zinc, polyesters, vinyls, acrylics, PVCs, epoxies, phenolics, and fluorocarbons. Many of these coatings can be supplied in multicolored printing or embossing of the surface, including woodgrain effects.

Fabrication of precoated metals

Precoated steel sheet can be sheared, punched, stamped, press-braked, roll-formed, deep-drawn, and spun with the same tools as uncoated steel. Die life is generally longer due to the coating's internal lubricants. The flat metal stock (0.002 in.–0.090 in. thick, 0.5 in.–70 in. wide; 0.05–2.29 mm thick, 12.70–177.8 cm wide) comes in coils.

The use of precoated metal places some demands on the designer. For example, the bend radius must be at least equal to $2t$ and it is better if it can be $3t$ or $4t$. If drawing is required, the radius should be extremely generous.

Exposed edges caused by slitting, notching, or punching presents two problems: appearance and corrosion. Quite often the raw edge can be turned under, as shown by the examples in Fig. 22-16(a). Another method is to cover the edges with a molding or snap-on trim (b).

Preventing corrosion along the exposed edges is often done by using a galvanized-steel substrate. When the material is cut, the zinc coating tends to "wipe" in the direction of the cutting motion, giving protection to the exposed edge. Where this is not possible, lockseaming provides another alternative. Also, exposed surfaces can be treated if corrosion resistance is essential.

Welding. Precoated steels can be welded. In general the temperature should be low (consistent with good welds). The weld cycle should be short. Resistance spot welds and projection welds are the most used (Fig. 22-17).

Fasteners. A wide variety of fasteners can be used. The designer can usually select a fastener that will meet the needs of the design and still be hidden from view or used to accent a feature of the product (Fig. 22-18).

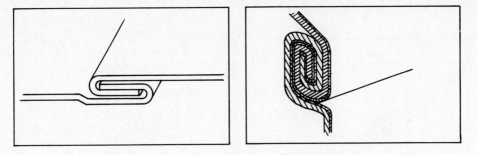

(a)

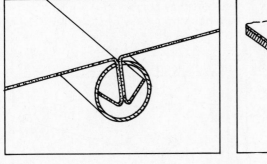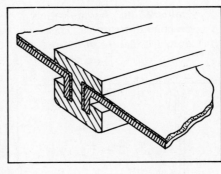

(b)

Figure 22.16. Exposed edges of precoated materials may be folded under, as in the seams shown at (a), or covered with snap-on trim, as shown at (b).

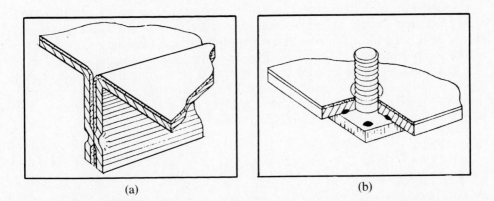

(a)   (b)

Figure 22.17. Precoated materials can be spot welded, as shown at (a), or projection welded, as shown at (b). In the spot weld the coating is burned off in the weld area.

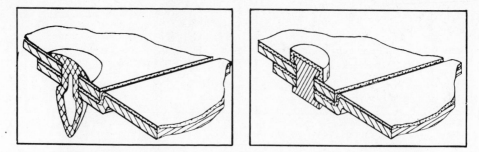

Figure 22.18. A wide variety of fasteners may be used to assemble precoated materials. They may be hidden from view or capped to accent a feature of the product.

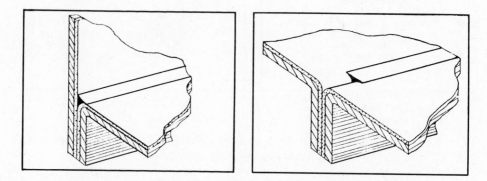

Figure 22.19. Adhesively bonded joints may have a PVC bead placed on top by hot-air welding to enhance the appearance.

Adhesives. Both thermoplastic and thermosetting adhesives are used to bond precoated metals. Since there are many adhesives available, the one used should be the one lowest in cost that meets strength and adhesion requirements, can be easily handled, and is compatible with the coated surface.

The surfaces to be bonded are first coated with adhesive, then brought together, often with heat and/or pressure. The resulting joint seals, fastens, insulates, resists corrosion, dampens vibration, and resists fatigue. Joints can be filled with PVC using hot-air welding for enhanced appearance (Fig. 22-19).

Problems

22-1. Can ultrasonic cleaning be used to remove oxides or tarnish? Why or why not?

22-2. What cleaning steps should be used for parts that are heavily contaminated by oils and grease?

22-3. Why are ultrasonic cleaners in the 20 kHz range or lower objectionable?

22-4. Why are conversion coatings called by that name?

22-5. How does a conversion coating prevent corrosion?

22-6. What is a big advantage of the electrostatic spray process?

22-7. What is the reason for the increased use of powder coatings in recent years?

22-8. (a) What three coatings are ordinarily used on die castings and how are they applied?

(b) What is the usual thickness of three electroplated coatings on die castings to be used outside?

22-9. Can chromium be used directly as a plating surface on any metal?

22-10. How thick a coat of chromium is required on a soft material to consider it a hard coat?

22-11. What is the big advantage of vacuum metallizing?

22-12. Why is the usual application of vacuum-metallized coating able to withstand wear and corrosion?

22-13. What is meant by the sputtering process?

22-14. What practical application can you think of for sputtering other than that mentioned in the text?

22-15. What is meant by electroforming?

22-16. A nickel electroformed part has a DPH of 130 to 250. Would you classify this as soft, medium-hard, or hard? Why?

22-17. Why is electrocleaning one of the best methods of cleaning?

22-18. How can large machine castings be cleaned prior to painting?

22-19. In what respect are ultrasonic cleaning and electrocleaning opposite?

22-20. How does the vibration frequency affect ultrasonic cleaning?

22-21. Is anodizing a conversion coating? Why or why not?

22-22. What is the main difference in use between phosphate coatings and chromium coatings?

22-23. Why can hydrocarbon solvents more easily meet the Clean Air Act requirements than an alkaline water wash?

22-24. What would be a good cleaning solvent for the following:

(a) oils on elastomer;

(b) ionic soils on plastics;

(c) water on metals?

22-25. What is meant by the expression, "Zinc is a sacrificial coating?"

22-26. Give four reasons why powder coatings have become so popular in recent years.

22-27. What system is best for galvanizing small parts?

22-28. In what way is the electrostatic fluid bed better than the dip system?

Bibliography

Adams, Frank. *Taking the Mystery Out of Metallizing.* New York: Technical Publishing Company, 1976.

AISI Committee. "Cost Effective Precoated Steels." *Mechanical Engineering* 96 (February 1974):37–42.

Curry, David T., ed. "Sputtering Comes Out of the Clean Room." *Machine Design* 51 (January 1979):95–99.

Editor. "Mechanical Galvanizing: An Alternative to Hot Dip." *Manufacturing Engineering* (October 1980).

———. "Powder Coating of Steel: Where It Stands, Where It's Going." *Manufacturing Engineering* 85 (June 1980):98–99.

Feinberg, Bernard, ed. "Electrostatic Powder Coating."*Manufacturing Engineering and Management,* 74 (June 1975):32–35.

Gabe, D. R. *Principles of Metal Surface Treatment and Protection.* New York: Pergamon Press, 1972.

Metal Finishing. Hackensack, N.J.: Metals and Plastics Publications, Inc., 1974.

Murphy, E. M. "High Solids, Here Now!" *Products Finishing* (September 1980).

Ramsey, R. B. "The Niche for Fluorinated

Solvents." *Metal Progress* (April 1975).

St. John, Warren. "Powder Coating." *Me-* *chanical Engineering* 96 (February 1974):23–29.

Appendix

Table A. Properties of metals

Metal or Alloy	Symbol (Main Elements)	Melting Point °F	Melting Point °C	Tensile Strength ksi	Tensile Strength MPa	Yield Strength ksi	Yield Strength MPa
Aluminum 99%	Al	1220	660	13	89.63	5	34.47
Aluminum 2011-T3	Al			55	379.2	43	296.5
Brass, yellow	Cu Zn		900–1100	40–110	275.8	18	331.0
Beryllium	Be	2340	900–1000	70–200	70–200		
Chromium steel (5100)	Cr	3430	1890				
Copper	Cu	1981	1083	32	220.6	10	68.95
Cast iron							
Grey	Fe C	2065–2200	1130–1204	20–60	172.4		
Malleable (pearlitic)	Fe C			60–105	413–724	40–90	275–620
Ductile	Fe C			65–150	448–1034	30–60	206–413
Iron, wrought	Fe			40	275.8	27	186.2
Lead	Pb	621	327	3	20.68	1.9	13.09
Magnesium	Mg	1202	650	25	172.4	13	89.63
Molybdenum (4150)	Fe Mo C	4760	2625	105–124	724–855	70–100	490–690
Silver	Ag	1760	960	23	158.6	8	55.16
Steel	Fe		1510				
Low carbon	Fe C	2798		50–75	413.7	40	275.8
Medium carbon	Fe C	2600–2730		70–120	579.2	52	351.6
High carbon	Fe C	2600		100–200	675.7	72	496.4
Steel, nickel (2515)	Fe Ni C			90–105	724.0	69–75	579.2
Steel, cast	Fe C			72	496.4	40	275.8
Steel, stainless (304)	Fe Ni Cr			85	586.1	35	241.3
Titanium (Ti-55A)	Ti	3300	1820	35	241.3	25	172.4
Tungsten	W	6170	3387	500	3447		
Zinc alloy	Zn	727		41–47	282–324		

Table A—*Continued*

Metal or Alloy	Elongation in 2 in. Percent	Modulus of Elasticity 10⁶ PSI	Modulus of Elasticity MPa	Thermal Conductivity BTU/Sq Ft Hr/°F/In.	Thermal Conductivity Cal/ Sq Cm/Sec/Deg. C/Cm	Coefficient of Expansion (In/°F) × 10^{-6}	Coefficient of Expansion (Cm/°C) × 10^{-6}
Aluminum 99%	35	10.2	70330	1570	0.53	12	25.3
Aluminum 2011-T3	15	10.2	70330	1570	0.53	9.8	25
Brass, yellow		14.0	96530	2680	0.94	11.0–11.6	12.4
Beryllium						9.4	
Chromium steel (5100)		7.1	4891			9.8	
Copper		15.0	103400	2680	0.94	9.2	16.7
Cast iron							
Grey	0–1	13	89630	310	0.106	6.0	10.2
Malleable (pearlitic)	10–25	12.5	86125			6.6	11.88
Ductile	1–26						
Iron, wrought	3–20	29.0	199900	418	0.143	6.7	
Lead		2	1379	240	0.083	16.4	29.1
Magnesium	15	6.5	5128	1090	0.37	14.3	12.06
Molybdenum (4150)	21–16						
Silver		10.5	72345	2900	1.00	10.6	18.8
Steel							
Low carbon	27–38	30.0	206800	460	0.18	6.7	12.06
Medium carbon	12–32	30.0	206800	460	0.18	6.7	12.06
High carbon	2–23	30.0	206800	460	0.18	6.7	12.06
Steel, nickel (2515)	27–24	30.0	206800	400			
Steel, cast		30.0	206800				
Steel, stainless (304)	50	29.0	199900		0.137	9.3	16.74
Titanium (Ti-55A)	24	12.1	83369				
Tungsten	10–16	51	351600	783–754	0.34	2.6	4.5
Zinc alloy		18.5	127460		0.264	15.2	26.3

Table B. Suggested cutting speeds for lathe work

Material	Cutting Speed for High-speed Steel Tools, Surface Feet per Minute
Aluminum and alloys	200–400
Brass and bronze, soft	100–300
Bronze, high-tensile	70–90
Cast iron:	
Soft	100–150
Medium	70–100
Hard	40–60
Copper	60–150
Malleable iron	80–90
Steel:	
Low-carbon	90–150
Medium-carbon	60–100
High-carbon	50–60
Tool-and-die	40–80
Alloy	50–70

Table C. Application of carbides. (*Courtesy* Metalworking.)

Material	Operation	Grade	Speed (sfpm)
Cast iron	Roughing	C-1	100–250
	Milling	C-1, C-2	150–300
	Precision finishing	C-3, C-4	200–500
Aluminum	Roughing	C-2	400-up
	Finishing	C-2, C-3	400-up
	Grooving	C-3	400-up
Brass	Roughing	C-2	150–300
	Finishing	C-2, C-3	250–400
Carbon steel	Roughing	C-5	125–200
	Medium cuts	C-6	150–325
	Milling	C-5, C-6	175–400
	Finishing	C-7, C-8	250–350
	Precision finishing	C-7, C-8	300–500
Alloy steel	Roughing	C-5	100–175
	Finishing	C-6, C-7	150–200
	Precision finishing	C-8	250–500
Heat-treated steel	Roughing	C-6	80–100
	Finishing	C-7	100–125
Steel castings	Roughing	C-5, C-6	100–200
	Finishing	C-5	150–225
		C-6	175–375

Drilling speed is 80% of turning speed.

Table D.   Materials machinable by ceramic. (*Courtesy* Metalworking.)

Material	Tool Life	Best Average Finish (mu-in.)	Usual Speeds (sfpm)
Steel—Soft to R_c 30	Good	35	1000–2400
—R_c 30 to R_c 45	Excellent	25	500–1500
—Hard (R_c 45 to R_c 65)	Excellent	20	75– 800
Cast iron	Excellent	35	300–2500
300 series stainless steel	Poor	35	400–1000
400 series stainless steel	Good	35	600–1500
Carbon-graphite	Excellent	—	200–1000
Plastics	Good	—	300–3000
Tungsten	Excellent	—	150– 800
Stellite	Excellent	—	60– 600
Tantung	Excellent	—	60– 140
Molybdenum	Fair	—	500– 800
Copper	Excellent	—	600–2000
Titanium	Poor	—	Not recommended
Brass	Good	—	To 3000
Bronze	Good	—	To 3000
Aluminum bronze	Poor	—	Not recommended
High-temp alloys (iron base)	Poor	—	Not recommended
High-temp alloys (nickel base)	Poor	—	Not recommended
High-temp alloys (cobalt base)	Good	—	60– 600

Table E. General properties, cost, and fabricating methods of plastics. (*Courtesy* Machinery)

Thermoplastics

Name of Plastic Material	Cost	Typical Applications	Maximum Tensile Strength psi	Maximum Service Temperature, °F	Significant Material Characteristics	Fabricating Methods
ABS	medium	pipe, wheels, valve bodies, refrigerator parts, automobile parts, pump impellers, tool handles, football helmets	4000 to 8000	175 to 212	High impact strength—2 to 10 foot-pounds per inch; tough to −60°F; resists acids, alkalis, and salts	Injection molding, extrusion, calendering, vacuum forming
Acetal	medium	instrument clusters, carburetor parts, gears, bearings and bushings	9000 to 10,000	185 to 220	Very tough and rigid with good fatigue properties; resists most solvents	Forming, molding, extrusion, machining
Acrylic	medium	aircraft canopies, camera lenses, automobile tail-light lenses, pump parts, sprinkler heads, signs, and lighting diffusers	6000 to 10,000	140 to 200	Good impact strength—.4 to 2 foot-pounds per inch; optical clarity; weather resistant; tough at low temperatures	Extrusion, casting, injection molding, compression molding, machining
Cellulosics (Cellulose Acetate)	medium	rigid packing, recording tape, combs, shoe heels	8000	120 to 170	Resists moderate heat and weak acids, alkalis, many solvents	Injection, compression or blow molding, vacuum forming, machining
(Cellulose Acetate Butyrate)		packaging, pipe, tubing, tool handles, and steering wheels	7000		Resists heat, weather, and weak acids, alkalis, many solvents	Blow injection and compression molding, extrusion, machining and drawing
(Cellulose Acetate Propionate)		packaging, steering wheels, appliance housings, telephone cases, and pens	6000		Resists heat, weather, and weak acids, alkalis, many solvents	Injection and compression molding, extrusion, machining

Table E—*Continued*

Name of Plastic Material	Cost	Typical Applications	Maximum Tensile Strength psi	Maximum Service Temperature, °F	Significant Material Characteristics	Fabricating Methods
(Ethyl Cellulose)		cabinet edge molding, flashlights	7000		Resists heat and weak acids, alkalis, and many solvents	Injection and compression molding, extrusion and drawing, machining
(Cellulose Nitrate)		billiard balls and fabric coatings	8000		Resists moderate heat and weak acids, alkalis, and many solvents	Machining
Ethylene—Vinyl Acetate	medium	syringe bulbs, tubing, gloves, dampening pads, pool liners	3600	150	Flexible, high impact strength; useful to −150°F	Injection and blow molding, extrusion
Fluorocarbons (TFE, PTFE) (FEP, PVF$_2$)	high	cooking utensil coatings, gaskets, piston rings, lining, low-friction surfaces	6500 / 7000	500 / 400	Very high chemical resistance; useful to cryogenic temperatures; very low friction	Molding and extrusion; hot or cold formed; machined
Ionomer	medium	skin packaging, toys, bottles, wire insulation, trays, covers			Transparent, tough at low temperatures, good insulator	Injection or blow molding, extrusion, thermoforming
Nylon	high	washers, gears, fasteners, motor parts, zippers, electrical parts and supports, rollers and wheels, bearings	8500 to 12,500	250 to 300	High impact strength, wear resistant, resilient, useful down to 0°F, resists all common chemicals	Injection, compression and blow molding, cold forming, extrusion, machining

Material		Uses		Properties		Fabrication
Parylene	medium	insulating and protective coatings for paper, cloth, ceramics, metals, microcircuits		Best insulating plastic known at temperatures near absolute zero; resists all organic solvents	200 to 240	Coating
Phenoxy	high	electrical coatings and insulation	8500	Clear, tough, rigid, hard; impact strength—2 foot-pounds per inch	170	Blow and injection molding, extrusion, coating
Polyallomer	medium	automobile panels, pipe fittings, hinges, shoe lasts	4000	Good impact strength—1.5 foot-pounds per inch	200	Injection molding, extrusion, vacuum forming
Polycarbonate	medium	armor for small arms fire, structural parts, insulating housings, safety helmets, rocket launcher handles, pump parts, gears, bearing balls		Extremely good impact strength—16 foot-pounds per inch; rigid and transparent	250	All molding techniques, machining
Polyethylene	low	kitchen ware, refrigerator parts, pipe, containers, seal rings, structural housings	1000 to 5000	Tough, flexible or rigid, resists weathering and common chemicals	175 to 250	Molding, extrusion, calendering, coating, casting, and vacuum forming
Polyphenylene Oxide	high	electrical insulation, battery cases, medical instruments, hot-water plumbing equipment, nose cones	11,600	High impact strength—1.2 foot-pounds per inch; useful to −275°F; good dielectric; high modulus and low creep	225	All conventional techniques

Table E—*Continued*

Name of Plastic Material	Cost	Typical Applications	Maximum Tensile Strength psi	Maximum Service Temperature °F	Significant Material Characteristics	Fabricating Methods
Polypropylene	low	helmets, pipes and fittings, valves, appliance parts, hinges	5000	230	Excellent electrical insulator over a wide range of temperatures	Injection and blow molding, extrusion
Polystyrene	low	toys, wall tile, food trays, appliance housings, battery cases, instrument panels	3000 to 8000	125 to 165	Hard, rigid, poor impact resistance; resists most foods, household fluids	Injection and compression molding, extrusion and machining
Polysulfone	high	automobile parts, power tool and appliance housings, pipe, computer parts	10,000	300	Rigid and ductile with good electrical properties over a wide temperature range down to −150°F	Injection and blow molding, extrusion and thermoforming
Vinyl	medium	raincoats, upholstery, hose, wire insulation, gaskets		130	Abrasion and wear resistant; resists most common chemicals	All molding methods, extrusion, casting, and calendering (machining for PVC)

Table F. Thermosetting plastics

Name of Plastic Material	Cost	Typical Applications	Maximum Tensile Strength psi	Maximum Service Temperature °F	Significant Material Characteristics	Fabricating Methods
Alkyd	low	automobile starter parts, vacuum tube supports, insulators	8000	300	Resists weak acids and most solvents; good electrical insulator	Compression molding
Allylic	medium	lenses, aircraft windows, electronic parts, laminates	6000	350	High dielectric strength; resists most solvents; optically clear	Injection, compression or transfer molding and extrusion
Amino	medium	melamine tableware, distributor caps and counter tops; urea appliance housings	5000 to 10,000	170 to 210	Colorful, hard, scratch resistant, resists detergents, cleaners, alcohol, oil	Compression, transfer and plunger molding
Casein		buttons, toys and novelty items			Smooth surface, rigid, tough, resists gasoline, organic solvents and chemicals; poor temperature, humidity resistance	Machining
Cold molded	low	small gears, jigs, dies, plugs, handles			Resists alkalis, solvents, water, and oils	High pressure molding and post curing
Epoxy		adhesives, tanks, laminated tooling	12,000	500	Flexible and resistant to many chemicals	Molding, extrusion, casting, and potting
Phenolics	medium	distributor caps, telephones, insulators, appliance parts, grinding wheel bonds	4000 to 10,000	250 to 450	Hard, good thermal insulator, electrical insulator and good chemical resistance	Injection, compression, transfer or plunger molding, and casting
Polyester	low	impregnant for cloth, paper, etc., and for reinforced automobile bodies, luggage	1000 to 17,000	150 to 300	Tough, resists most solvents, acids, bases and salts	Molding, casting

Table G. Unclassed plastics

Name of Plastic Material	Cost	Typical Applications	Maximum Tensile Strength psi	Maximum Service Temperature °F	Significant Material Characteristics	Fabricating Methods
Polyimide	high	aerospace parts, valve seats, bearings, seals, retainer rings, hose and tubing, piston rings, compressor vanes	10,000 to 12,000	425 to 450	High heat resistance, low friction and good wear resistance. Has thermoplastic structure and nonmelting character of a thermoset	Machining
Polyurethane	medium	foams, coatings, auto parts, tire treads, aircraft structures	over 5000		Tough, impact resistant, resists most common chemicals	Molding, extrusion, casting, and calendering
Silicone	high	switch parts, coil forms, motor and generator insulation, aircraft radomes and ductwork, electronic component encapsulation	4000 to 6000	over 500	High heat stability; highly resistant to mineral acids and corrosive salts; good dielectric. Either thermoplastic or thermosetting forms are available	Compression and transfer molding, extrusion, calendering, casting, coating and foam

Table H. Metric conversion factors

Property	Unit	Symbol	Exact Conversion			Approximate Equivalency
			From	To	Multiply by	
Length	metre	m	inch	mm	2.540×10	25 mm $= 1$ in
	centimetre	cm	inch	cm	2.540	300 mm $= 1$ ft
	millimetre	mm	foot	mm	3.048×10^{-4}	
Mass	kilogram	kg	ounce	g	2.835×10	2.8 g $= 1$ oz
	gram	g	pound	kg	4.536×10^{-1}	kg $= 2.2$ lbs (35 oz)
	tonne (megagram)	t	ton (2000 lb)	kg	9.072×10^{2}	1 t $= 2200$ lb
Density	kilogram per cu metre	kg/m^3	pounds per cu ft	kg/m^3	1.602×10	16 kg/M^3 $= 1$ lb/ft^3
Temperature	deg Celsius	°C	deg Fahrenheit	°C	$(°F - 32) \times 5/9$	$0°C = 32°F$ $100°C = 212°F$
Area	square metre	m^2	sq inch	mm^2	6.452×10^{2}	645 mm^2 $= 1$ in^2
	square millimetre	mm^2	sq ft	m^2	9.290×10^{-2}	1 m^2 $= 11$ ft^2
Volume	cubic metre	m^3	cu in	mm^3	1.639×10^{4}	16400 mm^3 $= 1$ in^3
	cubic centimetre	cm^3	cu ft	m^3	2.832×10^{-2}	1 m^3 $= 35$ ft^3
	cubic millimetre	mm^3	cu yd	m^3	7.645×10^{-1}	1 m^3 $= 1.3$ yd^3
Force	newton	N	ounce (Force)	N	$2,780 \times 10^{-1}$	1 N $= 3.6$ oz
	kilonewton	kN	pound (Force)	kN	4.448×10^{-3}	4.4 N $= 1$ lb
	meganewton	MN	Kip	MN	4.448	1 kN $= 225$ lb
Stress	megapascal	MPa	pound/in^2 (psi)	MPa	6.895×10^{-3}	1 MPa $= 145$ psi
	newtons/sq m	N/M^2	Kip/in^2 (ksi)	MPa	6.895	7 MPa $= 1$ ksi
Torque	newton-metres	Nm	in-ounce	Nm	7.062×10^{3}	1 Nm $= 140$ in oz
			in pound	Nm	1.130×10^{-1}	1 Nm $= 9$ in lb
			ft pound	Nm	1.356	1 Nm $= 75$ ft lb
						1.4 Nm $= 1$ ft lb

Table I. Basic welding symbols and their location significance

Basic Welding Symbols and Their Location Significance

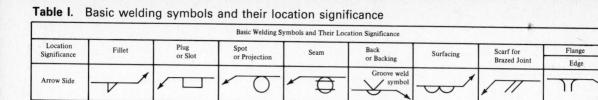

Location Significance	Fillet	Plug or Slot	Spot or Projection	Seam	Back or Backing	Surfacing	Scarf for Brazed Joint	Flange / Edge
Arrow Side					Groove weld symbol			
Other Side					Groove weld symbol	Not used		
Both Sides		Not used	Not used	Not used	Not used	Not used		Not used
No Arrow Side or Other Side Significance	Not used	Not used			Not used	Not used	Not used	Not used

Supplementary Symbols Used with Welding Symbols

Convex Contour Symbol
Convex contour symbol indicates face of weld to be finished to convex contour

Finish symbol (user's standard) indicates method of obtaining specified contour but not degree of finish

Weld-All-Around Symbol
Weld-all-around symbol indicates that weld extends completely around the joint

Joint with Backing
With groove weld symbol

See note

Note: Material and dimensions of backing as specified

Joint with Spacer
With modified groove weld symbol

See note

Double-bevel groove   Note: Material and dimensions of spacer as specified

Melt-Thru Symbol
Any applicable weld symbol

1 mm

Melt-thru symbol is not dimensioned (except height)

Flush Contour Symbol
Flush contour symbol indicates face of weld to be made flush. When used without a finish symbol, indicates weld without subsequent finishing

Finish symbol (user's standard) indicates method of obtaining specified contour but not degree of finish

Multiple Reference Lines
First operation shown on reference line nearest arrow — 1st

Second operation, or supplementary data — 2nd

Third operation, or test information — 3rd

Field Weld Symbol
Field Weld symbol indicates that weld is to be made at a place other than that of initial construction

Complete Penetration
Indicates complete penetration regardless of type of weld or joint preparation

Location of Elements of a Welding Symbol
Finish symbol

Groove angle; included angle of countersink for plug welds

Contour symbol

Root opening; depth of filling for plug and slot welds

Length of weld

Effective throat

Pitch (center-to-center spacing) of welds

Depth of preparation; size or strength for certain welds

Field weld symbol

Specification, process, or other reference

Arrow connecting reference line to arrow side member of joint

Tail

Tail (Tail omitted when reference is not used)

Weld-all-around symbol

Basic weld symbol or detail reference

Number of spot or projection welds.

Reference line

Elements in this area remain as shown when tail and arrow are reversed

F A R
T S (E) L-P
(N)
(BOTH SIDES) (ARROW OTHER SIDE / SIDE)

Supplementary Symbols

Weld-All-Around	Field Weld	Melt-Thru	Backing, Spacer	Contour		
				Flush	Convex	Concave

Basic Joints–Identification of Arrow Side and Other Side of Joint

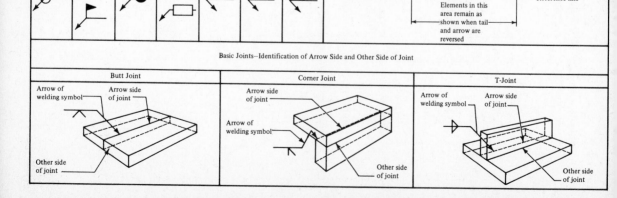

Butt Joint
Arrow of welding symbol

Arrow side of joint

Other side of joint

Corner Joint
Arrow side of joint

Arrow of welding symbol

Other side of joint

T-Joint
Arrow of welding symbol

Arrow side of joint

Other side of joint

Table I—*Continued*

Basic Welding Symbols and Their Location Significance

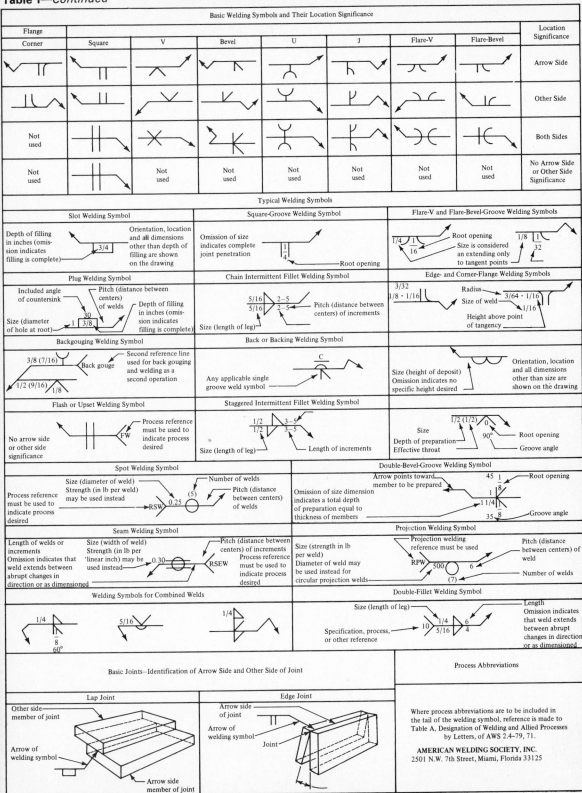

Flange									Location Significance
Corner	Square	V	Bevel	U	J	Flare-V	Flare-Bevel		
									Arrow Side
									Other Side
Not used									Both Sides
Not used	Not used	Not used	Not used	Not used	Not used	Not used	Not used		No Arrow Side or Other Side Significance

Typical Welding Symbols

Slot Welding Symbol

Depth of filling in inches (omission indicates filling is complete)
3/4
Orientation, location and all dimensions other than depth of filling are shown on the drawing

Square-Groove Welding Symbol

Omission of size indicates complete joint penetration
$\frac{1}{4}$
Root opening

Flare-V and Flare-Bevel-Groove Welding Symbols

1/4 · $\frac{1}{16}$
Root opening
Size is considered an extending only to tangent points
1/8 · $\frac{1}{32}$

Plug Welding Symbol

Included angle of countersink
Pitch (distance between centers) of welds
Depth of filling in inches (omission indicates filling is complete)
Size (diameter of hole at root)
1 · 30 · 3/8

Chain Intermittent Fillet Welding Symbol

5/16 · 2–5
5/16 · 2–5
Pitch (distance between centers) of increments
Size (length of leg)

Edge- and Corner-Flange Welding Symbols

3/32
1/8 · 1/16
Radius
Size of weld
3/64 · 1/16
1/16
Height above point of tangency

Backgouging Welding Symbol

3/8 (7/16) Back gouge
1/2 (9/16) 1/8
Second reference line used for back gouging and welding as a second operation

Back or Backing Welding Symbol

C
Any applicable single groove weld symbol
Size (height of deposit) Omission indicates no specific height desired
Orientation, location and all dimensions other than size are shown on the drawing

Flash or Upset Welding Symbol

FW
No arrow side or other side significance
Process reference must be used to indicate process desired

Staggered Intermittent Fillet Welding Symbol

1/2 · 3–5
1/2 · 3–5
Size (length of leg)
Length of increments

Size
Depth of preparation
Effective throat
1/2 (1/2) · 0
90°
Root opening
Groove angle

Spot Welding Symbol

Size (diameter of weld) Strength (in lb per weld) may be used instead
RSW 0.25 · (5)
Number of welds
Pitch (distance between centers) of welds
Process reference must be used to indicate process desired

Double-Bevel-Groove Welding Symbol

Arrow points toward member to be prepared
Omission of size dimension indicates a total depth of preparation equal to thickness of members
45 · 1
$\frac{1}{8}$
1 1/4
35 · 8
Root opening
Groove angle

Seam Welding Symbol

Length of welds or increments Omission indicates that weld extends between abrupt changes in direction or as dimensioned
Size (width of weld) Strength (in lb per linear inch) may be used instead
0.30
RSEW
Pitch (distance between centers) of increments Process reference must be used to indicate process desired

Projection Welding Symbol

Size (strength in lb per weld) Diameter of weld may be used instead for circular projection welds
Projection welding reference must be used
RPW 500 · 6 · (7)
Pitch (distance between centers) of weld
Number of welds

Welding Symbols for Combined Welds

1/4
8
60°
5/16
1/4

Double-Fillet Welding Symbol

Size (length of leg)
Specification, process, or other reference
10 · 1/4 · 6
5/16 · 4
Length Omission indicates that weld extends between abrupt changes in direction or as dimensioned

Basic Joints–Identification of Arrow Side and Other Side of Joint

Process Abbreviations

Lap Joint

Other side member of joint
Arrow of welding symbol
Arrow side member of joint

Edge Joint

Arrow side of joint
Arrow of welding symbol
Joint

Where process abbreviations are to be included in the tail of the welding symbol, reference is made to Table A, Designation of Welding and Allied Processes by Letters, of AWS 2.4–79, 71.

AMERICAN WELDING SOCIETY, INC.
2501 N.W. 7th Street, Miami, Florida 33125

Index